# EXPLANATION IN THE BEHAVIOURAL SCIENCES

# EXPLANATION IN THE BEHAVIOURAL SCIENCES

*Edited by*

## ROBERT BORGER

*Senior Lecturer in Psychology, Brunel University*

## FRANK CIOFFI

*Senior Lecturer in Philosophy, University of Kent*

## CAMBRIDGE

### AT THE UNIVERSITY PRESS

1970

Published by the Syndics of the Cambridge University Press
Bentley House, 200 Euston Road, London N.W.1
American Branch: 32 East 57th Street, New York, N.Y.10022

© Cambridge University Press 1970

Library of Congress Catalogue Card Number: 71-105497

Standard Book Number: 521 07820 2

Printed in Great Britain
at the University Printing House, Cambridge
(Brooke Crutchley, University Printer)

# PREFACE

The behavioural sciences abound in explanatory programmes over whose nature and value there is intense disagreement. It seemed to the editors that a good way of contributing towards the clarification, if not the resolution, of these disagreements would be to confront advocates of particular points of view with comments and criticisms directed at specific statements of their position, and to give them the opportunity of equally specific replies.

Some disagreements are due to divergent estimates of the heuristic promise of particular approaches; several of the contributions to this volume take the form of eliciting the presuppositions behind such estimates and attempting to assess their tenability. Others have a more intractable source in fundamentally different conceptions of what would constitute an explanation, of the kind of investigation certain explanatory problems require and of the type of consideration that is pertinent to their solution. We have therefore included exchanges on these more general issues, such as the relation of reasons to causes and of purposive to mechanical explanations, etc.

In 1775 the French Academy of Sciences decided that it would no longer examine contributions purporting to have squared the circle. Many philosophers and other workers in the humanities have adopted an attitude towards the more general aspirations of the behavioural sciences rather like that of the French Academy towards squared circles. This attitude is not merely a manifestation of 'impatience at the somewhat ridiculous swagger of the programme, in view of what the authors are actually able to perform' evinced by William James towards the pretensions of the psychology of his day, but of doubt as to the very coherence of the objectives in terms of which the rationale of behavioural research is often stated. It finds its classic expression in Wittgenstein's remark that in psychology we have experimental methods and conceptual confusion and that problems and methods pass one another by.

This issue sometimes takes the form of a query as to the extent to which discoveries in the behavioural sciences could supplement, supplant or transform our workaday notions of why human beings behave as they do. Could they effect as radical a reorientation of our conception of human affairs as the researches inaugurated by Galileo, or Newton, or Einstein, or Planck, did that of the physical world? Or is there a distinctive immutability in the notions which we traditionally employ in the characterization of human actions? Attitudes towards this issue range from Professor Hamlyn's insistence that we would count nothing as making an action intelligible which did not set it in the familiar, customary framework of ends and rational means to those ends, to Professor Sutherland's suggestion that only when we have an adequate understanding of the mechanisms that produce sequences of movements will we get really clear about the meaning of concepts such as 'action' which now we see through a glass darkly, as it were, but then face to face.

Heuristic justifications of particular lines of enquiry tend, however, to get stated in terms which themselves raise questions of logical appropriateness. Thus, the neat dichotomy between issues of heuristic promise on the one hand, and of conceptual propriety on the other with which we began, breaks down and it has not proved possible to confine discussions of the latter to those sections of the book which were specifically commissioned to deal with them.

The task of assembling the material of this book has proved a more protracted one than the editors in their innocence had anticipated. We should like to express our gratitude to our contributors, many of whom have had to wait a long time for the appearance of their papers, for their patience and forbearance but above all for their willingness to co-operate in a rather novel and exacting venture.

R.B.  F.C.

# CONTENTS

# CONTENTS

# BIOGRAPHICAL NOTES ON CONTRIBUTORS

BANNISTER, D. Medical Research Council External Scientific Staff and Head of Clinical Psychology Department, Bexley Hospital, Dartford Heath, Kent.

Formerly visiting Professor at Ohio State University, Columbus, Ohio, U.S.A.

Educated: Universities of Manchester and London.

BLACK, MAX. Professor of Philosophy, Cornell University.

Director, The Society for the Humanities, Cornell University.

Formerly Professor at University of Illinois; lecturer and tutor, London University Institute of Education.

Editor: *Contemporary Philosophy Series, Philosophical Review.*

Educated: Universities of Cambridge, Göttingen, London (Mathematics, Logic, Philosophy).

BLAU, PETER M. Professor of Sociology, University of Chicago.

Director, Comparative Organization Research Program, University of Chicago.

Pitt Professor of American History and Institutions at the University of Cambridge (1966–7).

Former editor: *American Journal of Sociology.*

Educated: Elmhurst College, Illinois, University of Cambridge. Columbia University.

BOAKES, R. A. Lecturer in Experimental Psychology, University of Sussex.

Educated: University of Cambridge and Harvard University.

BORGER, ROBERT. Senior Lecturer in Psychology, Brunel University, England.

Educated: University of Wales (Mathematics), University of Cambridge (Statistics), University of Oxford (Psychology and Philosophy).

BROWN, ROBERT. Senior Fellow, Institute of Advanced Studies, Australian National University, Canberra.

Formerly visiting Lecturer at the University of Wisconsin and the University of Massachusetts.

Educated: University of Chicago (Anthropology) and University of London (Philosophy).

ix

CHOMSKY, NOAM. Professor of Modern Languages and Linguistics, M.I.T.
>Formerly visiting Professor at the Universities of Columbia, California and the Institute of Advanced Studies, Princeton.
>Educated: Universities of Pennsylvania, Chicago and London.

CIOFFI, FRANK. Senior Lecturer in Philosophy, University of Kent.
>Formerly Lecturer in Philosophy, University of Singapore (1956–65).
>Educated: University of Oxford (Philosophy and Psychology).

DONAGAN, ALAN. Professor of Philosophy, University of Illinois.
>Former Chairman, Department of Philosophy, at both Indiana University (1961–5) and the University of Minnesota (1957–61).
>Educated: Queen's College, University of Melbourne, and Exeter College, Oxford (Philosophy).

EYSENCK, H. J. Professor of Psychology, University of London.
>Director, Psychology Department, Institute of Psychiatry.
>Formerly visiting Professor at the University of Pennsylvania, and at the University of California, Berkeley.
>Editor in Chief: *Behaviour Research and Therapy*.
>Editor: *International Monographs of Experimental Psychology*.
>Educated: University of London.

FARRELL, B. A. Fellow of Corpus Christi College, Reader in Mental Philosophy, Oxford.
>Educated: University of Cape Town and Oxford.

GRUNDY, J. H. Lecturer in Philosophy, University of Keele.
>Educated: Universities of Keele and Oxford.

HALLIDAY, M. S. Lecturer in Psychology at the University of Sussex.
>Educated: University of Cambridge (Psychology).

HAMLYN, D. W. Professor of Philosophy, University of London.
>Head of the Department of Philosophy, Birkbeck College, University of London.
>Formerly Research Fellow at Corpus Christi College, Oxford, and Lecturer at Jesus College, Oxford.
>Educated: University of Oxford (Lit.Hum. and P.P.P.).

HOMANS, GEORGE C. Professor of Sociology, Harvard University.
Formerly visiting Professor at the Universities of Manchester, Cambridge, and Kent.
President, American Sociological Association (1963–4).
Educated: Harvard University (English Literature).

JARVIE, I. C. Professor of Philosophy, York University, Toronto.
Director, Graduate Programme in Philosophy.
Formerly taught at London School of Economics, University of Hong Kong, Tufts University, Boston University.
Educated: University of London (Anthropology, Scientific Method).

PETERS, R. S. Professor of Philosophy of Education, University of London Institute of Education.
Formerly Reader in Philosophy, Birkbeck College, University of London. Visiting Professor Harvard Graduate School of Education and visiting Fellow to the Australian National University.
Editor: Philosophy of Education section of Routledge & Kegan Paul College of Education Library.
Educated: Oxford University (Classics), University of London (Philosophy).

PRIBRAM, KARL H. Professor Departments of Psychiatry and Psychology, Stanford University.
Formerly Director, Neurophysiological Research and Laboratories, Institute of Living, Hartford, Conn.
Visiting Scholar and Lecturer: Massachusetts Institute of Technology, Clark University, Harvard University, Haverford College, University of Southern California, Beloit College, Center for Advanced Study in Theoretical Psychology, University of Alberta.
Editorships: *Neuropsychologia; Brain Research; Experimental Brain Research; International Journal of Neuropharmacology; Brain, Behaviour and Evolution; Science of Behaviour Series.* (Penguin Books.)
Educated: University of Chicago.

SUTHERLAND, N. S. Professor of Experimental Psychology, University of Sussex.
Chairman, Laboratory of Experimental Psychology.
Formerly visiting Professor at Massachusetts Institute of Technology and Fellow of Magdalen College and Merton College, Oxford.
Educated: Magdalen College, Oxford (Ancient Greats and Psychology, Philosophy and Physiology).

TAYLOR, CHARLES. Professor of Political Science, McGill University.
Professor of Philosophy, Université de Montreal.
Formerly visiting Professor at Princeton University.
Educated: McGill University and Oxford University.

TOULMIN, STEPHEN. Professor of Philosophy, Michigan State University.
Formerly Fellow of King's College, Cambridge.
Professor of Philosophy, University of Leeds, and Director, Unit for the History of Ideas, Nuffield Foundation (London).
Educated: King's College, Cambridge (Mathematics, Physics, Philosophy).

WATKINS, JOHN W. N. Professor of Philosophy, University of London.
Convener, Department of Philosophy, London School of Economics.
Educated: L.S.E. and Yale.

WATSON, A. J. University Lecturer, Department of Psychology, University of Cambridge.
Senior Tutor, Fitzwilliam College, Cambridge.
Editor: *The Quarterly Journal of Experimental Psychology*.
Educated: University of Oxford (Psychology and Philosophy).

WINCH, PETER. Professor of Philosophy, University of London King's College.
Formerly visiting Professor at the University of Rochester.
Editor of *Analysis*.
Educated: St Edmund Hall, Oxford.

WISDOM, J. O. University Professor of Philosophy and Social Science, York University, Toronto.
Formerly University Professor at the State University College of New York at Fredonia and distinguished visiting Professor at the University of Southern California.
Formerly editor of *The British Journal for the Philosophy of Science*.
Educated: University of Dublin, Trinity College (Philosophy, Mathematics).

# REASONS AND CAUSES

## *by* STEPHEN TOULMIN

All behaviour is explicable: some actions are also justifiable

### 1. INTRODUCTION: THE PROBLEM

Ever since Galileo and Descartes laid down the guiding strategy for modern natural science, men's attempts at self-understanding have repeatedly stumbled over the concept of *reasons*. On the one hand, scientists have extended the scope of categories like *body* and *matter*, *physical* and *mechanical*, into ever larger areas of the natural world; and each new extension has seemingly subjected further kinds of phenomenon and system to the reign of 'causal necessity'. On the other hand, all men— scientists and non-scientists alike—have continued to think and act, take stands, criticize each other and justify themselves; and have cited as the factors relevant to an understanding of their conduct, not the physical or mechanical *causes* underlying their actions, but rather the *reasons* for which they acted as they did.

Inevitably the question has been raised, again and again: 'How are these two modes of understanding to be reconciled?' On the face of things, the 'reasons' for which we act have no analogy to 'causes'. As Wollaston put it, in an objection quoted by Joseph Priestley in his *Disquisitions Relating to Matter and Spirit*:

When I begin to move myself, I do it for some reason and with respect to some end. But who can imagine matter to be moved by arguments, or ever ranked syllogisms and demonstrations among levers and pulleys?[1]

And certainly: the way in which reasons serve to determine our conduct appear very unlike that which we associate with the most typical 'causal mechanisms' and 'physical processes'. Yet are we to conclude from this that the human reason can retain a sphere of influence only by holding causality at bay? On that supposition, we shall be putting ourselves in an embarrassing situation. For, then, every step forward in the neuroscientific analysis of brain-mechanisms will shrink the area of operation within which *mind* and *thought*, *mental* and *rational* categories, have any application. Like the God of the natural theologians, the Human Reason will find itself

---

[1] Priestley, 1777, Sec. 8, Objections to the system of materialism considered, *sub* Objection 3. See Passmore (ed.), 1965, p. 121.

on a diminishing sandbank, with the tide of Science rising all around it. The intellect—operating in its scientific mode—will thus be in danger of undercutting its own position, and so discrediting itself. No wonder Alexander Pope located the central mystery of human self-knowledge where he did: 'Chaos of Thought and Passion all confus'd' (he declared) Man was

> In doubt to deem himself a god or beast;
> In doubt his Mind or Body to prefer . . .
> Sole judge of Truth, in endless error hurl'd;
> The glory, jest and riddle of the world.
> (Alexander Pope, *Essay on Man*, Ep. II, 1, 8–9, 17–18)

The mystery has become no less teasing with the passage of time. It played a major part in Descartes' decision to adopt a psycho-physical dualism. In his system, the mind-substance was primarily concerned, not with the passions and feelings that humans share with animals, but rather with language, thought and reasoning, in which men are apparently unique. Writing before the major advances of modern physical science, Descartes could afford to take this easy way out. It cost him little to draw a hard-and-fast line between 'mental' phenomena, such as reasoned thoughts and choices, and 'material' phenomena, characterized by causality and extension. A century and a half later this escape-route was less attractive. Being rightly impressed by the far-reaching success of Newton's physics, Kant hesitated to treat the contrast between reasoning and causation as a distinction between different kinds of phenomena. Within the phenomenal world (he argued) we were bound to regard *all* happenings as causally determined: the 'rational self' must therefore be understood, not as a 'phenomenon', but rather as a 'transcendental' idea belonging to the world of 'noumena'. Its task was not to 'hold causality at bay', since causality belonged only to the world of phenomena: rather, it operated on quite another level. Yet Kant's 'transcendental self'—like its intellectual offspring, Schopenhauer's 'Will'—was scarcely a solution to the central problem: 'How can the explanation of human behaviour, in terms of reasons, be reconciled with the explanation of natural phenomena, in terms of causes and mechanisms?' On the contrary, it merely restated the *problem*; and did so in a less tractable form since, on Kant's own confession, there was very little at all that we could *say* about 'noumena', and still less about the precise relations between 'noumena' and 'phenomena'.

In consequence, the question of reasons and causes is still very much alive today. It arises as an analytical problem in philosophy, as a methodological problem in psychology, and in neurophysiology as a source of paradox. For instance: in his Auguste Comte Memorial Lecture for 1964,

A. J. Ayer sets out to rebut the views of Gilbert Ryle and Ludwig Wittgenstein, whose arguments (he says) have made it 'almost a commonplace among philosophers that motives are not causes' (Ayer, 1964, p. 12)—Ayer himself ends by tentatively 'assimilating motives [and this term embraces also reasons] to causes' (Ayer, 1964, pp. 24–5). The same continued disagreement about the distinction between reasons and motives on the one hand, and causes on the other, is illustrated in recent books by Richard Peters[1] and Charles Taylor (Taylor, 1964: see especially Chaps. 1 and 2). In Taylor's book, indeed, the distinction is erected into an absolute dichotomy—the 'teleological' or 'intentional' explanation of human actions being sharply contrasted with the 'mechanistic' explanation of bodily movements. Taylor's arguments are criticized by Stuart Sutherland elsewhere in this volume (Sutherland, 1970, pp. 137–8), and we shall return to the topic later in this essay.

In psychology, the dispute about the status of motives and reasons is central to the argument between the followers of B. F. Skinner, who claim to rid psychological explanation of all 'mentalistic' terms, and those psychologists for whose theories terms like 'reason' and 'motive' are indispensable and irreducible. If it makes sense to talk of reasons and motives at all (Skinner's followers would argue) these terms must simply refer to 'conditioned verbal responses'.[2] That being so, such terms will not serve as tools for the explanation of psychological phenomena: rather, the verbal behaviour-patterns will themselves become additional phenomena, requiring to be explained in their turn.

Finally, in the neurophysiology of higher mental functions, the problem of reasons and causes reaches—depending on your viewpoint—the ultimate point of acuteness, or of absurdity. Many neuroscientists believe that we are at last within sight of explaining, in neurophysiological terms, all the basic causal interconnections and influences involved in the operation of the brain and central nervous system. And when that day finally arrives—as Charles Townes likes to remind us[3]—the scientists concerned will certainly wish to *take credit* for their intellectual feat. 'Take credit for what intellectual feat?', we may ask: 'For the scientific discovery that strictly causal brain-mechanisms underlie all rational thought-processes— *including* the scientific discovery that strictly causal brain-mechanisms underlie all rational thought-processes.'

[1] Peters, 1958: The argument of the present essay at certain points goes over ground already covered by Professor Peters in this book.

[2] Cf. Honig (ed.), 1966: esp. the essay 'Conditioning human verbal behaviour' by Holz & Azrin, pp. 790–821.

[3] Verbal discussion at the Neurosciences Research Program's Intensive Study Program (July–August 1966), Boulder, Colorado.

This neat restatement of our central problem I shall call *Townes'*
*Paradox*. On the one hand, scientists have increasingly good evidence that
neurophysiological processes in the brain conform to the same general
principles as other 'physical' or 'mechanistic' processes; and that such
processes underlie, are associated with, correlated with or counterparts of
*all* our thoughts and ideas, even the most 'rational'. On the other hand,
those same scientists claim personal responsibility for, and feel a justified
pride in, *their own* thoughts and ideas. How are we—and they—to reconcile
those two attitudes? If all our thoughts and ideas are subject to causality,
what place is there for claiming personal responsibility for any of them?
For are they not all alike *necessitated*? This is the form in which the
problem of 'reasons' presents a continuing challenge to the scientists
of the late 1960s.

## 2. THE STRUCTURE OF THE ARGUMENT

So much for the problem to be explored in this essay: our strategy envisages three successive attacks. We shall begin by looking at some questions
of 'linguistic analysis' or 'logical grammar' in the hope, first, of isolating
some representative types of situation in which one would *prima facie*
allow that a man was 'acting for reasons' and, secondly, of marking those
situations off from others in which such a description would on the face of
it be quite out of place. This linguistic enquiry will, however, be merely
preliminary. The main thrust of our attack will come in the second phase,
when we follow up the hints suggested by the logical grammar of reasons,
and look at the *behavioural* distinctions that underlie (and give a point to)
those differences in our *linguistic* practices and usages.

This behavioural enquiry will concentrate on one feature: namely, the
contrast between actions performed as applications of some learned procedure, ritual or technique, and those types of behavioural response which
are unlearned or autonomic. Acting rationally—doing things 'for reasons'
—is an art, or rather a complex set of arts; and these arts differ from simple,
automatic responses to stimuli, not least, in having to be acquired. Accordingly, we shall have to consider, in a general way, how the contrast between
unlearned responses and acquired skills comes progressively to be *established*, in the course of the individual's lifetime. As we shall have occasion
to remind ourselves, the 'factors' or 'considerations' which serve as
'reasons' for a man, can do so *only* if he recognizes them as 'carrying
weight': in fact, such considerations can carry weight with him at all only
to the extent that he does so recognize them. From this we shall go on to
infer that, in one legitimate sense of the term, it is a necessary condition

4

of a man's doing something 'for a reason' that he should recognize the consideration in question as being a possible reason for so acting. And then the question will arise, whether his *recognition* of the reason may not be counted among the factors which are *causally* relevant to an understanding of his action.

We shall not be driven to the point of calling the 'reason' for a man's action a 'cause' of his acting as he does; still less to the point of treating his 'reason' for acting as imposing a 'causal compulsion' on him. Both these doctrines will remain barbarisms. On the contrary, we shall have to distinguish between talking *directly* about an agent's 'reasons' for acting, or about the acceptability of those reasons; and talking *indirectly* about his recognizing, being moved by, acknowledging the force of, or seeing the weight of those 'reasons'. Causal questions will not be found to arise so long as we confine ourselves to appraising 'reasons for acting' directly; yet this will leave entirely open the question: 'May not a man's *accepting* a consideration (*recognizing* its force, *acknowledging* it as a reason) cause him to act as he does?' This distinction, between 'reasons' for acting and 'having reasons' for acting, will be a conceptual one; as contrasted with the alleged *empirical* distinction, between 'genuine actions' and 'bodily movements', whose validity we shall have to reconsider.

In the third and final phase of our attack, we shall return to our initial problem, with that fundamental distinction in mind. As we shall see, both these two modes of speech (talking directly about reasons, and talking about the recognition of reasons) are perfectly legitimate in appropriate situations, and no satisfactory argument has yet been advanced for supposing that they are incompatible. It is only if we slide unthinkingly between the two terms of this distinction—between our 'reasons' for acting and our 'having reasons' for acting, between the 'considerations' in the light of which we act and the 'weight' those considerations carry—that we may be driven into philosophical perplexity about the relation between reasons and causes; and only if we allow this to happen need there be any doubt about the propriety of introducing causal enquiries and explanations into the behavioural sciences, or about the special terms and conditions which we must recognize in doing so. On the central question at issue here—we shall discover—Immanuel Kant was not far from the essential truth: the first steps towards seeing how reasons and causes (rationality and causality) are to be reconciled in the sphere of human conduct are (i) appreciating the fact that they are not in competition, and (ii) understanding why this is so.

### 3. GIVING REASONS FOR OUR ACTIONS

What, then, is involved in acting 'for reasons'? We must begin by eliminating certain irrelevant examples. We shall be concerned here with only some of those cases in which one may speak of the 'reasons' why a man acted as he did: namely, those cases in which the reason in question can also be called 'his' reasons for the action. For, in a very broad sense, the word 'reason' is frequently interchangeable with the word 'explanation'—in this broad sense, we even speak of the 'reason why' it snowed yesterday. (It doesn't have a 'reason for' snowing!) This sense can also be applied to certain types of human behaviour: e.g. when we speak of the 'reason why' a girl blushed, viz. someone mentioned the young man she secretly loves. In such a situation, the girl herself may have every reason, if possible, to avoid blushing: thus, the reason *why* she blushed was not her reason *for* blushing. Similarly, with the reasons why a man is driving dangerously—by reason of intoxication—or the reasons why he is unfit to govern—by reason of physical or mental incapacity: here too, *the reasons why* are clearly not to be understood as *his reasons for* that behaviour or condition.

These cases are outside the scope of 'reason', as philosophers have discussed it. But, to come nearer home, explanations of a man's behaviour may still fall short of stating 'his reasons' even in some marginally rational examples. For instance, in cases of inadvertence, oversight or emotional stress, we act in ways that can be *explained* by citing reasons for our behaviour; though not in terms of our reasons for acting as we did: 'Why did you give the porter a $100 bill as a tip?', 'Why did you blurt out that unfortunate remark?' These cases, too, will be irrelevant to our main discussion.

Still, it is not enough to dismiss these examples as irrelevant: we have to understand also what makes them irrelevant. If, in such cases, we do not regard the agent as acting 'for a reason' (or, at any rate, not for *the* reason cited in our explanation) that is because these examples differ in crucial respects from those in which we would most typically account for someone's conduct by citing *his reasons for* acting as he did. So let us now turn and look at some more representative examples, in the hope of getting some clue to the special character of the types of behaviour which concern us directly here. We may usefully consider three classes of examples in turn.

### 3.1 *Giving reasons as justifying*

If the specification of a 'cause' is typically the outcome of a diagnosis, then the specification of a 'reason' (in one required sense) is typically the response to a challenge. If we are asked to explain the reasons for which we adopted some attitude, belief or course of action, we commonly understand this request as demanding a justification. According to the nature of the context, this justification may be of several different kinds. In some situations, we would understand the request as calling for a moral or legal justification—in this sense, the challenge 'Explain yourself, Sir!' has the force of 'Excuse yourself, if you can!' (Often the precise force of the challenge is apparent from the situation alone: we have to recognize implicitly just what moral, legal or other accusation our 'reasons' are required to rebut.) In other situations, we would understand such a challenge as an intellectual one: the implicit demand is then for an *argument*, whose effect must be to demonstrate that our belief, attitude or policy is realistic, properly-thought-out and/or effective.[1]

The request, 'Give your reasons . . .', sometimes presupposes that the action in question was decided on calculatedly, following on some explicitly-thought-out train of argument or deliberation; but this is not invariably presupposed. Although explicit reasoning may take place in some cases— e.g. 'What new medical calculations, statistical arguments and/or strategic considerations lead you now to discount the danger from nuclear ballistic missiles?'—this is not universally necessary, as a condition of conduct's being 'rational' or 'justifiable'. While we do quite frequently pause before acting and 'rehearse' possible defences to foreseeable challenges, much of our rationally-defensible conduct is in practice habitual, without thereby ceasing to be justifiable. Where prior rehearsal is demanded, the statement, 'I had a reason for acting as I did', has the force, 'I anticipated the objections to my action, and satisfied myself that they could be met'. Where prior rehearsal is *not* demanded, as with habitual actions, it is enough if we can articulate a justification for our conduct *after* it has been challenged. For the purposes of our present argument, however, the distinction between cases where prior deliberation is demanded, and those where it is not, will not be of central importance.

Where we fail to anticipate a challenge to our actions, on the other hand, we may find the demand for 'reasons' mysterious. Suppose that a casual tourist is sunning himself beside the Cam, and a College porter enquires about his 'reasons for' lying on the sacred lawns. He may not know what to make of the enquiry: 'What do you mean, "reasons"? I had no particular *reasons* for lying on this particular patch of grass: it didn't occur to

---

[1] Mr K. Kipnis has raised with me the question, whether we can apply similar distinctions in the case of our reasons for liking or disliking people or things. At first glance, 'reasons for liking' appear less tractable than 'reasons for acting', and it will be as well to begin our analysis by considering the latter only. But we shall return to the question of 'reasons for liking' below: see note 1, page 16.

me that I *needed* reasons!' In that case, the demand for reasons draws a blank, because the tourist had not foreseen, and so anticipated, any objections to his course of action.

### 3.2 *Giving reasons as signalling intentions*

Turning now to a second class of examples, we shall find that 'reasons for' actions are not always concerned (at any rate on the surface) with justifications; while enquiries about 'reasons' are not always (at any rate not obviously) to be understood as challenges, calling for arguments or defences. In these cases, explanations of the reasons for our behaviour come much closer to explanations of other happenings: they are designed to explain anomalies, make sense of seeming incongruities, to show how some puzzling element in our conduct can be fitted into an intelligible pattern.

One way in which they do this is by registering the fact that the action in question is not self-contained or self-explanatory, but is rather an *instrument* or *means* of achieving something beyond itself. In these cases, to specify one's 'reason' for an action is to imply that the action in question was instrumental, and to indicate or 'signal' the goal, destination or 'intention' towards which it was instrumental. 'Why did you whistle *The First Nowel* as we passed the Post Office van?'; 'To let the postman know that there are some outgoing letters in our mailbox . . . it's a private code we use.' (Contrast the answer: 'I didn't have any particular reason—I was simply whistling because I was happy.')

The incongruity which gives rise to the enquiry about 'reasons' arises here, just because the action is not self-contained, and appears puzzling when taken in isolation. In this context, the very function of 'declaring your intentions' is to resolve such uncertainties and incongruities. 'What are you doing that for?', here means, 'Where are you headed? What direction are you going?'; and to signal your intention, or give your reasons, is like putting up a destination-board. ('Young man, what are your reasons for paying so much attention to my daughter? Tell me your intentions: are you set on/headed for/aiming at matrimony?')

### 3.3 *Giving reasons as classifying or redescribing*

Alternatively, to specify the reasons for an action may have the effect of characterizing the action, not as an instrument, but rather as one instance of a ritual or formalized mode of behaviour. 'Why did you put your three middle fingers to your brow as we passed the Post Office van?': 'It's the

Boy Scout salute—the postman is my junior Scoutmaster.' In such cases, the question, 'What is your reason for doing that?' differs little from the question, 'What are you doing?' Indeed, there is no very sharp line between these two questions. As Professor J. L. Austin used to emphasize, the 'natures', 'consequences', 'effects' and 'results' of human actions merge into one another, and an action will often be described or classified in terms of its effects. In an appropriate situation, to hold three fingers of the right hand to your brow just *is* to give the Scout salute; and the arguing about your 'reason' for doing so is met by pointing out this fact.

In appropriate circumstances, for that matter, to whistle *The First Nowel* may just be to pass on a message. Wherever an instrumental action depends for its effect on some understood convention—as with a code— the dividing line between intentions and rituals cannot be an absolute one. Even actions involving no such quasi-linguistic element can be explained in either of these ways, alternatively. The question, 'What is your reason for dropping purple crystals into that bottle of water?', can be answered either in instrumental terms or by a redescription—'It's permanganate of potash: I'm adding it to the water in order to disinfect it', or alternatively, 'It's permanganate of potash: I'm disinfecting the water'.

Although the justificatory element characteristic of our first class of examples (i) is not obvious in the case of classes (ii) and (iii), no very clearcut line can be drawn between the first class and the others. With human behaviour, the concept of 'expectations' has a complexity missing in the case of sticks and stones. In the case of our fellow-men, the behaviour we expect always balances up two things: the behaviour that past experi- ence has led us to *expect from* them, and the behaviour we are entitled to *expect of* them. Correspondingly, people can catch us out either by acting unpredictably (like objects whose workings have defeated us) or else by acting out of line (by deviating from accepted norms): their actions may be either surprising, or open to objection, or both at once. In our original class of examples (i), our attention was first caught by the element of deliberation, justification or ratiocination they involve, and so by the implicit appeal to norms; but these same examples frequently involve also instrumental or formalized elements which link them to classes (ii) and (iii). ('What long-term goal underlay the deliberations of which that action was the outcome?' 'In what capacity did you enter on the deliberations of which . . . ?') Meanwhile, the latter examples may invoke 'reasons', not merely so as to make puzzling behaviour more intelligible, but also in part with the aim of justification. The fuller question, 'Why did you whistle *The First Nowel* as we passed the Post Office van? After all, it's mid- summer!', brings to the surface the *breach of custom* implicit in the original

9

question. (Carols are tunes for the wintertime.) And the same kinds of consideration may arise equally well in other cases: 'Why did you put your fingers to your brow like that? It looked like an obscene gesture!'

### 3.4 *Actions not done for reasons*

To round off this preliminary collection of examples, we must ask: 'In what cases would we *suspend* the use of the term "reasons", and rebut any presumption that an agent had acted "for a reason"?' To answer this question, let us return to the examples that we dismissed at the outset as irrelevant.

We may properly discuss the reasons why a girl blushes in third-person terms, but it would be inappropriate to ask about 'her reasons for blushing': she would rightly retort, 'I can't help blushing'. Similarly: if a man is unfit to drive his car, or unfit to carry out the duties of his office, this fact may be explicable—there may be 'reasons why'—but to ask the man to 'give his reasons' would betray a misunderstanding: in retrospect, he could only say, 'I was incapacitated'. Nor do we insist on 'reasons for', when someone acts in the heat of the moment: 'I had no reason to blurt out that unfortunate remark, I was simply carried away.' Nor do we, in the case of a purely inadvertent action: 'I had no reason to give the porter a $100 bill—the light was so bad, I just couldn't tell it from a $1 bill.' Nor do we, in the case of action which, though surprising or objectionable, can be explained or excused on analogous grounds: e.g. 'I was dog-tired'.

Half a dozen such examples will begin to indicate what marks off actions over which the demand for reasons is out of place from those about which we do press—and are ostensibly justified in pressing—the demand for the agent's reasons. For in all these 'irrelevant' examples, the behaviour in question has reverted, in one respect or another, from being deliberate, 'high-grade' conduct, to being autonomic, incompetent, uncontrolled, inattentive or otherwise 'lower-grade' conduct. The significance of this point will soon be apparent.

To draw together the threads of this argument: all four general types of behaviour distinguished so far can be illustrated by considering one single kind of action—namely, *blinking*. Ordinarily, blinking is autonomic: it happens without our having to deliberate or form the habit, without our using it instrumentally to achieve some end, and without our performing it as a formalized or ritual action. Normal blinking can of course be explained, but it must be explained in purely causal terms: by citing the physiological factors which immediately determine the moment of blinking, or the organic functions which the blinking protects and maintains, or the evolutionary advantages an animal species derives from possessing a blinking-mechanism. The one way *not* to explain a man's normal blinking is by specifying his *reasons* for blinking.

Yet, in some exceptional situations, we may have 'reasons' for blinking. After certain ear-operations (e.g. fenestration) the patient experiences severe attacks of giddiness, which are slow to clear up of themselves. The best means of stabilizing the visual field, and so controlling the giddiness, is for the patient to keep on blinking—i.e. to keep on blinking deliberately, and to do so more often than normal. (And, in fact, the first exercise in the established programme of physiotherapy following a fenestration operation is precisely that.[1]) So if we see such a patient in hospital, blinking away doggedly, we may well ask him about his 'reasons' for blinking so much— 'I'm blinking like this to control the dizziness'. The implication is clear. If we use a mode of behaviour which is normally autonomic as a deliberate instrument or technique, questions about 'reasons' which would normally be irrelevant at once arise: thereupon the behaviour ceases to be explicable in purely causal terms alone.

When we blink on purpose, blinking moves from the reasonless class (iv) to the intentional class (ii). But cases can equally be imagined in which it might count as a ritual class (iii), or calculated class (i) action. (Little girl to mother, in church: 'Mummy, why are you kneeling down and blink-ing?'—'Hush, darling; I'm *praying*.' Onlooker to research physiologist: 'Why do you keep blinking at half-second intervals?'—'Well, I calculated the rate of evaporation of mucous fluid from the eyeball, and concluded that, in poor illumination, one's acuity of vision may be adversely affected, if one's blinking-rate drops below 1-a-second [etc.]') The scale running from actions clearly 'done for reasons' to behaviour clearly *not* 'done for reasons' accordingly ranges between calculated, reasoned-out actions, at one extreme, to purely autonomic behaviour at the other. Passing from carefully-thought-out actions, through instrumental and ritual acts, to inattentive or impulsive ones, and finally to pure reflexes, we move pro-gressively from cases in which a complete explanation necessarily involves the agent's 'reasons' for his actions, to cases in which complete explanation can be undisguisedly and undilutedly 'causal'.

The adverbial qualification—'undisguisedly and undilutedly'—is en-tirely indispensable here. Too often, this difference is presented as an *unqualified* contrast, between 'rational' actions and 'causal' bodily move-ments—a formulation which begs our central question. We must not assume from the very outset that actions can be 'rational' only if they are

---

[1] My wife has first-hand experience of this treatment, and drew my attention to the facts here reported.

not 'causal' (and vice versa): on the contrary, we must turn now and consider the question, 'Are explanations of an agent's reasons for his actions also—but only disguisedly—causal?' This will be our central question in the next phase of the argument.

## 4. RATIOCINATION AND OTHER LEARNED PROCEDURES

Language (as Wittgenstein insisted) acquires meaning only in the stream of life. Linguistic examples such as we have collected here need to be understood in relation to their deeper behavioural contexts: 'language-games' are behavioural syndromes from which speech-acts draw their sense. So can we now find some more general basis for our preliminary distinction, between 'high-grade' behaviour—about which the demand for 'reasons' is appropriate—and 'lower-grade' behaviour—about which no question of 'reasons' arises? And can we show how the special character of high-grade actions (actions done for reasons) determines how questions about an agent's 'reasons' are commonly to be answered?

To come straight to the point: all the examples considered here of actions done for reasons can be regarded as *applications of procedures* (methods of calculation, techniques, rituals or other formalized modes of behaviour) that we have *learned during our lifetimes*. If we pursue the implications of this fact, that will take us a long way towards unravelling the central tangle over 'reasons' and 'causes'.

(i) Let us return, first, to the most elaborate class of examples: namely, those in which the reasons we offer for our actions, attitudes or beliefs involve 'ratiocinative' procedures of deliberation, calculation or justification. Whether we are engaged in a piece of scientific research, in playing a game of chess, appraising alternative political strategies, or working out a legal manoeuvre: in all such activities, we are acting deliberately, in the primary sense of *deliberating*—applying procedures of deliberation which involve the consideration of arguments and, in a broad sense, of calculations. To engage in such activities at all, we must have a mastery of the relevant procedures; and no one could question that these procedures, at any rate, have to be acquired during our lifetimes. A physicist calculates the energy of a particle whose track appears in a bubble-chamber photograph; a chess-master considers how each of two alternative moves would be likely to affect his position four or five moves ahead; a statesman ponders a possible *démarche*, balancing up the merits of a show of strength against the risks of alarming his opponent; a company lawyer explores the foreseeable objections to raising new capital by debenture, rather than on bank loan or through a share issue. In each case—manifestly—the agent goes through complex procedures of argument which he acquired only during his lifetime, over an extended period of time. And notice: when we agree to be 'answerable' for performances of these 'high-grade' types, we do so just because we are applying rational procedures of types that we have learned to perform. Having mastered quantum mechanics, chess tactics, or the arts of politics or law, we are ready to make use of these arts and techniques; and it is then up to us to do so in an effective and justifiable manner.

Indeed: over these particular kinds of behaviour, we normally do not hesitate to take credit or blame. A perceptive explanation in particle physics, an elegant end-game play in chess, a striking political manoeuvre or powerful legal dodge: it is in such things as these that the 'craftsman of rational deliberation' (so to call him) finds his satisfactions. And conversely: scientists and chess-players, politicians and lawyers alike, would most readily blame themselves for using their rational procedures carelessly, ineffectively or unimaginatively. An accountant who added a column of figures too hurriedly, a chess-master who blundered into an elementary checkmate, or a lawyer who overlooked some straightforward objection, would in each case regard himself—quite properly—as 'answerable', for having carried through *inadequately* procedures of a type he had been trained to perform to a far higher standard.

(ii) To move on, next, to ritual and customary actions: these also involve 'reasons', just to the extent that they are *learned*. A new recruit, on joining the Boy Scouts or the U.S. Marines, has to pick up the customary, rule-governed actions with which he must respond to different situations, in his capacity as a Boy Scout or a Marine. Such formalized actions commonly become habitual, and cease any longer to require explicit ratiocination or calculation. Yet they, too, are the outcome of learning-by-drilling; and they, too, can typically be performed, only by an agent who has learned to recognize what response is called for in a given situation. Nor is this pattern confined to the specialized rituals of limited organizations: as A. J. Ayer demonstrates in his Auguste Comte lecture, such formalized patterns do much to determine the character of all social life (Ayer, 1964, pp. 7–8, 23–4). Only a man who has learned how the fluttering of a handkerchief can serve as 'waving goodbye' will be in a position to recognize how the words, 'I'm seeing my mother off to the seaside', can serve to answer the question, 'Why are you flapping your handkerchief like that?'

At first sight, it is true, this argument is in danger of becoming too intellectualistic. Where the 'reasons' for an action (e.g. holding three fingers up to your brow) do not go beyond the descriptive level—'I'm a Boy Scout, and this is the Boy Scout salute'—it seems somewhat pretentious to talk of the action as 'determined' by 'considerations' which 'carry weight', and so on, and so on. Yet, even here, to know the reasons for a man's actions enables one to see what considerations—in a weakened sense—weighed with him in acting as he did. For, within an organization like the Boy Scouts, or within society at large, one permanent consideration having a bearing on your conduct will be your standing within that social group. Of course, you will not normally stop and think *before* saluting of any decisive consideration *in favour of* saluting; since, for one who is an active Scout, it will be a standing rule to salute when occasion demands. On the other hand, an active Scout who deliberately *refrains from* saluting *will* be answerable for that dereliction; and his standing in the organization may therefore be an implicit consideration deciding him to salute, even where it is paradoxical to describe him as saluting 'in order to preserve his standing'. (Contrast the situation where somebody is in two minds whether to break with an organization: then we might indeed say, 'He is continuing to salute *merely* to protect his status in the organization, until he finally decides whether to break with it or not—it doesn't follow that his heart is in it'.) After all, then, formalized and ritual actions are performed in the light of considerations, and *learned* considerations at that: the difference is merely that in these cases the considerations carry weight implicitly, rather than being explicit.

(iii) Finally, let us turn to instrumental actions. We behave in many ways that have beneficial effects on us. Some we do not have to learn— normal blinking, for instance; others we have learned to perform, as ways of bringing about those effects. And, once again, the dividing line between those instrumental actions for which we would agree to be answerable, and so specify our 'reasons', and those beneficial actions and modes of behaviour about which no question of 'reasons' arises, runs between the *learned* and the *unlearned*. A man who 'just blinks' as we all normally do, autonomically, could not give 'his reasons for' blinking, and blinks without having learned to do so. But a patient who blinks deliberately, to stabilize his visual field after an ear-operation, has learned to use blinking as a means to the end: the onset of giddiness serves as a 'consideration', which lends great 'force' to his decision to start blinking. In learning how to use deliberate blinking for this purpose, he learns *ipso facto* to answer the question, 'Why are you blinking like that?' So here, too, the ability to give 'reasons' for an action—indeed, the very relevance or meaningfulness of those 'reasons'—is directly associated with the fact that the action is an application of a learned procedure.

What is the significance of this point? Recall Townes' Paradox: that brain-physiologists will wish to take credit for neurophysiological discoveries which—on a naive view—destroy the whole basis for 'responsibility' and 'credit'. For are not verbal responses and behaviour-patterns causally-produced, as much as any others? And how can we take credit for the causal outcomes of the neurophysiological processes in our brains? A classic neurosurgical case (reported by Foester, Gagel and Mahoney[1]) will make clear both the force and the weakness of this naive view.

A German patient occupying a responsible public office, complained of periodical disablement, of apparently neurological origin. Under direct surgical inspection, using a local anaesthetic only, a small tumour was discovered attached to the hypothalamus of his brain; and, when mechanical traction was applied to this tumour, the patient switched into a manic mood and burst out with a stream of verbal obscenities. Now, on the one hand, we might be tempted, first, to infer from this case that the patient's obscene language was released in a purely causal manner, by the local surgical handling of his hypothalamus; and, secondly, to extend this inference further, to the hypothesis that the release of all linguistic behaviour is dependent on appropriate cerebral causation. On the other hand, one small detail of the case is worth noting:[2] the patient immediately apologized for his bad language, saying, 'I can't think what *made* me say that!' The obscenities were something which he regarded as 'happening despite himself': they were not part of a behavioural syndrome for which he was ready to be answerable, and their occurrence—whose causal stimulus he was in no position to observe—filled him with embarrassment and confusion. This case, accordingly, illustrates *both* the probable universality of causal mechanisms *and* the abiding distinction between responsible and internally-caused behaviour.

In practice, then, we can still distinguish between actions which involve the use of learned procedures—for whose skilful application we are prepared to answer, and so, in the most literal sense, accept 'responsibility'— and behaviour which short-circuits all deliberation, like that stimulated by local handling of the brain. So, while it is legitimate enough to describe the verbal obscenities of the hypothalamically-stimulated as the effects of *causal compulsion*, it is entirely questionable to infer from this that *all* verbal responses are similarly the effects of 'causal compulsion', even in cases which manifestly involve ratiocination. Probably, indeed, causal mechanisms are present in all cases. Yet, where the phenomena themselves are so different, we must surely suppose that these mechanisms are operating in correspondingly different ways. Perhaps, again, there may be a kind of 'compulsion' present in some ratiocinative situations. Still, on the face of

[1] In 'The hypothalamus', *Res. Publ. Assoc. Res. Neur. Ment. Dis.* (1939–40).
[2] Reported verbally to Professor W. Nauta of M.I.T. by the late Dr Mahoney.

it, this will be merely the 'compulsion' of *compelling reasons*—reasons whose weight overbears that of all relevant counter-reasons—and this 'compulsion' does nothing to impugn our 'responsibility'. When it comes to bowing before compelling arguments, we do not thereby become automata, devoid of all responsibility. On the contrary, faced with unanswerable arguments, we show our responsibility in yielding to them.

The first outlines of a solution to our central problem will now be becoming apparent. The actions we perform 'for reasons' may all be subject, in a suitably qualified sense, to 'causality'—but it is to a causality all their own. When, e.g. in a ratiocinative, class (i) example, we specify the 'reasons' for which a man acted as he did, we note the key-features of his situation that figured in his deliberations: that is to say, the considerations which *carried weight* with him. In such a context, the phrase 'carry weight' is neither empty, nor merely figurative. The aspects of a man's situation which carry weight in his deliberations may indeed determine what he does, and they may do so as effectively as any 'internally causal' factor. Yet such aspects or 'considerations' are capable of carrying weight only because the agent in question has learned to recognize their relevance to the deliberative procedures in which he is engaged. (Only a man who knows how to integrate can recognize the significance of boundary conditions; only a man who understands tax-law will know how to respond to changes in interest-rates; and likewise for other modes of reasoning.)[1]

To generalize: actions can be done 'for reasons' only if they are actions of types which we *learn* to perform. That we do learn to act in these different ways—that our behaviour is modified by learning—is a manifest fact about human beings which psychology and neurophysiology may attempt to account for, but must not presume to explain away. And, whether we consider complex, ratiocinative conduct, or formalized customs and rituals, or the instrumental actions we use as means of achieving certain results—in each case, our actions can be accounted for completely only by

[1] Here we may take up once again the question about 'reasons for liking or disliking' raised above in note 1, page 7. To deal with this question briefly: The question 'Why do you dislike him so violently?' may be taken in either of two ways, or in both, depending on the example. It may mean 'What do you take such grave exception to (find open to such objection) in his attitude, behaviour or manner?'; or it may mean 'What is it about him that triggers off so uncontrollable a distaste in you?'; or it may mean 'What is the nature of your objection to, or your distaste for him?'

In the light of our discussion in this essay, it can now be argued that 'taking exception' is yet one more instance of applying a learned procedure, while 'experiencing an uncontrollable distaste' approximates rather to autonomic behaviour. To the extent that this is so, we can give reasons *for* our dislikes, when these are based on objections; but our sheer distastes can be accounted for only in terms of reasons *why*. At the same time, it must be noted that this distinction is normally harder to draw in the case of likes and dislikes than it is in the case of actions—likes and dislikes are only marginally *disputanda*.

referring to the 'considerations' which 'carry weight' with us: to invoke inner 'causal mechanisms' alone is not enough. This is not to question that the things we do on account of our customs, techniques and deliberations, somehow involve our brain-mechanisms, just as much as the things we do on a pure stimulus–response basis—if not more so. It would, indeed, be surprising if the rational skills and capacities which we learn so effectively, and put into action so regularly, did *not* have some neurophysiological basis. Yet the existence of such an internal neurophysiological basis, if considered in isolation from the 'weight' which different external considerations 'carry' for us, fails to make fully intelligible the actions which we do 'for reasons'. Our capacity to recognize features of the situation in which we have to act as providing 'reasons' for acting in one way or another, and to act accordingly, is something which cannot be ignored, even in a complete *causal* account of our actual behaviour. And just how the 'reasons' which account for our actions come to 'weigh' with us, and so 'move' us, is a question we must face directly, if we are to understand from what source they derive their peculiar kind of 'causal efficacy'.

### 5. THE CO-EXISTENCE OF REASONS AND CAUSES

René Descartes' dichotomy between Mind and Matter, and Immanuel Kant's dichotomy between *noumena* and *phenomena*, were both introduced —among other things—to deal with the perplexing contrast between ratiocinative conduct and purely causal responses. Yet this perplexity (we can now argue) might have been forestalled, instead of explained, if only the philosophers had paid enough attention to the *learning* on which all rational conduct depends.

The basic source of our problem is this: That the philosophical discussion of 'reasons' and 'causes' in human behaviour too often begins, and ends, with the consideration of *adult* behaviour. Learning and education are taken for granted. Professional training, character-formation, personality-development, all the refinement of skills and responses, discriminations and perceptions: the philosopher does not pause to look at these, but rather considers them as completed—and is then puzzled that the adult, waking end-product of this long process displays behaviour which differs so strikingly from that of the unconscious, the intoxicated or the brain-stimulated. And of course: if we shut our eyes to all those steps by which the differences between behaviour done 'for reasons', and behaviour triggered off in a 'purely causal' way, are *progressively established* within an agent's lifetime, we may well be perplexed by the magnitude of these differences, and be tempted as a result to consider the distinction *absolute*.

Yet the contrast between a jungle and a garden is not an *absolute* one. The same mechanisms of growth operate in each case; but in the garden systematic selection, transplantation and training have imposed an orderly pattern on their results. Nor are we compelled to treat the contrast between actions done for reasons and responses to causal stimuli as an absolute one, either; if only we are prepared to acknowledge how, in this case too, selection, discrimination, reinforcement and training have had their effect. (It is no accident that the vocabularies of gardening and education overlap.) In a sufficiently extended sense, education, too, is surely a 'causal' process. Considerations weigh with us because we have learned to appreciate their 'force': they come to carry weight with us (that is) in the course of our experience and education. An educated and experienced man necessarily reacts to the situations he encounters in a different way from a primitive or an infant; and his understanding of those situations is shown, not least, by his acquired ability to recognize relevant factors—i.e. those factors capable of 'moving', or 'having force for' him in his ratiocination.

To return to Wollaston's argument, as quoted from Priestley in our introduction:

When I begin to move myself, I do it for some reason and with respect to some end. But who can imagine matter to be moved by arguments, or ever ranked syllogisms and demonstrations among levers and pulleys?

Wollaston continues:

Do we not see, in conversation, how a pleasant thing will make people break out into laughter, a rude thing into a passion, and so on. These affections cannot be the physical effects of the words spoken because then they would have the same effect, whether they were understood or not. It is therefore the sense of the words which ... produces those motions in the spirits, blood and muscles (Priestley, in Passmore (ed.), 1965, p. 121).

Since Wollaston understands 'the sense of the words' as 'an immaterial thing', he banishes it from the world of 'physical causes'. Yet, as Priestley immediately retorts, this argument begs the fundamental question:

Since it is a fact that reasons, whatever they may be, do ultimately move matter, there is certainly much less difficulty in conceiving that they may do this in consequence of their being the affection of some material substance, than upon the hypothesis of their belonging to a substance that has no common property with matter ... To say that reasons and ideas are not things material or the affections of a material substance, is to take for granted the very thing to be proved (Priestley, in Passmore (ed.), 1965, pp. 121–2).

The role of reasons in our actions may not be crudely physiological, yet it does not follow that 'reasons' and 'causes' must be distinguished so absolutely as to become incompatible. We learn to understand the sense of the words we hear, just as we learn to respond to other external factors; and the fact that men respond to a language intelligently only if they understand it can prove only this—that having learned the language is *one* of the necessary causes, without which they would not have responded as they did.

Similar arguments can be brought to bear on the current philosophical debate. Let us start from the position stated in A. J. Ayer's Auguste Comte Lecture. Ayer attributes to Ryle and Wittgenstein the thesis that motives —and this includes reasons—'are not causes' (Ayer, 1964, p. 12). In the light of a variety of examples, he is led to question this thesis, and concludes himself that—on the contrary—motives and reasons can quite legitimately be 'assimilated to causes' (Ayer, 1964, pp. 24–5). Yet, one must ask, are we really obliged to settle on one or other of these two positions? Can we describe the relation between reasons and causes adequately in either of these two forms? We must, of course, make some allowances for the ambiguities of the term 'motives', which is so much more comprehensive than the narrower term 'reasons': still, if one poses Ayer's question in the sharpened-up form, 'Are *reasons* causes or not?'— 'Can *reasons* be assimilated to causes or not?'—the implications of our analysis will be clear. Statements about a man's reasons for his actions are not in competition with statements about the causes of his behaviour. The two sorts of statements are made on quite different levels. The operative question is, not whether 'reasons' can or cannot be *equated* with 'causes': it is, rather, whether justifying a man's actions (in terms of 'reasons') is *compatible* with explaining it (in terms of 'causes'). And, up to this point, we have found no adequate argument for supposing that they are not compatible.

Talking, first, on a causal level: what may be 'assimilated to causes' is not a man's *reasons* for an action, but rather his *having reasons* for that action—i.e. his *recognition of* those reasons as having weight for him in his particular situation. (Ayer's illustrative examples suggest that he himself need insist on nothing more than this (Ayer, 1964, pp. 23–4).) For what can serve as a 'motive', and 'cause' a man to act, is not a bare verbal argument, but his acceptance of that argument as 'compelling'; and the task of diagnosing the causes of his action involves considering his entire situation as a totality—including his prior attitudes and skills, and the factors he will accordingly recognize. Considering such a situation from the outside, we might well have occasion to conclude that the effective cause of a man's action was the coming-to-his-attention of some new 'consideration'. ('Finding his wife's love-letters moved him—i.e. gave him a compelling reason—to act as he did.') In itself, a reason neither has, nor fails to have, causal efficacy: it is not a fact or event, but a consideration. What *can* have causal efficacy is the fact or event of a man's having, being given, recognizing, or acknowledging the force of, that reason.

By contrast, taking up the 'rational' point of view involves considering a man's action, not *from the outside*—from a clinical, diagnostic standpoint —but *from within the context* of the action itself: enquiring, 'How far can this action be justified, in the light of factors/considerations/possibilities which the agent could recognize and respond to?' In discussing an action from the rational point of view, we need be understood *neither* as asserting *nor* as denying anything about its 'causes'. We may infer, diagnostically, that the agent was 'caused' to act as he did, by accepting some reason-for-acting as a conclusive one; but this has no bearing either way on the justificatory question, whether that compelling reason was as uniquely appropriate as he supposed to the actual demands of his situation.

In short: the reasons for which we act play an indispensable role within the overall network of causes by which our behaviour is determined. But the *statement* of our reasons for behaving as we do is never, in itself, a *statement* of the causes determining that behaviour. The two kinds of statement are neither equivalent nor inconsistent. They simply by-pass one another.

Granted: once we have recognized the authentic differences between statements about the 'reasons for' behaviour, and statements about the 'causes of' that same behaviour, we may be tempted to register their authenticity by asserting baldly that *reasons are not causes*. But this particular formulation can well prove gravely misleading; since it may all too easily lead others to conclude that having reasons for actions is incompatible with there being causes of those same actions. Alternatively: once we have recognized how a man's acceptance of reasons as conclusive may serve as a necessary and sufficient condition of his actions, we may be tempted— equally—to assert baldly that, after all, *reasons are causes*. Yet this, too, can be just as misleading a formulation; since it may all too easily conceal the differences between *statements* of reasons and *statements* of causes— between appraisals of actions, in relation to norms and standards, and diagnoses of those same actions, considered as phenomena within a causal economy.

It was, indeed, one of Kant's great insights that reasons were neither the same as, nor necessarily incompatible with, causes; but that, on the contrary, the rational justification of human action was potentially compatible with the belief that *acting-for-reasons* takes place within a complete causal economy. Kant's final account of the matter, given in terms of *noumena* and *phenomena*, may have been unhappy; but that was for extraneous reasons, since the other implications of the *noumena/phenomena* dichotomy were so obscure. Taken by itself, however, his proposed resolution of the problem of reasons and causes—though programmatic—appears quite legitimate. For, considering this problem in isolation, the world of *noumena* reduces to the 'universe of discourse' in which we appraise the *justification* of human actions; while the world of *phenomena*, by contrast, refers to that other 'universe of discourse' concerned with causally-interpretable associations in the natural world. So re-stated, Kant's view that the rational world of *noumena* and the causal world of *phenomena* 'co-exist' implies precisely what we have tried to establish here: namely, that the giving of reasons in justification of our actions is distinct from, yet compatible with, the discovery of causes in explanation of those actions.


## 6. TELEOLOGY AND PHYSICAL SYSTEMS

All behaviour is explicable: some actions are also justifiable. Everything we do has causes: some of the things we do have reasons, too. When our actions are done for reasons, those reasons enter into the causal explanations of the actions. But they do so only indirectly, by way of the 'rational arts'—moral reflection, practical deliberation, intellectual calculation—which are inculcated in us through education and experience. And they do so without losing their rational character, of 'having force for us' rather than 'forcing us', of 'carrying weight with us' rather than 'overpowering us', of being 'compelling' rather than 'compulsive'. That, in five sentences, is the outline solution to which this discussion of our problem has led us.

Some will protest that this so-called solution is excessively tame. Surely —they will reply—the differences between acting-for-reasons and behaving-in-a-causally-determined-way cannot be reduced to a difference between *façons de parler*. Yet such a retort will miss our central point. The contrast between talking about 'compelling reasons' and talking about 'causal compulsions' is not the main substance of our conclusions: it is merely an index, reminding us of the more fundamental contrast between certain learned modes of behaviour (ratiocinative, instrumental and formalized) and those unlearned responses which operate autonomically, regardless of all learned procedures and techniques. In the contrasts between 'reason-talk' and 'cause-talk' there are preserved, and made explicit, behavioural distinctions which philosophical prejudice might otherwise tempt us to deny. The virtue of focussing attention on the linguistic differences is simply this: that they can inoculate us against the temptation to oversimplify the inherent variety of human behaviour.

Even so, many readers will object that the present essay does not tell the whole story. Yet that, of course, is something it does not pretend to do: all it aims to do is to locate an essential starting-point for any fuller account. This being so, different readers will wish to press onward in different directions. Some will claim that our conclusions still understate the gulf separating rational, intentional, goal-directed actions from causal, mechanical, reflex bodily movements. (Indeed, Charles Taylor's book on *The Explanation of Behaviour* (1964) sets out to demonstrate that this gulf is absolutely unbridgeable: if ostensibly genuine human actions did prove to be explicable in terms of underlying causal mechanisms—he concludes —then all discussion of 'intentions' and 'reasons' would lose its meaning. In that case, our actions would cease to be genuinely 'teleological', and would become mere reflex bodily movements—devoid of all intention, significance or rational justification.) Others, meanwhile, will pursue our conclusions in a rather different direction. Our distinctions are all very well so far as they go (these readers will argue) but, since they are largely semantic, they may lead us to exaggerate differences that lack all deeper scientific relevance. For, once we concede that all talk about 'reasons' is compatible with our behaviour's being causally determined, the last real intellectual barriers separating men from mechanisms will have been dismantled. (This—as I read it—is the force of Stuart Sutherland's argument in reply to Charles Taylor's position.)

Yet both these continuations will—in different ways—undo the modest good this essay was designed to achieve. And the result is to leave Taylor and Sutherland at cross-purposes, in consequence of their reliance on certain ambiguous words, which we have taken care to avoid. For instance: Charles Taylor's argument, to demonstrate that a genuinely 'intentional' action could not have an underlying 'mechanistic' explanation, rests on a definition of 'mechanistic' explanation taken, not from a study of actual scientific practice, but rather from the writings of logicians. Taylor shows that the formal patterns used by logicians to characterize 'causal explanation' make nonsense when applied to 'intentional' or 'teleological' conduct; and thereupon regards his point as sufficiently made (Taylor, 1964, Chaps. 1 and 2, esp. Sec. 2.3, pp. 37–45). Yet the arguments by which he distinguishes 'intentional human actions' from 'causal processes' (in the formal logicians' sense) are not such as to be applicable to human actions alone. On the contrary, Taylor's arguments take precisely the same form as those which other writers advance to demonstrate that the logicians' account of 'causal processes' misrepresents the real-life 'mechanistic' explanations of physical science also. So the difference between 'teleological' and 'mechanistic' explanation does not correspond—as Taylor claims—to the difference between a medieval physics based on Aristotle and a modern physics based on Galileo, Descartes and Newton. On Taylor's definition of the term 'teleological', both pre-Galilean and post-Galilean physics employ 'teleological' explanations.[1] As a result, the idea of a 'mechanistic system' capable of 'intentional action' is nothing like the contradiction Taylor supposes it to be.

Sutherland's argument is open to question in the opposite direction. He may turn out to be justified in his *negative* claim: that it would be a mistake to rule out beforehand, absolutely and in principle, all application of 'ratiocinative' terms to machines of sufficient complexity—in the belief that this would destroy the essential significance of 'rationality'. But his *positive* thesis, that the brain is 'a physical system', simply trades on the ambiguity of the word 'system' to conceal all those genuine differences which have traditionally been invoked to justify the distinction between deliberate and autonomic behaviour, and so between reasons and causes.

In one sense, of course, the brain *is* 'a physical system': viz. that we need not suppose that any of the physical and chemical processes taking place within it violates the general principles of physics and chemistry in doing so. However, once we begin to study the brain as a functioning organ involved in (say) vision, we are compelled to treat it, not merely as 'a physico-chemical system', but also as 'a physiological system': or rather, as *part* of such a physiological system, in which it is linked up into a complete sensory–motor circuit, together with the eyeball, retina, optic nerves, motor nerves, and the muscles attached to the eyeball. It is the overall functioning of this entire circuit that physiologists must in the long run explain, and the integrity of the whole system is essential to this functioning: frustrating its operation at any point may frustrate this functioning, so that any impulse to press the question where exactly 'the real seeing' goes on should probably be resisted. (The muscles on the eyeball, for instance, may seem of secondary importance; yet by now it is well established that, if eye-movements are prevented and the retinal image is thereby stabilized, the visual discriminations which constitute normal 'seeing' become impossible.[2]) When we go further, and consider the role of the brain in deliberation and action, we must widen our span of attention again, so as to treat the entire nervous network—sense-organs, brain and all—as forming part of a still wider 'system': one which associates the agent with all those aspects of his environment that can serve as

[1] See, for instance, Mischel, 1966, pp. 40–60, esp. pp. 43–51; and also cf. Toulmin, 1961, Chap. 3, and Toulmin, 1966, pp. 129–33.

[2] Out of an extensive body of work on the visual system, in all its complexity, one may refer particularly to the work of Ditchburn on the stabilization of the retinal image: cf. Ditchburn & Ginsburg, 1950, p. 36.

'relevant considerations' in the course of his deliberation. This done, anyone is at liberty to call the brain 'a physical system' if he chooses; since the physical and chemical properties which the brain manifests when studied in isolation will then be entirely compatible with the more complex roles it performs within wider physiological and cognitive systems.

What is liable to lead to misunderstanding is, first, insisting that the brain is a physical system, and then operating with an oversimplified model of what 'a physical system' is; and here the manner in which Sutherland describes the operation of the central nervous system gives rise to legitimate hesitations. For it is essential throughout to do full justice to the *internal* systemic character of the CNS, rather than considering the peripheral nerves merely as chains of neurons passing on linear cause-and-effect influences up to, and away from, a central computer-style brain;[1] and it is essential also to recognize fully the systemic *interaction* between the entire CNS within, and the environment outside, without which deliberate behaviour must be physiologically unintelligible.

In fact, any picture of the nervous system as operating by the linear transmission of 'impulses' along cause-and-effect pathways, like bicycle-chains, represents a hangover from an out-dated physics.[2] Within the framework of the seventeenth century 'mechanical philosophy', which attempted to explain all natural phenomena in terms of the transmission of 'motions' or 'vibrations' on contact or impact, there may have been no alternative. (To that extent, Hartley's neurological 'vibratiuncles' made the best of an oversimplified view of matter.) But, by now, the natural philosophy of contemporary science has abandoned the simplistic mechanical models of seventeenth-century physics, and is 'mechanistic' only in a much more extended sense. All the way up the scale from elementary quantum mechanics to the neurophysiological bases of cognition, twentieth-century scientific ideas require us to consider the systems we study— whether physical or physiological—as integrated functional totalities; and the organization of these systems is such that their overall structures and activities need to be *analysed together*.

The neurosciences, then, can hope to explain the causal mechanisms

[1] Cf. Sutherland, 'Is the brain a physical system?', in this volume, pp. 97–122. Note especially the emphasis Sutherland places on the role of the brain as a central processing-organ, which acts as the channel by which microscopic 'impulses' from the receptors *cause* similar 'impulses' in the efferent nerves. Apart from the fact that these 'impulses' are electrical rather than mechanical, the implicit model is identical with that underlying Hartley's 'vibratiuncles' in the early eighteenth century.

[2] The argument presented here is set out much more fully in another essay: Toulmin, 1967. That essay analyses the changes which have affected the idea of 'mechanism', and the consequential ideal of 'mechanistic explanation', over the last 350 years; and throws serious doubt on any suggestion that physical scientists have been operating with a single, unchanging ideal of explanation throughout the period since Galileo.

underlying human actions as involving physical—and physiological—systems; but they can do so only on certain very definite conditions. For any adequate account of human action and understanding necessarily starts with a totality: comprising the human agent (central nervous system and all) as shaped by genetics, education and experience, together with the environmental situation in which he finds himself, embracing all the thousand external 'factors' or 'considerations' which the agent may either respond to, or ignore. The resulting account can claim to be 'mechanistic' in this sense: that all the specific brain-processes it invokes conform to the same general principles as the rest of physics and physiology. But it cannot usefully be based on too crude a model of what a 'physical system' is, and is capable of doing; for all deliberate human actions depend in part on external, historically-conditioned 'factors' and 'considerations'. So neurophysiological brain-processes will always be part of a much larger story, comprising also rational and historical elements.[1]

A neuroscience capable of doing justice to the part which the central nervous system plays in human action must, therefore, give a physiological account complex and sophisticated enough to allow for all those 'high-grade' types of behaviour which are shaped by history, and by the development of the individual's intellect and personality: viz. the employment of deliberative procedures, the formulation and execution of intentions, the making of scientific discoveries, the conception of works of art. Given these qualifications, the ambitions of the neuroscientists need not be unrealistic ones. Rather, it is the fears which prompt philosophers to distinguish 'intentional actions' *absolutely* from 'causal bodily movements' which are unrealistic. For, although human behaviour may be subject to causality as much as all other phenomena, causal accounts of rational, intentional conduct will be plausible, only provided they include references to the learned procedures—ratiocinative, instrumental or formalized—which are applied in such conduct.

Given these qualifications, too, no paradox need follow from Townes' hypothesis: that strictly causal brain-processes may be found to underlie all rational thought-processes, *including* the discovery that strictly causal brain-processes underlie all rational thought-processes. Had scientific thought been a mere 'secretion' of the brain—if a scientist's writings flowed from his hand compulsively, like obscenities from the mouth of a hypothalamically-stimulated patient—then scientists would indeed be left

---

[1] The reconciliation of 'rational' and 'causal' explanations in psychology proposed in this essay closely resembles, in form, the reconciliation of 'physicochemical' and 'vital' explanations in physiology worked out by Claude Bernard. This is no accident, for the methodological issues involved are very similar in both cases: on this point, see Goodfield, 1960, esp. Chaps. 7 and 8.

with nothing in which to feel personal pride. Yet, if the present essay has demonstrated anything, it is this: that the discovery of neurophysiological correlates to both autonomic and ratiocinative behaviour need do nothing to destroy the distinction between them. And since, by itself, the existence of neurophysiological correlates leaves our rational thought-processes as 'rational' as ever, the neuroscientists are welcome to discover these correlates, and to take credit for their discoveries as well.

One last caution is necessary. Suppose that, for the time being, no neurophysiological models can be devised capable of providing correlates of deliberation, intention and the rest: then, in the meanwhile, scientists must possess their souls in patience. To use the insufficiency of existing physiological models as a ground for denying the undeniable—namely, that we *do in fact* employ deliberative procedures, formulate and execute intentions, make scientific discoveries, and conceive works of art—would simply be to falsify the evidence. As Kant saw clearly, the existence of rationality is as much an indispensable axiom of all science and philosophy as the existence of causality. The only problem is, to make explicit the terms on which they can co-exist.

### REFERENCES

Ayer, A. J. 1964. *Man as a Subject for Science*. London.
Ditchburn, R. W. & Ginsburg, B. L. 1950. Vision with a stabilized retinal image. *Nature* **170**.
Goodfield, G. J. 1960. *The Rise of Scientific Physiology*. London.
Holz, W. C. & Azrin, N. H. 1966. Conditioning human verbal behaviour. *Operant Behaviour: Areas of Research and Application*. Ed. W. K. Honig. New York.
Mischel, T. 1966. Pragmatic aspects of explanation. *Philosophy Sci.* **33**.
Passmore, John A. (ed.) 1965. *Priestley's Writings on Philosophy, Science and Politics*. New York.
Peters, R. S. 1958. *The Concept of Motivation*. London.
Priestley, J. 1777. Objections to the system of materialism considered. *Disquisitions Relating to Matter and Spirit*.
Priestley, J. 1965. *Priestley's Writings on Philosophy, Science and Politics*. Ed. John A. Passmore. New York.
Taylor, C. 1964. *The Explanation of Behaviour*. London.
Toulmin, S. 1953. *Philosophy of Science*. London.
Toulmin, S. 1961. *Foresight and Understanding*. London.
Toulmin, S. 1966. Are the principles of logical empiricism relevant to the actual work of science? *Scient. Am.* **214**.
Toulmin, S. 1967. Neuroscience and human understanding. *Neuroscience: A Study Programme*. Eds. G. Quarton, F. Schmitt & E. Melnechuk. New York.

# COMMENT

## *by* R. S. PETERS

### INTRODUCTION

I have sympathy with the lines along which Toulmin has attempted both to retain and to reconcile the dichotomy between reasons and causes. To distinguish the 'language game' in which actions are justified from that in which they are explained and to point out that they are distinct from but compatible with each other certainly throws light on some of the puzzles in this area. But it does not throw quite enough light; for some of the murkiest problems are located precisely at the points where these 'language games' intersect and these are the regions which Toulmin leaves too much in the shadows. His attempt, too, to make explanatory discourse dependent on justificatory discourse does not really work when a wider range of examples is taken. But it is ingenious and interesting nevertheless.

Looked at in a global way, for instance, one could say that the cluster of problems connected with free will and responsibility are located at the point where explanatory theories of human behaviour intersect with justifications and excuses. Similarly in ethics generally it is accepted that generalizations about human behaviour cannot provide reasons of a justificatory sort for conduct. But, nevertheless, such generalizations are relevant in various ways to justification; the problem is to see the kind of relevance which they have. It is easy enough to refute ethical naturalism, but it is a much more difficult and important task to show the points in ethical theory where generalizations about 'human nature' must have a place.

Toulmin, in dealing with the explanation of human behaviour, makes a move corresponding to that made by naturalists in the justification of conduct. He gives the inner track to justificatory discourse just as they give it to explanatory and descriptive discourse. There may be something to be said for this move in some areas of behaviour; but it is incumbent on anyone who makes it to consider a wide range of behaviour and not to confine himself to a contrast between involuntary behaviour and highly institutionalized behaviour as Toulmin tends to do. To put my point more historically: Toulmin's treatment is much influenced by Kant's. He puts in a more modern way Kant's distinction between 'the world of noumena' which 'reduces to the "universe of discourse" in which we appraise the justification of human actions' and 'the world of phenomena', which 'refers to that other "universe of discourse" concerned with causally-interpretable associations in the natural world'. But what Kant never succeeded in showing is how respect for the law or for rational beings as the source of law could ever induce anyone to act. I think there are similar problems in Toulmin's refurbishing of the Kantian approach.

Let me get out of the way, first of all, the notion that anything depends on the use of the word 'cause', in relation to the type of explanation given. I agree with Toulmin that 'causal questions will not be found to arise so long as we confine ourselves to appraising "reasons for action" directly' but that once they do arise 'accepting a consideration (recognizing its force, acknowledging it as a reason)' may cause a man to act just as much as movements in his brain or chemical changes may cause other things. But I do think that a closer examination is required of notions like 'accepting the force of a consideration' in an explanatory as distinct from a justificatory context, and especially their relationship to the concept of 'wanting', which Toulmin more or less omits.

I propose therefore to examine (1) Toulmin's account of justificatory discourse, (2) his account of reasons as causes, (3) his views about the importance of learning in the sphere of reasons for action and (4) his account of the relationship between mental and physiological causes. My conclusion will be that what he says about (1) is acceptable but incomplete, about (2) is ingenious but untenable, about (3) largely superfluous and about (4) true but irrelevant to the main point at issue.

## I. GIVING REASONS FOR OUR ACTIONS

Toulmin starts his account of acting 'for reasons' by eliminating certain irrelevant examples. He first of all dismisses cases like that of a girl blushing where we speak of the 'reason why' she blushed, viz. someone mentioned the young man she secretly loves, which could not be described as her reason for blushing. He links this with a case of a man driving dangerously —by reason of intoxication. These cases, however, are very dissimilar and a consideration of the first case opens up wider issues.

They are dissimilar first because 'intoxication' explains not the driving but the manner of the driving whereas it is simply the blushing that is explained in the first case. Secondly, blushing is not something that we do, like driving; it is something that happens to us. But it does not happen to us, like the lapses from skill in the case of driving dangerously, because of chemical conditions induced by alcohol; on the contrary it happens to us because of the way in which we view or understand a situation. The mild impropriety of talking about the girl's reason for blushing does not derive, in other words, from the absence of 'considerations'; rather it derives from the fact that we do not usually talk this way when we are in the realm of the involuntary. But the impropriety is surely only mild; for we might easily say to the girl: 'Look, you have no reason to blush. What is wrong with being in love with such an irresistible man? He isn't married, is he?' We would, in other words, be trying to alter her view of the situation, her estimate of relevant considerations. Cases like these are palpably not outside the scope of 'reason'. If they were we would have no basis for making the distinction, which we do make, between justified and unjustified emotional reactions. What Toulmin really means is that cases like these are not cases of actions, but of reactions. His initial dictum: 'All behaviour is explicable: some actions are also justifiable' is a bit too epigrammatic. For he ignores the possibility of reactions as well as actions being justifiable or unjustifiable.

In most cases, where taking account of considerations can be linked with reactions, it can also be linked with actions. The appraisal, in other words, can form part of a motive for action as well as of an emotional reaction (for distinction see Peters, 1961–2). The girl, for instance, might have walked away when she heard the young man's name mentioned. We would then say perhaps that she acted out of shame. And there could then be arguments about the appropriateness of her action as well as of the appropriateness of her understanding of the situation. We might similarly argue about the appropriateness of Othello's jealousy both in respect of his understanding of Desdemona's situation and in respect of his actions which were done in the light of his understanding. But supposing someone maintained that 'jealousy' was linked with ways of viewing situations that were never appropriate. Presumably a lengthy discussion might ensue about the appropriateness of notions to do with rights, ownership, exclusive relationships, etc., and still remain in a more or less self-contained realm of justificatory discourse. But could the same sort of discussion ensue for long if fear were taken, instead of jealousy, as the appraisal that was being questioned? There might be preliminary discussions about what things were or were not dangerous, i.e. about what the appropriate objects of fear were. But supposing someone said 'But you never ought to feel fear'. What could the reply be? Surely only that some things really are dangerous—i.e. they are likely to hurt or to lead to injury or death. What could be said, then, to a person who replied 'What if they do?'.

We would not, of course, be so bewildered by a person who did not regard this as a decisive consideration when viewed alongside other aspects of a situation. For many people regard pain or even death as less undesirable than other alternatives open to them. But what would bewilder us is a person who did not consider this a relevant consideration at all. This would not be just because it seemed very cogent to us personally; for we are aware of individuals who have idiosyncratic and seemingly arbitrary stopping points in the citing of considerations. It might not necessarily be because of our lack of acquaintance with anyone who was not moved by the thought of danger or death. For it might be that we had never met anyone who was a stranger to envy or jealousy but we could nevertheless imagine people living in a different type of society who might never look at situations in these ways. We would be bewildered, surely, because it would be difficult for us to conceive of a being who would not be, to a certain extent, moved by the consideration that he would be hurt or killed. If this does not count as a reason for avoiding something, we might think, what kind of inter-subjective content could we ever give to the notion of a negative consideration? Without some common anchorage such as this how could we ever understand the sorts of considerations adduced by people living under such a variety of institutions? In other words some very general notions of 'normal human nature' are necessary to give content to the notion of a reason for acting. Some very general notions of what men *qua* embodied men *tend to accept* as reasons for action is necessary to give content to our justificatory discourse. It is also probably the case that it is in the context of considerations such as that of pain and injury that the conceptual scheme of reasons for action is learnt.

The question now arises about the status of generalizations of this sort in relation to justificatory discourse. Generalizations about human nature often, of course, feature in justification when they are made relevant by a normative principle. For instance the fact that the breaking of promises tends to make people suffer is accepted as a reason against breaking promises if the principle is accepted that what makes actions right or wrong is their effect on human welfare. But the generalization with which we are here concerned does not enter justificatory discourse in this way; for it is postulated as underpinning in some way the fundamental principle connected with the avoidance of suffering, which makes other empirical generalizations relevant. Its status is much more that of a *presupposition* of justificatory discourse. Without a grounding in such universal features of human nature how could justificatory discourse ever function as a public form of discourse for getting people to act?

Another interesting point, too, about this type of generalization is that it has a peculiar status as a generalization. It is not an established empirical generalization that people invariably tend to avoid pain. Some kind of 'other things being equal' clause has to be added which indicates the need of a special explanation if somebody does not avoid pain. Neither is it a straight contradiction to say 'he realized that it was going to be painful, but he took no steps to avoid it'. The point is that the tendency to avoid pain is connected in our minds with our notion of a normal human being and this notion is developed to a large extent by our own experience and by our experience of what others tend to say and do. We need special explanations for people like masochists because they do not seem to accept the view that, if anything is a reason for not doing something, the fact that it will involve pain is a paradigm case of a reason. The avoidance of pain thus occupies a kind of twilight status between our understanding of what causes people to do or to avoid things and our convictions about what, in justificatory terms, are weighty considerations. Toulmin, as a Kantian, must surely look favourably on the suggestion that Kant's problem of the connection between rational acceptance and action might be dealt with by making explicit a few presuppositions of this sort.

Considerations relating to pain and death have been introduced into the argument so far in order to locate a region where justificatory and causal types of discourse intersect. They are also a particular case of 'compelling reasons' of a sort that have no niche in Toulmin's positive account. With what he actually puts into this account there is little cause to quarrel. Indeed, as he observes, much of it is an elaboration of distinctions that I tried to make within what, in *The Concept of Motivation* (1958), I called 'the purposive, rule-following model', though even in this Toulmin leaves on one side complexities connected with concepts such as 'motive' and 'need' which, as I then tried to point out, convey both justificatory and explanatory suggestions. On his view the giving of reasons can relate to straightforward cases of moral or legal justification, to cases where intentions are signalled, perhaps to demonstrate the instrumentality of a piece of behaviour, and to cases where an action is classified as being in conformity with a rule. He also distinguishes these types of cases from others where there is no appropriate appeal to reasons because the action is autonomic, uncontrolled, inattentive or otherwise 'lower-grade' conduct. But what he does not dwell upon is the 'compelling' features which a few reasons within this general family have which are not shared by others. Toulmin has, of course, a place for the notion of a 'compelling reason'. It is one 'whose weight overbears that of all counter-reasons'. The 'weight' derives from the fact that the agent 'has learned to recognize their relevance to the deliberative procedures in which he is engaged'. And there is a very large number of cases that can be dealt with in this way. As Toulmin himself puts it: 'Only a man who understands tax-law will know how to respond to changes in interest-rates.' The question, however, is whether this notion of relevance within a deliberative procedure exhausts the notion of a 'compelling reason'.

Consider, by contrast, cases like those of a man who has not eaten for a long time, or of a man deprived of water in a desert. When we talk of a 'compelling reason' in such cases we are not simply thinking of courses of action such as moving towards a restaurant, or an oasis in so far as they are perceived as relevant in a deliberative process, concerned with attaining obvious goals. The 'weight' of considerations, in other words, does not derive simply from the logic of the situation; it derives much more from what might be called the 'peremptoriness' of the goals in question. It is in relation to goals such as these that the classical theory of 'drive-reduction' has some plausibility. We are justified in using terms such as 'peremptoriness' and 'drive' about such forms of behaviour partly because of the felt over-ridingness of the goals in question rather than of means estimated to lead to them, and partly because of our general knowledge about the peculiar features of these goals under certain conditions. We know, for instance, that when men are well advanced in these deficit states no other types of goal will seem equally attractive. When in such states we not only grasp the force of considerations to do with—e.g. food; we also know that they will *as a matter of fact* outweigh other considerations. In respect, too, of how a man sees his situation these states of mind have a special character. It is not just that they are obsessive and crowd out other forms of sensitivity; for jealousy, envy, and ambition can do this. It is also that the way in which the world is perceived is intimately connected with the satisfaction of the specific goals in question in a way in which it is not in other forms of motivated behaviour which are not so closely tied to deficit states. An envious or jealous man, for instance, may well tend to see situations overwhelmingly in terms of others having things which he wants or to which he thinks he has a right. But what he can do which is appropriate to these appraisals is not always very clear or specific. With hunger, thirst, and with some cases of sexual desire, on the other hand, what has to be done is obvious and specific. For the appraisal of the situation is tied to the lack of the state of affairs that is wanted. Sensitivities and responses are sucked up in a jet of appetite.

In cases like these it is because the scope for effective deliberation is so circumscribed, rather than because the conclusions of deliberation are so obvious, that we speak of 'compelling reasons'. And these sorts of compelling reasons are very pertinent when questions of responsibility for actions are raised. Toulmin rightly says that our responsibility is not impugned when we bow before compelling arguments. But surely it is very much in these other drive-reducing cases of compulsion? In such cases of diminished responsibility, compelling reasons of this sort might, of course, excuse rather than justify a piece of behaviour. Or they might not even excuse it. But the particular type of assessment of behaviour does not affect the postulated connection between our assessment of people's reasons for acting and our understanding of what a normal human being would do who is in a state of this sort. Indeed the dependence of types of assessment on such understanding very much reinforces the main point that I am making. In appraising, therefore, the behaviour of people who act with compelling reasons of this sort, our notion of what counts as a weighty reason for acting depends to a large extent on our understanding of what we and people generally tend to do in situations like these. Again, as in the case of fear of pain or death, what counts as a justificatory reason depends in part on a particular type of empirical generalization.

In my *The Concept of Motivation* (Peters, 1958) I maintained that the concept of 'motive' itself is used in justificatory types of context when somebody is up for assessment and we ask for knock-down explanations of the untoward. I suggested that we are prepared to rank 'jealousy', 'ambition', etc., as 'motives' because of their explanatory status in our culture. They are ways of appraising situations of an interpersonal sort issuing in generalized tendencies to act in typical ways which most people tend to acquire in varying degrees as part of their initiation into our culture. Reference, therefore, to such widespread and predictable tendencies is a convincing type of explanation of behaviour that deviates in various ways from that which is conventional and institutionalized. It also is relevant when we wish to explain more personal, as distinct from institutional, types of emphasis and involvements; for the sorts of things that count as motives such as envy, jealousy, ambition, selfishness and benevolence depend upon appraisals of interpersonal relationships which are all-pervasive in a society and not tied down to particular activities or institutions. We do not learn what such motives are by consulting psychological textbooks. Rather we learn them by acquiring them in our personal relationships with others who already act out of them. Our learning to act in the light of such considerations develops *pari passu* with our learning to interpret the behaviour of others. Although, as has already been pointed out, the appraisals which are definitive of these different motives lack some of the features of these connected with deficit states such as hunger and thirst, they nevertheless issue in reasons for acting which are so universal and predictable that they provide paradigm cases for the notion of 'moving to action' which the term 'motive' conveys. Indeed in some cases such as those of sexual jealousy, motives can be so powerful that they enter the realm of psychologically compelling reasons and are cited as excusing conduct or as providing extenuating considerations. Again, behind our assessments of people and of conduct lurk generalizations deriving from our general understanding of how normal human beings tend to act in the light of certain types of considerations.

## 2. REASONS AS CAUSES

So far I have tried to show that justificatory discourse does not, as it were, stand quite on its own feet. Behind the language game of justification lurk certain very general presuppositions about human nature which are relevant to what counts as a justification and what does not. I have also shown that the notion of the 'force of a consideration' which does a lot of work in Toulmin's thesis is an ambiguous one. When, however, Toulmin tries to make such notions do an extra stint of work in his account of reasons as causes, his position becomes increasingly untenable.

Toulmin argues that reasons can be regarded as causes if an agent accepts the weightiness or relevance of considerations. In causal explanations the direct appeal to reasons, which is involved in justificatory discourse, gives way to the indirect appeal; the transition is mediated by recourse to the psychological notion of accepting a consideration. This thesis has a *prima facie* ingenuity and attractiveness about it; for certainly when asked what caused a man to do something such as change his job we can say things like 'He saw that he was no good at it' which is a case of accepting the weight of a consideration. But there is one obvious objection to this general thesis which suggests the necessity for further analysis. The objection is that a person can accept the relevance or force of a consideration and not act. I have elsewhere (Peters, 1958, p. 9) distinguished *a* reason for acting from a man's reason for acting. It is only if *a* reason is also *his* reason that accepting a consideration counts *prima facie* as an explanation of a person's action. Accepting the relevance or force of a consideration can just be a reason for acting which, though accepted as such by the agent, leads to no action of the appropriate sort. A man can say 'I accept that I am no good at this job. This consideration has considerable weight for me. But I am not going to give it up. I enjoy it too much.' There is a further difficulty, too, relating to self-deception and rationalization. I knew a girl of a missionary disposition who indulged in promiscuity. When asked 'Why?' she said 'In order to cure men of homosexuality'. This was not simply a reason; it was her reason. But would we accept it as an explanation? Even 'his reason' does not always count as an explanation (see Peters, 1958, pp. 57–61).

Reasons for action, which count as explanations, must surely relate in some way to what a person wants or does not want, likes or dislikes, takes pleasure in or finds painful and so on. Toulmin nowhere deals with the function of these 'pro' and 'con' words in the explanation of human actions. He gets near it when confronted with the difficult case of 'liking' or 'disliking'. But he slips out by assimilating this case either to that of 'experiencing an uncontrollable distaste' or to applying a learned procedure. He omits to point out that 'learned procedures' (e.g. that of deliberation between alternatives) only get off the ground in the context of action because the alternatives, with reference to which we deliberate, are states of affairs which we want or do not want, like or dislike, etc. Toulmin's examples of the 'rational arts' are altogether too sophisticated and intellectual ('Only a man who knows how to integrate can recognize the significance of boundary conditions; only a man who understands tax-law will know how to respond to changes in interest-rates; and likewise for other modes of reasoning.' p. 16). His concept of 'reasons for action' in terms of considerations capable of carrying weight is taken from complicated social situations in which the ends of the procedure are not made explicit. We need to take a more mundane example such as saving money in order to buy a house to bring out what is missing. In cases like these there is a 'learned procedure' all right, but its 'relevance' is to be understood in the light of its instrumental function in relation to getting something that is wanted. Without institutionalized purposes such as these, together with the more fundamental ones connected with motives and deficit states, it is difficult to see how notions like 'weight' and 'relevance' could have application as determinants of action. For deliberation presupposes both things that are wanted and courses of action deemed 'relevant' in relation to their instrumentality to such wants. In brief there is a conceptual connexion between 'reason for action' and 'wanted for its own sake'.

What this seems to suggest is that in the explanatory sphere notions like 'recognizing the force of a consideration' presupposes reasons which count as explanations, whereas Toulmin attempts to explicate the explanatory function of reasons in terms of notions like 'weight' and 'force' of considerations. If a person scrambles through a hedge in order to avoid an enraged bull 'his reason' for indulging in this 'learned procedure' is to avoid pain, death, etc. If this is not explanatory in the realm of human action, what is? Where, then, does recognizing the force of a consideration come in? It would come in if he had thought about getting out over the gate but had noticed that the gate was very high. That would have been a consideration that had force—in relation to possible means of escaping what he feared. Similarly if he had been deliberating about whether to walk on or to retrace his steps, and had observed the bull making rapidly towards him, that too would have been a consideration that weighed with him as relevant to a choice between different ends that he could pursue, namely that of reaching his destination and that of avoiding harm. But in all such cases considerations have weight because of existing reasons for action. It is not the acceptance of the force of considerations that explains a man's actions; rather it is because his actions are made explicable by his purposes that his acceptance of some considerations rather than others is made intelligible.

To put these points in a more formal way: in explaining human actions we assume, as a basic postulate, that if a person wants (or has an aversion to) something and has or acquires information about how to attain it (or avoid it) he will take the necessary steps, other things being equal. What Toulmin calls 'considerations', which are seen as having weight or force, usually refer to the beliefs which a person has or acquires about matters affecting his attaining what he wants. We speak of 'force' or 'weight' when we are assessing such information in contexts of deliberation about alternative courses of action. But the acceptance of the weight of considerations only has *explanatory* force because of our acceptance of the basic postulate of rationality which is presupposed in our explanation of actions. And it is the 'wanting' part of this which makes our beliefs the effective *causes* of actions.

Toulmin neglects this crucial point because he rests his case on the contrast between deliberation in 'the rational arts' and responses of an unlearned sort issuing from the autonomic nervous system. He nowhere considers the status of actions of a less institutionalized sort involving desires—e.g. for company, approval, security, or achievement. Such desires may or may not be learned, but they are certainly different in character from whatever 'causes' autonomic blinking. Toulmin is perfectly correct in arguing that a great number of human actions are to be explained by reference to what Popper calls 'the logic of the situation' in institutionalized contexts which provide a complicated framework of deliberation. Furthermore, the purpose or purposes of activities like chess, playing the stock exchange, science and politics (to take his examples) are not explicitly stated in giving explanations in terms of 'the force of considerations' within such contexts, because they are taken for granted by those who are initiated into these activities. But they are there nevertheless; for how would we ultimately make any sense of a scientist's activities unless we assumed not only that he was concerned with finding out what was true but also that this purpose was one that actually explained what scientists did? We might, of course, be sceptical about this particular purpose and regard scientific activity as a form of sublimated sexual curiosity, or something like that. But that would be only to give a dubious psychological explanation either of how this institutional purpose was learned or of its psychological potency. It would be merely to complicate the story about the purposes underlying scientific activity; it would not be to dispense with the necessity of postulating purposes.

Within such institutional contexts, too, we appeal to more general purposes such as those built into 'motives' such as ambition, envy, and benevolence to explain individual differences in behaviour. But the 'force of considerations' in explanations such as these is even more obviously a function of their relevance to what is wanted than in the case of explanations which presuppose institutionalized purposes. And, of course, there are a great number of explanations in completely non-institutional contexts such as the case of avoiding the bull cited above, where the 'force of considerations' has palpable reference to purposes. If a man is seen jumping about slapping his coat pocket which has smoke coming out of it, we might say that he accepted the force of the consideration that his pocket was on fire. But such a consideration would only have 'force' on the assumption that he did not want to get burnt or to buy a new jacket.

So far I have argued that Toulmin's notion of accepting the force of a consideration, which he tries to make do the explanatory work in his account of reasons for action, only in fact does such work because of purposes which are presupposed. Could he argue that his notion could help to explicate the notion of wanting something, not instrumentally, but for its own sake? Could this notion throw any light either on learned wants such as 'owning a house' or unlearned ones such as avoiding pain? What help would it be to ask in an explanatory context 'Is owning a house a consideration that has weight with you?' or 'Is avoiding injury a consideration that has weight with you?' Surely all it would do is to draw attention to an aspect of a state of affairs which might be brought about and to ask whether this is an aspect under which it might be wanted? I cannot see that such a cumbersome circumlocution would add anything to our understanding of the notion of 'wanting' except an invitation to the agent to reflect whether he really wants a state of affairs under a certain aspect or not. It is, of course, possible to find equivalents in the language of 'considerations' for states of affairs that are ultimate objects of desire or aversion. 'That wire is dangerous' involves a non-neutral appraisal of an ultimate sort and functions a consideration very much like its equivalent 'you will get hurt (or killed) if you touch that wire'. But nothing is added by way of explanation by talking about this as a consideration which has weight. For pain and death just are, other things being equal, states of affairs that human beings want to avoid. The weight depends on the want.

If the main line of my argument against Toulmin's thesis is accepted, I think that what it shows is that his thesis is round the wrong way. His case, as I understand it, is that a man grasps that, e.g. the avoidance of injury is a justification for, e.g. scrambling through a hedge with a bull behind him and that it is because he accepts the force of a consideration such as 'a bull is after me' that he adopts this 'learned procedure'. To which my reply is that it is often the case that a man accepts the force of a consideration deriving from a normative principle, but does not act in the light of such acceptance. A reason has also to be his reason, not just a reason that has 'weight' for him. Furthermore, as I argued in the first section, justificatory discourse does not stand quite on its own feet. It has to be underpinned by some notion of normal human nature which, in the end, amounts to some general assumptions about reasons for action which are in general operative as causes of behaviour. Usually in human life, as Bentham remarked about pleasure and pain, reasons indicate what we ought to do as well as determine what we shall do. But this coincidence is not invariable and it is also important to assign proper priority to these two ways of looking at reasons. My case is that Toulmin has reversed the proper order of priority in his treatment of explanatory discourse and that, in his treatment of justificatory discourse, he has failed to appreciate the importance of assumptions about human nature.

### 3. LEARNED PROCEDURES

Toulmin has a whole section devoted to showing that all his cases of actions done for reasons involve learning. This is undoubtedly true; but I cannot really see what it adds to his argument. Perhaps he is led to think that it shows something because of his tendency, which I have already noted, to contrast actions done for reasons with phenomena like blinking which are as near to unlearned reflexes as one can get in the sphere of human behaviour. To use his own words his fundamental contrast is 'between certain learned modes of behaviour (ratiocinative, instrumental, and formalized) and those unlearned responses which operate autonomically, regardless of all learned procedures and techniques'. (p. 21) Supposing, however, that he had contrasted actions done for reasons with cases like those of people with free-floating anxiety, or with phobias. These all involve learning, often of a very complicated sort; they are progressively established within an agent's life-time. Yet they manifestly are not cases of things done in the light of relevant or appropriate considerations. Indeed a complicated story is told about learning in order to account for the manifest irrationality of the actions. But what the story reveals is that an irrelevant or inappropriate type of learning took place.

38

It might be argued that such cases really tend to support Toulmin's case. For it must be the case that, if a being is capable of high-grade behaviour involving complicated learning, he must also be capable of being in error and making faulty generalizations which lead him to act in irrelevant or inappropriate ways. But this is an over-simplification; for what operates in cases such as these is probably a different type of learning which gains ascendancy over the more normal type postulated by Toulmin. I do not myself hold much brief for conditioning theory in so far as it purports to give a convincing explanation of how most forms of human behaviour are learned. But I think it may well apply to limited ranges of behaviour—e.g. the development of phobias, attitudes, and the early learning of simple non-intelligent acts. I also suspect that many of the goals which come to be operative in rational action are first learnt by some process which approximates to conditioning.

Now the point about conditioning, if interpreted in a strict sense, is that it shows how certain sorts of things can be learnt, whether they are reactions as in classical conditioning, or non-intelligent acts, as in operant conditioning, without the agent *grasping* that they are in any way 'appropriate' or 'relevant'. For situations are artificially constructed in which the possibility is eliminated that intelligence could be employed, that something could be seen as a way of obtaining some desired end. It may be odd, of course, to call these 'learning' situations, but then psychologists often interpret 'learning' as referring to any established change in behaviour that is not due to maturation. If, however, some forms of behaviour, especially complicated forms like those involved in the training of circus animals, can be 'learnt' in this way, then what becomes crucial in Toulmin's thesis is not learning as such but particular types of learning, which have important similarities with the type of behaviour that develops—i.e. they involve notions like 'seeing the point', 'grasping what is relevant', etc. It will not do to say that rational behaviour can be characterized as that which involves learning; for practically all forms of behaviour involve this. What is important is the type of learning involved. In human behaviour there are all sorts of different things to be learnt—skills, information, principles, attitudes, reactions, wants. Psychological theories and educational theories derivative from them have been for too long obsessed by the notion that there is one type of learning and by the consequent tendency to extrapolate one form of learning to explain the learning of very different forms of behaviour. It looks as if Toulmin is presupposing a theory of learning which involves concepts connected with consciousness such as 'consciousness of a task', 'seeing that something is relevant', 'succeeding', etc. This is, no doubt, to talk about 'learning' in a precise way as distinct from the more general notions connected with 'change in behaviour' favoured by some psychologists. But if this is the case then the emphasis on learning does no extra work for him in his account; for the burden is carried by the conceptual scheme connected with consciousness, which is presupposed already in his characterization of 'reasons for action'.

There is, too, a further point about the relevance of learning in connection with the status of wants in Toulmin's theory which has already been found to be unsatisfactory. Does he wish to argue that 'reasons for action' must relate to wants that are learned? Is it only the politician, the stockbroker and the chess-player who have reasons for action and not the lad brought up by monks who beholds a girl for the first time in his teens and begins scheming to waylay her under the influence of a seemingly unlearned want? In psychological theory there is a perennial debate about which wants are innate and which are acquired. Are we to be in doubt about whether we are right in ascribing reasons for action to people in particular cases until these difficult matters about learning are resolved by psychologists?

### 4. PSYCHOLOGICAL AND PHYSIOLOGICAL CAUSATION

I come now to the final section of Toulmin's paper in which he makes a few remarks about 'Teleology and Physical Systems'. About this section I have even more radical doubts than about his preceding sections. It seems to me that in his preceding sections he has developed some important points about the distinction between justificatory and explanatory discourse and has shown that in a wide sense of 'cause', accepting reasons, etc., can count as causes and that therefore there need be no incompatibility between 'reasons' and 'causes'; for justification, assessment, etc., are not incompatible with explanation. But surely those who have been worried by 'reasons' and 'causes' in the past, who have tended to interpret the concept of 'cause' in a mechanical, or quasi-mechanical sense, have really been worried about the relationship between *different levels of explanation.* They have been worried about the relationship between physiological explanations and explanations in terms of 'his reasons', to use my own previous distinction. No amount of talk about the difference between justification and explanation touches this problem and Toulmin nowhere comes to terms with the relationship between these different levels of explanation.

Toulmin objects to the crude model of a physical system employed in the past by those who have given an account of the brain process underlying human behaviour and stresses the need for a more sophisticated account which will be appropriate to 'high-grade' types of behaviour. He therefore concludes that 'no paradox need follow from Townes' hypothesis: that strictly causal brain-processes may be found to underlie all rational thought-processes, including the discovery that strictly causal brain-processes underlie all rational thought-processes'. (p. 25) But nowhere does he chance his arm on what precisely is meant by 'underlie'. No light is thrown on this problem by pointing out that scientists can still take credit for their discoveries because such remarks occur in justificatory rather than in explanatory discourse. Of course explanatory and justificatory discourse can co-exist. But the traditional problem of 'reasons' and 'causes', which is really a particular problem (perhaps loosely expressed) within the general problem of the relationship between the mental and the physical, is in no way illuminated by making this move. My final verdict on this part of Toulmin's paper can thus be stated by adding to his critical epigram 'All behaviour is explicable: some actions are justifiable', the question 'But what does this show about the appropriateness of and relationship between different levels of explanation?'.

REFERENCES

Peters, R. S. 1958. *The Concept of Motivation*. London.
Peters, R. S. 1961–2. Emotions and the category of passivity. *Proc. Arist. Soc.* **26,** pp. 117–34.

# REPLY

## *by* STEPHEN TOULMIN

The apparent dichotomy between reasons and causes provides the fuel for half a dozen groups of philosophical problems, and my own paper was addressed to only one of these. Richard Peters has focussed attention on an overlapping, but different, group of questions; so he finds my treatment partly convincing, partly irrelevant, and largely incomplete. In return: his comments strike me as useful and interesting, if considered as a supplement to my own argument, but as criticisms or objections I find them wide of the mark. Under the circumstances, therefore, the best way of rebutting his criticisms will be to deal with them indirectly—showing how they can be circumvented, if one only recognizes how the two approaches are interrelated and dovetail together.

The distance between our respective starting-points can be measured in historical terms. (i) David Hume, on the one hand, presupposed that thoughts and reasons have some practical relevance to our conduct, but was puzzled to know how anything so bloodless as a 'rational consideration' can ever be causally *efficacious*. To this question, he answered (like Richard Peters) that 'rational considerations' can have real psychological effects only if every such consideration is annexed to a suitable 'passion'— i.e. 'want' or 'drive'. (ii) René Descartes and Immanuel Kant, on the other hand, were preoccupied with another question, whose answer was taken for granted in Hume's initial presupposition. 'Within a material world governed by physical causality,' they asked, 'can any place be found for rational deliberation *at all*?'; and both of them ended—though in very different ways—by answering, in effect, 'No'. Rational thoughts and physical processes must be kept rigorously apart: either by treating them as manifestations of different 'substances' ('mental' and 'material') or by relating them to distinct, but complementary aspects of experience ('noumenal' and 'phenomenal').

Evidently, these two areas of enquiry overlap; and, within the framework of an exhaustive analysis of the relations between phenomena, behaviour and conduct, both groups of problems would have to find an adequate solution. But the main focus of attention is rather different in each case, and so also are the types of example which bear directly on the points at issue. So let me repeat straightaway that I was concerned in my paper with the Cartesian problem, rather than with the Humean one. At points, my treatment of the issues facing Descartes and Kant became confessedly schematic; and I could not hope to tackle the Humean problem also within the limits of the space available. Richard Peters, for his part, is concerned with this other group of questions—which (as I understand them) begin to arise only at the point where my own discussion leaves off.

The element of cross-purposes between us is already clear early in his comments, where he remarks:

What Kant never succeeded in showing is how respect for the law or for rational beings as a source of law could ever induce anyone to act. I think there are similar problems in Toulmin's refurbishing of the Kantian approach.

Given his own particular problem (I believe) Kant would have retorted that it was not incumbent on him to prove that a rational consideration, e.g. conforming to the law, is a sub-species of 'inducement'. On the contrary: the ostensible differences between rational considerations and inducements were the very things that created a problem for him, and he was determined to see them respected. If David Hume (or Richard Peters) proposes to allow rational considerations into the causal world, only at the price of inventing a special type of 'psychological cause'—in the form of a '*drive* to respect the law' or '*wanting* to conform to the law'—that will raise further questions. But Kant's own problem is the prior one: to prevent the universality of causal necessity within the physical (or 'phenomenal') world from making complete nonsense of all rational (or 'noumenal') considerations whatever.

The same is true in my case also. The very point of posing Townes' Paradox as I did is to spotlight the Cartesian question, 'If and when neuroscientists master the entire material structure and mechanism of the brain, what room will remain to talk about rational considerations or deliberation *at all*?' Of course, having established that we still have all the elbow-room we need for the idea of 'rationality', even after all possible neurophysiological enquiries have been completed, we can then go on to ask, '... and what further conditions must be satisfied, if "rational considerations" are to be causally efficacious *in fact*?' But this question—which I take to be Richard Peters' question—is a subsequent one, which arises only when the prior Cartesian question has been dealt with satisfactorily.

The same cross-purpose underlies also most of Peters' detailed reservations about my argument and illustrations. To summarize: he finds (1) my account of justificatory discourse 'acceptable but incomplete', because the examples I analysed were irrelevant to the Humean problem—having been chosen with the Cartesian problem in mind. He regards (2) my account of reasons as causes 'ingenious but untenable', because he understands me as claiming that a 'reason' can possess causal *efficacy* on its own—whereas I was interested only in the weaker question, on what conditions rational considerations can have causal *relevance*. And he dismisses my discussions of (3) learning and (4) physiology as either 'largely superfluous' or 'irrelevant to the main point at issue', because I do not show their bearing on the analysis of motivation, or psychological causation. Yet my remarks about (4) in fact bear directly on the contemporary neurophysiological form of the Cartesian problem, which was the main point at issue for me; and my outline discussion of (3) is simply a preliminary sketch, indicating how we might set about mapping the area of 'higher mental functions' that neurophysiologists need to come to terms with.

These responses can be supported from Richard Peters' own text. To begin with his major criticism (2): he considers my examples of rational deliberation 'altogether too sophisticated and intellectual' and recommends us to consider, instead, 'a more mundane example such as saving money in order to buy a house', or that of a man 'scrambling through a hedge with a bull behind him'. These examples enable him to argue that, even in the most extreme 'rational' cases, the 'force' of considerations does not by itself provide the 'weight' needed to make them causally effective. In his pithy restatement of Hume's dictum about Reason being the Slave of the Passions, he declares: 'The weight depends upon the want.'

For Descartes and the neo-Cartesians, by contrast, the difficult problems arise precisely over the 'sophisticated and intellectual' cases, where there is no evident 'want' to add 'weight' to the reason itself. Descartes (after all) had no compunction about analysing human conduct arising from 'the passions of the soul' in the same terms as animal behaviour, and so assimilating it to the mechanical processes of the causal/material/extended world. He was puzzled, rather, by the capacity of human beings for creative language-use, rational judgment and argument—the things which mark men off from other animals. And these same activities remain the major source of puzzlement and controversy today: whether we consider them in purely behavioural terms (Chomsky *v.* Skinner on language-learning) or in terms of their neuroanatomical basis, as those which by-pass the seat of normal 'drives' or 'wants' in the 'limbic system', and depend rather upon 'cross-modal connections' in the 'association cortex'.

Evidently, then, Richard Peters' preferred example—of the man being chased by a bull—misses the point of Descartes' puzzle. There is no real problem for Cartesian mechanists or twentieth-century neurophysiologists about actions done from fear, since in such cases the role of reasoning is manifestly a secondary one. It is precisely over the 'sophisticated and intellectual' example that the Cartesian difficulties begin, and that is the reason why I selected them for discussion. Recall Wollaston's way of posing the problem: the question for him is, how so 'immaterial' a thing as 'the sense of the words' (i.e. language) can give rise to 'motions in the spirits, blood and muscles'.

Who can imagine matter to be moved by arguments, or ever rank syllogisms and demonstrations among levers and pulleys? . . . Affections cannot be the physical effects of the words spoken because then they would have the same effect, whether they were understood or not.

And it is at this very point that the relevance of topic (3), viz. learning, becomes apparent. For any answer to Wollaston must include a reminder that what permits the sense of words/arguments/demonstrations to *become* causally relevant to an explanation of our behaviour is the fact that we have *learned* to understand them.

This is not, of course, to contradict Peters' own points. The very act of meeting Descartes' and Wollaston's objections may simply bring us face-to-face with Hume's problem, and it may well be desirable, subsequently, to make the move that Peters makes: i.e. to insist that rational considerations can themselves become causally *efficacious* (not merely *relevant*) only if a man is 'open to reason', and open also to the specific *kind* of reason in question—ready, as he puts it, to make that reason 'his' reason. But the first step was to show, in a way that has some relevance to twentieth-century physiology, that the category of 'reasons' and the category of 'causes' can be brought into an intelligible relation, without destroying the essential character of either. That was all I set out to do.

Let me now go back to topic (1). There is, in fact, some virtue in drawing a completely sharp contrast between reasons and causes—at any rate, at the outset. One can, no doubt, find plenty of intermediate examples, which blur the lines between rational conviction and causal compulsion, justification and explanation: these intermediate cases mix together elements that are exemplified in isolation only in more extreme cases. Yet it is, surely, advisable to consider the extreme cases first, so as to circumvent the additional complications involved in the mixed ones. Furthermore: as we move across the spectrum of examples from one extreme to the other, we shall find many of our key-terms showing systematic ambiguities, and Richard Peters' own analysis of emotional reactions and behaviour seems to me to suffer through overlooking these ambiguities.

For example: he takes me to be questioning whether 'reactions as well as actions' can be 'justifiable or unjustifiable', and he claims in return that even a girl's blush might—to stretch a point—be discussed in rational, justificatory discourse. Yet one must turn the question back to him and ask, 'In what sense could it be so discussed?' The very term 'reaction' covers examples of many distinguishable types, and we commonly discuss the 'justifiability' of such 'reactions' in correspondingly different ways. At one extreme, we may say, e.g. 'Mr Chamberlain's reaction to Hitler's threats was to appease him by misplaced concessions'—and in this case we shall include under the word 'reaction' an entire course of action, or policy, whose justifiability can indeed be challenged in strictly rational terms. At the other extreme, a girl may be subject to such uncontrollable blushing that 'rational criticism' is entirely out of place—the only sense in which we can describe her blushes as 'unjustified' is the weaker sense of 'inappropriate', 'disproportionate', or 'without a relevant object'. If it was certain beforehand, indeed, that no such criticism could do anything to moderate her blushing, we might well doubt whether the question of 'justifiability' or 'unjustifiability' arose at all in this context.

The Cartesian, at any rate, is entitled to resist any argument which runs all these different kinds of reactions together, and so blurs the differences which are the very source of his puzzlement. For he well understands that a 'passionate' reaction may be 'justified' or 'unjustified', in the weaker sense of being disproportioned to the actual character of an object or situation. What he questions is the supposed analogy between this weaker sense of 'justifiable', and the stronger sense which is relevant to more 'conscious' or 'intellectual' reactions—i.e. the sense in which 'justifiable' means 'warranted', 'well-founded' or 'rationally-based'.

Let me turn now to question (3). Here again Peters' preoccupation with the Humean problem distracts him from the questions my own paper was concerned to discuss. Thus, he asks, rhetorically:

Does Toulmin wish to argue that 'reasons for action' must relate to wants that are learned? Is it only the politician, the stockbroker and the chess-player who have reasons for action and not the lad brought up by monks who beholds a girl for the first time in his teens and begins scheming to waylay her under the influence of a seemingly unlearned want?

(This example presses home his earlier objection that 'it is the "wanting" which makes our beliefs the effective *causes* of action'.) Yet, given that my own aim was merely to show how reasons can be causally relevant at all—not to prove that they are necessarily 'effective causes' on their own—this example, too, can be turned against him. His 'lad brought up by monks' who 'schemes to waylay a girl' is, even on his own description, involved in a course of action that is in part 'deliberative': not, to be sure, in so far as it springs from a nascent sexual drive, but at any rate in so far as it calls for 'scheming'. And scheming is just the sort of instrumental behaviour which has to be developed by learning, and which I cited as one typical variety of actions done-for-reasons.

My own basic distinction then applies unchanged to Peters' chosen example. Contrast the two questions, 'Why did the lad hide behind the hedge at dusk?', and 'Why did he waylay the girl at all?' The answer to the first question ('So as to take the girl by surprise') does indeed indicate the lad's *reason-for* acting in the specific way he did; and it does so by referring to the element of deliberate 'scheming' involved in his behaviour. By contrast, the answer to the second question ('It was his newly-awakened sexual drive') tells us rather the *reason-why* the lad embarked on this course of action at all; and the factor cited may well be neither 'conscious' nor 'deliberate'—still less, 'learned'. Characteristically, Richard Peters' example is just the kind of case in which an agent might say, of his entire course of action, 'I can't think what came over me'. It illustrates the idea of 'reasons-for-acting' only to the extent that it involves learned 'deliberative' behaviour: to the extent that we explain the behaviour as the expression of a sexual drive, this may tell us the 'reason why' the lad acted in this way, but hardly his 'reason for' so acting.

I have attempted to meet Richard Peters' comments in no merely defensive spirit—hoping to patch up the holes he apparently punched in my argument—but rather constructively, hoping to use the cross-purposes between us to clarify further the philosophical problems of human action. For, in this clarification, the task of distinguishing clearly between the Cartesian and Humean questions, and explaining how they bear on one another, is an indispensable first step. In the long run, a complete account of 'psychological explanation' must relate together all the half a dozen or more different modes of discourse (types of 'language game') that apply to human behaviour—ranging between the purely neurophysiological and the purely justificatory, by way of the propaedeutic and the psychopathological, the dispositional and the diagnostic. Within such a broader account (as Peters rightly guessed) I believe that we must

talk about 'learning' in a precise way as distinct from the more general notions connected with 'change in behaviour' favoured by some psychologists,

and I believe that, if we do so, we shall end by recognizing how far *rationality* is one of the end-products of what may be called 'the education of consciousness'.

This task has both scientific and philosophical aspects, and these need to be looked at in conjunction. About the scientific aspects, I myself have learned much from the work of Soviet psychologists, e.g. from L. S.

Vygotsky's *Thought and Language*. (Russian work on the role of speech in the development and regulation of thought and behaviour ties in much more closely and naturally than the more fashionable kinds of American psychology with the results of recent analytical work in philosophical psychology.) At the same time, in so far as the task is still a philosophical one, it remains *analytical,* and on this account Peters' further objection:

If this is the case then the emphasis on learning does no extra work for Toulmin in his account; for the burden is carried by the conceptual scheme connected with consciousness, which is presupposed already in his characterization of 'reasons for action'

misses the mark again. For I drew attention to the importance of learning in my own paper, simply in order to make explicit one essential aspect of 'the conceptual scheme connected with consciousness'. For reasons both of space, and of the extreme difficulty of the subject, my resulting remarks were (I concede) schematic and inconclusive. But Peters himself would allow (as I know from his other work) that there are essential relations between the concepts of education, rationality and consciousness, and that the character of these relations is still largely in need of analysis. This is the joint task to which, in their different ways, both my paper and his comment should be read as complementary contributions.

# THE EXPLANATION OF PURPOSIVE BEHAVIOUR

## *by* CHARLES TAYLOR

### I

'Explanation' has many meanings, and there would seem to be no common ground between its use in some ordinary contexts, and its role in science. But the scientific sense is continuous with at least one common meaning in ordinary speech. Here 'explain' very often means 'to make what appears strange and outlandish understandable'. This is often done by relating what appears to stand outside the normal course of events to this course in another way. Thus, if someone does something which is strange or shocking, we 'explain' it by describing the context in a way which makes the action understandable given current conceptions of human motivation. Someone suddenly gets angry in the course of a discussion; we ask why? The event is explained when it is pointed out that a subject mentioned is taken as a reflection on the man's honour, or that his convictions are very strong on this subject and he cannot stand its being spoken about lightly, or something of the kind.

Explanation involves here, therefore, bringing the strange back to a place in the normal course of events. Something analogous holds of explanation in a scientific context. But here what corresponds to the 'normal course' is established by the explanation and not already received in ordinary conceptions. We usually claim to have explained an event when we give an antecedent from which it follows; not indeed *any* antecedent, but one which has a certain saliency: either because it is the one which we can alter most easily, or because it is the one which varies most often, the other conditions being standing conditions, or for some other reason. But in a scientific context, we have come to demand something more: we expect the antecedent condition to be singled out in concepts which show the connection between this outcome which we are explaining and a host of others. Typically, for instance, the antecedent conditions might be expressed in a statement attributing certain values to key variables, which had they had other values, would have determined other predictable outcomes. This element is perhaps always present, even in explanations of the most rudimentary kind. If we explain that a bridge has collapsed because of the weakness of one of its supports, we obviously understand that a stronger support would have kept it up, *ceteris paribus*. What the progress of science adds here is exactness, the ability to define the antecedent conditions of breakdown more exactly, and also to predict a great number of other outcomes more finely.

But what it also patently has added is explanation at greater depth. To pursue the above example, the progress of scientific enquiry not only adds exactness to the common sense explanation by giving a measure of strength, but also sheds light on the determinants of strength. This accomplishes more than simply adding to our criteria for measuring strength, though it certainly does that. It also *explains* why the original criteria are criteria of strength; that is, it accounts for the correlation between strength and the properties by which we originally assessed it. For instance, let us say that we start off estimating the strength of possible materials roughly as we go about building the bridge, by, say, the type of material: this alloy is stronger than that metal, and so on. When, however, we come to understand something about molecular structure, then we are in a position to explain the original rough correlations by which we operated. But, in line with what was said above, we consider this explanation a scientifically satisfactory one partly because it shows the connection between the outcomes we are explaining—here the correlations between certain materials and certain strengths—and a whole host of others. We know more than that material $M$ has molecular structure $S$ and therefore a certain degree of strength. We know also more exactly what modifications in it would alter its strength, more exactly therefore how it differs from other materials whose strength is different. And this means, of course, that we can go beyond the original repertoire of materials with which we were operating at first; for the new language of molecular structure enables us to give the formula for new materials as yet non-existent.

Scientific explanation therefore has two important properties: it gives the antecedent conditions of the explicandum in terms of a set of factors which make evident its connection with others, which makes clear with some exactitude what would need to be changed for other outcomes to eventuate; and it also is capable of building in tiers, that is the correlations which explain at one level can be taken as explicanda and explained at another. The result is that the progress of science sees more and more outcomes connected by the network of explanatory theory. For, as a set of connected correlations are themselves explained at a deeper level, they are connected with a wider class. Scientific theory thus becomes more wide-ranging as research progresses; it brings under the same framework widely separated phenomena; it homogenizes, one might say, what at first appears different.

Thus the laws of Kepler by which one can explain on one level certain aspects of planetary behaviour are explained in turn by Newton in terms of the law of attraction and certain initial velocities. In this way the regularities that Kepler charted are shown to be one case of a broader class of regularities to which Newton's laws also apply. Newton can be said to have explained Kepler's laws because he has shown the connection to a broader range of phenomena by singling out the factors alteration in which would alter the behaviour of the system in known ways; or, in other words, because he has singled out the variables such that, where they are found to have other values, a different sort of system is predictably found to be in operation. The same set of variables which enables us to account for the movement of the planets—mass, distance, velocity—enable us to account also for the revolutions of the moon, of artificial satellites, and so on. They also help us to account for the movements of bodies in systems of a quite different kind. Newton's account homogenizes, as it were, what at first appears different.

Similarly, the regularities described in Boyle's law and Charles' law are explained at a deeper level by the Kinetic Theory of gases. These phenomena are linked up with a host of others and the connection made clear through the laying bare of a set of variables concerning molecular movement, variations in which account for a great number of different outcomes, including the ones in question. Similarly, modern atomic theory has brought together phenomena previously segregated under the domain of supposedly separate disciplines, physics and chemistry.

Scientific explanation thus connects and links phenomena with a wider range. But it is a mistake to think that this connection is like the subsumption of a generalization under a broader one. This has been the view of some philosophers influenced by logical empiricism. Thus Braithwaite (1953, pp. 302–3) cites, as an imaginary example of an explanatory relationship, 'animals are mortal' as an explanation for 'men are mortal'. The latter can, of course, be deduced from the former. But this will not do, it is just the kind of connection which is useless for purposes of explanation. In the examples above, we have seen that the phenomena to be explained are connected to others by being linked with antecedent conditions which in another form also condition these other outcomes. But if we explain that men die by the fact that animals die, what is the antecedent condition? 'Being an animal' is what we have to reply here, and this certainly does connect us with other animals. But not in such a way as to account for the differences, as to delineate in what the difference consists. Contrast the Newton–Kepler case above. Here the movements of the planets around the sun are connected with, say, the movements of satellites around the planets. But the connection is made not simply by the affirmation that they are to be explained by the same laws, but by singling out the important common variables and stating what results follow from different values of these variables. The explanation connects by *accounting for the difference*. But giving 'animals are mortal' as an explanation for human mortality does none of this. It simply affirms that human mortality must be accounted for as part of a more general phenomenon. One can say that it simply *affirms* the connection without *showing* it.

Whether or not to call this 'explanation' is, of course, ultimately a stipulative matter. But it is certainly essential to distinguish the two types of answer to the question 'why?'. Both can make some claim to the title in virtue of their connection with the ordinary context sense above, in that both can be used to show an event or class of events to be part of a broader 'normal course' of events. Both have their utility. But they are nevertheless very different. The explanation which subsumes an event or a class of events under a broader generalization can serve to set the stage for an explanation of the kind which shows the connection. It circumscribes the phenomenon, as it were, which we have to explain. To subsume human mortality under animal mortality tells us to look for the explanation of human mortality not in something peculiar to man but in something which he shares with all animals. This kind of subsumption can thus be a useful preliminary to explanation in the strong sense. But it obviously is not equivalent to it.

Explanation in the strong sense therefore builds in tiers in a way which cannot be understood simply as the subsumption of less under more general laws. Rather the correlations on one level are explained by those on a deeper level in a way which shows their relation to other possible outcomes. The first can be seen as a special case of the second; they are shown as exemplifications of the deeper correlations under some initial conditions. Thus the laws of Kepler are the laws of Newtonian mechanics exhibited in the relevant initial conditions holding for our solar system. We can thus often say of two explanations of a given phenomenon, that they are compatible but that one is *more basic* than the other, for it appeals to deeper level laws which themselves can be used to explain the laws invoked in the shallower account.

One important practical respect, therefore, in which the two forms of explanation differ is this: that the stronger kind lays the basis for an advance in technology since in showing the differences between outcomes, it also gives us information which can be used, if the techniques are available, to control the outcome: because it tells us what antecedent conditions are relevant, and how alterations in these conditions affect outcomes, it tells us what would have to be altered to alter the outcome. It enables us also to bring about new, as yet non-existent outcomes, as when, for instance, we create new elements with the aid of the theory of atomic weights. It is evident from the explosive development of technology which has accompanied modern scientific progress that explanation has not all conformed to the model Braithwaite presents us with, but has been of the strong variety. If all that science had realized were subsumptions, we would have no more knowledge of how to control our environment.

But is it not far-fetched to link scientific explanation in the strong sense to explanation in the context of ordinary behaviour? It seems to me that there is a continuity between the two. Explanation, say, of outlandish behaviour in ordinary life brings the strange back to what is considered normal for the class of events under consideration (say, human acts, or emotions, or whatever). It shows it to partake of the same nature as other events in this class. Similarly, scientific explanation shows an event to partake of the 'norm' in the sense that it is accounted for by the same set of factors which account for all events of this class. It shows the event concerned in its relation to all the others. But in the case of scientific explanation, the 'norm' is discovered, not given. And this is, of course, a central fact about the scientific endeavour, that the factors in terms of which events are to be accounted for and thus related to each other are not necessarily immediately in evidence. They have to be discovered, and their discovery involves our taking the phenomena to be explained at the right level of analysis and with the right conceptual framework. For it is only as identified and characterized in a certain way that the phenomena can be exhibited as all being variations in the same fundamental dimensions. The search for a conceptual framework, as a search for the definition of 'normalcy', is therefore central to all scientific enquiry.

This idea of a 'normal course' of events sometimes enables us to make a non-arbitrary distinction between necessary and sufficient conditions. As was mentioned above, this distinction is sometimes made in ordinary contexts: one condition is singled out as the 'cause' or explanation, and the others are seen as necessary conditions. The 'cause' or sufficient condition is singled out either because it is the one we can bring about, or because all the other conditions are considered to be part of the normal state of affairs, and only this one different (e.g. a bridge falls down because of a weak support; but it also falls because of a car going over it; this, however, is part of the 'normal' conditions for the bridge, and therefore counts as a necessary condition), or because the rest are standing conditions and the 'cause' is an event (from this point of view the cause of the bridge falling is the car that drove into it). But in the case of scientific explanation, we often have a less arbitrary criterion. The class of events which are explained by a given set of factors and thus connected may form a system or systems, in this sense that the set of correlations involved may hold only if certain boundary conditions are fulfilled. The boundary conditions may be simply the relative independence of the system from outside interference: thus Kepler's laws apply only in the absence of some large foreign body traversing our solar system, such as is regularly depicted in science fiction novels. Or they may be some positive state conditions, such as, for example, an organism's requiring such and such atmospheric, climatic, etc., conditions to function. But in either case the outcome concerned is contingent on the boundary conditions' holding as well as on the antecedent's being present. But we distinguish the antecedent here as cause or sufficient condition since it is what embodies the explanatory force; it connects the outcome concerned to others, and allows us to predict and create precise alterations in outcome. The boundary conditions, on the other hand, are connected indiscriminately with a whole host of outcomes, all those which are explained by variations in antecedent conditions within the system. We thus speak of these conditions as necessary. The title 'cause' or 'sufficient condition' of a given outcome is reserved for the antecedent which selectively explains this outcome, although within the bounds of the necessary conditions, which are, however, shared with others.

## II

The search for a conceptual framework, for a concept of the 'normal' course, is no less vital in the sciences of behaviour than it is in the physical sciences. Only in the former field it seems to have met with much less success. One can describe the state of disarray and contention in which we find the sciences of man—whether in the field of political science, sociology, anthropology or psychology—as arising from deep disagreements over the conceptual frameworks which are appropriate. Each of the above fields is the scene of several rival 'approaches', no one of which seems to be able to establish itself to the satisfaction of all workers in the field as the definitive framework.

In psychology, one of the major questions concerns the validity for scientific purposes of the type of framework implicit in our ordinary concept of action, an explanation in terms of concepts of the range of 'purpose', 'goal', 'meaning'. Concepts of this range are central to some forms of explanation (e.g. psychoanalysis), but have been vigorously rejected by others (behaviourism).

The form of explanation which we can call explanation in terms of purpose can perhaps be more closely defined in terms of two principal features: (1) explanation of behaviour takes a teleological form, and (2) explanation is sometimes cast in terms of the way the agent sees himself and his situation, of 'the meaning' that they have for him. Let us examine each of these features in turn.

It is fairly obvious that our ordinary form of explanation of behaviour in terms of action is teleological in form. This is not to say that all ordinary explanations of behaviour as action point to some end at which the action is aimed, but that all such explanations carry the implication that acting for the sake of an end is a fundamental feature of human behaviour (for further discussion of this cf. Taylor, 1964, Chap. 2). Now a teleological form of explanation can be characterized in this way: where the behaviour is explained by the goal to which it is aimed, in other words, where it is explained by that 'for the sake of which' it occurs.

Now it seems to be a widespread belief among researchers in the field of academic psychology that this form of explanation is inherently non-empirical, that is, untestable; that it involves an absurdity, like explaining an event by another one subsequent to it (which would certainly make prediction rather difficult), or postulating an unobservable force. They have been encouraged in these beliefs by philosophers of science of the positivist persuasion. But a little reflection should show that this view is ill-founded. That behaviour is a function of what it serves to bring about, rather than some other unrelated factor, is a fact (if it is one) about the form of the antecedent; it is not a claim to dispense with an antecedent altogether or the invocation of an unobservable one. An explanation is teleological if the events to be explained are accounted for in this way: if $G$ is the goal 'for the sake of which' events are said to occur, $B$ the event to be explained, and $S$ the state of affairs obtaining prior to $B$, then $B$ is explained by the fact that $S$ was such that it required $B$ for $G$ to come about. In other words, a teleological explanation is defined as such by the form of the antecedent, a form in which the occurrence of the event to be explained is made contingent on the situation's being such that this event would bring about the end in question.[1]

---

[1] One can perhaps add: that this event *and this event only* would bring about the goal. This stronger expression is equivalent to the formulation above: a situation where $B$ is *required* for $G$. But although the stronger form is usually appropriate, since we are dealing with beings with a limited repertoire (as we shall see below), the weak form must be mentioned as a possibility: if there are several possible occurrences in a given situation which would bring about the goal, then a teleological explanation making reference to *this* goal cannot account for the fact that one is chosen rather than another. The teleological explanation can thus be only partial; it yields us a necessary but not sufficient condition. Of course, the selection among possible 'routes' to the goal may be accounted for by another goal which one serves more than the others (parsimony of effort, or custom, or dignity, or whatever). Because with animate beings, it can plausibly be claimed that a unique selection of this kind usually takes place, through limitation of repertoire, or through some norm, I shall go on using the strong formulation (that $B$ is required for $G$) in what follows.

Now it is evident that an antecedent of this kind is both open to observation, and open to observation *ex ante* (thus permitting prediction). Of course, it makes *reference* to the end which will occur later, and reference, too, to the explicandum (and to a relation between it and the end). But this is not the same thing as making the end the antecedent, or defining the antecedent in terms of the consequent, or any other such vices of explanation. That the situation is such that $B$ is required for $G$ can be established independently of discovering whether $B$ has occurred. It can be objected that, defined in this way, teleological explanation ceases to be a distinct form; for if the teleological antecedent is an observable state of affairs, that $B$ is required for $G$, then surely it can also be described in other terms, terms which made no reference to the event explained or the end. Let us call this type of description an 'intrinsic' characterization. Then for any state of affairs which one can describe as requiring $B$ for $G$ an intrinsic characterization can be found. But if the situation so characterized is taken as the antecedent, then we no longer have a teleological explanation. Hence we cannot claim that this latter is a distinct form with a different empirical content.

But this objection is fallacious. The antecedent condition in any given causal sequence can be described in an indefinite number of ways. But that does not mean that we can indiscriminately characterize the situation described in any of these ways 'The antecedent'. First, and most obvious, there are descriptions which pick out the features of the situation which are causally irrelevant. An accident occurs because the driver is drunk. But the state of affairs where the driver is drunk is also that state of affairs where he is talking too much. But this does not mean that the state of affairs where he is talking too much is the antecedent of the accident. We could imagine a case in which we would call it the antecedent, where he was talking too much and thus not paying attention and thus didn't see the sign, and so on. But this is a separate case from that where his reaction time was impaired through drunkenness or where drinking too much made him foolhardy, and *this* is how the accident occurred, although his being drunk also had other results such as making him tell you his life story. In this case his loquaciousness, although causally linked to drunkenness as an effect, is not causally relevant to the accident.

But, of course, those who propose to translate out teleological antecedents undoubtedly have in mind to choose intrinsic characterizations which are causally relevant. But this is where a problem arises: not all characterizations of an antecedent which pick out its causally relevant aspects are on the same footing. For they differ in the degree to which they yield a functional relation valid in general or over a wide range of instances. Thus the state of affairs where the driver is drunk is also the state of affairs where he has imbibed the greater parts of the contents of a bottle with a label marked 'Canadian Club'. This latter description undoubtedly characterizes the antecedent in a causally relevant way, unlike the reference to loquaciousness above. But it is not as adequate as a description of the antecedent. For it describes this situation in a way which is not as widely generalizable as the original description in terms of drunkenness: in the next village housewives use old Canadian Club bottles to keep fruit juice in, other men get drunk and have accidents after drinking gin or scotch, some men can drink this much and not get drunk, and so on, and so on.

But what is important here is not the generality. This is the corollary of something else, namely that some descriptions offer a more basic explanation than others. An explanation $A$ is more basic than another $B$, as we have seen in the previous section, when the correlations invoked in $B$ can in turn be explained by those invoked in $A$ together with certain initial conditions. The correlations of $B$ are shown thus to be a special case of the correlations in $A$. Thus with the accident case above: the explanatory correlation involved in the description of the antecedent as 'his having drunk the contents of a bottle labelled "Canadian Club"' must itself be explained by the explanatory correlation involved in the description of the cause as 'his being drunk'. In other words, his drinking the contents of this bottle led to his having the accident only because of his being drunk led to the accident (more basic correlation) and his drinking the contents etc., made him drunk (initial conditions).

Thus we can see that not all descriptions of a causal antecedent, among those which pick it out as causally relevant, are on the same footing; some are more basic than others, and we can say, for instance, that the event characterized as $B$ (less basic characterization) is only an antecedent of the explicandum because in the circumstances its being $B$ is also its being $A$ (more basic characterization). Thus we can hardly claim that the two descriptions are interchangeable. The fact then that a teleological antecedent also bears an intrinsic description, even an intrinsic description which picks it out as causally relevant, proves nothing by itself. We still have to know which description is more basic. The intrinsic one may apply to the antecedent of a given event in one case, or even in a wide class of cases, but it may nevertheless be that this is only the case because in these circumstances having the intrinsic characterization amounts to having the teleological one. This may be the case, or it may be the other way around. But the important point is that this cannot be determined *a priori*. Thus although we might admit it to be true *a priori* that for any state of affairs which we can describe as requiring $B$ for $G$ an intrinsic characterization be found, this is utterly irrelevant to the point at issue. What is relevant is which explanation is more basic, and this can only be discovered empirically.

Which of the two is more basic is a matter of which yields a really general functional relation, one in terms of which the other can be explained as a special case. It may be possible to establish this by seeing that one applies to cases where the other does not. But even where the limits of experimentation make a discovery of this kind impossible, we may establish their relationship by seeing that the set of correlations to which one belongs and from which it can be derived can be used to account for (and perhaps also manipulate) the phenomenon over a wide range, whereas the set to which the other belongs yields only a random collection of *ad hoc* relations. In other words, the first is part of a system which can be used to account for variations in the phenomena which the second cannot explain; this is enough to show that the former offers the more basic explanation. For the set of *ad hoc* regularities parallel the individual correlations of the more basic system of laws. But since these latter should be seen as special cases of this system the former must as well.

After weighing this objection, we are in a better position to define what is at stake in the question about frameworks of explanation in psychology. The question whether we should use a framework based on explanation by purpose involves the question whether we should allow a teleological mode of explanation. But this means surely, whether we should allow that the most basic explanation is teleological. No one can quarrel with the thesis that out of human and animal behaviour we can abstract regularities at *some* level which are teleological in form. This is what we do all the time in ordinary speech and have been doing for some hundreds of thousands of years. The question is whether these regularities can in turn be accounted for by others of a non-teleological sort. And this turns on the question whether the teleological regularities can be explained by non-teleological laws which account for a wider range of phenomena, for instance by allowing us to predict similar outcomes even when the normal teleological antecedent isn't present, but the 'intrinsic 'one is (e.g. a stimulus, or condition of the brain or nervous system), or else by offering a system permitting greater prediction and control, from which the teleological regularities can be derived as disconnected *ad hoc* connections. Or whether, on the contrary, the relations are not reversed, and the teleological laws provide a wider explanation.

Of course, in applying teleological explanation to animate beings account has to be taken of the fact that all species, even man, are limited in some way, that for each species there is a certain shape of the universe of possible behaviours which we can call its repertoire. Thus, in saying that $B$ is required for $G$, account is implicitly taken of the limitations of the repertoire. In the abstract, there may be some more efficient way of getting $G$, but $B$ is the way of getting it which lies within the capacities of the being in question. This cannot be made the ground for an objection against teleological explanation, that it inadequately characterizes the response. In an abstract sense this is true. But there is no incoherence in the claim that once the repertoire is given we can only account for the selection of the response which occurs by invoking an antecedent of teleological form. In terms of the distinction of the previous section we can see the behaviour of the animal in question as a system, where the limits of the repertoire constitute part of the boundary or necessary conditions, and where in consequence that $B$ is required for $G$ is a sufficient condition for $B$'s occurring. And, indeed, we have no choice but to look at things this way, if no other law-like correlations can be found to account for the occurrence of $B$ and other behaviours.

But there is a second and just as important restriction to be made in applying teleological explanation to animate beings, and that is that we have to take account of the way that the agent sees the situation. We come then to the second main feature of explanation by purpose. Plainly, explanations of human behaviour, for instance, can only be teleological if we interpret 'requiring $B$ for $G$' as 'requiring $B$ for $G$ in the view of the agent'. That an action is required in fact for a given goal will not bring it about unless it is seen to be such; and many actions can be accounted for in terms of the goals of the agents concerned which in no wise really serve these goals. Thus an important part of our explanations of human behaviour consist in making actions intelligible by showing how the agent saw the situation, what meaning it had for him.

It is obvious that on a lower level an analogous principle holds for animal behaviour, or at least for that animal behaviour which is phylogenetically high enough for us to be tempted to account for it in teleological terms. But there is still a qualitative difference in the importance of this principle when we come to the human level. Men through language can differ in culture to the point that recognizable human goals, which are in a sense universal—not only hunger, sex, etc., but also the need for self-respect, human companionship, and so on—take on characteristic forms in different cultures. Their meaning is different for men from different backgrounds. Thus, the set of values which are important and in terms of which men see their situation and themselves are different and often untranslatable from one culture to another. Translating words from a foreign language as 'honour', 'integrity', 'sainthood' and so on may create more misunderstanding than comprehension of another people.

Explanation in terms of purpose therefore involves taking into account the conceptual forms through which agents understand and come to grips with their world. That people think of their environment in certain concepts, that is, use certain modes of classification, is an element in accounting for what they do. Indeed, it can be said to define what they do. For if we think of actions as defined by the purposes or intentions which inform them, then we cannot understand man's actions without knowing the concepts in which they frame their intentions.

This principle is valid whether or not we explain men's behaviour in concepts which they would understand or accept. We may show that much of what men think of their own behaviour is not only untrue, but the result of a powerful drive towards self-deception or repression. But this is to appeal from how men say that they see their situation to how they really see it; it is not to make this dimension irrelevant. From this point of view, therefore, psychoanalysis is a development of explanation by purpose. For recognizing that much thought and motivation is unconscious, it accounts for this in terms of goals; indeed, repression itself is accounted for teleologically; and thus a central aspect of psychoanalytic explanation concerns the meanings that events, situations, symbols, have for the agent, even if these are unconscious.

This accounts in part for the suspicion with which psychoanalysis is regarded among academic psychologists. For there is a strong resistance to explanation by purpose in these circles. This springs from a deep-lying complex of views about scientific method, observation and explanation, according to which only explanations by efficient causes can be accepted, and these must not make use of any intentional properties (properties of the way in which a reality is seen by a subject). This outlook thus rejects both teleological explanation and any reference to the meaning of the situation for the agent; it rejects explanation by purpose root and branch.

The historical roots of this view are complex; modern academic psychology draws its hostility to teleology partly from the scientific tradition of modern times. Behaviour theorists seem to be hypnotized by the victory of Galileo and Newton over the teleological physics of Aristotle, and seem to aspire to achieve the same thing in philosophy.[1] But they are also opposed to explanation by purpose because it seems to involve having recourse to the unobservable. Here we can see the influence of positivism and its

---

[1] But it may be that the lesson which the seventeenth century has to teach us is precisely the opposite: not to try to cram all reality into the same mould; then men learnt not to treat inanimate nature as though it was animate, now perhaps we have to learn the inverse. More generally, we can learn from the opposition to Galileo how readily men can fail to grasp empirical reality because their philosophical system hasn't allowed a place for it.

ancestor empiricism. For empiricism backs up the rejection of explanation by purpose in two ways: the notion that all knowledge is based on impressions which come to the mind from outside leads first to a dualistic notion of body and mind as in causal interaction (for the body, too, is 'external' reality *qua* spatial), and second gives rise to a notion of observation according to which it is difficult to give a sense to observing an action as against observing the corresponding bodily movement (for they cannot be thought to cause distinct 'impressions' on the mind). In both these ways empiricism gave support to the view that the mental could not be directly observed; it could only be inferred from physical external behaviour.

This is the dichotomy from which behaviourism starts. Once we have a dualist notion of body–mind interaction, we only need to suppress one term and we have behaviourism. At this point, all the different sources of the behaviourist orientation give each other mutual support. How can we take account of intentional properties (how the agent sees things) if this involves taking account of totally unobservable entities? Moreover, entities whose only function can be to interfere with the only form of law-like regularity we can observe? 'There is no separate soul or life-force to stick a finger into the brain now and then and make neural cells do what they would not do otherwise' (Hebb, 1949, p. xiii). In short one cannot accept intentional properties because this would involve some strange interactionism. But if one cannot accept intentional qualities, then obviously one cannot accept teleological explanation, for this can only be applied to the goals and situation as envisaged by the agent, as we have seen above. Thus all causality is efficient. But then reciprocally, if all causality is efficient, any attempt to explain behaviour by purpose can only be interpreted as the introduction of another efficient cause, viz. 'purpose'. But this is not among the efficient cause that we observe, therefore it must be unobservable; moreover, it must be operating in addition to the causes we observe so it must be in interaction with them. But this is necessarily an unverifiable hypothesis, and of no interest to science. It follows that scientific psychology cannot take account of purposes, and hence of the goals that agents seek and the ways that they envisage them; that is, it has no use for intentional properties. 'If one is to be consistent, there is no room . . . for a mysterious agent that is defined as not physical and yet has physical effects . . .' (Hebb, 1949). We are back where we started.

The behaviourist view of science is a kind of closed circle, a self-induced illusion of necessity. For there is no self-evidence to the proposition that the mental is the unobservable. In a perfectly valid sense I can be said to observe another man's anger, sadness, his eagerness to please, his sense of his own dignity, his uncertainty, love for his girl, or whatever. I can find out these things about another sometimes by just observing him in the common sense of that term, sometimes by listening to what he says. But, in this latter case, I am not leaning on the fruits of some dubious and uncheckable 'introspection' on his part. For what people say about themselves is never in principle and rarely in practice uncheckable. Nor, as we have seen, does any necessity attach to the proposition that teleological explanation is non-empirical, unverifiable or absurd. Quite the contrary. It follows that teleological explanation in terms of intentional properties requires no appeal to an interactionism involving the unobservable.

In fact the idea of interactionism is closely linked to behaviourism. It is the alibi, the only other ontological hypothesis which is admitted, which by its very absurdity gives behaviourism its unchallengeable metaphysical credentials. The hidden assumption of behaviourism is that only these two possibilities are open. But this cannot be accepted *a priori*. The premisses from which this assumption could be derived are far from self-evident, as we have seen. The choice lies rather between a form of explanation in terms of efficient causes and non-intentional properties, such as is exhibited by the different behaviourist theories, and some form of explanation by purpose as we see, e.g. with psychoanalysis. The issue between these two can only be decided by the evidence; it cannot be foreclosed by *a priori* considerations about scientific method, which are so reminiscent of the pleas of the schoolmen of the seventeenth century that the new physics just *couldn't* be right.

### III

Let us examine more closely the debate between these two forms of explanation in modern psychology, and see at what stage it has arrived. As was said above, academic psychology, in Anglo-Saxon countries anyway, has largely been working on the hypothesis that explanation must be in terms of efficient causes and non-intentional properties (let us call this the 'mechanist' orientation, for short). This has been the case for at least the last half-century.

But this approach has taken a particular form, known generally as 'behaviourism', whose aim has been to establish such an explanation at the level of 'molar' phenomena, that is, gross elements of the environment and bodily movement. This may seem strange in that the original impulse for mechanism, in the seventeenth and eighteenth centuries, springs rather from seeing the human organism as a machine, and therefore tends naturally to a mechanistic physiology, an attempt to trace out the connections on the level of the nervous system. If behaviourism eschewed this path, it was partly because of the great difficulty of the enterprise since so little was known (and is known) about the finer workings of the nervous system and the brain. To which was added the additional incentive that what came to be known contradicted the simpler mechanistic models, like that of the 'reflex arc'. For instance, the work of Lashley (e.g. Lashley, 1929) seemed to show that for many functions the brain operated in a complex way which couldn't be reduced to the linking of an afferent to an efferent stimulation. Moreover, many functions couldn't be identified with a specific set of connections in the brain. In face of this, it was perhaps normal that psychologists should abandon the track of mechanistic physiology as premature, or at least as not likely to yield immediate results, and should attempt to establish a mechanistic behaviour theory on the gross observable level of environment and movement. It remained, of course, an article of faith that this would eventually be shown to be derived from a set of mechanistic laws on the physiological level, but meanwhile psychology would be getting on with the job of bringing behaviour into the domain of Science.

The enterprise was undertaken with a quite extraordinary lack of doubt as to its feasibility. There were, of course, differences of approach. Not all theorists went as far as Skinner in eschewing all reference to inner processes. Some preferred the line of Hull who allowed recourse to 'intervening variables', which, however, remained unrelated in any but the most incidental and speculative way to any physiological embodiment. But all were convinced that a mechanistic science of behaviour on the molar level was possible. I say that the lack of doubt was extraordinary, for a minute's reflection will show that even if we grant the behaviourists' *a priori* thesis concerning the rightness of mechanism (and we have seen in the last section that there is no reason to grant this), there is still no guarantee that the mechanistic laws can be discovered *at a specific level*. There is no rule in force in the universe which says that the laws governing a given range of things must be discoverable at any level of analysis, and using any range of concepts. Indeed, precisely the opposite is the case, and the crucial step in the development of any science is the discovery of the conceptual framework in terms of which explanatory laws can be discovered. It may well be that what we know about intentional properties can be fully accounted for in laws applying exclusively to non-intentional properties: but it may also be that this can only be done on the level of neurophysiology. Even the mechanist faith cannot assure us that it can be done just by watching rats in a maze.

And the sequel has shown this confidence in a molar mechanist science to be misplaced. The result can only be described as a failure. Whatever the ultimate truth of the mechanist thesis, it is now becoming pretty clear that it cannot be established on this level. The attempt was to link the environment, characterized non-intentionally as the 'stimulus', and behaviour, characterized non-intentionally in terms of movement as the 'response', in a series of law-like correlations. Since obviously the relation between stimuli and responses changes over time, notably with learning and with changes in motivation, stimulus–response connections were in turn to be shown as functions of behaviour history. In the case of learning this meant a function of the responses to stimuli the animal had emitted in the past (or a function of the responses to stimuli in a context of reward, according to the law of Effect). The motivational condition of the animal was held to be a function of his history of gratification and deprivation. In this way, all the different dimensions of behaviour could be coped with while remaining in the purview of efficient causation and non-intentional properties, as Hull put it, 'colourless movement and mere receptor impulses as such' (Hull, 1943, p. 25).

But the trouble was that this programme broke down at just about every point.

1. The phenomena of 'insight' as reported by Gestaltists, and also by a large number of other students of normal behaviour, and the phenomena of 'improvisation' in a new situation, as shown, e.g. by the experiments of Tolman and his associates, make it just about impossible to exhibit the behaviour resulting from learning as a function of past responses to stimuli. Of course there are attempts to account for the element of adaptive novelty by means of *ad hoc* hypotheses, of which the best known are Hull's 'stimulus acts'. But apart from the fact that no consistent role was ever devised for these unobservable stimuli which could account for all the phenomena concerned,[1] the frequency with which recourse must be had to them begins to raise questions about the criteria for success and failure of molar science. For being without any known physiological basis, inner stimulus acts are fully as unobservable as the 'ideas' of introspectionist yesteryear, that is we come to know of their operation in exactly the way in which we ordinarily see that an animal has 'caught on' or 'learnt to get around the maze', and so on. The set of hypotheses about these inner stimuli add in no way to our ability to explain and predict the phenomena; indeed, the need to have recourse more and more to hypotheses of this kind is itself a symptom, as with the Ptolemaic system of old, that we are on the wrong track. 'Insight', that is the ability to see new relationships and behavioural possibilities in the situation, and some generalized knowledge of the environment seem to be irreducible facts of human and (higher) animal behaviour on the molar level.

---

[1] For the confusions involved, cf. Deutsch, 1960, and also Taylor, 1964, Chap. 8.

2. The notion of 'stimulus' itself cannot fill the role for which it is cast. It is meant to be a physically defined feature of the environment, or perhaps a physiologically defined 'receptor discharge'; but in fact the matter cannot be left here. For, as one would suspect, animals can be trained to react differentially to all sorts of features of the environment, not only to its elements, but to its configuration, to the number of elements rather than their shape, size, etc.,[1] and even to complex relations.[2] Thus the animal's response to a given environment is not a function of the same elements all the time, but the way in which it is relevant to his behaviour can change. Or, as we would say in ordinary speech, the way the environment is seen by the animal changes. Of course, behaviour theorists have responded to these 'discoveries' by a set of hypotheses concerning the organization and selection of stimuli.[3] But these, too, raise questions concerning the viability of molar theory. Since even a casual inventory of the problem will show that the operation of selection and organization involved in perception is immensely complicated, and connected with so many other dimensions of mental function, motivation, interest, past experience, novelty, and so on, it is clear that some simple mechanism[4] will not do. Moreover, we are unlikely to unearth the mechanism (if there is one) by free invention, without careful study of perception itself, and some knowledge of the physiology involved. For we are unlikely to be able to account for the operation of this mechanism in molar terms as we could, for instance, a simple mechanism of selection between different elements. It is obvious that the problem of accounting for 'the way the animal sees the environment' goes way beyond the selection of different parts for saliency to the very structuring of the perceptual field into 'parts' in the first place, and this cannot be a simple function of the environment, but involves an important contribution of central processes. Besides, what we know of the operation of the brain confirms that 'sensory signals invade *ongoing activity*, which is capable of distorting these signals to favour or work against their further penetration of the central nervous system, all in accordance with the concurrent and antecedent events taking place simultaneously in many regions of the nervous system' (Livingston, 1962, p. 71). Thus it would seem clear that the hope for mechanism lies in some 'centralist' theory, the fruit of physiological enquiry, and that at a molar level, intentional language concerning the way the animal sees the situation is unavoidable.

3. Analogous problems arise for the 'response'. This, too, cannot be treated as a simple 'colourless movement'. On the contrary, we only derive

---

[1] Cf. experiments of O. Koehler, reported in Thorpe, 1956.
[2] Such as the matching experiments of Nissen show, e.g. Nissen, 1953.
[3] Cf. Nissen's own comments, 1953; also Broadbent, 1958 and Deutsch & Deutsch, 1963.
[4] Like, e.g. Wyckoff's learned response of 'attention', cf. Wyckoff, 1952.

a law-like relation when we define it in terms of its goal. It is obvious that animals have a more or less flexible repertoire, and that they can substitute one behaviour route for another to the same goal, if the first is inappropriate; so that to say that an animal has 'acquired a response' is often to say that he has learnt to bring about some outcome. This, together with the evidence on improvisation, points strongly to the conclusion that, on the molar level, we have to think of behaviour as directed to certain goals. The *ad hoc* hypotheses which have been devised to cope with this suffer from the same disabilities as those mentioned above,[1] and point again to the conclusion that the hope for mechanism must lie on the molecular physiological level.

4. The field of motivation theory also raises insuperable problems for behaviourism. The aim of behaviourism is obviously to avoid the teleological notion of behaviour as directed to goals, and therefore the concept of 'drive' was devised as an activator of behaviour which was nevertheless 'directionless', that is, didn't give behaviour its direction and shape. This latter was thought to be the function of stimulus–response connections, either innate, or built up in learning history.

But this position is hard to maintain in face of the evidence (see 3 above) that behaviour on the molar level must be understood as purposive, as directed to certain goals. Some semblance of plausibility can perhaps be maintained for instrumental behaviour, such as finding the way through a maze to get food; this behaviour could be seen as the result of $S–R$ connections 'stamped in' by previous training. But when it came to 'intrinsically' motivated activity, like exploration, exercise, not to mention more complex human ambitions, the scheme obviously breaks down. The attempt to put human desires, such as for money, esteem, power, etc. into the 'drive as directionless activator' mould[2] merely produces some ingenious verbal adaptations.

Behaviourism is being forced, therefore, more and more towards a 'molecular' or 'centralist' approach. For it becomes clearer and clearer that the operations of selection and organization which are crucial to behaviour cannot be accounted for (mechanistically) in terms of properties of the environment and gross skeletal behaviour alone; that reference has to be made to the properties of central processes in the brain and nervous system.

It would seem, in other words, that the molar level, the level in which we speak of human or animal acts and the features of their situation which provide their context, is irretrievably teleological and purposive. Indeed,

[1] Cf. MacCorquodale & Meehl, 1954 and discussion in Taylor, 1964, Chaps 8 and 9.
[2] Cf., e.g. Dollard & Miller, 1950 and discussion in Taylor, 1964, Chap. 10.

it may be asked why so much effort has been spent arriving at this conclusion. Future generations may well look on molar behaviourism with the awe and wonder reserved for some incomprehensible cult of a previous epoch. For the facts of insight, improvisation, goal-direction, and intentionality are so obvious in human behaviour and that of the animals which are most similar to us, that it was only by special efforts that they could be overlooked or minimized to the point of appearing tractable to $S-R$ approach. Behaviourists had to go some distance down the phylogenetic scale (not too far, though—the rat is still intelligent enough to refute the hypotheses of his tormentors) and above all devise experiments of a certain rigidity (discrimination and maze experiments) to keep a semblance of plausibility to the ambition of a mechanistic molar science. Those who tried to track animals in their normal environment (ethologists, like, for example, Thorpe) soon revealed the real capacities of animals of different species. The $S-R$ language then only remains appropriate for more rigid instinctual routines of animals lower on the scale,[1] and even there requires extensive changes.

The hope for mechanism thus lies on the neurophysiological level. But it might be thought that here its victory was assured; and this in virtue of another *a priori* argument. Either we admit that behaviour, thought, etc., can be accounted for in terms of physiology, or we have to claim that there is in addition some non-physical reality lying behind behaviour. But this latter hypothesis leads us back to the type of interactionism which we considered in the last section and which is an impossible basis for scientific advance, since it involves reference to a factor which cannot be observed. Besides this, interactionism is unplausible in itself: where exactly in the flow of physiological happenings is there an interruption from the mental? Where, indeed, is the mental? Can we really believe that there are states of the mind for which there is no expression (or, if this is preferred, correlate, or aspect) in neural terms, so that the change from mental state $A$ to mental state $B$ could occur without any alteration in neural state? We thus seem driven to adopt the other possibility, viz., that behaviour, thought, etc., can be accounted for in terms of physiology. Thus Hebb (1949, p. xiii): 'Modern psychology takes completely for granted that behaviour and neural function are perfectly correlated, that one is completely caused by the other. There is no separate soul or life-force to stick a finger into the brain now and then and make neural cells do what they would not otherwise.'

Now this conclusion doesn't follow, not, that is, unless we already assume in deriving it that all explanation is mechanistic.

[1] The kind of behaviours which, e.g. Tinbergen and Lorenz have uncovered.

For this assumption is needed to step from the thesis that all mental states have a neural expression to the thesis of mechanism. Let us call this first the expression thesis. According to it, any condition which we describe in intentional language, e.g., that a person is thinking, or desiring, or suffering, or purposing or intending, or enjoying something, and so on, must have (can only come to be if mediated by) a neural expression. The word 'expression' is chosen here in a perhaps vain attempt to preserve neutrality between two views, one of which would use the term 'correlate' to mark the view that there are two separate entities here, the other of which might use a term like 'aspect', 'facet' or description, thus underlining the identity of the two. Neutrality may be all the more desirable in that these two views may not jointly exhaust the possibilities, and may be both misleading, as we shall see below. To say that all thoughts, etc., must have a neural expression is not necessarily to say that a given thought has a characteristic neural expression which always pertains to it whenever it appears in the mind of any human being, or even whenever it appears in the mind of a given human being. The expression thesis need involve no guarantee that the criteria for identity of mental states will parallel the criteria of identity of brain states. What is required for the thesis, however, is that there be no disembodied thought. But if each thought doesn't have a characteristic neural expression, how can we verify whether a given thought is embodied or not in the neural events contemporary to it? How, in other words, can the thesis be given a content? We want to claim that no mental state can come to be unmediated by neural expression, without claiming anything about the nature of this expression (although empirical discovery *could* reveal regularities which would allow us to judge 'same thought' on neural criteria). The thesis then must be to the effect that no change can occur from mental state $A$ to mental state $B$ without there being a change in neural state which can be seen as its expression.[1]

Now this thesis in no way entails the thesis of mechanism. For it is quite compatible with the view that explanation by purpose is the appropriate model for human and animal behaviour. If the thesis of expression is true then we can certainly trace the pattern of mental life in the pattern of excitation in the brain, and we can establish laws and correlations governing the latter's function; but it does not follow that these correlations must make no reference to purpose, that is, that they be neither teleological in form, nor themselves susceptible to more basic explanation in terms of intentional properties. The expression thesis leaves the two possibilities

---

[1] This last clause is there to rule out of consideration irrelevant neural changes which plainly have no relation to the mental state change in question, and also to give a place to those correlations between mental and neural states which can be established.

open, and the issue between purpose and mechanism is untouched by it. We can only derive the thesis of mechanism, if like most theorists of the behaviourist persuasion, we assume it beforehand. Once more the *a priori* argument is shown to be bogus. There is just no substitute for examining the facts, that is, discovering the nature of the laws which hold in this field. Let us turn then to see if we can define more exactly what the issue is about in physiological psychology.

The trouble is that not enough is known about the workings of the brain and nervous system from this point of view. The simpler earlier views which saw the nervous system as a set of afferent–efferent connections, set up through changes in synaptic structure, have been shown to be inadequate. So have views which ascribed specific behaviours to definite anatomical pathways. Any normal function of animal or man in a waking condition involves a complex integration involving different parts of the nervous system. How exactly this comes about to produce the behaviour selected is as yet not clear.

But the dimensions of the problem can be clarified. An animal or man has a repertoire of possible behaviours, a set of capacities for action. Out of this repertoire, somehow the appropriate response is (often) selected. The question is, how does this come about? It cannot be by some connection 'stamped in' by past experience as was once thought, for behaviour shows improvisation and novelty. We are typically not dealing with a set of fixed movements, but with plastic capacities which can be deployed to many effects. The selection determines how they are to be deployed. Thus I am thirsty: I reach out my hand for a glass of water, or, I go to get a glass at the sink, or I go to a fountain or to a restaurant, or ring for room service, or call for the waiter, and so on, *ad infinitum*. I select somehow the behaviour which is appropriate (or which I believe is appropriate). It is not necessary that I should have done this particular act before in quenching my thirst. It is not even necessary that I have done this particular act before at all. It may be the deployment of a newly-acquired capacity in a novel way: it may be a motor capacity: I may have to execute some as yet unprecedented series of movements to get what I want (though this is not likely if all I want is a drink); or it may be another type of capacity: someone may have told me about phrase books, and I go to get one and look up the phrase for 'give me a drink' in the language of the country. But in spite of the element of novelty we are usually equal to the task.

The fact that much of our behaviour is the exercise of plastic capacities of this kind, motor or other, shows the inadequacy of much traditional learning theory. Learning seems to be much more the acquisition of a capacity of this kind, which is available then for a host of different accomplishments, than it is the mastering of specific performances and achievements. The child may learn to reach and grasp while trying to get a rattle to put in his mouth, but this motor capacity is then available the rest of his life for whatever other goals he has where reaching and grasping are appropriate.

The problem is thus to account for how the appropriate response gets selected. We must discover, that is, in virtue of what a response is chosen among many possible. We are looking for an antecedent condition of this form: a property of the response which earns its selection. Now a mechanistic solution to this problem must select a property which provides us with a non-teleological antecedent, that is, one which makes no mention of that property of the response which is its being appropriate, that is, required for the goal in question. Thus theorists of this persuasion generally look for some mechanism which would select the response in virtue of some anatomical or chemical or electrical property, the formula for which would already be encoded in the nervous system as a result of the need-state which the response relieves; so that as a result of the need-state, just this response is activated. Let us call this intrinsic property $I$. It would thus be the case that the appropriateness of the action would be explained on a mere basic level by the fact that the behaviour having $I$ is selected, and $I$ is appropriate. And the fact that need-state $N$ gives rise to appropriate action $A$ is accounted for by the fact that $N$ gives rise to a state of nervous system, $RI$, which is such as to activate behaviour the neural structure of which has the property $I$. Or $A$–$N$ is explained by $RI$–$I$.

Some explanation of this kind may turn out to be the correct one. But it is an unjustifiable assumption that the true explanation *must* be of this form. It may be, that is, that we can only account for the selection of the response in terms of its appropriateness. In other words, it may be that there is no other property $I$ which determines which behaviour is selected —the antecedent condition of its selection is just that a behaviour is $A$ (within the limits of the repertoire); there is, in other words, no further explanation for $A$'s occurring on the grounds that it is $I$. On this hypothesis, possessing a given capacity would just mean having a nervous system and cortex which within the limits of the capacity sets up the pattern of excitation required by the behaviour which is appropriate, which suits the agent's goals. If I am thirsty, and there is a glass of water to hand, and no reason not to drink it, etc., then I must reach for it, that is, the pattern of excitation in the 'sensory', 'motor' and 'association' cortex is set up which is involved in my reaching for the glass.

In this case, we are accounting for the neural developments in teleological terms. But the expression thesis still holds. I have the goal of reaching for the glass, and a pattern of excitation is set up which is involved in my reaching for the glass. But these are not two events in causal relation one with another. The second is the neural expression of the first. Of course, it may only be part of that expression. For human beings have many ways of having goals, besides the most direct primitive expression which is trying to encompass them. They can, for instance, contemplate them without acting. This is by means of another, more complex capacity, that of verbal thought; and the display of this is not without neural expression. Thus what is involved in my reaching may be only part of the neural expression of my having this goal. But it can be the whole of it, and often is in what we call unreflective action, and moreover, even where it is part, it can only be called the *result* of my having the goal if this latter is identified with something, e.g. thinking about the goal, which has another neural expression. That the pattern of excitation involved in reaching for the glass is the neural expression of the goal of reaching for the glass is thus not a fact to do with a supposedly ghostly inner cause, but rather the fact that we explain this event teleologically, that it arises when it is appropriate in the situation (as understood by the agent).

What is it for the situation to be appropriate? It can be a fact of body chemistry, as in the case of thirst.[1] Or it can be a much more complicated fact about the attitudes, outlook, fears, ambitions, etc., of the person. I am about to make a speech. I am nervous, am thinking over what I am going to say. I can sense that I am going to have a frog in my throat as soon as I start to talk, thus embarrassing me severely. I reach for the glass of water (or go through any of the other routines mentioned above). Here the goal being appropriate is something to do with the way I see my situation, with other goals and desires. The neural expression of this must be very complex indeed. But in either case the question is whether the link between chemical state or neural pattern on one hand and the appropriate behaviour (or the neural pattern involved) on the other can be explained by a non-teleological mechanism of selection, or whether what is crucial in the two cases is just the appropriateness of the goal.

But this last example brings us to a second point. A teleological approach of this kind would also make use of intentional properties in explanation. If I explain that the neural pattern expressive of nervous thoughts and rehearsings before the speech gives rise to the pattern involved in reaching for the glass in a teleological way, i.e. by making the behaviour of reaching for the glass appropriate, then the properties in terms of which the first

---

[1] *This* is, of course, subject to manipulation on mechanistic principles, the question is whether the link between the chemical state of deficit and the appropriate behaviour can be explained mechanistically, through some non-teleological selection mechanism of the kind described above.

makes the second appropriate become relevant for explanation. But in order to explain how the first condition makes the action of reaching for the glass appropriate, we have to make reference to how I see my situation, how I judge a certain possible outcome as embarrassing, and so on. In other words, on a teleological view, what makes the first neural condition an antecedent of the second is something to do with what it is a neural expression of, viz. a certain set of thoughts and feelings about my situation. And thus intentional properties also enter into the explanation on this view.

But once again we would be wrong to see this as a form of interaction. That the content of thoughts enters into the explanation even of neural events doesn't mean that we are making appeal to another sphere. The intentional content of thought is only in another sphere from its neural embodiment for us if we already accept a dualist hypothesis. But what we are saying is that the neurological level would be the wrong level on which to find an explanation of behaviour of this complexity, if this hypothesis is right; that for this range of behaviour, the key level of explanation is a psychological one, and that the shape of the developments in the cortex can themselves not be fully explained without reference to this level. In other words, for this range of behaviour, the most basic level of explanation is psychological.

This is not a striking or outlandish thesis. Nor is it in conflict with the thesis of expression. It was only thought to be because of a confusion between two theses: (1) that mental function is not a set of extra entities or a process taking place in another medium which could occur without neural function but is in interaction with it (thesis of expression); and (2) that mental function and the neural events which express it can be explained on the most basic level in terms of concepts of the neurophysiological level (which make, therefore, no mention of intentional properties), and mechanistically at that.[1] Now this second thesis is not self-evidently true, if it is true at all. On the contrary, there is no ontological guarantee, as we have seen, that a given range of phenomena are explicable in concepts of one level rather than another. The progress of science would, indeed, be much easier if there were; for we would be spared the difficult search for the right conceptual framework, the right notion of 'normalcy', in terms of which the phenomena can be presented as functions of the same set of factors. We cannot say *a priori*, therefore, what concepts will prove adequate for the most basic explanation of human behaviour; and there is no greater implausibility in the thesis that we shall have to account for the

---

[1] It is necessary to add this last clause because it cannot be held to be true *a priori* that all physiological explanations are non-teleological, although behaviour theorists generally assume them to be.

sequence of certain neural events by appeal to psychological laws governing intentional properties and purposes, than there is in its converse.

But nor does thesis (2) follow from thesis (1): that there is no disembodied mental life tells us nothing *per se* about the laws governing this life. The derivation can only be made if we assume the thesis of mechanism. The blithe way in which this assumption is made can be seen in the quote from Hebb above: 'Modern psychology takes completely for granted that behaviour and neural function are perfectly correlated, that one is completely caused by the other.' It seems that, for Hebb, the second half of the sentence is just another way of saying what is conveyed in the first half. But this is far from being the case. The first half can be taken as an assertion of the thesis of expression; but the second half, that behaviour is completely caused by neural function (we are safe in assuming that Hebb didn't mean the converse), amounts to *a priori* legislation about the type of concepts which will figure in an adequate explanation. Whether this is so can only be decided by the facts, by discovering what explanations actually succeed. The implied inference from (1) to (2) is therefore based on a *petitio principii*.

But it is the easier to make in that it introduces a certain simplicity. For explanation by purpose introduces a complexity not found in mechanism. For it marks a distinction between levels where mechanism sees none. Obviously some aspects of our behaviour must be accounted for in concepts appropriate to a neurological level of description. For one thing the limits of our repertoire must be, for another certain determinants of desire —hunger, thirst, sexual desire most notably—are under the control of our body chemistry, for a third, many forms of breakdown must be accounted for in neurological terms. Explaining our behaviour by purpose, therefore, involves a view of behaviour as existing on several levels. This poses the problem of the relations between them.[1]

[1] We can thus see the debate about the supposed identity of sensations and brain processes in a new light (cf. J. J. C. Smart, 1963; Hilary Putnam, 1961; and U. T. Place, 1956). The whole question is badly put because it involves a confusion between theses (1) and (2). We are asked to consider a given class of mental and a given class of neural states as entities and decide whether they are identical. Put in this way, we have to say, no; for there is no guarantee that a given mental state will always be expressed by the same (qualitatively) neural state. Indeed, there is some reason to believe that this is not the case, given the multi-functional nature of much of the brain. But that does not mean that mental states are separate entities from neural states. Given the thesis of expression, we can say that mental function is not some set of extra entities apart from neural function. But put this way, the thesis of identity loses all its interest. The really interesting question is whether mental function can be explained by laws whose concepts are drawn from descriptions only appropriate on the neurological (or more radically the physical level), or whether it must be explained by concepts of its own range. (In other words the question whether thesis (2) is true.) This is the question of 'reduction', and this is what materialists like Smart and Place appear really to want to establish. But to

I say 'several' levels, and not just two, because we are not dealing with the dualism of traditional Cartesianism or empiricism where the 'mind' can be thought to make use of a number of automatisms in order to encompass its ends. No instrumentalist view will do. In fact there is no clearcut demarcation between the levels, on any hypothesis which accepts the thesis of expression.

We can in fact see two types of relations between the levels. On one hand the 'higher' behaviours can obviously only exist if certain conditions, specifiable in neurological terms, are met. In this way the existence of the neural conditions necessary to capacity and normal function can be seen as boundary conditions in the sense discussed earlier. These conditions must hold for any normal function, but once this is assured sufficient conditions can only be discovered for behaviour in teleological terms.[1]

On the other hand, there is plainly a continuation between behaviours which are more 'automatic' and those which are more flexible. The 'automatic' behaviours are the ones which are more rigid, are set off by a limited set of releasing conditions and are performed in a more or less stereotyped way; this probably corresponds to a less complex and widely integrative neural pattern (something more closely approaching the classical theory of the 'arc'); in any case, automatic behaviours are those which can most plausibly be accounted for in mechanistic terms. These behaviours dominate low in the phylogenetic scale, as we can see from the rigid instinctual patterns described by ethologists (cf., e.g. Tinbergen, 1951), and also early in ontogeny. What is striking is how plasticity develops as we advance both phylo- and ontogenetically. More plastic behaviours are more adaptive, involve intelligence in judging their appropriateness and in adapting means to the end. Thus men, who start their lives with an automatic routine, sucking, as their only food-getting behaviour later develop an almost unlimited repertoire which can be directed to this end.

But in the case of man, and perhaps some of the higher animals, there is an interesting development in the other direction. Men are capable of learning and as it were entrenching routine behaviours. Indeed, this procedure even seems to be essential to their achievement. Because secondary functions can become 'second nature' to us, as we say, we can concentrate on other things and extend our scope. Those who know how

put it in terms of identity is to confuse the issue: because reducibility is not a question about, e.g. sensations and brain states as entities, but about the laws governing mental and neural functions in general. The question is not: are sensations and brain states identical; but, can we account for sensations in terms of laws mentioning only brain states and their appropriate description (that is, without any mention of psychic properties)? This, if I understand him, is the principal point of the interesting article by Richard Rorty, 1965, which restates the identity thesis in a defensible form.
[1] Cf. the distinction between 'necessary' and 'sufficient' conditions in Sec. 1.

to drive cars will slam their foot on the brake when the 'releasing stimulus' of something running across the road impinges on their receptors, very much as geese will fly from the shadow shaped like their species predator. And in either case, the reaction may happen where it is inappropriate, for instance on an icy highway.

But with man, the routine is potentially under higher control in the immediate situation, and is subject to training in or out in the long run. It is thus rightly called 'second nature'. On this view, therefore, there can be no clearcut line of demarcation between 'higher' behaviours whose necessary conditions are psychological in form and 'lower' ones where an explanation may be adequate in exclusively neurological concepts.

Thus the question between the two modes of explanation can perhaps be put in this way: are there certain non-teleological neurophysiological mechanisms which can account entirely for the selection, integration of functions and organization necessary for behaviour, or must we explain some behaviour in terms of its appropriateness for the goals of the agent concerned? In the latter case, the 'higher' behaviour is subject to certain necessary conditions which are defined in purely neurophysiological terms, and is embedded in a complex of 'lower' behaviours which approach at the limit a degree of rigidity which makes mechanism plausible; but sufficient conditions for these higher behaviours cannot be given without using purposive and intentional concepts. The question is, in other words, whether all purposive behaviour can be explained on a more basic level mechanistically, or whether on the other hand, different aspects of the stream of behaviour must be seen as taking place at different levels, albeit not rigidly separated, for some of which the most basic explanation remains psychological and hence in terms of purpose.

It is impossible to say at this stage whether this latter thesis or the mechanist one offers the best hope for a science of behaviour. All that we can say is that what is known of the operation of the brain is compatible with explanation in terms of purpose. We have already seen that for many functions the brain operates by a complex integration involving many regions rather than on fixed paths. It is also clear that sensory inputs are themselves organized by ongoing activity (Livingston, 1962). In addition the evidence of lesions in both animals and man is that the brain is in part a locus of general capacities which can serve a great many ends, rather than a collection of specific functions. This was the conclusion of Lashley (1929, pp. 175–6), for rats, and of other studies for man.[1] Lashley's principle of mass action, that impairment in performance is proportional to the size of the lesion, but much less dependent on the locus, tells in the same direction.

[1] E.g. the classical work of Goldstein quoted and discussed in M. Merleau-Ponty, 1945.

So much is well established. But some evidence tells specifically in favour of the conception of behaviour on different levels. Lashley (1929) and others have discovered that lesions can induce loss of habits already learned, but that the same habits can be relearned after. Which might imply that behaviour once learned may become more stereotyped and involve less general neural activity; lesion might therefore abolish it. But the capacity to learn, although perhaps less efficient because of the lesion, is a general capacity and survives. There is also some encephalographical evidence (cf. Jasper, 1958) that learning requires a contribution of a large part of the brain, but that a behaviour, once learnt, requires only a segment to carry it out. This might be thought as well to support the notion that some behaviour requiring greater effort of intelligence involves the display of more general capacities, and that, once the fruit of learning is routinized, it becomes more localizable and identifiable with a specific neural pattern.

Another indication that routine functions may become identified with specific patterns and pathways, but not irrevocably, comes from Ivo Kohler (reported in Attneave, 1962). Kohler carried farther Stratton's experiment with inverting lenses. After three or four weeks, the wearer comes to see the world upright again. So that what starts off as a rigidity of function of certain nerves (a 'rigidity' highly justified in terms of their normal role in our perception) is gradually corrected when it becomes mal-adaptive.

Thus the picture of behaviour on different levels, straddling more complex, intelligent, conscious and neurally diffuse performances on one end and more automatic, rigid, unreflective and neurally localizable performances on the other, which seems to arise from the purposive view, is not an implausible one in the light of the neurological evidence.

But the evidence is far from decisive in either direction, and the debate remains open between mechanism and a purposive theory. This is perhaps a meagre conclusion to so much argument. But it has some importance in the light of the widespread belief in the *a priori* necessity of mechanism as the only 'scientific' form of explanation. We can, moreover, draw some further conclusions from this discussion about the way to proceed in behaviour science.

Once we recognize that the question is open between mechanism and purposive explanations, some doubt is cast on the utility of speculations about neural mechanisms of selection and organization which are not neurophysiologically based. This is especially true when the proposed mechanisms are extremely simple. For example a well-known work in this field in recent years, *Plans and the Structure of Behaviour*, by Miller, Galanter & Pribram (1960), presents a feed-back mechanism, the TOTE (test-operate-test-exit). But this concept plays no part in the interesting and often highly valuable reflections in the greater part of the book, which are expressed in terms of the concept of 'Plan'. These reflections, shorn of the obeisance to mechanism, turn around a conception of behaviour as directed and hierarchically organized, features which fit easily into a conception of behaviour as purposive. The TOTE mechanism only regains its relevance, after the introductory chapters, in connection with some very sketchy 'neuropsychological speculations' at the end of the book. Its principal function seems to be that of a kind of logical talisman: as long as Plans are seen as potentially reducible to TOTE hierarchies, it is permissible to use the concept to make all sorts of teleological discoveries about behaviour. But the real value of the book lies in the remarks on the structure of behaviour.

Again, Hebb (1949) in his *Organization of Behaviour* presents mechanisms, the 'cell assembly' and 'phase sequence', which although expressed in neurological terms are not based on any direct supporting neurological evidence. They represent avowedly speculations on the basis of behaviour. But it is hard to see what the utility of this might be. Hebb hopes to cope with some of the phenomena invoked by Gestalt theorists and which made trouble for molar behaviourists. He succeeds, in a rather loose way in 'saving the phenomena'. But it is very unlikely that a theory like his based on specific connections and lowered synaptic resistance can cope with the form of learning which involves acquiring a general capacity, like walking, swimming, talking, using a library catalogue, etc., which can be used for an indefinite number of behaviours once it is mastered. Of course, we could try to save it by a special set of *ad hoc* hypotheses of an already depressingly familiar kind about 'generalization' of one performance to others. But since we are dealing with a purely speculative mechanism, whose operation we can in no wise verify directly, *ad hoc* additions of this kind have a largely gratuitous nature. We are free to add and modify them at will to 'save' any phenomena that turn up. But this raises the question what use this serves. In these conditions, any theory could 'survive'.

Hebb's speculations fall between two stools. They yield neither a verifiable neurophysiological hypothesis, nor a really valuable psychological reflection. For this attempt to save the phenomena can be an immensely futile operation. Since the 'phenomena' concerned are largely those of other psychologists, gleaned in experiments which in effect greatly narrow by their very design the behaviour which can be exhibited, the game is played within fairly narrow bounds. In fact it would have no point at all, if it weren't for the great interlopers, such as the Gestaltists, and the Tolman group, who having different views about animal behaviour, design different experiments which show capacities which animals could never reveal in the behaviourist's maze, but which everyone except academic psychologists knew they had all along. An attempt is then made to fit these findings within the purview of mechanistic explanation by inventing *ad hoc* hypotheses of an unverifiable kind about inner events. In this regard, the phase sequences of Hebb are on all fours with the 'stimulus acts' of Hull, except that the former sound more neurophysiologically plausible. But since so often the only new phenomena allowed are that small range of intelligent behaviours which these cognitive theorists with access to the journals have managed to demonstrate experimentally, the game is still pretty remote from reality and a shadow victory has to be won again after each new discovery. Thus Spence (1937) painstakingly tries to account for one result of Köhler showing relational perception by a specific *ad hoc* hypothesis, when the evidence of relational perception of many other and diverse kinds is there in abundance in everyday human behaviour (and some of it has even been reported in the journals, too).

What is really needed is to throw open the doors, and examine the real world, take account not only of the more surprising performances of the white rat, but examine also what men can do. What is needed is a reflection on behaviour in its own terms and a classification of its different varieties, a study of its structure, which will reveal the full range and limits of flexibility and intelligence. We need to see what has to be explained to get an idea of what it would mean to explain behaviour. We can no longer go on the behaviourist assumption that we can grasp the simple lower behaviours first, and then build from there to the higher ones. This assumes that there are no differences of level in behaviour between more rigid and more intelligent, and thus begs one of the most central questions of psychology. All it can lead to is an examination of those behaviours which have already been made to fit the theory. In order to make progress the science of behaviour has to knock down the walls of the maze and look afresh at the real world.

### REFERENCES

Attneave, F. 1962. Perception and related areas. *Psychology: A Study of a Science*, vol. 4. Ed. S. Koch. New York.
Braithwaite, R. B. 1953. *Scientific Explanation*. Cambridge.
Broadbent, D. E. 1958. *Perception and Communication*. London.
Deutsch, J. A. 1960. *The Structural Basis of Behaviour*. Cambridge.
Deutsch, J. A. & Deutsch, D. 1963. Attention: some theoretical considerations. *Psychol. Rev.* **70**, 80–90.
Dollard, J. & Miller, N. E. 1950. *Personality and Psychotherapy*. New York.
Hebb, D. O. 1949. *The Organization of Behaviour*. New York.
Hull, C. L. 1943. *Principles of Behaviour*. New York.

Jasper, Herbert. 1958. Reticular and cortical systems and theories of the integrative action of the brain. *Biological and Biochemical Bases of Behaviour*. Eds. H. F. Harlow & C. L. Woolsey. Madison.

Lashley, K. S. 1929. *Brain Mechanisms and Intelligence*. Chicago.

Livingston, R. B. 1962. Neurophysiology and psychology. *Psychology: A Study of a Science*, vol. 4. Ed. S. Koch. New York.

MacCorquodale, K. & Meehl, P. E. 1954. Study of E. C. Tolman. *Modern Learning Theory*. W. K. Estes *et al*. New York.

Merleau-Ponty, M. 1945. *Phenomenologie de la Perception*. Paris.

Miller, G. A., Galanter, E. & Pribram, K. H. 1960. *Plans and the Structure of Behaviour*.

Nissen, H. W. 1953. Sensory patterning versus central organization. *J. Psychol.* **36**, 271–87.

Place, U. T. 1956. Is consciousness a brain process? *Br. J. Psychol.* **47**, 44–50.

Putman, Hilary. 1961. Minds and machines. *Dimensions of Mind*. Ed. Sidney Hook. New York.

Rorty, Richard. 1965. Mind–body identity, privacy and categories. *Rev. Metaphysics*. **19**, 24–54.

Smart, J. J. C. 1963. *Philosophy and Scientific Realism*. London.

Spence, K. W. 1937. The differential response in animals to stimuli varying within a single dimension. *Psychol. Rev.* **44**, 430–44.

Taylor, Charles. 1964. *The Explanation of Behaviour*. London.

Thorpe, W. H. 1956. *Learning and Instinct in Animals*. London.

Tinbergen, N. 1951. *The Study of Instinct*.

Wyckoff, L. B. 1952. The role of observing responses in discrimination learning. *Psychol. Rev.* **59**, 431–42.

# COMMENT

## *by* ROBERT BORGER

Taylor's central proposal is that of 'teleological' as an alternative to 'mechanistic' explanations. Most of the time mechanism is treated as synonymous with an essentially peripheralist, stimulus–response model—and this is shown to be unsatisfactory for dealing with the kind of question about behaviour that we normally answer in terms of someone's views and intentions. It also seems to be incapable of accounting for the flexibility and adaptiveness that characterizes goal-directed behaviour, in humans and in (some) animals. When mechanistic explanations are conceived more broadly as being 'centralist' and physiological, it is argued that there is still no guarantee for the adequacy of explanation in such terms, since the assumption of physiological correlates for mental phenomena does not ensure being able to avoid *references to purpose*, even within this framework. This leaves us with 'teleological explanations'. But before we consider such explanations, as well as the rejection of mechanism, in more detail, I must mention a somewhat idiosyncratic difficulty that Taylor's conclusion presents me with.

Over the years, I have engaged repeatedly in what would be described as goal-directed activities of various degrees of complexity—just now, quite deliberately, I made a cup of coffee, involving the usual component steps of kettle filling, coffee grinding, etc.—and some of these activities have been discussed or explained, both by myself and by others, in terms of objectives, purposes and motives, both conscious and unconscious. Such accounts appeared to be entirely appropriate and it seems to me that in the context in which they occurred, no other form of explanation would have satisfied the states of uncertainty or puzzlement that invited them. Yet—I have to reveal it—I am a mechanism. Mechanistic explanations would therefore seem to be relevant *in some way* to all aspects of my behaviour. Does this mean we made a mistake in talking about my behaviour in 'terms of concepts of the range of purpose, goal, meaning'? Or is it possible that there are aspects of the behaviour of (some) mechanisms that can only be discussed in purposive and not at all in mechanistic terms?

I want to argue that explanations in terms of purpose, and mechanistic explanations, are not rivals, provided they are considered in the context of two different types of enquiry; and to speculate that Taylor may have arrived at (what seems to be) his position because,

(i) his conception of mechanism is too much influenced by linear systems, of which the various *S–R* models are examples, and which are in fact inadequate for *both* purposes; and,

(ii) because, perhaps through lack of interest, he may not regard answers to one of the questions as constituting *explanations* of behaviour.

The two—by no means exhaustive—types of enquiry I have in mind are concerned with,

(*a*) the identification of goals, and the tracing of relationships between them, conducted within a conceptual framework—and the corresponding language—that is based on the experience of having goals and taking what is judged to be appropriate action. In experience there is no gap between the execution of goal-directed activity and 'having the goal'; in the language of experience, the relationship between wanting to do *X* and doing *X*, in the absence of constraints, is a non-contingent one, and does not therefore admit further explanation;

(*b*) the development of models, or blueprints for models, that are *capable* of exhibiting a given range of (goal-directed) behaviour.

Enquiries under (*a*) are concerned with what we usually call 'understanding' or 'making sense of' someone's behaviour; those under (*b*) are a stage on the way to *building* a system with certain characteristics and potentialities for behaviour. The large differences between the behaviour (and appearance) of people and that of (most) machines, has led us to reserve the language of purpose and intention for the former; the only way in which we normally 'understand' machines is in terms of their structure and modes of operation. In my own case, the type (*b*) problem would consist in attempting to reconstruct, probably by very slow stages, the original specifications, or their equivalent. If it has made sense, in the context of (*a*) enquiries, to talk about my intentions, or about my believing, rightly or wrongly, that a certain course of action would lead to a particular desirable result—then the specifications under (*b*) will have to make possible the kind of behaviour that evokes such discussions. If over the years I have learned from experience and have developed in a variety of ways, the specifications will have to allow for this also.

All this is not to suggest either that the quest for a complete blueprint will be rapidly, indeed ever, successful, or that references to specifications would provide a substitute for explanations in terms of purpose on those occasions where the latter have hitherto been appropriate. The quest for such specifications in the broadest sense—it may for example attempt to incorporate physiological data, or avoid direct reference to 'hardware' altogether—is the expression of one kind of puzzlement generated by behaviour. I regard *this* as the search for mechanistic explanations, using 'mechanistic' in much the same sense as Deutsch has used the term 'structural' (Deutsch, 1960), and whereas the relative success of any particular formulation is something to be determined empirically, this is not true of the *possibility* of success of the entire enterprise.

I now want to consider stimulus–response models and Taylor's alternative to mechanism in relation to these two types of enquiry.

Stimulus–response models of behaviour are rooted in what is really a third type of preoccupation, expressed most explicitly by Skinner: that of discovering how behaviour may be influenced and controlled. For such a purpose, it will be useful to have a set of probabilistic statements directly linking particular environmental conditions and behaviour, together with methods for modifying the probability values involved—useful even if it should turn out to have a limited range of applicability. Even if most of the time complex processes inside the organism rule out simple linear relationships between environment and behaviour, between 'input' and 'output', it may be possible to manipulate circumstances in such a way as to override internal complexity and eliminate it from our calculations. Reducing an animal to 80% of its normal body weight is one way of achieving such simplification—starvation is a good method for 'emptying the organism'. The trouble starts when a system of relationships which is adequate for prediction and control subject to very specific boundary conditions, comes to be used as a basis for *explanation* in either of the senses indicated above. Even though $S$ and $R$ may be all that is observable, $S$–$R$ does not necessarily provide a satisfactory prototype for the mechanism—or organism—whose behaviour may at different times bear a whole variety of different relationships to its environment. Despite a great deal of ingenuity that has gone into the elaboration of $S$–$R$ models, they do not seem to be capable of producing the sort of behaviour that we would wish to describe as purposive. Basically they remain linear, rather than feedback models, and it is an empirical question to discover whether or not a given type of model can or cannot generate a given type of behaviour.

The failure of mechanistic explanations in general to satisfy the other sort (type ($a$)) enquiry is a different matter. To say how a system works is not intended to answer a question about what was happening on a given occasion. The observable output might have been achieved in a whole variety of ways. More fundamentally, though we might attempt to phrase a hypothesis about a particular action—my action say—in terms of the operation of sub-systems and their inter-relationship at the time, such an explanation apart from probably being enormously cumbersome, could not answer a question about purpose, posed within a framework in which reference to mechanism is reserved for those cases for which explanations in terms of purpose are not considered appropriate: in which to talk of mechanism therefore seems to question the very phenomenon it is called upon to explain.

Taylor presents the prescription for his teleological alternative in two stages. The first stage (p. 55) makes an explanation teleological if *B is explained by the fact that S was such that it required B for G to come about.* My recent activity in grinding coffee, boiling water, etc., would be accounted for by pointing out that the initial situation was such that it required ground coffee and boiling water, etc., to be brought together, if a cup of coffee with particular characteristics was to be the result. The explanation apparently invokes a feature of the situation, *S*. But there is obviously something incomplete about this, for the situation was also such that it 'required' a quite different series of events to produce a cup of tea, or a ham sandwich—yet somehow only the coffee-producing events occurred. All situations have unlimited potential for developing into others, and there is nothing at this stage of the formulation which accounts for the selection of any particular facet of this potential.[1]

The basis for the selection is introduced on p. 59 as a 'second restriction' on applying teleological explanations; *B* must be required for *G in the view of the agent*, it also being understood, implicitly, that *G* is something the agent wants to bring about. 'That an action is required *in fact* for a given goal will not bring it about unless it is seen to be such.' But this emphasis on the agent would seem to be a first and only, rather than a second restriction; for as far as what is required *in fact* is concerned, no account is given either of how the situation influences the choice of one among many potential goals, or, once the goal is chosen, how it predisposes the agent to view it one way rather than another. The only contribution that it makes is that of specifying the physical limitations imposed by the situation—one cannot boil a kettle with hallucinated heat.

When Taylor introduces *teleological antecedents* as a separate feature of teleological explanations, separate that is from references to the agents' view, he seems to commit the same error for which he berates *S–R* behaviourism: the attempt to explain goal-directed behaviour as a product of the stimulus situation. The explanation differs apparently from the behaviourist position in that it invokes special 'non-intrinsic' features of this situation. I shall attempt to show, however, that the only distinction between teleological and 'intrinsic' descriptions (p. 57 of Taylor's article) consists in a covert reference to the agent. If this reference is

---

[1] Indeed, the definition of a teleological explanation at this stage is such that *any* sequence of connected events could be explained teleologically. Yet 'the condition of the ground was such that it required a lot of water to turn it into mud' is hardly satisfactory as an explanation of heavy rainfall.

separately allowed for, the case for talking about teleological antecedents disappears.

Coffee can be described as dark brown, as being of a given density, as having a variety of chemical properties. It can also be described as having the property of being turned into powder by particular types of mechanical action, of being then partly soluble in water to form a liquid of a particular kind and so on. These are all quite straightforward 'intrinsic properties'— for when we are *listing* the properties of a substance (or situation) $S$, rather than singling out one of them in the context of some explanation, it makes no difference as far as the claims about the nature of $S$ are concerned, whether we say of $S$ 'it requires $B$ to turn it into $G$' or '$S$ has the property of being turned into $G$ by $B$'. All the potentially unlimited number of sequences of events, in which $S$ can be an element, provide the basis for possible descriptions of $S$, and these could be expressed in *equivalent* intrinsic or teleological forms.[1] Thus when we consider the *general description* of $S$, it is not a question of substituting for a teleological description *another* one in terms 'which make no reference to the event explained or the end' (Taylor, p. 57). At this stage, there is no particular $B$ to be explained, only *possible B*'s, and statements about the effects of different $B$'s on $S$. These are, in fact, what Taylor calls intrinsic descriptions, irrespective of the grammatical form that is used to convey the information. To show, as Taylor does in his 'Canadian Club' example, that there is a variety of possible *intrinsic* descriptions of any situation, only some of which may be appropriate as explanations in a given context, does not provide an argument for there being two *types* of property, teleological and intrinsic—yet he claims implicit support for the distinction when he concludes 'the fact then that *a teleological antecedent also bears an intrinsic description* (my italics), even an intrinsic description which picks it out as causally relevant, proves nothing in itself'.

The statement '$S$ was such that it required $B$ to bring about $G$' *works* as an explanation for the occurrence of $B$, because typically the questioner is ignorant of the relationship between $S$, $B$ and $G$. The gap in this understanding of the agent's behaviour is completed by a piece of information about the (intrinsic) properties of things and situations. Taylor, however, uses this form of explanation to fill a different gap—a puzzle about the nature of goal-directed behaviour as such. *This* is not solved by a con-

---

[1] The properties that would normally be included in a descriptive list would be a selection from this infinite population, limited by the current state of discovery, and by considerations of what would be found relevant to situations likely to be encountered. But although we do not normally say anything about the interaction of coffee with sulphuric acid, or the effect of putting it into a car engine, this omission in no way affects what would in fact happen in the appropriate circumstances.

sideration of intrinsic properties as conceived above. A dummy confronted by these same properties would not have generated $B$, nor a human agent unaware of the relationship, nor one having a different goal. The clue to this problem is to be found in, rather than outside, the agent and has to do with his having intentions, views, objectives and all the rest. As Taylor says (p. 55) 'That behaviour is a function of what it serves to bring about, rather than some other unrelated factor, is a fact (if it is one) about the form of the antecedent.' The critical part of that antecedent, or antecedent condition, is, however, decidedly a function of properties of the agent. It may be that Taylor wants to put part of the burden of explanation on special teleological properties of *situations*, because like $S$–$R$ theorists, he feels the need for explanation in terms of antecedent conditions, yet like them is not prepared to regard views and intentions as constituting such conditions. Instead of invoking ghostly stimuli 'to make neural cells do what they would not do otherwise', he seems to be hanging special properties on to quite ordinary situations—properties that achieve their effect in virtue of language rules and thereby avoid the absurdity of dualism.

A very explicit appeal to rules of usage is made by Norman Malcolm in a recent paper entitled 'The conceivability of mechanism' (Malcolm, 1968), to argue an incompatibility between explanations involving purpose or intentions, and mechanistic explanations. The relationship between brain states and behaviour must be contingent, whereas that between behaviour and intention is to some extent non-contingent because of the way in which we use words like 'intend'; hence we must be dealing with different types of explanation, and there can be no equivalence between 'brain state' and 'intention'. Taylor in *The Explanation of Behaviour* (Taylor, 1964, p. 33) writes 'This is part of what we mean by "intending $X$" that, in the absence of interfering factors, it is followed by doing $X$. I could not be said to intend $X$ if, even with no obstacles or other counter-vailing factors, I still didn't do it.' But the term 'intention' is involved in two kinds of relationship—its relationships with other terms within normal language, where it is subject to constraints of usage, and its relationship to events, where it draws attention to something that goes on where it makes distinctions between situations in which we feel the need to use this kind of terminology, and those in which we don't. It seems to me that Taylor, Malcolm and many others who have written about purpose and mechanism are using constraints within one system of conceptualizing—that inherent in our normal way of talking about behaviour—to argue for the in-admissibility (inconceivability in Malcolm's paper) of *another* system, related to the same phenomena, but constructed from a different point of view and for a different purpose. The fact that we use—have tended to use —the term mechanism in a restricted sense which makes it incompatible with the term purpose, should not be taken to imply that the systems which exhibit purposive behaviour have no principles of construction.

For both Taylor and Malcolm, mechanism constitutes a denial of purpose. 'The viewpoint of mechanism thus assumes a theory that would provide systematic complete non-purposive causal explanations of all human movements not produced by external forces' (Malcolm, 1968). But how does the man whose movements have just been given a purposive explanation *function*? In virtue of what sort of property is he able to do things which lead us—rightly—to ascribe to him purposes and points of view? What sort of instructions should we give to our technicians to produce one like him? If at this stage I abandon my earlier conceit (if that is the right word) of being an assembly of micro-miniaturized circuits, this does not make any fundamental difference. If 'building' a human being is an impossibility, it is not a logical one. The pursuit of that goal, or type of goal, is a perfectly respectable interest, and the formulations arising from it constitute one approach to the explanation of behaviour. As an approach it must take account of purposive explanations, as drawing attention to phenomena that are to be reproduced. Reference to the appropriateness of an action to a goal, or of a goal to other, higher order goals, is in this context a starting point, rather than the end of the line. The problem is, as Taylor says (p. 70), '. . . to account for how the appropriate response gets selected'. But the fact that a mechanistic explanation may not *mention* a response's appropriateness, i.e. that it may not use this particular term, does not necessarily mean that it therefore ignores this feature of the situation. Unless appropriateness *can* be translated into terms that can form the basis of construction and simulation, the building of a system exhibiting purposive behaviour is an impossibility. Unless this impossibility is insisted on, the translation of 'appropriate' and related concepts at whatever level of complexity into structural terms is an aim whose gradual achievement has only practical obstacles. There is no point at which 'the attitudes, outlook, fears, ambitions, etc., of the person' *take over* from body chemistry —or whatever system of concepts may be appropriate to a 'person' constructed from some other materials. They belong to parallel frameworks of reference.

Explorations within the mechanistic framework may not, at any given time, contribute significantly to what we know about the range of phenomena ordinarily discussed in terms of attitudes, ambitions, etc. They do not at the moment. What is important about contributions like that of Hebb's 'Neuropsychological model' (Hebb, 1949, 1959) or the TOTE

(Miller, Galanter & Pribram, 1960) is that in their very different ways they mark a step forward within this framework compared to the S–R system of Hull and its derivatives. When it has become increasingly apparent, that a particular type of model that has dominated the mechanistic approach, is inconsistent with a wide range of phenomena, then a promising reformulation, even if it is not followed up very much, produces relief, and opens up a whole variety of topics that have been ignored with varying degrees of embarrassment. The embarrassment is the result of a restricted view of what a mechanistic approach involves and the discovery that there is no room for purpose within it. This point has been amply developed by Taylor. It seems to me, however, that his exclusion of mechanism is the reverse side of the same fallacy.

Although I am almost entirely in agreement with Taylor's introductory analysis of scientific explanations, I do not think that it supports his later conclusions. Two main points are made: that scientific explanations are arranged in tiers, with relationships at one level being *accounted for* by translation into the terms of another; and that scientific explanations make possible innovation of control and construction. The example of reference to molecular structure, to account for differences in the strength of materials, is quoted. And Taylor writes (p. 52) '. . . the stronger kind (of explanation) lays the basis for an advance in technology, since in showing the difference between outcomes, it also gives us information which can be used, if the techniques are available, to control the outcome . . . it enables us also to bring about new as yet non-existent outcomes, as when, for instance, we create new elements with the aid of the theory of atomic weights.' But this is precisely what Taylor's proposed form of 'basic' explanation of behaviour does not do. What technological advance in the area of goal-directed behaviour can result from regarding the *appropriateness* of a response as an ultimate explanatory concept? In what sense does this type of explanation *account for* the common characteristics of all the instances of behaviour that are grouped together as purposive? The stage corresponding to an appeal to the molecular level is missing. To explain the strength of materials by reference to molecular arrangements, or their temperature by molecular movement, does nothing to invalidate the concepts of strength or temperature. To attempt the specification of systems capable of different forms of goal-directed behaviour does not deny the concept of purpose. The terms in which such specifications would be made need enter into the ordinary discussion of behaviour no more than molecular concepts into discussions of the strength of bridges.

REFERENCES

Deutsch, J. A. 1960. *The Structural Basis of Behaviour*. Cambridge.

Hebb, D. O. 1949. *The Organization of Behaviour*. London.

Hebb, D. O. 1959. A psychoneurological model. *Psychology: A Study of a Science*. Ed. S. Koch. London.

Malcolm, N. 1968. The conceivability of mechanism. *Phil. Rev.* **77**, 45–72.

Miller, G. A., Galanter, E. & Pribram, R. H. 1960. *Plans, and the Structure of Behaviour*. London.

Taylor, Charles. 1964. *The Explanation of Behaviour*. London.

# REPLY

## *by* CHARLES TAYLOR

Reading Borger's comment on my piece made me realize how far I was from communicating what I wanted to say clearly. For in fact I have no quarrel with him over one of the major points he makes. I don't believe that there is any argument in principle which can show that mechanistic explanation is impossible, that in other words such an explanation is 'inconceivable'.

In this I disagree with Norman Malcolm's paper ('The conceivability of, mechanism', *Philosophical Review*, January 1968) which Borger mentions, and with which he would want to associate me. Rather I think, with Borger, that there is no necessary incompatibility between our describing and explaining behaviour by purpose in ordinary life or in the context of scientific theories of teleological–intentional type (like, for example, psychoanalysis) on one hand, and our being able to give a mechanistic neurophysiological account of them on the other. Like him, I believe that such a mechanistic explanation would be related to our ordinary purposive ones on the model of the 'deeper level' explanations which I talk about in the first part of my paper and which Borger mentions with approval towards the end of his comment.

But I feel I should say more about this, for Borger's error in aligning me with Malcolm is a very understandable one: in the *Explanation of Behaviour*, I use a set of arguments in a rather confused way which seem to have the same purport as Malcolm's. And I think this argument is well worth examining.

An essential stage in Malcolm's reasoning is the point that we cannot have two explanations which purport to give us necessary and sufficient causal conditions for the same event, unless they are co-ordinated in some way (cf. Malcolm's article, p. 56, although he doesn't make this point in the way I do). Thus the hope that we can resolve the philosophical problems of mechanism by some 'double-truth'-type doctrine, and hold, for instance, 'that there could be two different systems of causal explanations of human movements, a purposive system and a neurophysiological system' (Malcolm, 1968) is an utterly vain one, if we think of these systems as being without relation to each other.

For a causal explanation always has subjunctive and counterfactual implications. It implies something about what would have happened if other conditions had obtained. We could think of two separate and unrelated laws stating necessary conditions for a given event, where both had to be met to provide a sufficient condition; there could be two unrelated laws stating sufficient conditions, where neither is necessary. But one cannot state a necessary and another a sufficient condition without there being some link between the two. Since for both to be true it would have to be that the antecedent of the sufficient condition law couldn't be fulfilled without the antecedent of the necessary condition law being fulfilled also. And this necessary concomitance could itself be expressed as a statement. We can then put the case this way: we can only hold to both explanations if this linking law holds good; should it turn out to be false, then one of the two would have to be rejected.

Now with our hypothetical mechanistic neurophysiological theory we have a system which is meant to provide us with both necessary and sufficient conditions for all behaviour ('sufficient' here means 'sufficient within the system'; we take for granted that a whole host of external conditions for the normal functioning of the organism: enough air, absence of cosmic catastrophe, temperatures between certain limits, and so on, are met). And with our ordinary purposive explanations we often offer, in a quite unsystematic way it is true, necessary and/or sufficient conditions (where 'sufficient' has the same sense as the above).

There is thus no room for two such systems in a single range of phenomena unless they are linked systematically. One can imagine their being held provisionally as independently true because for a time we cannot set up a crucial experiment; but if they are really independent such an experiment must be possible, and it will decide the issue. Or one might imagine that the system will not admit of a crucial experiment in principle; but that would mean that the system will not admit of a situation in which the predictions of the two deviate; and this is just another way of affirming their systematic connection.

To say this is to say that purposive and mechanistic accounts are potential rivals, that is they *can* be in a relation such that to offer one for a given phenomenon is to rule out the other. And this is what happens for instance when we try to show that a given piece of behaviour was not action and cannot be explained by desire or intention on the grounds that it was a reflex or heteronomously-controlled movement.

But in order to show that mechanism is 'inconceivable' one has to go further and establish that these accounts are always rivals. For in this case, only one could be correct. One could, of course, argue that this need cause no embarrassment to mechanism: if there is a fight to the death why should it not be the victor, and the sloppier, less 'scientific' purposive explanation go under? But there is something preposterous about this suggestion, for it implies not only that all our ordinary explanations in terms of purposes, desires, intentions, feelings and the like are mistaken and have always been, but even that the very language we use for action and feeling, which is impregnated with the logic of these ordinary explanations, that this language is systematically inappropriate and misleading. And this is just too much to swallow. (Malcolm's inconceivability argument goes further in trying to explore the paradoxes that arise from someone's stating the mechanist thesis on the assumption that it excludes the possibility of intentional behaviour; for stating is a form of intentional behaviour. But this further turn of the screw as it were presupposes that we have already caught mechanism in the vice by showing that it is incompatible with purposive explanations.)

The key step in the argument is that which shows that these two accounts are necessarily rivals. But above we saw that they were rivals unless systematically co-ordinated. And in fact the model for this co-ordination is readily available: it is that of the reduction of one theory or set of laws to a more basic one, along the lines I mention in the first part of my paper. And it is just this kind of more basic level theory which proponents of mechanism promise us.

The crucial step in Malcolm's argument is thus that in which he purports to show that this type of relationship between the two modes of explanation is impossible. Malcolm starts from the proposition that the general form of purposive explanations is of this type: let us say that we are explaining why $O$ emitted behaviour $B$, then we might say that $O$ had goal $G$ and believed that $B$ was required for $G$. This is a clearly recognizable form of purposive explanation.

But this explanation may seem incomplete as it is. For those who hold that all explanations must contain at least one general statement as well as particular statements of fact, it will inevitably appear so. What is missing would be some general proposition to the effect that 'whenever an organism $O$ has goal $G$ and believes that behaviour $B$ is required to bring about $G$, $O$ will emit $B$' (cf. Malcolm, p. 47). Malcolm in any case reconstructs purposive explanation with this as major premiss.

Now an argument might break out right here, as to whether this was not a misleading reconstruction of ordinary explanation by purpose. True, it seems necessary if we want to set this up in parallel to explanations of events in inanimate nature, or mechanistic explanations in general, for these seem always to involve reference to general propositions. But one might question whether this parallel presentation doesn't have precisely the effect of obscuring the difference between the two.

But let us waive this objection and see Malcolm's argument through. His point is that this major premiss of purposive explanations, with the addition of a *ceteris paribus* clause, becomes a non-contingent proposition. It is part of what we mean by someone having goal *G* that he will take action appropriate to encompass it 'other things being equal' (or, better, says Malcolm: 'provided there are no countervailing factors'). Now not only does this have no analogue in the major premisses of mechanistic explanations, but it introduces an obstacle in principle to the explanation of purposive behaviour on a more basic level by a mechanistic neurophysiological theory.

For it is a property of a more basic law $L_2$ that it allows us to express the conditions in which the less basic law $L_1$ holds. But the major premiss of purposive explanation above is not contingent, and hence 'cannot be contingently dependent on any contingent regularity' (Malcolm, p. 50). It follows that it cannot be explained on a more basic level by a mechanistic theory.

But then if we rule out this kind of systematic co-ordination between mechanistic and purposive explanations of behaviour where the former plays the role of more basic explanation, we are forced to consider them as always in rivalry. And if they are rivals, then one must be wrong. Since it is preposterous to hold that our ordinary explanations are *all* mistaken and have been for millennia, we have to conclude that the mechanist thesis cannot be true.

I think the flaw in this argument lies in the step in which we rule out more basic mechanistic explanations of the major premisses of purposive explanation on the grounds that these are *a priori*. Malcolm provides the wherewithal to invalidate this step on the very next page (p. 51). For since he wants to show the odd and untenable consequences of mechanism which flow from the fact that it is incompatible with the principles of our ordinary purposive account, he must give some sense to this notion of incompatibility, and hence to the idea that 'purposive explanations... would be refuted by the verification of a comprehensive neurophysiological theory of behaviour'. Now this cannot mean that the purposive principles are shown to be false; for they are *a priori*; they are conceptual truths. We can, however, show that such principles have no application to the world, or more particularly to behaviour; and this is what would be established by a comprehensive mechanist theory on Malcolm's view.

But this same move can be made in connection with a more basic mechanist explanation. The principles of a neurochemical theory cannot account for the purposive principle's being true, for this it is in virtue of the sense born by 'goal', 'desire', 'intention', and so on. But what it can explain is why these principles have application to the world.

And when we think of it, we can see that this is exactly what such a theory would claim to explain. The cortical state which corresponds to my desiring $X$ is not characterized in the theory as a condition which *ceteris paribus* issues in the doing of $X$. So that the state-description (let us call it '$S$') has a quite different logic from that of 'desiring $X$'. But this does not prevent it from figuring in a more basic explanation. No one would claim that $S$ helps us to explain why when people desire $X$ they tend to encompass it, for this correlation is not contingent. But we can say that it is because there are such states as $S$ which are linked in the theory, contingently but with great regularity, with certain overt behaviours that we can have and apply our ordinary concept which transforms this link into an *a priori* one.

Malcolm's argument, which crucially depends on showing that purposive and mechanistic explanations are always rivals, thus fails at this point. And I would like to say generally that I do not believe that an in principle argument can be constructed to rule out the possibility of a more basic mechanistic explanation of behaviour. Where do I then disagree with Borger?

Simply here: that Borger seems to want to go to the other extreme, and oppose to the thesis that mechanism is in principle impossible, the claim that it must be true. At least this is what he seems to be saying (p. 82) where he declares that in 'the search for mechanistic explanations', 'whereas the relative success of any particular formulation is something to be decided empirically, this is not true of the possibility of success of the entire enterprise'.

But there is no more reason to accept this *a priori* thesis than there is to accept that mechanism is inconceivable. What do we know at this stage about the potential more basic level explanations of behaviour? How can we be sure that they are mechanistic? It is by no means clear that any deeper level explanation must be mechanistic; psychoanalysis, for instance, offers us an account more basic than our ordinary everyday one. Can we assert today that some version of this theory cannot be more firmly established?

Borger says baldly (p. 80) 'I am a mechanism.' But how does he know? Isn't there perhaps a confusion between two different theses here? There may be good in principle (philosophical) arguments to show that I am a material object, that is, that a dualistic notion of man in which an immaterial substance is thought to be combined with a bodily one cannot really be sustained. In this case whatever happens in us which we can describe with mental predicates will have bodily concomitants. But—and this was the burden of the argument in my paper—nothing follows from this about the plausibility or implausibility of mechanism.

93

That it seems to, reflects a narrow imagination in philosophy of science. From the fact that everything that happens in me has a bodily concomitant, it may seem plausible to conclude in the light of our mechanistic traditions in philosophy of science that everything can be explained in the scientific languages which have been found applicable to bodies in general, i.e. physics and chemistry. But this inference is far from forced on us. Certainly some ranges of, in the broadest sense, 'behaviour' of organisms must be so explicable—at the limit, it is evident that in falling off a high building I follow the same laws as any stone. But it does not follow that all our behaviour is to be so explained.

Of course, we all have our hunches about what a future science of human behaviour might be. If asked for mine, I would be bound to answer that I look forward to something at present unimaginable: I don't believe that we will have a neurophysiological science of any of the recognizably mechanistic forms that are now being canvassed; but I don't flatter myself that I have an alternative model. People of basically anti-mechanistic bias like myself will get a jolt, but so will the present generation of practising academic psychologists (if we all live to see real progress in this domain). It will be a theory, I imagine, which will only be 'reducible' to a physics and chemistry that are themselves greatly extended and enriched in their biological applications.

But all this is just hunch. My main point here, and in the paper, is that nobody knows now or can know. And this is where I disagree both with the position that Borger attributes to me and with his own.

But this is not all that can be said on the subject. If for the ultimate outcome all bets have equal odds, we can still discuss what short-term research strategies seem most promising. And here it does not seem that attempts to adumbrate what an overall mechanistic theory might be like are of very much use. Borger seems to agree to this when he says (p. 86) that explorations within the mechanistic framework are not at the moment significantly contributing to our understanding of 'the range of phenomena ordinarily discussed in terms of attitudes, ambitions, etc.'.

But the case can be made stronger than this. What emerges from the argument above against the thesis that mechanism is inconceivable is that mechanist and purposive explanations can co-exist to the extent that they are related as more and less basic explanations. And this means conversely that we should be able to derive from an eventual true mechanistic neurophysiological account of behaviour those purposive explanations which we have established as valid. It follows that it would help towards the discovery of such a global mechanist theory, if this is possible, to explore further the nature of behaviour and to expand our repertory of explanations of a teleological and intentional kind.

I would go further and claim that talk about a global mechanist theory tends to be of a very far-fetched speculative kind, and is likely to remain such until we get a great deal more knowledge either about the operation of the brain and CNS, or about the structure of those higher behaviours (such as speech, or certain typically human emotions) which are central to human life—or about both. That academic psychology both in Britain and America should look with such reverence on these speculations while accepting very bad epistemological arguments for regarding teleological and intentional theories (like psychoanalysis) with at best suspicion is more than a pity. It is a great source of intellectual sterility in the sciences of man.

# IS THE BRAIN A PHYSICAL SYSTEM?

## *by* N. S. SUTHERLAND

### I. INTRODUCTION

It seems odd to raise the question 'Is the brain a physical system?' since the natural temptation is to reply unequivocally 'Yes'. The question is posed here because it has been fashionable for certain philosophers in recent years to put forward arguments implying that the brain is not a physical system. Thus some philosophers have argued that because movements are not actions, behaviour cannot be explained in terms of the workings of the brain and its interaction with the physical environment; others have argued that there is a dimension of unpredictability that applies to human behaviour which does not apply to the behaviour of purely physical systems. If these arguments are correct, then it would be impossible in principle to explain and predict behaviour in terms of the physical workings of the brain. If, however, the brain and the body constitute a physical system in which matter behaves according to the same laws as in other physical systems but differs only in the complexity of its organization, then behaviour could in principle be explained in terms of physical events.

Moreover, if the brain is a physical system, there is no room for the operation of any additional non-physical influence on behaviour such as volition, wants or motives where these terms are construed as referring to anything other than very complex categories of physical events. The reason for this is that if the brain is a physical system, then human behaviour will be predictable to within the same limits of accuracy as the behaviour of any other physical system. If it is predictable to within these limits, then there is no room for any systematic departures from predictions based on the organization of matter in the brain, though of course there will be random departures due to the inherent difficulties of making completely accurate predictions about any physical system. Hence there is no room for any non-physical influence on behaviour.

It should be stressed that even if the brain is a physical system, we may need to make use of new concepts and principles in explaining behaviour. This is equally true in the explanation of other physical systems. Thus the explanation of the behaviour of a gas, of a star or of a thermostat involves us in developing concepts appropriate to the way the matter in each system is organized. These three systems are, however, all mechanistic systems

97

and the principles used in explaining the behaviour of the whole system can be inferred from a knowledge of the laws governing the component parts of the system together with a knowledge of how these component parts are organized. Our claim is not that the explanation of behaviour does not involve new principles but only that if the brain is a physical system then these new principles are reducible to the laws governing the behaviour of elementary particles in the same way that the laws governing the molar behaviour of a gas are reducible to laws governing the behaviour of the molecules that compose it.

If we interpret the question 'Is the brain a physical system?' to mean 'Does matter at the atomic and molecular level in the brain obey the same laws as matter in other physical systems?' then the question is clearly an empirical one to which we do not yet know the answer with certainty. For reasons of scientific economy, many scientists working on the problem of the brain and behaviour assume that the brain is a physical system in this sense. It is important, therefore, to decide whether this belief can be shown to be false on *a priori* grounds as some philosophers imply.

The remainder of this article falls into four parts. First, the question of what is the empirical evidence for regarding the brain as a physical system is discussed. Secondly, an attempt is made to specify the fallacy that under-lies all efforts to show on *a priori* grounds that the brain is not a physical system. Thirdly, six of the detailed arguments that have been brought against regarding men as physical systems are considered. Finally, the question is raised of what empirical evidence there could be against the brain being a physical system and what would be the implications for scientists of its not being one.

## 2. EMPIRICAL EVIDENCE

In what follows the phrase 'physical system' will be used to mean any system in the space–time continuum in which matter and radiation conform to the same lower level laws to which they conform in those parts of space time that we would all agree constitute inanimate systems. For the purposes of study it is convenient to break up the universe into parts which have some sort of continuity of form through space and time. Examples are the sun, the solar system, an atom, a wireless set, a car. It is characteristic of such physical systems that they have some principles of internal organization which we seek to discover; we also try to discover how they are affected by physical influences from outside, and systems vary in the extent to which they are subjected to external influences. In the case of animals and man the interplay between system and environment is extremely complicated and we cannot get very far in understanding the system itself without studying in detail its interaction with the environment.

One obvious difficulty with the above account is that we do not yet know and perhaps never will know all the laws governing the behaviour of matter and of energy in other forms. There is, however, no reason to suppose that the matter composing the human body obeys different physical laws known or to be discovered from the laws governing matter in inanimate systems. It may be that it differs only in the way it is organized and if this is true we may hope to explain and understand the behaviour of the system by studying the very elaborate organization of its components. With a system of such complexity, one of the main methods of studying the organization of the parts is to study input–output relationships and infer the internal organization, rather than merely to study the separate parts. The first approach is that of the experimental psychologist, the second that of the neurophysiologist or neuroanatomist and both are necessary to acquire an understanding of the system. An analogy may help to make this clear: we came to understand the behaviour of gases by subjecting them to various operations and studying the effect on the gas as a whole, rather than by attempting to follow the detailed behaviour of individual molecules. However, it was possible having studied the molar characteristics of gaseous systems to infer how the behaviour of the individual molecules composing the systems gives rise to these characteristics.

It is, of course, impossible to demonstrate that the human brain and body constitute a physical system as defined above. Nevertheless, it is plausible on inductive grounds to suppose that they do. Unless we have strong evidence to the contrary, we assume for the sake of economy that the same fundamental laws apply to matter in different systems. This is the method by which science proceeds. To make a rational decision on the plausibility of the induction would involve a careful scrutiny of all that is known about the behaviour of matter in animate systems and particularly in man. It is not possible to undertake such a task here. However, the successes of biologists in explaining such things as the nervous impulse in terms of the known properties of matter have increased the plausibility of the proposition very considerably over the last few decades.

On a more macroscopic scale, considerable evidence has accumulated recently that the movements that animals and men make are themselves attributable to physical causes. We shall be concerned in this section only with physical movements and will leave the question of the relations between such movements and actions until later; the problem of consciousness will also be deferred.

# IS THE BRAIN A PHYSICAL SYSTEM?

We can regard the human being as a physical system into which is fed a certain stimulus input, namely, those physical energies which affect the receptors—the eyes, the ears, taste buds, etc. Studies of receptor organs have shown not surprisingly that unless a receptor organ is affected by a particular stimulus, it is not possible for the organism to respond to that stimulus: the converse of this is not true—that is, some stimuli may affect receptor organs differentially but it is impossible for the organism to respond to them differentially. This is because some of the differences detected by peripheral organs are not transmitted over further parts of the nervous system. Thus the cat retina is sensitive to variations in the wavelength of the light falling upon it, but this sensitivity is not matched by the ability of the whole animal to discriminate colours.

The output from the system consists in the movements of various organs such as the arms, legs, and vocal chords. There are other outputs such as blushing and crying which themselves are caused by movements of a less macroscopic variety. These outputs are determined by muscle contractions, which in turn can be shown to be caused by impulses running down efferent nerves. There is good evidence to show that the impulses in peripheral efferent nerves are themselves caused either by further impulses originating in receptors and transmitted via the spinal cord (spinal reflexes) or by nerve impulses in the brain itself. Stimulation of the brain at certain loci produces predictable movements.

Our present state of knowledge of brain organization and functioning is of course still very rudimentary, but it is possible to produce by electrical stimulation certain complex learned series of movements: thus if the thirst centre in the hypothalamus is stimulated, a goat will give a series of responses which it has previously learned to perform in order to obtain water. Moreover, stimulation of the brain and lesions to different parts of the brain produce changes in psychic functions in a predictable way. Thus damage to a fairly restricted set of centres interferes with speech and language, including the capacity to use language internally. Under certain circumstances stimulation of the temporal lobe leads to the appearance of very vivid auditory and visual imagery which patients themselves describe as reliving past experiences.

The plausibility of the proposition that all behaviour is determined by the state of the brain and nervous system depends upon evaluating the evidence so far accumulated that this is true. Such an evaluation would involve detailing all the discoveries made by the sciences of experimental psychology, neurophysiology and neuroanatomy. Now to show that some mental functions and some responses depend upon brain mechanisms is not to prove that all do. The latter proposition depends upon induction from what is already known, and as in any inductive argument one can always deny that the evidence is enough to support the generalization: the evidence can at best only make the generalization more or less plausible. However, no one who has not been through the evidence carefully is in a position to evaluate the plausibility of this generalization, and it will be argued in the next section that it cannot be ruled out of court on *a priori* grounds based on the way we apply concepts in everyday language.

Although we have not been able to present the evidence in detail, an attempt has been made to show on what type of evidence the belief that men are physical systems rests. We must now consider whether there is any evidence suggesting that men are not physical systems. The most obvious evidence lies in the fact that there are many functions carried out by human beings which are not carried out by any known inanimate physical system. If men behave in totally different ways from inanimate systems, then this suggests that matter in the human body differs in some way from matter in inanimate systems. There are two possible ways in which the matter could differ and they are not mutually exclusive. Either matter obeys different laws in the human body from the laws it obeys in inanimate systems or it obeys the same laws but is organized in very different ways. The first alternative would mean that men were not physical systems, the second would imply merely that they were unusual physical systems. A hundred years ago it might have been plausible to believe the first alternative. Today, however, it is clear that many of the functions at one time thought to be unique to animate beings can in fact be performed by machine. This strongly suggests that the differences between men and machines depend upon the way matter is organized rather than on differences in the laws governing its behaviour in the two types of system. To evaluate the plausibility of this assumption involves a knowledge of what functions can at present be performed by machines.

If we understand the behaviour of a physical system, it is always possible to produce a second physical system which behaves in some sense in the same way as the first. The behaviour of a gas or of the solar system can be modelled on a computer. If we know what are the elements composing a system and what their relationships to one another are, we can build a second system whose elements are put in correspondence with those of the first preserving the relationships which hold between the elements of the first system. Proofs of this proposition have been attempted by McCulloch and Pitts and by Kleene: for our present purpose it is not necessary to prove it, but only to envisage it as a possibility. One test for discovering whether we have understood the workings of a physical system is in fact to simulate the system by a second: we can then see whether the effects of a given input on the second system are the same as on the first. If they are the same we take this as evidence that our model is correct, particularly if it responds to inputs whose effects were not envisaged in the original design in the same way as the system we are trying to simulate.

Because of our limited knowledge, it is clearly not possible at the moment to simulate all of human behaviour in this way or even to simulate a very substantial part of it. Because of the complexity of organization of the parts of the human body, and the vast amount of input, the problem is a very formidable one. However, some aspects of human behaviour can be simulated in this way. Simon and Newell have written a program for a computer which makes the computer simulate the ways in which beginners find proofs in elementary logic. The computer prints out an account of what it is doing at each stage, and these accounts resemble the introspections of human beings engaged on the same problem. The computer was able to solve problems not taken into account in the design of the program, and from the protocols seemed to solve them in the same way as people. It is obvious that computers and machines are today able to carry out many functions which one hundred years ago would have been thought to be functions that only man was capable of performing. Thus computers can carry out mathematical calculations, some of which could not be performed by any human being because of the limitation of the life span; automata can already recognize a variety of simple patterns and there can be little doubt that within a few years we shall have devised machines that can perform complex recognition tasks such as identifying faces and 'reading' handwritten characters; computers can find proofs, they can play chess, though not as well as the average club player; they can store and retrieve information and hence have at least an analogue of memory; and they can exhibit goal-directed behaviour.

It should be noted that where a computer can perform a function which we previously thought of as being performed only by man, it does not necessarily perform it in the same way as man: unless the elements of the computer and their relationships can be set in one–one relationship with the physical elements mediating a function in man, the machine is not a model of the human brain. Thus numerical calculations are normally performed by a computer in a different and more efficient way from the way in which they are performed by our own brain.

Two further points should be made about machines which perform functions formerly mistakenly thought to be performed only by man. First, no matter how many such functions are performed by existing machines, it will for a very long time always be possible to point to further functions which cannot be performed by machine. No machine can at present classify tastes in the same way as man, nor has any computer yet written a great symphony, though they have produced some very tolerable melodies. In view of the number of typically human functions which can at present be performed by machine, it is quite pointless to argue about what further functions may or may not similarly be performed in the future: we can only guess or wait and see, but if we are going to guess we should at least examine the achievements of machines to date rather carefully. It is often forgotten how much the situation has changed in the last two hundred years. Two centuries ago, nobody would have had any idea how to go about building a machine which could perform any of the mental functions performed by man. Until very recent times, engineers concentrated on transferring energy: it is only within this century that the problem of transferring information and analysing it by machine has been seriously tackled. Two hundred years ago nobody would have known how to set about building a machine which exhibited an analogue of purposive behaviour: today it is a commonplace that there are mechanisms designed to achieve a certain goal which correct deviations from a target by systematically varying their output in such a way that over a wide range of input variations they still achieve the same goal. The existence of such mechanisms has put purposive behaviour in a new light and removed some of the mystery from it: although not all purposive behaviour can be simulated, at least we have an idea how to go about it. The same is true about the processes involved in recognition or in calculating or proving theorems.

Secondly, in describing what functions existing computers carry out, I have used language which many philosophers would perhaps prefer to restrict to human beings. Machines were said to 'recognize', 'learn', and 'have a memory'. This sort of language is today used by people engaged in automata studies without hesitation; precisely because many of the functions carried out by machines are so closely analogous to functions carried out by man, it is very convenient to use the same words for both. Thus when we say a machine can recognize handwritten words we mean that it can consistently give the same output to a large number of written instances of the same word, and it will systematically give different outputs to different words where there is a large range of variation in the way the actual characters forming the words are written. This is at least very close to what we mean when we say of a man that he can recognize words, and certainly the only way we have of knowing that anyone else can recognize a class of things is to discover whether he can systematically give one response to that class of things and other responses to other classes.

The arguments of this section may be summarized as follows. From studies of the behaviour of matter in animate systems there is some evidence that it obeys the same lower level laws in such systems as in inanimate systems. There is also evidence that human behaviour considered as a series of movements is caused by physical events in the muscles and nervous system. It might be thought that the difference in the functions carried out by animate and inanimate systems was evidence that matter behaves differently in the two systems. However, our ability to simulate by machine many functions formerly thought to be the prerogative of man alone suggests that it is differences in the way matter is organized rather than differences in the laws governing the behaviour of matter that is important.

### 3. INVALIDITY OF 'A PRIORI' ANSWERS TO THE QUESTION

We now turn to the question of the underlying fallacy in the attempts that have been made to argue on *a priori* grounds that human behaviour must be explained in terms of different principles to those used in the explanation of inanimate physical systems.

Perhaps the main reason why some philosophers deny that human behaviour can be explained in terms of the interaction between the physical elements that compose the brain and body is that this notion conflicts with certain concepts which are enshrined in our everyday language. The following quotation from Geach is an extreme instance of this type of argument: 'Machines do not think: for thinking is one of the special activities of living things, and machines are not alive. A machine—any machine—is a paradigm of what is not alive, as a man or a cat is a paradigm of what is alive. It is idolatrous superstition for men to ascribe superior wisdom to their own inanimate artefacts.' What Geach appears to be saying is that in ordinary language the word 'thinking' is only applied to men and higher animals. It is not normally applied to other physical systems. However, the concepts that are enshrined in ordinary language merely reflect the conceptual system which has been evolved by our ancestors and which has been found useful to them. As well as reflecting the wisdom of our ancestors, language also reflects their muddles and mistakes. The fact that there are certain words that can be applied to men but which are not normally applied to inanimate physical systems does not in itself prove that there is a difference in the way men regarded as physical systems behave and in the way in which all non-animate physical systems actual and possible behave: it merely shows that our ancestors believed that there was such a difference. To be accurate it shows slightly more: it shows that it was useful to them to hold this belief. A concept, as reflected in language, will only survive provided that it has some usefulness to society: but it is easy to see that the concepts which are useful to society at a given period of time do not necessarily have any objective validity. Thus no one would argue that God exists because the concept of God exists, nor that winds are alive because some primitive societies supposed them to be.

Geach suggests that the word 'thinking' is only applicable to things that are alive and are not artefacts, and this proves that we have some criterion for what is alive: it does not prove that thinking cannot be understood and explained in terms of the organization of matter in the human nervous system, nor does it prove that we could not build a machine in which matter is organized in the same way. These are empirical questions and cannot be solved by an appeal to ordinary language. In the extreme case, it is possible that one day we might succeed in stringing atoms and molecules together in such a way that the resultant artefact could not be told apart from a human being by any test other than that of how it came into existence.

There is a slightly more subtle form of the argument put by Geach. There are certain words that we can apply consistently to different phenomena: this proves that there is a real difference between the phenomena. Two such words are 'voluntary' and 'involuntary': there is a large measure of agreement about which actions are voluntary and which are involuntary, and this shows that there is a real distinction between such actions. It does not, however, show the nature of the distinction. The man in the street, if he thinks about it at all, would probably suppose that voluntary acts are in some way not determined, whereas involuntary acts are. Certainly some philosophers have supposed this. But they may well be wrong about what the criteria for the use of these words are. It might be that voluntary acts are simply acts which are determined in much more complex ways than involuntary acts. It is impossible to use such distinctions to give information about empirical questions. The fact that two sets of phenomena have different names which can be consistently applied by different observers proves only that there is a real difference between the two sets of phenomena: it does not of itself establish what the difference is; moreover, people may use the words consistently and themselves be mistaken about what the difference is.

Now there are a whole series of words that are normally only applicable to man or at best to man and some animals: examples are such words as 'recognizing', 'free will', 'voluntary', 'calculating', 'punishment'. There is a strong temptation to believe that because such words are not normally applied to anything else except man, human behaviour must be determined by principles quite different from those which operate in any other physical systems whether actual or possible. This conclusion clearly does not follow: there must be some differences between men and inanimate objects or we would not consistently classify them apart. What these differences are, however, is a matter for empirical investigation. Thus the question of whether human behaviour can be understood in terms of the workings of the brain and body regarded as physical systems is an empirical one: it cannot be settled by fiat, and it cannot be settled by appealing to any distinctions built into ordinary language. The only way in which it can be settled is by investigating the brain and behaviour and seeing just how far we can explain behaviour in terms of the workings of the brain.

#### 4. EXAMINATION OF DETAILED 'A PRIORI' ARGUMENTS

We shall now consider in more detail six of the arguments that philosophers have produced in order to show that it is improper to regard men merely as physical systems.

I. One of the objections most commonly raised is that no matter how much we were able to explain the actual physical movements made by a human being in terms of the workings of the nervous system we would still not have explained his actions since an action is more than a set of physical movements. It is certainly true that in categorizing actions we take into account very much more than a person's physical movements: we have to consider the relation between his body and the external world, his knowledge about the external world, and his intentions and much else besides. Supposing that we knew the state of the brain completely at one time, and that this knowledge allowed us to predict that when a certain stimulus input occurred a particular set of muscle movements would result, we would still not be in a position to categorize the action, and there are many questions about it which we would not have answered.

A case that has been raised recently in this connection is that of a car driver who gives a signal to the car behind that he is slowing down. If we knew the state of the brain of someone driving a car at a particular moment, we might be able to specify a class of stimuli which if they occurred would lead to his making a complex series of extensor and flexor movements with the muscles of his right arm. The set of stimuli would correspond to the car in front slowing down and to a variety of other contingencies and the set of muscle contractions would correspond to the action described as giving a hand signal for slowing down. It is clear that in so far as we have only considered in isolation from everything else the brain state that gave rise to this set of muscle movements we have not given anything like a full explanation of the action of signalling. An everyday explanation of the act of signalling would take into account all sorts of things about the relation between the driver who signalled and the external world both at the time the signal was made and in the past. Philosophers who argue that actions cannot be explained in physical terms seem to forget, however, that it is open to the physical scientist also to take all these relations into account: *of course* actions cannot be explained by taking just the state of the brain at a given time and nothing else into account, but there seems to be no reason at all why a full explanation in physical terms cannot be given if we consider the state of the brain in relation to the external world.

Part of an everyday explanation for the action of signalling might be that the man who signalled was invariably courteous. If we consider not just the state of the brain, but what movements the brain would give rise to under a large variety of different inputs and if it is the case that it tends to give rise to bodily outputs which please other people, and if it were the case that if his brain had not been arranged in this way then he would not have signalled, then all that is involved in saying 'He signalled out of courtesy' will be taken account of in the explanation in terms of brain processes. In order to do this, however, we have had to consider the relationship between brain processes and the external world, but there is no reason at all why the physical scientist should not do this and indeed if he wants to have a full explanation of human behaviour he must do so.

An analogy may help to clarify this point. We shall consider how to describe the behaviour of a missile that homes on a target. If the missile is off course, one of the motors may fire more strongly and bring it back on course towards the target: we might describe this movement as 'the missile correcting its course'. Let us suppose that we have a detailed knowledge of the mechanism inside the missile, and from its state at a particular time we have deduced that if a particular input occurs a particular motor will fire more strongly, and this will result in just the movement that has occurred. This knowledge would not justify us in saying that the movement was a corrective movement made to bring the missile back on its course towards the target. We could only make the latter claim after we had considered not merely the mechanism on its own but the mechanism in relation to a large class of possible inputs to that mechanism. If the mechanism is such that when the missile is off course, under a wide range of different conditions an input will occur that results in an output that tends to bring the rocket back on course, then we will be able to make statements about the target of that rocket and about a particular movement being a corrective movement. But to do this we have had to consider not merely the mechanism in the rocket but the relation between the rocket and the external world and we have had to consider not merely what it does but what it would have done under other circumstances.

To claim that actions cannot be explained in physical terms is rather like claiming that the rocket's behaviour cannot be explained in physical terms on the grounds that a knowledge of its mechanism at a particular time does not of itself guarantee the truth of a statement such as 'it is making a corrective movement'. The truth of this statement *can* only be guaranteed by examining the mechanism of the rocket in relation to possible inputs and outputs, in other words taking into account the relation between the mechanism and the environment. Moreover, we can make statements about the rocket making corrective movements without having examined the mechanism at all: if we observe enough rockets of a similar type, and under different circumstances they all make movements which bring them back on a course towards a target, we shall infer that they are built in such a way as to home on a target and to make corrective movements when they get off course. The fact that we can make such statements about the rocket without knowing what the mechanism is does not mean that a full explanation of the rocket's behaviour cannot be given in physical terms provided we include in the explanation both details of the mechanism in the rocket and details about the relation between the mechanism and the rocket's environment.

It is, of course, true that the mechanism of the human brain is of a very different order of complexity to that in the rocket, and the relationships between man and environment that have to be taken into account in explaining actions are very much more complicated than the relationships that have to be considered in determining whether a rocket has made a corrective movement. Even in the case of such an apparently simple action as giving a hand signal when driving, the complexities are enormous. To describe something as a signal we have to consider not only one human being but others as well—a series of muscle movements can only serve as a signal between drivers when they belong to a class of movements which have specific effects (mediated by the brain) on other drivers. We would also have to take into account the driver's knowledge of his environment. As a result of past experiences, human beings have a representation of parts of the external world in their brain. Examining the organization of the brain at a particular time is not enough to tell us what is represented there without doing much further calculation: we have to consider what is there in relation to possible stimulus inputs, and in turn the relation between these inputs and the physical objects which give rise to them must be examined. We also have to take into account what goals are represented in the brain—again we must consider not just its organization at a particular time, but its organization in relation to potential inputs. And all these things interlock.

It may be said that it is not at the moment possible to give an account of any but the simplest actions in purely physical terms, and this, of course, is true. All that is being argued is that this does not prove that actions are not ultimately explainable in physical terms nor does it show that they involve anything more than extremely complicated arrangements of the particles composing the physical world, though in classifying actions we have to take into account unfulfilled conditionals, i.e. when we say some-body ran to catch a train we are committed to the belief that he would not have run if he had not thought he was late, and he would have run in a different direction if his beliefs about the streets had been different. This sort of thing, however, is equally involved in our statements about the rocket correcting its course and indeed is involved in any classification of events where some of the criteria used in the classification reside in the causes that have led up to the event we are classifying.

2. A second argument that has been advanced against human behaviour being explainable in terms of physical causes is very similar to the first. The argument runs as follows: to apply many of the words describing mental functions we have to be sure that the behaviour to which the description is applied meets some sort of norm. Since knowledge of the brain can never tell us whether the behaviour does or does not conform to a norm, we can never have a complete account of behaviour in physical terms. Hamlyn has argued in this way in his book on the *Psychology of Perception*, and other philosophers have used similar arguments.

This argument seems to rest on the same fallacy as the first argument on the difficulty of describing actions in physical terms. Hamlyn argues that in the normal use of the word 'see', whatever is seen has to be seen in an appropriate way: but clearly a knowledge of the change in brain state set up by a visual input will not be enough on its own to tell us whether something has been correctly seen or not. This argument, however, again only shows that we cannot fully explain behaviour in terms of the brain without taking into account the relation between the brain and the external world: once that relationship is taken into account, we can see that it might be possible to have an explanation of seeing in purely physical terms provided we leave consciousness out of account for the moment.

If we show someone a square and say of him 'He sees a square', we normally mean both that there is a square there, and that light from the square is affecting his eyes, and that because of this he will have certain behavioural dispositions which are in some sense appropriate to a square being there. That is, if asked to say what the object is he will describe it as a square, if he is asked to draw it he will draw a square, if he is asked to say whether it contains any curved lines he will say 'no' and so on. Now what constitutes appropriate behaviour will depend upon the goals a man has in view and upon his knowledge of the world. If for some reason a man wants to lie about what he sees, then he will clearly not give any of the responses to the square instanced above, though it is still true that if he had not wanted to lie he would have given these responses if he had seen it. If it is true that our goals are represented in the brain and our knowledge or rather beliefs about the world are also represented, then a knowledge of the organization of the brain at a given time taken in conjunction with a knowledge of the relations between the brain and the external world, would enable us to give an explanation of seeing in physical terms—again still leaving consciousness out of account. We would know how different physical objects were represented in the brain, and if a representation occurred corresponding to a particular object being in a particular position in space at a particular time and if this change occurred as a result of light falling on the retina and transmitting a message to the brain via the optic nerves, we would have an explanation for why someone had seen something correctly. If the light from the object produced a change in the brain corresponding to a representation of something other than what is in front of the eyes, then we would say the object had not been correctly seen, and we would be able to explain why this was so. However, all this can only be done if we consider the brain in relation to the surrounding physical world and not if we consider the functioning of the brain on its own divorced from the effect of different possible inputs.

Thus, the argument from norms does not prove any difficulty in principle about explaining behaviour in physical terms and the argument involves the same fallacy as the argument from actions. Both arguments are based on supposing that those who seek to explain behaviour in terms of brain functioning can only study the brain as an isolated phenomenon and are not to be allowed to study it in relation to the environment.

3. We shall next consider some arguments that are intended to show that human beings are not completely predictable: some philosophers have supposed that if there were a sense in which human behaviour was not completely predictable this would be enough to differentiate human beings from other physical systems, but this is plainly false since there is no evidence to suggest that other physical systems are completely predictable and there is some evidence suggesting that they are not. To establish a difference along these lines between men and machines, it would be necessary to show that men are unpredictable in a way that machines are not.

The first argument runs as follows: if we predict what someone is going to do, this prediction, if he is informed of it, may well influence what he does, so that he may always choose to falsify it. In this form the argument does not prove very much: it might still be possible in theory to predict accurately what someone was going to do, provided we did not inform him of what the prediction was. Moreover, it is not true that we can never make accurate predictions when we inform a person of the prediction: there is no reason why anyone making a prediction about someone else's behaviour should not have taken account of the consequences of informing him about the prediction before delivering it. It will then still be possible in some instances to make and deliver a correct prediction. There is no mystery about this, and if we could build a machine with which we could communicate there would be no problem in building it in such a way that its behaviour would sometimes be altered when a prediction was made to it so that the prediction would be falsified.

There is a stronger form of the same argument, whereby it can be argued that in principle individuals cannot always predict correctly what they themselves will do. If a man makes the prediction to himself his own internal state is changed and this change in his own internal state may affect his subsequent behaviour and so lead to the falsification of the prediction. A man cannot take into account the changes that the prediction will make in himself because, if he does, the taking into account of those further changes will itself influence his future state and he can still never be sure that this will not lead to behaviour falsifying the prediction. Once again it is easy to envisage a machine which exhibited similar properties, and indeed it is impossible to envisage a machine not having these properties. No matter what calculations a machine performs about its own internal states, it cannot take into account the changes caused by the calculations and therefore could not predict its future states with complete accuracy. Thus this sort of attempt to square freedom of the will with predictability does not involve the postulation of any basic difference between men and machines.

The second attempt to introduce some sort of indeterminacy into the operation of the brain was set off by the discovery of Heisenberg's uncertainty principle: Eddington and others, including some eminent neurophysiologists, have tried to find a loophole for freedom of the will by supposing that events in the brain and therefore behaviour may be influenced ultimately by the behaviour of very small particles like electrons. Since their behaviour is only predictable in principle to within certain limits of accuracy, the behaviour of the whole brain can never be predicted to within these limits of accuracy.

If in the operation of the brain some uncertainty due to the operation of Heisenberg's principle occurs, this again does not constitute any difference between the operations of the brain and the operation of a machine. In fact quite the reverse is true: if such an uncertainty appears in the behaviour of any physical system and therefore of any machine, the brain would in fact only be unlike a machine if such uncertainty did not appear in the operation of the brain. It may be said, that perhaps the brain is so arranged that the effects of uncertainty at certain points are magnified and produce very big differences in behaviour. In many physical systems, although we may not be able to predict with complete certainty the future positions and velocities of individual electrons, we can predict with considerable certainty what the state of the whole will be at some subsequent time since random variations tend to be self cancelling. However, in the brain it might be so arranged that at certain points what happened was determined by the behaviour of a very few electrons, and that what happened at these points was crucial for what subsequently happened in the rest of the brain. It is in fact a simple matter to build a physical system of this sort. For example, we might provide a source of radioactivity to be detected by a Geiger counter and connect the output from the counter to a computer: the program the computer executes at any given moment can be made to depend on the occurrence of unpredictable inputs from the Geiger counter.

Curiously enough one of the main problems in the design of automata is to *minimize* the effects of very small causes rather than to maximize such effects. Indeed one of the biggest differences between the behaviour of computers and men is that men appear in some ways to be more reliable. We can predict with some accuracy the types of errors men are likely to make in performing a given task; despite the fact that computers are much less complex, we cannot in general predict what types of error they will make. It is in fact certain that there are devices built into the human nervous system to prevent random fluctuations influencing the final output, and the problem facing the behavioural scientist is more to discover what makes behaviour so predictable than to think of physical mechanisms which would lead to a lack of predictability of behaviour.

Two further points are worth making about the argument from Heisenberg's uncertainty principle. First, if behaviour is determined by random events this does not leave any loophole for a non-deterministic account of freedom of the will. Random causes could only lead within a physical system to random behaviour and the actions most characteristically described as voluntary are highly systematic—for example good deeds. Secondly, if it were found that physically unpredictable events occurred in such a way as to lead to highly systematic behaviour, this would be evidence for denying that the matter in the brain obeys the same laws as matter in inanimate physical systems. The behaviour of saints is often very predictable. Supposing we found that their good actions were determined by events such as the position and velocity of an individual electron that could not be predicted on physical grounds. In some instances we would be able to predict the action from a knowledge of the saint's character. We could therefore form a knowledge of his character and of his brain state and the environment at a given moment. This means that the electrons in the brain are not conforming to the laws to which we believe these particles conform in inanimate systems. We would be completely unable to stimulate such behaviour in an automaton, and we could no longer regard the matter in the brain as obeying the same lower level laws as matter obeys in inanimate physical systems. However, there is no evidence that matter does behave in this fashion in the brain. We shall return later to a consideration of the implications of such evidence if it were found.

4. The fourth argument against thinking of man as a physical system is the argument from Godel's theorem. This has been expounded by Kemeny and by Nagel and more recently by Lucas. Lucas' argument runs somewhat as follows: Godel has shown that within some consistent systems of logic, there are propositions which can be seen to be true, but which are not provable in the system. Machines are logical systems, therefore there will always be some true propositions about the logic of a given machine which the machine cannot prove. Men, however, can always see the truth of these propositions and therefore we can always go one better than any existing machine. The argument is obscure, but it seems worth making three brief points about it.

(*a*) In the first place, the argument does not prove that all men are superior to machines but only that some mathematical logicians are. It would be fairly easy to build a machine today of such complexity that most people could not possibly see, no matter how much it was explained to them, that some propositions about the logic of the machine were true where the machine itself could not prove these propositions.

(*b*) There may well be true propositions about the logical system of our own brains that we ourselves cannot know to be true. The difficulty in predicting what we ourselves will do has already been instanced. It may be said that this is an inductive prediction, not a deductive proof, but Lucas identifies a machine with its logic, and if this identification is made, then a proposition about what someone will do can be regarded as a proposition in the logical system and cannot be proved true by a person because he cannot take into account any changes in his own system brought about by having gone through the proof. Thus it is very far from clear that the limits which Lucas ascribes to the operation of a machine do not also apply to mankind.

(*c*) The third and final point about the argument from Godel is taken from a recent article on this problem by Jack Smart. He points out that men only discover the logic of their own system by induction. If a machine is allowed to make discoveries about its own syntax in the same way, it could, given the storage capacity, store its own syntax and operate on this in a higher order language: at any one moment there would be propositions about the language it was using which it could prove, but given the storage capacity it could in turn store the higher order language and operate on this in a yet higher order language. Thus although at any one time there would be some propositions about the machine's own language which the machine had not proved there would be no limit to what it *could* prove about itself. Moreover, there is no reason why storage capacity should set such a limit since we would have to allow the machine to store information in the external world, just in the same way as people do when they resort to pencil and paper.

5. C. Taylor has recently put forward a new argument suggesting that behaviour cannot be explained in physical terms. Unlike many other philosophers who have written on this topic, he believes that the issue is an empirical one. He argues, however, that teleological explanations and causal explanations are not compatible with one another—some systems (physical systems) require an explanation in causal terms, whereas the behaviour of higher animals and man requires an explanation in teleological terms. It is not valid to apply both types of explanation to the same phenomenon. Taylor's argument for this proposition is as follows.

In teleological explanations the occurrence of a response ($R$) is explained by showing that it is required for $G$ where $G$ is a goal of the system. Causal explanations attempt to specify a set of initial conditions ($C$) such that when they occur the response ($R$) will be made and the joint occurrence of $C$ and $R$ lead to $G$. If there is an infinite set of conditions ($C$) which produce $R$, it will be impossible to specify them and hence for any system or organism satisfying this condition, mechanistic explanations are bound to fail. There appear to be two flaws in this argument.

If we include in the features of the situation ($C$) producing a response ($R$) those features which are not relevant to whether or not the response is produced, there will clearly be an infinite number of situations which will lead to $R$: there are an infinite number of situations in which when a match is struck it will light, but this does not mean that we cannot isolate and specify the causal conditions for the match lighting. Taylor does not discuss this point and his use of the word 'situation' is extremely vague, but the claim that people will make the response required by their goal in an infinite number of situations is plausible only if we include irrelevant variations in the situation.

Secondly, there is something wrong with Taylor's account of teleological explanation: there is no known system of which it is true to say 'an event's being required for G is a sufficient condition of its happening' where G is the goal of the system, yet this is how Taylor characterizes teleological explanation. No man invariably does what is required to reach his goal.

Taylor himself is aware of the problem: he writes 'to say that anything is a goal for animals of a given species is to say that they will do whatever is necessary *within the limits set by their motor capacities* to attain it'. Even this statement is plainly false and elsewhere we find 'an animal will do whatever is necessary *within certain bounds* to encompass G' (my italics) but Taylor never resolves the confusion and elsewhere he writes 'in short the lion will charge from a point where, all things considered, the prey can best be seized', thus implying that lions are omniscient.

If Taylor were right and it were true that an attempt to specify the situations under which an animal will do things involves us in non-teleological explanations, then he would have shown that no teleological explanation can explain behaviour since unless we can specify these circumstances we cannot explain why given that a person has a certain goal he makes the appropriate response under some circumstances and not under others. This point can be brought out by an analogy: it is possible to explain the behaviour of a missile directed by a servo-system in teleological language by saying 'Its target was New York and therefore it corrected its course to allow for strong winds over Nova Scotia'. Now although this seems a sensible explanation of the rocket's behaviour Taylor would presumably say that it was invalid because there was an alternative mechanistic explanation which would specify under what circumstances the rocket would change course and under what circumstances it would not. According to Taylor, only when it is *impossible in principle* to specify the circumstances under which a system gives a certain output, can we use teleological explanation; this implies that we can never have a detailed explanation teleological or otherwise. If *both* types of explanation are to be incomplete, however, there is no case for saying that the teleological explanation is the only valid one: only in the non-existent case where a system *always* did what was required to achieve a goal would the teleological explanation be superior to the non-teleological.

Thus in the absence of any system that always does what is necessary to achieve a goal, Taylor's argument falls to the ground, and if he were right about the nature of teleological explanation it would never be valid to use it. Taylor goes on to argue that teleological explanations of human and animal behaviour have been more successful to date than non-teleological explanations and suggests that this is evidence that it is impossible to give causal explanations of behaviour. However, the limited success of attempts at causal explanation so far cannot be used to show that such explanations can never be successful, and since Taylor has failed to show that teleological and causal explanations cannot both be validly applied to the same system the argument would in any case not prove that there is some inherent impossibility in giving causal explanations of behaviour.

As suggested above, it is not possible to infer any empirical proposition about differences between animate and inanimate systems from the fact that in ordinary language teleological explanations are far more often given of human behaviour than of the behaviour of inanimate systems. It is in fact impossible to give an adequate account of the behaviour of any complex servo-system without introducing words like 'target' or 'goal' and this does not prove that such systems are different from other physical systems except in the way in which matter is organized in them. Moreover, to someone unsophisticated about machines, the easiest way to give some understanding of the behaviour of a rocket is to use a teleological explanation, and such explanations are certainly useful and appear to be perfectly valid.

6. The last of the arguments under review against regarding human beings simply as physical systems is the argument from consciousness. Supposing it were the case that men were physical systems. If this were true, we might be able to understand and explain behaviour in terms of the organization of the matter in the human body and in particular the organization of the elements composing the nervous system. Thus, if we knew enough about the human brain and nervous system we might be able to build a super computer which although built out of electronic components would work in the same way as the human brain. That is to say, the input elements would each correspond to the different receptors, and the output elements could be put in one–one correspondence with the muscles of the human body. We now subject the machine to a series of inputs which correspond to the inputs a human being gets in a life-time, and we keep a detailed record of the machine's output: we discover that if protocols of these outputs are examined together with protocols of the outputs given by men exposed to the corresponding series of inputs, it is impossible to say which protocols were taken from the machine and which from people. Some people would argue that there would still be little temptation to say that the machine was conscious. Moreover, although we might talk about the machine 'seeing', 'hearing', 'having a memory', 'recognizing', 'discriminating' and even 'thinking' it could be claimed that when we applied these words to the machine we were using them in a slightly different sense

from the sense in which we use them of human beings since it could be argued that all these words have some reference to consciousness.

If we could build such a machine, we would have in some sense an explanation of consciousness but it would not help us to say what consciousness is. We would have an explanation of consciousness in the following way. A man in two different states of consciousness could always in theory do things which would reflect the difference in his states of consciousness. We could not recognize that we were in two different states of mind at different times without being able to give expression to the fact in behaviour, and the only information any of us can have about other people's states of mind comes from their behaviour. If, however, behaviour itself is determined by the workings of the brain, then it follows that different states of consciousness must be correlated with different states of the brain. If to talk of two different states of consciousness implies that under the right circumstances two different responses would have been made, then the brain itself must be in different states when we are in different states of consciousness. If the *brain* is a physical system, it must be true that its organization at a given time will determine what a man does at that time, and it must also be true that we can deduce from it what he would have done if a different input had been given—that is if it is true of a person that at two different times he would have done two different things if a given event had occurred, then at those two times his brain must have been in two different states.

To summarize, the argument proceeds as follows: first it is assumed that the brain and body are a physical system, and differences in behaviour must be accompanied or immediately preceded by differences in the state of the brain; any disposition to behave differently even if not actualized must be accompanied by a difference in the state of the brain; therefore since differences in conscious states are accompanied by differences in behavioural dispositions, differences in consciousness will be accompanied by a difference in brain states. This argument leads to two conclusions. If different states of consciousness can always lead to differences in behaviour, the behavioural scientist who studies behaviour is not leaving something out of his account. Secondly, any machine which simulates the brain as a physical system does have a representation of consciousness in so far as its internal state will differ from time to time; its output is determined by its internal state, and differences in its internal state will represent differences in conscious states in men. Only some differences in its internal state will, of course, be relevant since not all potentially different responses in men are accompanied by different states of consciousness. Incidentally, whether or not we would want to use the word 'conscious' of the machine we shall certainly need a word like it. At some times the machine will be able to report on its own internal states, at other times it will not, and this will be a useful distinction to apply to it.

The above argument shows that different states of consciousness might be represented in a machine: this, of course, does not answer the question of whether the machine is conscious. Differences in consciousness would be explicable in terms of the functioning of the nervous system, but we would be no nearer to saying what consciousness is. Whether or not we would ever want to ascribe consciousness to a machine built out of electronic components no matter how accurately it simulated human behaviour, there is no question but that if we could construct a physical entity indistinguishable from a human being, the accident of not having been born of woman would not make us want to deny that it (or he) was conscious. The above argument is intended to show that if the brain is a physical system we cannot look on states of consciousness as influencing behaviour if we think of consciousness being a non-physical entity. It would, however, be consistent to believe both that the brain was a physical system and to believe in some form of psychophysical parallelism if any meaning can be attached to this doctrine.

## 5. THE POSSIBILITY OF FINDING NEGATIVE EVIDENCE

The above arguments are intended to show that the question of whether or not the brain is a physical system is an empirical question. If the brain is a physical system then it should be possible to find explanations of behaviour in physical terms. We have tried to show that the question cannot be settled by appeals to ordinary language and that arguments attempting to show on *a priori* grounds that physical explanations of behaviour are not possible are fallacious. We have also tried to show that the proposition that the brain is a physical system is plausible and that there is some evidence in its favour although like any other inductive generalization it cannot be taken as proven.

At this point we should enter a caveat. Even if human behaviour can be understood in terms of the chemistry of our brains and bodies, this does not mean that in our explanations exactly the same concepts will be used that have been used to explain other physical systems with which we are familiar. The fact that matter in the human body is organized in a different way from that in which it is organized in most familiar physical systems means that we will need and have needed to develop different concepts in the explanation of human behaviour to those used in the explanation of a heat engine or the solar system. Some of these concepts may resemble those used in everyday explanations of human behaviour; thus, in talking about the artefacts with which they deal, computer programmers and others engaged on automata studies unhesitatingly use concepts normally applied only to higher animals and man. Nor do they do this in jest. They do it because much the most convenient way of talking about some automata is to apply words like 'recognizing', 'calculating', 'memory', and even 'thinking'.

The study of human and animal behaviour and of information-processing machines has already led to the development of new concepts and even of new branches of mathematics such as information theory, theory of games, and the branch of mathematics suitable for describing feedback processes. Nor is there anything new in this; the concepts needed to describe the behaviour of a star are not the same as those needed to describe the behaviour of the individual particles of which it is composed. It can be expected that many such new concepts and principles will be required in the explanation of human behaviour. The important point is that such concepts need not refer ultimately to anything other than physical events, and to the extent to which we are successful in explaining human behaviour in these terms we will be able to devise machines which also can only be understood in the same terms.

Even such a simple system as a thermostat requires new concepts and new mathematics in order fully to understand its working; we do not properly understand the behaviour of a thermostat until we understand the concept of negative feedback, and in order to understand and predict its workings we need to apply a branch of mathematics specially developed for the purpose—control theory. The question of whether the higher level laws of behaviour can be completely reduced to the lower level laws governing the behaviour of the elementary particles in our brain will not be raised here. We have argued that such reduction is possible only to the same extent as it is possible in treating of other physical systems. Just as we cannot understand the behaviour of a gas without introducing concepts such as pressure, we cannot understand the behaviour of a man without introducing concepts such as 'goal'. Neither of these concepts are needed to describe the behaviour of a single molecule whether it is in a gas or in the brain. It is for this reason that experimental psychology and neurophysiology are likely to remain distinct sciences.

An attempt has been made to demonstrate that it is plausible to regard man as a physical system. Part of the evidence for this belief is the success that has been achieved so far in simulating typically human functions by machine. Extrapolating from the success obtained within the last twenty years, it seems likely that within the next century machines will be constructed with which it is possible to carry out a normal conversation and which will be capable of performing most if not all of the remaining functions that on this planet can at present only be carried out by man. In explaining the behaviour of such machines we shall, of course, need to use terms and *concepts* similar to those required for the explanation of human behaviour.

There is one further argument that could be raised by philosophers anxious to prove that there is something non-physical about the human mind. It could be argued that the existence of such machines could never show that words like 'intention' and 'purpose' are merely ways of describing extremely complex categories of physical events since the machines themselves will be built by man and it is only the fact that man is a rational being that will enable him to produce machines that exhibit rational behaviour. Such machines could not have come into existence through the blind operation of physical forces. The answer to this argument is that there is evolutionary evidence that man himself came into existence through the operation of such blind forces, aided, admittedly, by the lucky accidents that made possible the formation and stable existence of the complex molecules that compose our bodies. The main criterion for what sort of organism could be evolved by chance is that it should have survival value. Our own selfishness and irrationality (as well as our rationality) stems from this. It is most unlikely that we will breed computers for survival value. There will, therefore, be no reason for automata to exhibit selfish behaviour and within a century or a millennium, computers may well be very much more rational than man himself. They will certainly be more saintly.

If the question of whether the brain is a physical system is an empirical one, then it must be possible in principle to look for evidence that the brain is *not* a physical system as well as evidence that it *is*. Any evidence that suggested that the matter in the living brain did not obey the same laws as matter elsewhere would count against the brain being a physical system. On a really close examination of the brain, we might find that the matter in it simply defied the laws obeyed elsewhere. We might find that, by examining the matter in the brain, we could not predict its future states by the same laws that we use to predict the future states of inanimate systems.

It might also be possible to obtain evidence that the brain is not a physical system by the study of behaviour alone. If it were established that certain people could recognize the future, and that this was not done by any process of induction based on present knowledge, we would suspect there was some sort of causal gap in the functioning of the brain. Its state at any one time could never be understood in terms of its state immediately beforehand since its states would in some way be systematically influenced by future events. We shall refer to such a breakdown in the laws of causation as a 'causal gap'. If the precognition were a prediction based on knowledge of the present there would be no problem—that knowledge would be represented in the state of the brain, and we could understand the prediction made in terms of the brain circuits performing the necessary inductions and giving rise to the pattern of movements of the vocal cords involved in enunciating the prediction. However, if someone had been shielded from all physical channels through which he might have acquired the knowledge necessary to predict a future event and he was still able to foresee the future, then the state of the brain must be changing from one moment to the next in a way which is arbitrary with respect to its own physical workings. We would have no idea how to go about building a machine which could simulate this behaviour.

It is within the bounds of possibility that such causal gaps in brain functioning do exist, though on present evidence it is rather unlikely. It is worth considering what we could do about it if such causal gaps were found.

When we examined the ways in which future states of the brain failed to correspond with predictions made about them based on present states of the brain, we would find either that they did so in a systematic way or that no matter how hard we looked they did so in a very unsystematic way. We could, of course, never be certain of the latter alternative—the ways in which they departed from our predictions might be systematic even though we had not discovered any system. Moreover, if behaviour is systematic, as it appears to be, and if ultimately behaviour is governed by the state of the brain, then of course the brain must operate in a systematic way. The alternative that no system would be discovered is thus both implausible and not very interesting, since all we could do is to keep on looking for a system.

If the departure from predictability were systematic, we might start building up new laws governing the behaviour of particles in the brain: such laws might have to be of a very different kind from the laws of physics—for example they might have to incorporate things like action at a distance without our being able to detect any changes in the physical media between the two points. In addition to working out in this way new laws concerning the behaviour of matter when incorporated in animate systems, we might also try to look for some more fundamental laws which would enable us to subsume both the behaviour of matter in living systems and in non-animate systems under the same general laws. Thus the scientist who wanted to explain behaviour in terms of the workings of the brain would not be doomed to failure just because the brain did not work according to the same laws as inanimate systems: he would only be doomed to failure to the extent to which no system was discoverable in its workings, and on the assumption that motor activity is directly caused by events in the efferent nervous system this implies that the workings of the brain cannot be less systematic than behaviour itself.

If we did find that matter in the brain obeyed different laws from matter elsewhere, we would have two alternatives open to us: we could either say that the brain was not a physical system or that there were two very different kinds of physical system. Irrespective of which of these lines we took, the question of whether the brain is a physical system in the same sense in which other parts of the space–time continuum constitute physical systems remains an empirical issue.

# COMMENT

*by* J. H. GRUNDY

I

Sutherland's question, 'Is the brain a physical system?', is indeed, as he himself suspects, an odd one. It is surely common ground between those who assert and those who deny that all human activity can be explained in physical terms that the brain *is* a physical system; it is because the brain is a physical system on the one hand, and because on the other hand it appears to have a great deal to do with the explanation of human activity, that neurological findings figure so much in these debates. And the assertion that 'it has been fashionable for certain philosophers in recent years to put forward arguments implying that the brain is not a physical system' can only be justified if one takes the word 'brain' to mean 'that, whether physical or non-physical, in terms of which human activity is to be explained'. That this is a use of the word which is in Sutherland's mind is indeed suggested by several of the things he says at the beginning of his paper: for example, that if the brain is a physical system, there are no non-physical factors to be taken into account in the explanation of behaviour, and that 'if the brain is a physical system, then human behaviour will be predictable to within the same limits of accuracy as the behaviour of any other physical system'.

To use the word 'brain' in this way can, however, be muddling and dangerous. The danger is, of course, that one might take 'brain' in one sense, that sense in which it means 'that in terms of which human activity can be explained', then show that the brain, in the other sense of the mass of nerve tissue in our heads, is a physical system; and finally conclude that one has shown that behaviour must be explained in terms of the states and mechanisms of the brain in the second sense, in which it is uncontroversially a physical system.

What Sutherland presumably means is that *people* are physical systems; and, in fact, he avoids the danger mentioned in the previous paragraph. Perhaps the only point at which his argument is influenced by his eccentric use of the word 'brain' is in argument 6 of §4 where he argues that, since different states of consciousness can always be 'reflected' in a person's behaviour, then 'if the brain is a physical system we cannot look on states of consciousness as influencing behaviour if we think of consciousness being a non-physical entity' (p. 118). But even here this illicit move does not seem to do more than slightly reinforce Sutherland's approval of the conclusion that consciousness is to be explained in physical terms, a conclusion for which he offers other arguments.

## II

Sutherland is surely right in thinking that the very general *a priori* arguments like the one put forward by Geach do not have any force against his position; even if it is conceptually inappropriate to talk of artefacts thinking, being conscious and the like, this shows nothing about whether artefacts can simulate the processes by which we think and are conscious. (And in any case, it is by no means obvious that we could never attribute thought and consciousness to artefacts.) But although Sutherland is clearly right in saying that it is only empirical research, and not *a priori* argument, which can settle the question of whether the processes of the human brain can be simulated in machines, he is open to the criticism that he dispenses with *a priori* considerations too quickly. This shows itself in two ways: first in the lack of analysis of the specific concepts in terms of which we describe human behaviour, and secondly in Sutherland's failure to acknowledge the mind–body problem. The lack of detailed analysis of what the human attributes and activities are that one may be attempting to duplicate in machines is the only lack of *a priori* consideration that directly affects a programme of *empirical* research such as Sutherland envisages; but any claim that such a programme of research, well designed according to the detailed analysis of the concepts used in talking about human behaviour, shows that human behaviour can be explained in purely physical terms is a claim which must acknowledge the mind–body problem.

## III

(i) It is worthwhile spending quite some time showing the importance, even to a programme of empirical research, of *a priori* consideration of our talk about human behaviour. If it is to be argued that human attributes can be duplicated in machines, one must know what these attributes are. As Sutherland himself points out, to show that two systems have the same output with a given input is not enough to show that both systems can be described in the same way. To take an example which Sutherland himself does not discuss, the question of whether computers can be described as intelligent. Someone, seeing a computer make a correct calculation might be tempted to say that it is intelligent; after all, when presented with certain data it does get the calculation right, which a man presented with the same data might well not do; and it most likely does the calculation in an infinitesimally small fraction of the time it would take a person to do so. As Michael Scriven (1953) points out, however, to call the computer intelligent would be misguided; like an abacus or a slide-rule, it is an instrument which is more or less useful, but not more or less intelligent. The intelligence is shown by the person who programs the computer, and even the more sophisticated description of computing machines as 'high-speed idiots' is, strictly speaking, wrong; they are simply not to be placed anywhere at all on the genius–idiot spectrum.

But, it might be objected, this is to talk of computers as they are at present, and to ignore the advances which electronic engineering will make possible. It would, for example, be a fairly modest advance to build a computer which incorporated in the design of its mechanism all the rules of logic and of evidence which we use in solving our various problems; we would no longer have to instruct the machine by means of the program to proceed according to these rules. Would we then be able to say that computers are intelligent? Well, the question is not simply one of what advances in electronic engineering are possible. Nor is it a question simply of allowing or not allowing computers the epithet of 'intelligent'. What is taking place when one envisages advances in computer design and scope and asks how much nearer they bring computers to being intelligent, is a conceptual exploration of what intelligence is, what the attributes are possession of which puts something into the intelligent–unintelligent, as distinct from the simply non-intelligent, category. Until the notion of intelligence has been clarified, no questions as to whether intelligence has been simulated or duplicated in machines can be adequately answered, and in the meantime simple comparisons between different input–output situations are likely to be misleading.

(ii) Another case, which Sutherland does discuss, is that of goal-directed behaviour in machines. Here again, the conceptual exploration is stimulated but not completed by noting certain similarities between the input–output situations of men and of machines. A missile may consistently alter its course in such a way that it hits its target even when the target changes its position, just as a person changes his behaviour when a change occurs in what is needed for his ends to be gained. It would perhaps be unduly restrictive to refuse to allow the use of such terms as 'goal-directed' and 'making corrective movements' in describing the missile's behaviour, though it would be as well to use them with cautionary quotation marks. For before any conclusions can be drawn about whether the missile has displayed behaviour which can be likened in any full-blooded way to human purposive behaviour, we need to know just how variations in movements to suit new goal-achieving conditions fit into the pattern of notions which make up our purposive explanation of actions. For example, when a man's behaviour changes in such a way that a certain result occurs which would have been the result of his previous behaviour if circumstances had not changed, we should regard his change of behaviour as purposive if he believes the new behaviour to be a means to the result and if he wants to achieve the result. But the missile cannot be said to believe that its change of course will result in its hitting the target; nor can it be said to want to hit the target. And how essential is it for something to be said to be acting purposively that it can decide to do what it does, change its mind as to whether to do it or have a purpose in doing it?

## IV

(i) This type of point must be made again, and at some length, in connection with Sutherland's treatment of the explanation of actions in physical terms. He says, rightly, that to explain movements in physical terms is not to explain actions in physical terms. 'It is certainly true that in categorizing actions we take into account very much more than a person's physical movements; we have to consider the relation between his body and the external world, and his intentions and much else besides' (p. 106). Again he says (p. 107), '*Of course* actions cannot be explained by taking just the state of the brain at a given time and nothing else into account, but there seems to be no reason at all why a full explanation in physical terms cannot be given if we consider the state of the brain in relation to the external world'. Even a detailed knowledge of the mechanisms inside a missile would not entitle us to say that it had made a corrective movement; to be entitled to say that we should have to have observed its behaviour under a wide range of different conditions. If it had consistently behaved in such a way as to put itself back on course each time its target moved, then we could say (perhaps without knowing anything at all about its internal mechanisms) that it made corrective movements. Similarly, Sutherland argues, with actions; to have a physical explanation of actions we must study the state of a person's brain under a wide range of inputs, taking into account its relation to a very great deal of the man's environment, both at the time and in the past.

Now there are strong grounds for thinking that Sutherland is right in saying that it is possible in principle to give physical explanations of actions; but to do so full account must be taken of the differences between the various sorts of 'mere bodily movements' on the one hand and the various sorts of actions on the other. The exploration of these differences is both an empirical and an *a priori* task. Where the programme is the initial one of finding states and mechanisms in the brain which 'correspond' to the various kinds of behaviour (and leaving till later questions of the identity or non-identity of physical and psychological states) the nature of much of the research must be empirical. It will consist in neurological research on the one hand and, on the other, research into what, as a matter of contingent fact, the nature of the activities to be correlated with the neural events is. The distinctions philosophers make between the different sorts of human behaviour—including the broad distinction between mere bodily movements and actions—must be based on differences in what as a matter of fact takes place; indeed, the drawing of these distinctions is largely a matter of calling attention to the facts of what goes on—facts which have for some time often been neglected by psychologists anxious to deal with human phenomena in the same way as natural scientists so successfully deal with inanimate things. Thus it is a matter of fact, indisputable for all that our knowledge of it is largely prescientific, that much of our behaviour is done because we want to act in the way we do, because we have purposes, motives, wants and intentions, and not because some physiological causal sequence, of which we may know nothing, produces some movement such as flinching or starting at a sudden loud noise. The fact that philosophers often talk in terms of the vocabulary used in ordinary-language talk about and explanations of behaviour does not alter the fundamentally factual nature of the point that the explanation of behaviour like reflex-actions is quite different from the explanation of purposive behaviour. Discussion of the vocabulary is largely worthwhile in so far as it points to differences in what is seen, as a matter of fact, to take place. (Indeed, the vocabulary is occasionally added to or changed in order to focus attention on these differences. The word 'action', for example, is used by philosophers in a technical, and from the ordinary language point of view, in an artificially restricted way, to mark the distinction between behaviour which is to be explained purposively, and behaviour which is not to be so explained: a distinction which is otherwise only rather diffusely reflected in our vocabulary.)

(ii) But there is a large and very important *a priori* task in this matter; analysis of what it is to want something, to have an intention, to make a decision and to have a purpose is *a priori*. And it is an essential step in achieving the programme Sutherland envisages. Unless all actions turn out to be non-actions because they are not in fact purposive—and this is a possibility which it is difficult to take very seriously—they must be explained in purposive terms at a psychological level; consequently, if a physical explanation of them is to be found, it must be in terms of the neural 'correlates' of the components of purposive explanations. And to do this, one must know what those components are. Looking for these neural 'correlates' is presumably the most fruitful and convenient way of investigating the workings of at any rate a large part of the brain; but even if we fully investigated the workings of the brain without correlating these workings with events and states at a psychological level, we should still have to do the *a priori* analysis of the psychological events and states in order to complete the explanation of activities like purposive action: a point which Sutherland himself makes in his section on consciousness. And it is worth saying that this correlating would not be a matter of linking a neurological story with a psychological one simply to make the neurological one more rewarding in that it would be shown to accommodate our ordinary explanations (with, perhaps, the presumption that these ordinary explanations would have been outmoded). To be an explanation of actions, the neurological talk would have to be shown to be talk about what was also the having of a reason or intention.

The fact which so many philosophers have stressed,[1] that actions must be explained in terms of reasons, purposes, wants, intentions, etc., is, then, not the end of the idea of explaining actions in physical terms; it can, and it also should, be the beginning. In his section on actions, Sutherland gives only one passing reference to intentions and another to goals. He does at one point talk about 'part of an everyday explanation' of an action: a driver's signalling might be explained by the fact that he is invariably courteous (p. 107). He continues:

If we consider not just the state of the brain, but what movements the brain would give rise to under a large variety of different inputs and if it is the case that it tends to give rise to bodily outputs which please other people, and if it were the case that if his brain had not been arranged in this way then he would not have signalled, then all that is involved in saying 'He signalled out of courtesy' will be taken account of in the explanation in terms of brain processes. In order to do this, however, we have had to consider the relationship between brain processes and the external world, but there is no reason at all why the physical scientist should not do this and indeed if he wants to have a full explanation of human behaviour he must do so.

Sutherland is right to cast his net widely; as he points out, philosophers who deny that actions are open to causal explanations often seem to suggest an unduly restricted area in which such causal explanations could be sought. Purposive explanations are often contrasted with causal explanations 'in the sense in which mechanical explanations are causal

[1] For example, Peters, 1958; Winch, 1958; Melden, 1961; Taylor, 1964; and Hamlyn, 1953.

explanations'.[1] To be fair, this is usually done when attention is being drawn to the difference between the purposive explanation of an action and the explanation of mere movement in terms of neural impulses and muscle contractions. But of course there is absolutely no reason why investigation of the physical on-goings which accompany action should be restricted to this simple level.

(iii) The treatment of actions quoted above is, however, inadequate, resting as it seems to do on an inadequate Rylean analysis of motives as dispositions. It is certainly true that much of our behaviour does consist in dispositions and habits, and the temptation to think of all actions along those lines is considerable because, as Kohler (1961) points out, retention in memory, habits and dispositions are not as such parts of the 'phenomenal' world, and *a fortiori* are not consciously motivated in the way that actions are. Consequently, they are much easier to fit into physical explanations of behaviour. But such habits and dispositions must be distinguished from fully purposive actions, and this Sutherland's analysis fails to do. That the driver's movements in signalling please other people may be an unintended consequence rather than the reason why the driver signals. It is, after all, possible to train quite lowly organisms like fleas to give rise, under widely varying inputs, to bodily outputs which please people; but no one would say that to have explained these performances in physical terms is to have explained an action in physical terms.

Consequently, although Sutherland makes many good criticisms of the *a priori* arguments against his view, he does not, it seems to me, make enough of the *a priori* investigation which would not only be a help in exploring the brain, but which is at some point an essential step in the achievement of his programme.

v

Finally, something must be said about the mind–body problem, the most general and recalcitrant problem which a programme like Sutherland's must face. Many of the philosophers who assert that, for example, actions must be explained in purposive terms would also assert that the gap between this type of explanation and explanation in physical terms cannot be removed either by further *a priori* analysis nor by any amount of neurophysiological investigation. Their view is that purposive explanations can neither be derived from nor reduced to physical explanations. So that however much was known about the brain, a physical explanation of actions would still be impossible; talk of reasons and purposes on the one hand and physical states on the other belong to quite different systems, and the

[1] MacIntyre, 1960, p. 91. See also Peters, 1958, p. 12.

latter can never be interchanged or substituted for the former.[1] Thus, no explanation of actions in physical terms can be given.

As against this view there are, broadly speaking, three others. One of these, Epiphenomenalism, is the doctrine that mental events are caused by the same physical events that cause actions, or whatever the activity is; the mental events themselves are always effects, and never the causes either of further mental events or of physical events. The doctrine has been criticized on the ground that it makes mental states redundant and to have no influence at all on what goes on. Surely, it is argued, mental states do affect our behaviour; if I am angry I am more likely to have an accident while driving; if I make a determined effort, I am more likely to finish my work on time. The Epiphenomenalist can of course reply that while the mental state *appears* to influence what goes on, it is in fact the corresponding brain state which causes both the mental state and the accident or the greater speed of work. But this still leaves the mental state redundant; one wants to know why we should ever have mental states at all. And the doctrine conflicts with the view that wants and intentions do produce actions, and with the view which Sutherland maintains in his section on consciousness, that different states of consciousness produce different behaviour.

One of the other two remaining doctrines, the Identity Hypothesis, avoids making mental states redundant; according to this view, talk of mental states and of brain states may belong to different systems and be logically irreducible to or derivable from one another, but this does not prevent them being, *as a matter of contingent fact*, different ways of talking about the same events. The two systems of talk may have different meanings; but that does not prevent their having the same referent, nor does it mean that they are conflicting or incompatible with one another. In just the same way, a physicist's talk, *qua* physicist, about music is about the same events as the talk of music critics even though the two are not derivable or reducible to one another; and the legitimacy of the one in no way impairs the legitimacy of the other.[2]

It is presumably the Identity Hypothesis which Sutherland favours, although this is by no means clear from what he says. Unlike the third of the three doctrines aimed at showing that human behaviour is open to physical explanation, Central State Materialism, it does not deny that there

---

[1] Strawson, for example, maintains this view very unequivocally in *Freedom and the Will* (1963).

[2] See Flew, 1965, where this view is forcefully argued, though not in the context of defending the Identity Hypothesis. Flew's argument is of particular relevance to the type of argument Sutherland discusses in argument 2, §4 of his paper. There is no reason why we should not give physical accounts of items of behaviour and also assess whether they conform to various norms. Each set of talk will not do what the other does, but there is no reason why it should.

are mental states which we are aware of and feel; nor, since the referrents of the terms 'brain state' and 'mental state' are identical, does it make mental states redundant, as Epiphenomenalism does. Moreover, in the case of the Identity Hypothesis one cannot ask why we should ever have had both brain states and mental states, since on this view there is only one thing to have. But this only reinforces the puzzle as to how two such different sets of concepts can denote the same state of affairs; or to put the same question in a different way, how a brain state which, for example, cannot be known without observation by the person whose brain state it is and which has spatial characteristics can be one and the same thing as a mental state which its possessor can know without observation and which has no spatial characteristics. And of course no retreat to a double aspect theory is possible, for there is no third thing of which both brain states and mental states are aspects. (To say they are both aspects of a person is to go back to square one.)

While this *a priori* problem is being unravelled, empirical research into the brain will continue to add to the intellectual pressure to find a solution to it. As Sutherland's paper assumes, it would be an extremely uncomfortable situation to know perhaps all there is to be known about the physical happenings that take place when an agent engages in any activity, to feel that all aspects of these activities are accounted for—and yet to have to deny that any physical explanation could be forthcoming. This is especially so since only some of those who maintain that no physical explanation could be forthcoming have any wish to revive the notion of minds as incorporeal substances.

## REFERENCES

Flew, A. G. N. 1965. A rational animal. *Brain and Mind*. Ed. J. R. Smythies. London.
Hamlyn, D. W. 1953. Behaviour. *Philosophy*.
Kohler, W. 1961. The mind–body problem. *Dimensions of Mind*. Ed. S. Hook. New York.
MacIntyre, A. C. 1960. Commitment and objectivity: a comment. *The Sociological Review Monograph No. 3: Moral Issues in the Training of Teachers and Social Workers*.
Melden, A. I. 1961. *Free action*. London.
Peters, R. S. 1958. *The Concept of Motivation*. London.
Scriven, Michael. 1953. The mechanical concept of mind. *Mind*, **62**, No. 246. Reprinted in *Minds and Machines*. Ed. A. R. Anderson, 1964.
Strawson, P. F. 1963. *Freedom and the Will*. London.
Taylor, C. 1964. *Explanation of Behaviour*. London.
Winch, P. 1958. *The Idea of a Social Science*. London.

# REPLY

*by* N. S. SUTHERLAND

Grundy's treatment of my arguments is on the whole fair and sympathetic —almost too sympathetic since I find it difficult to identify the points of disagreement between us.

I am slightly baffled by the argument contained in Section I of Grundy's article since he appears to suggest that I use the word 'brain' to mean 'that in terms of which behaviour is to be explained' though subsequently he withdraws this allegation. I in fact use the word 'brain' to refer to the matter inside our heads.

When I pose the question 'Is the brain a physical system?', I intend to ask whether this mass of matter is a physical system in the usual sense of the term—that is to say whether it obeys the usual laws of chemistry and physics. I argued that if it does so, it should be possible to explain human behaviour solely in terms of the workings of the brain and the physical inputs to it from the external world. There is unfortunately an ambiguity in the term 'behaviour' since this could either refer to a series of movements made by the muscles of an organism or to a series of actions. For the moment let us postpone the problem of actions and take behaviour to consist of a series of muscle movements. I argued that if behaviour is influenced by non-physical events such as wishes, intentions and fears where these terms are construed as referring to something other than complex categories of physical events, then the brain could not be a physical system in the same sense as other physical systems: to allow for some non-physical influence to be brought to bear on behaviour there would have to be points at which either one brain state taken together with the physical input did not determine the next, or occasions when behaviour was not determined by the state of the brain. Any philosopher who denies that behaviour can be explained in terms of physical events would therefore be denying that the brain is a physical system in the usual sense of the term.

It is worth noting parenthetically that I am puzzled by one of Toulmin's arguments (p. 23 of this volume): he writes that my 'positive thesis, that the brain is a physical system, simply trades on the ambiguity of the word "system" to conceal all those genuine differences which have traditionally been invoked to justify the distinction between deliberate and autonomic behaviour'. It appears that all Toulmin is saying in this passage is that the organization of the central nervous system is in some ways different and in many ways more complex than that of the organization of other known physical systems, as I am at pains to point out in my article. It is precisely because of the complex ways in which the brain is organized that it is useful to use words like 'purpose' and 'goal' of men but not useful to use them of most inanimate systems. If, however, the main difference between a human being as a physical system and other known physical systems is in the way they are organized, then this opens up the possibility of building artefacts organized in similar ways and it would then presumably be legitimate to use some of the words at present reserved for the explanation and description of human behaviour of these artefacts. Toulmin is mistaken (p. 24) in supposing that I think the brain can be understood by considering only the isolated transmission of impulses in individual nerve cells. I thought I had made it clear that the workings of the brain can only be understood by unravelling the logic of the whole system, and that, because it is such a highly organized and intricate system, this process would involve us in using explanatory terms such as the word 'goal' that are not applicable to most physical systems and would also involve careful consideration of the relation of the central nervous system to the environment.[1] Incidentally, there are many other systems apart from brains that cannot be understood solely in terms of isolated physical and chemical processes (cf. my example of a guided missile). To understand how a computer works in practice we need a whole vocabulary of terms such as 'stack', 'procedure', 'list' that refer to organizational features of a program rather than to isolated physical processes.

Some explanation of behaviour can certainly be given in terms of concepts such as wishes, purpose and intention. The behaviour of a gas can be explained in terms of volume, pressure and temperature and these terms refer to complex aspects of the behaviour of molecules. In the same way terms such as intention may refer to complex categories of physical events, though it should be borne in mind that these physical events must include not merely brain states but these states in relation to the external environment.

It is, of course, true that I cannot prove that such terms do refer to complex categories of physical events in the same sort of way as terms like mass and temperature. All I am arguing is that no disproof of this assumption can be given on *a priori* grounds. Moreover, I am painfully aware that my own attempts to analyse the meaning of such words as 'courteous' are inadequate. Unfortunately, mental terms have defied adequate analysis by

---

[1] This is not the place to enter into technical details of how the brain is in fact organized but a popular account that should be intelligible to philosophers will be found in Sutherland, N. S. Machines like men? *Science Journal*, Oct. 1968, pp. 44–8.

philosophers for over 2,000 years. It may be that such terms will be best elucidated ultimately by empirical work on behaviour and upon the brain. The presuppositions behind the use of such terms could after all be false or misleading and empirical work could lead us to change radically our conceptions about man.

For these reasons I am sceptical of Grundy's claim (III (i)) that '*a priori* consideration of our talk about human behaviour' is important to a programme of empirical research. A consideration of animistic thinking about the weather, in some cultures enshrined in ordinary language, would scarcely have helped the science of meteorology.

Grundy writes 'until the notion of intelligence has been clarified, no questions as to whether intelligence has been simulated or duplicated in machines can be adequately answered'. The problem here is that when the word 'intelligence' was coined and throughout the history of the use of the word until very recently, it was simply not envisaged that any system other than a man and possibly some higher animals could show behaviour remotely resembling intelligent behaviour. It could therefore be argued (and indeed it has been argued by Geach) that the word intelligent is only applicable to people. If this were right the existence of machines mimicking 'intelligent' behaviour could both throw light on what intelligence is and also make us decide to change the way in which we use the word.

I am not happy with Grundy's assumption that before deciding whether a machine is intelligent we must have conducted 'a thorough conceptual exploration of what intelligence is, what the attributes are possession of which puts something into the intelligent–unintelligent as distinct from the simply non-intelligent category'. Since this conceptual exploration has yet to be carried out, if Grundy were right, we would be unable to use the word with any certainty of human beings. If a computer scientist were to produce a computer that initially was fed with a very general-purpose program allowing it to make inductions, process a sensory input, and formulate demands for further information and if it were then exposed to a rich environment about which it could learn, and if subsequently it proved possible for men to hold intelligent conversations with it, and it was able to play games like chess and to solve problems, I doubt if we would wish to withhold the word intelligent from it. One of the reasons why existing machines are not regarded as intelligent is that they do very little more than was envisaged by the designers of the machines and the programs. At present any one program can solve only a very limited set of problems and machines are neither exposed to a wide range of data nor do they themselves to any extent seek out particular kinds of data or make any selection from what is input.

It is, of course, open to the philosopher to doubt whether such a machine could in fact ever be constructed: all that I have argued is that there are no valid *a priori* arguments that will show in advance that this is impossible. Moreover, the existence of such machines might well make us change our mind about what is involved in the notion of 'intelligence' and could hardly fail to sharpen up the concept for us.

Grundy is right in saying that self-guided missiles do not have 'purposes' in exactly the same sense that men do. Nevertheless I believe that the existence of such devices and a consideration of how they work has already thrown some light on the use of terms like 'motive', 'goal', and 'purpose' even although it does not yet come near to telling us the whole story. So far then from analysis of everyday terms being a pre-requisite for the progress of empirical work aimed at gaining a scientific understanding of how man behaves, it may be that a proper understanding of the loose terminology we use to describe human behaviour will only be achieved as the result of empirical work and model construction.

The same types of consideration apply to Grundy's arguments about actions. Some of the distinctions drawn by everyday language are by no means as clear as Grundy seems to believe. He writes 'It is a matter of fact, indisputable for all that our knowledge of it is pre-scientific, that much of our behaviour is done because we want to act in the way we do, because we have purposes, motives, wants and intentions, and not because some physiological causal sequence, of which we may know nothing, produces some movement such as flinching or starting at a sudden loud noise'. We can already assign physiological causes to certain wants and actions. A hypothalamic tumour may turn a woman into a nymphomaniac actively seeking sexual intercourse with almost every man she meets. The tumour is not, of course, the sole cause of her behaviour—the behaviour is governed by a great many other aspects of brain functioning. Even reflexes are more complex than appears at first sight and involve elements of goal-directedness. A man who is exposed to a bell paired with an electric shock to the finger-tips will come to withdraw his hand involuntarily at the sound of a bell. If his hand is placed in a new position when the bell sounds, a completely different set of muscles will be used to effect the withdrawal movement. Even reflexes have elements of goal-directedness in their organization.

The mere fact that it is appropriate to use words like action and motive in describing and explaining some segments of behaviour and not others does not necessarily mean that the two segments of behaviour are different in kind—they may differ merely in the complexity with which they are organized and in both instances the individual muscle movements made may be fully determined by the workings of the nervous system.

Grundy seems to accept the idea that it might one day be possible to have a physiological explanation of all muscle movements including those that are involved in actions, but he is not prepared to accept that this would constitute an explanation of actions partly because of the difficulty of analysing what is meant by an action. It may, however, be that only when we have an adequate understanding of the mechanisms that produce sequences of movements will we be able to get really clear about the meaning of words such as 'action'. As I pointed out, even in providing scientific explanation of behaviour we shall certainly need in our explanatory language terms such as 'goal' or 'end-directed' which refer to complex ways in which matter is organized and we may well require terms that correspond fairly closely to whatever is meant by 'action' though such terms will refer to complex categories of physical events and will be capable of much more precise definition than are most of the terms used in everyday talk about behaviour.

I did not incidentally mean to imply that all actions arise out of dispositions or habit (compare Grundy's Section IV (iii)), though I would have thought some do and are nonetheless actions for that. Clearly some actions are performed on purpose though whether this means much more than that (*a*) the organism's behaviour is being determined by a certain goal (in the same sense that missiles can be said to have goals) and (*b*) the organism is able to give a verbal description of this goal can be left to the philosophers to decide.

Early in his article Grundy seems to accuse me of having ignored the mind–body problem, but in his final Section on this (v) he appears to represent my views fairly and sympathetically. He is right in supposing that I incline to accept the 'Identity hypothesis'—that is to say, I believe that on further analysis mental terms may turn out to reflect extremely complex categories of physical events. It is of course quite impossible to prove this assertion at the moment, but I have only argued that it is quite impossible to disprove it. We all seem to be agreed that there is at the moment no adequate analysis of mental terms—this being so, I do not see how one can be certain that the correct analysis—or better perhaps *a* correct analysis—may not be given in terms of physical events, and my own article was directed against *a priori* arguments trying to show that such an analysis is in principle impossible.

I am not as unsympathetic to the philosophical analysis of mental terms as might appear from some of the above, and indeed I feel that we already know a lot more about these terms as a result of such analysis over the past thirty years even though we are a long way off satisfactory answers. Furthermore, the results of some of these analyses have been of help to psychologists. My main reservation is that we may be in a much better position to understand the meaning of mental terms after a great deal of empirical work has been conducted just in the same way that precise definitions of terms like 'light', 'heat' and 'weight' could be said to have been facilitated by the investigations of physicists.

In this connection, I rather resent Grundy's allegation (Section IV (i)) that 'psychologists anxious to deal with human phenomena in the same way as natural scientists so successfully deal with inanimate things have for some time often neglected the distinctions drawn by philosophers between such terms as movements and actions'. Discussions of what is a response (and what is a stimulus) are only too common in the psychological literature. One reason why psychologists looked at from the outside may appear to ignore the philosophical literature is that we are nowhere near at the moment to obtaining fully-fledged scientific explanations of human purposive action. We cannot even adequately explain how it is that a rat runs a maze nor the mechanism whereby when a rat has learned to press a bar using its foot it will continue to press the bar by sitting on it if necessary, if its foot is tied up. We do, however, have the beginnings of explanations for these types of phenomena and we are obtaining some insights into how many aspects of human perception and memory work. Until this type of problem is solved we are unlikely to get very far with explanations of the very complex phenomena involved in purposive actions. Philosophers like Taylor tend to take psychologists too much *au pied de la lettre*. Nobody supposes that modern learning theory provides answers to all the phenomena of learning let alone the theories long since abandoned such as Hull's with which Taylor actually deals in his book. This does not mean that there is no truth at all in the theories: one may have to start with theories that are gross over-simplifications and elaborate them bit by bit in order to build up a theory which is some approach to the truth.

Finally, I would like to repeat a point made in my chapter. Just as there is no proof that behaviour cannot be explained in terms of physical events, there is equally no proof that it can be so explained. I have tried to argue only that it is not implausible that it might be so explained and that the plausibility of this proposition has increased in recent years with increased understanding of how the brain works and with an increased understanding of how to make machines that behave in some respects like man. One reason why I object strongly to the arguments put forward by philosophers such as Geach and Taylor is that, if taken seriously, they might prevent the attempt to find scientific explanations of behaviour. (Grundy himself makes the same point on p. 124.) I believe we would then never know the answer to the question of whether human behaviour can be explained in terms of physical events. The attitude of some philosophers towards experimental psychology is reminiscent of the attitude of the Roman Catholic Church towards physics and genetics at certain stages in history. It is fortunate indeed that such philosophers wield spiritual but not temporal power.

# CONDITIONING AND BEHAVIOUR

## *by* D. W. HAMLYN

I am not sure whether my main thesis might be better put by saying that the concept of conditioning has outlived its usefulness or by saying that it never had any utility in the first place. To say that the theoretical pattern of Pavlovian conditioning is an artificial abstraction which was not really exemplified even in Pavlov's actual experimental situation would not be to say anything new. It would not be new to say even that Pavlovian conditioning never in fact occurs. These things were said by some of the classical exponents of *S–R* learning theory. Hull maintained the more radical thesis and Guthrie and Skinner have maintained the only slightly less radical thesis that Pavlovian conditioning, type *S* conditioning as Skinner called it in *The Behaviour of Organisms* (1938), is an artificial abstraction, so that it does not appear in a pure form. These doubts about the notion of conditioning have not, however, prevented the notion from being used in some form or other. Indeed, we now hear of sensory conditioning and conditioned imagery, notions which are a far cry from the original conditioned reflexes of Pavlov's dogs. Moreover, the applications claimed for conditioning, e.g. within psychiatry, are manifold.

There can be no objection simply to the extension and stretching of a theoretical concept. It might indeed be argued that it is in such extensions of the application of theoretical concepts that the progress of science lies. But the fruitfulness of possible extensions of the application of a concept (and thereby of the concept itself) depends on what underlying analogies exist between the phenomena which the concept is invoked to explain. The question, therefore, is whether there are any appropriate analogies between the phenomena in connection with which psychologists have spoken of classical conditioning, operant conditioning, sensory conditioning, etc. Of course, what they all seem to have in common is that there takes place some process whereby a connection *appears* to be set up between two factors in the situation—between stimulus and response, between response and reinforcement, or between stimulus and image. I say 'appears to be set up' because strictly speaking it is merely a matter of hypothesis that there is such a connection. All that can be observed is that the response, or whatever it is, takes place and becomes more probable in the given situation. Whether this is due to an underlying connection and of what kind is another matter. That is where theory comes in.

I make this point because it may seem to some that the term 'conditioning' is not a theoretical term at all; it is merely a loose term used to cover any situation in which the probability of a response increases due to its association in some way in the experience of the organism with some other factor in the situation. Thus if someone could be got to have an image of a certain kind whenever a situation of a certain kind obtained (e.g. on being given a signal by an experimenter) this would be enough to justify the use of the term 'conditioning', and saying that the image had been conditioned. But this cannot be so, or at least to say this would be proof of the unhelpfulness of the term 'conditioning', since the label would then cover a wide range of quite different phenomena, having in common only that a certain response, image, etc., has the probability of its occurrence in situations of a certain kind increased, as a result of what has happened during a certain period. Surely it is not just *this* that is meant by 'conditioning'. If it is, then the term has certainly no explanatory value; and psychologists have certainly behaved as if it did have such value.

In fact, I think it impossible to separate the concept of conditioning from the other concepts which form its theoretical background. Any reference to conditioning must imply recourse to a certain kind of explanation of the phenomena in question, and that is where the theoretical background becomes relevant. In Pavlovian conditioning what is supposed to happen is that there is initially an understood connection between stimulus (the so-called unconditioned stimulus) and reflex reaction. By repetitive presentation of another stimulus (the so-called conditioned stimulus) in the same context, the reaction is transferred to it, so that conditioning takes place. There are certain things to note about this, things which may seem very obvious and boring but which require attention all the same. First, I have spoken of an 'understood connection' between stimulus and reflex reaction. I do not mean by this that we have to understand completely why the connection exists, although we must have *some* idea of why it does so. We do know at least the general pattern of the neurological mechanisms responsible for reflexes. But, secondly, this determines the sense which has to be given to the terms 'stimulus' and 'reaction' in these circumstances. For it is implied that the application of some form of energy to a group of nerve endings produces in the end, and if it is not inhibited (i.e. in normal circumstances), certain muscular contractions or other physiological effects. In the case of the patellar reflex one can speak loosely of tapping the knee and of a subsequent knee-jerk, but this is loose talk all the same. Similarly, in the original Pavlovian situation one can speak of food producing salivation in the dogs, although it is presumably the smell or taste, or rather the chemical effects on the olfactory or gustatory nerve endings, of the food which produces the reflex, and then only if the reflex is not inhibited, because, say, the animal is satiated with food. However, it may plausibly be claimed that even if one is constrained in the interests of economy to speak loosely concerning the phenomenon, one may have some idea of what is supposed to be going on and one can always in principle provide the more exact account. (Though what I have said about inhibition and normal circumstances reveals the fact that the reflex arc cannot be treated as an isolated mechanism, and that the stimulation that takes place in producing the reflex is not without its effect on the rest of the neurological system; that should go without saying. That the reflex is an abstraction has often been recognized.)

We have so far a phenomenon which is explicable in terms of a relatively isolable system. The transfer of the reaction to another stimulus must, if the account is to be more than superficially descriptive of what is seen to take place, be intelligible in similar terms. The connection with the new, conditioned stimulus must be capable of being understood in terms of the same or an analogous system, and what is colloquially referred to as the stimulus must be capable of being interpreted in terms of effects on nerve endings. The sound which Pavlov caused to occur in the case of his dogs is so capable of being interpreted in terms of its effects on the ear, and, as is well known, Pavlov put forward certain theories about the kind of neurological system which could make possible the connection which he supposed to be set up. There are no problems in his case about the end effect, the salivation, since this again can be seen in terms of the physiological workings of certain bodily organs. The mechanics presupposed in all this are, in physiological and therefore more plausible terms, what was supposed to take place in the association of ideas, according to nineteenth-century associationist psychology. It is no coincidence that the notion of conditioning was taken up with a certain alacrity by the behaviourist successors of the associationists (cf. Taylor, 1964, p. 143); but it was not only taken up, it was also extended, and that in questionable ways.

Such an extension took place in a variety of ways, but these fall into two main categories. There is first the extension of the notion of conditioning to phenomena which are not interpretable in terms of the notions of stimulus and reaction discussed above, and in which a mechanism of the kind postulated is to say the least dubious. Thus J. B. Watson (1925) spoke of conditioning where what calls out the reaction is not a stimulus in the above sense, but what he called a 'situation'. It is true that he attempted to construe situations as complex groups of stimuli, but this is not a plausible move. To say this is not to deny that when we are in a given situation our sense-organs are being stimulated in an extremely complex way. But this pattern of stimulation is not necessarily related to the situation *as it is seen by the subject*; and it is the latter to which the reaction is made. What Watson tried to explain in these terms is a subject's responding in an habitual way to a situation; but in that case it is what the subject *sees* the situation *as* that is pertinent. To put the matter in another way, while the sound in Pavlov's classical case might be said to cause the salivation when conditioning has taken place, a situation cannot be said to cause a response in the same way. But I have here spoken of a response, rather than a re-action; and it is clear that one must speak in this way; or at any rate if one speaks of a reaction, it is a reaction *on the part of* the subject, not one *caused in* the subject. It is not a contingent matter that in the classical conditioning situation the reaction is a reflex movement, a muscle move-ment or the workings of a bodily organ, such that the subject does not necessarily have control over it. The reaction or response that might be made to a situation, if it is to be an intelligible one, must be one made by the subject in the light of that situation. One could always construe Pavlov's dogs as having responded to the sound of the bell in the light of their perception of it (i.e. in the light of their perception of it as connected with food). This might not be a very plausible interpretation, because the response appears so automatic. But an interpretation of this kind would be the only possible one in the case of what is recognizably a subject's reaction or response to a situation. For how could a reaction of an automatic kind, similar to that of salivation to its stimulus, be at all intelligible in relation to a *situation*? Or rather, how could any intelligible claim be made that the mechanisms in this case are at all like those which might be invoked to explain Pavlovian conditioned reflexes? In other words, the extension of the notions of stimulus and reaction, if applied to this case, affects drastically our conception of what mechanisms may conceivably be involved; so that the case may be seen to be so different from the Pavlovian case that it is very unlikely to be an instance of the same phenomenon. What I am saying is that it is not a trivial matter that the notion of conditioning has been extended from conditioned reflexes to conditioned responses and that the notion of a stimulus has been extended so that more or less any-thing with which an organism might be concerned can be called a stimulus. In the process of this extension the cases have altered fundamentally, so that neither the application of the notion of stimulus, nor that of the notion of reaction, nor the possible mechanisms involved, are at all the same. It is difficult to see why one should suppose that one is therefore concerned with the same phenomenon, and why there is any analogy between the cases beyond a merely verbal one. Given this, it is also difficult to see what ex-planatory role is played by the notion of conditioning in the extended case.

Another facet of this problem is that while in the case of a conditioned reflex the eventual movements of the animal can be seen to be superimposed upon a natural causal connection between a stimulus and a physiological reaction, the natural *response* to a given stimulus (and it is worth noting that we should normally speak of a response *to* a stimulus rather than one caused by it) is not so easy to determine. If it *can* be determined it is not on causal grounds; or rather, it is not our understanding of the physiological apparatus which makes it seem natural. Our understanding of, say, the withdrawal of a limb on receipt of an electric shock is based on our idea of natural reactions to pain rather than on any idea of the physiological connection between the application of an electric current to the skin and the subsequent movement of the limb in question. Thus when a movement of a limb is 'conditioned' to, say, the sound of a buzzer via the application of an electric shock, it might seem more plausible to say that what happens is that the subject takes the buzzer as a *sign* that pain will ensue, and for that reason moves his limb away from the source of the pain; such a response might indeed become more or less automatic. To say this serves also to meet a possible objection that the movement is not phenomenologically different from that which occurs in cases of genuine conditioning; or, in other words, that the movement looks and perhaps feels like one which is caused by a buzzer in the same way in which it was previously caused by a shock. Phenomenology is certainly unreliable in such cases; there are many instances of movements which are automatic or habitual, without being in any way like a reflex reaction. And such movements may feel to the subject just like a reflex reaction, at any rate in the sense that the movement took place without previous consideration by the subject and in a certain sense involuntarily. But this does not prevent the one being a movement *caused by* the stimulus, the other a movement *made by* the subject in response to the effect of a stimulus. And once we have said *that* our understanding of the two cases must be quite different.

These considerations affect also the conception of stimulus generalization, which is often taken to be part and parcel of conditioning; for it often happens that a response or a kind of response is as a result of conditioning called out not by a given stimulus but by stimuli of the same or similar kind. In the case of a straightforward conditioned reflex, where the reflex movements or movements are *caused* by a stimulus, the possibility of a range of possible stimuli with the same general effect might be put down to indeterminacy in the cause and/or in the mechanism involved. Indeed this kind of indeterminacy might be put down as a characteristic feature of physiological mechanisms. On the molar or macrological level we tend to think of a limb movement being produced by some gross cause like a tap on the patella, although on the molecular or micrological level the actual pattern of stimulation of nerves and the actual pattern of muscle movement may vary from occasion to occasion. It is this to which I referred when I spoke of indeterminacy above. It is not that the cause is in any disturbing sense indeterminate; it is that the exact pattern of stimulation and physiological response is only indeterminately given when the phenomenon is identified as a reflex; and the same applies to a conditioned reflex. Here again the stimulus which calls out the conditioned reflex can be specified only as one of a certain kind, if one is working on the macrological level; and this is compatible with a whole range of patterns of actual stimulation. (It is to be noted once again how slippery the notion of a stimulus is in all this.)

But in the case of a response *to* something, if the response is one that takes place because the subject takes the so-called conditioned stimulus as a sign of something else and it is in virtue of this that the movement is made, the notion of generality has relevance in a quite different way. It is not now a question of indeterminacy in the identification of the pattern of stimulation. What is now to the point is that the subject takes not just, say, a particular buzzer sound as a sign of ensuing pain, but anything of a similar kind. And it is important what he does take as similar, what he classes together. This classification can be a matter of habit, not necessarily something conscious; a person or animal may have come to see certain things in a similar way without there being any conscious classification on any specific occasion. Thus the generalization, if it takes place, must be in what the subject perceives the so-called conditioned stimulus *as*. I say 'if it takes place' only to indicate that it need not take place to any significant degree. Yet to see $X$ as anything at all involves inevitably *some* generality. Thus the extent to which a person may respond to different sounds, and not just to a buzzer, in ways appropriate to the belief that these sounds are signs of ensuing pain in a certain part of his body, depends on the extent to which he takes these sounds to be alike in this respect or in some relevant respect. What is actually relevant here is not easy to determine in any absolute way, since the similarities that a person takes to exist between things depend on such considerations as his past experience, his interests and his sensitivities. At all events, if this is the sort of thing which takes place in this kind of conditioning, to speak of stimulus generalization is at the least misleading, since it is not the stimulus or its effect that is generalized, but rather how the subject perceives the situation in which he finds himself. This reference to perception is vital and cannot be excluded.

It may be said at this stage that my discussion shows only that the kind of case with which I have been concerned is not one of classical Pavlovian conditioning, and that the necessity to make reference to it affords additional doubt as to whether Pavlovian conditioning really occurs. What I have been describing is a case of what Skinner has called operant conditioning or conditioning of type $R$, which functions in a quite different way from classical conditioning. This brings me to the second category of extension to which the notion of conditioning has been subjected. What is essential to this second class of conditioning is that the factor which plays the main part is not what Thorndike called the 'law of exercise' (the function of practice or repetition) but what he called the 'law of effect'. There is not a simple association between two stimuli in relation to a given response, but association between a response to a situation (or operant) and the stimuli which this produces and which reinforce that response. This is instrumental conditioning, in the sense that the person's or animal's behaviour is instrumental to some given effect which is desirable for that person or animal, e.g. the production of food by the pressure of a lever by a rat in a Skinner box.

I have spoken in this of a 'desirable effect', but I do not pretend that this would generally be taken by those who refer to conditioning in this context as a desirable way of speaking. It would be *theory* that it is this which provides the reason for the continuance of the form of behaviour in question in the context; all that we can observe is that the production of a certain effect, e.g. the provision of food, does in certain circumstances itself produce further instances of the behaviour which produces the effect. What these circumstances must be and why they are contributory causes is another matter. The natural thing to say would be that the animal is hungry, that it is thus in a certain state; but the nature of that state and of the way in which it contributes to the behaviour in question has been the subject of much dispute (obvious though the answer may be to some).

Skinner's notion of an operant is meant to cover all forms of behaviour on the part of an animal which are not reflexes simply elicited by a stimulus. Whether or not stimuli are impinging upon the animal (and it is clear that there will generally be such) the behaviour is not directly elicited by them in the way that a reflex reaction is directly caused by its stimulus. The stimuli are simply occasions for the behaviour in question. Whether or not it is happy to speak of such behaviour as consisting of responses which are emitted by the animal rather than elicited from it is another matter. It is at all events clear that by the notion of an 'operant' Skinner means to cover all ordinary behaviour. But in that case the similarity between what is referred to as operant conditioning and the type of conditioning involved in the Pavlovian-style conditioned reflex becomes very small indeed. What is the force of calling it 'conditioning' at all? The answer can be only that despite the differences between operant conditioning and classical conditioning they are thought of as essentially two species of the same genus. They differ fundamentally only in the relations which hold between the stimuli and the response. Whereas in classical conditioning the connection that is set up is supposed to be that between the conditioned and unconditioned stimuli, in operant conditioning the connection is between the response and the stimuli that this produces in the situation in question.

This, however, is not enough to make them just two species of the same genus; one's understanding of these connections must also be similar—the connections must also have some similarity. In the case of classical conditioning, if it ever occurs, the connection must be a fundamentally mechanical one, and to the extent that it seems that a phenomenon cannot be interpreted in that way so also it becomes doubtful whether it can properly be classified as a case of classical conditioning. One's understanding of operant conditioning, on the other hand, must surely be different, since the connection, if there is one, will be set up only in certain circumstances, and under conditions of a special sort. One would expect the behaviour in question to become established only if the effect of it means something to the animal in question. This is the sort of thing that I was suggesting when I spoke earlier of the effect being desirable to the animal. But it also has to be noted that the effect that I have in mind here is, e.g. the production of food—an occurrence which is clearly a desirable one for a hungry animal. This will, of course, have in turn further effects on the animal concerned, though the exact nature of those effects will depend on other things that happen as well as on the general circumstances. Mere olfactory stimulation need not have any reinforcing effect on the tendency of the animal to press the lever that delivers the food. The animal must be in the position to eat the food—to receive a partial satisfaction of its hunger —and one must therefore expect the production of the food to set up a complex pattern of stimulation, the exact nature of which will depend on the set-up and on the exact position and state of the animal. There is nothing here corresponding to the association between stimuli or groups of stimuli that is characteristic of classical conditioning. What is set up is a complex series of interacting events, dependent on a variety of circumstances but in particular on the kind of system which constitutes the animal's nervous system and on the state which it is in.

What I have said implies that it should be possible in principle to set out the kind of mechanism which makes possible the tendency of the animal to press the lever when hungry. I am not here denying this possibility, although I believe that we have little idea at present of the exact nature of the mechanism or even of its general character. It is, on the other hand, quite clear that its complexity makes it very unlike the mechanism implied by the concept of classical conditioning—so much so that speaking of conditioning in this case can only be misleading. Furthermore, the mechanism in question must be taken merely as one which *makes possible* the tendency of the animal to press the lever. For pressing the lever is something that the animal *does*. Whenever it presses the lever certain bodily movements take place which might indeed be explained in terms of the mechanism under consideration, given the general situation in which the animal is found. Reference to the general system which all this comprises explains why these bodily movements or ones like them tend to be repeated. To the extent that the mechanism explains these movements so in a sense it explains what the animal *does*. It does so because only if the movements take place can the animal be said to press the lever (although pressing the lever is not just having these movements take place); *a fortiori* it is only if the mechanism functions in whatever way it does that the animal can be said to press the lever and to repeat its action. The 'only if' reveals there-fore that the reference to the mechanism is reference to the *necessary conditions* of action, not the sufficient conditions. This is not the case in the classical conditioned reflex, where the reaction is produced by the condi-tioned stimulus, so that the story about the stimulus and the system which it interacts with is a story about the *sufficient conditions* of the resulting movements. If one wishes to give the sufficient conditions of the *action* of the animal in the lever-pressing situation, as opposed to the conditions of the movements of its body that take place, one must make reference also to the kind of factors which I mentioned earlier. One must, that is, say such things as 'The animal continues to press the lever because the effect of doing so is a state of affairs which is desirable to it.' The conditions that thus have to be mentioned include its recognition that a certain state of affairs is desirable to it and that this state of affairs will be brought about by what it does. None of this would be possible but for the mechanism which is affected by stimuli which impinge upon it, and it is in that way—as constituting necessary conditions of action—that the mechanism furnishes an explanation of the action—but only in that way.

At all events, to say simply that the animal's action in pressing the lever is to be explained by saying that its action is reinforced by its getting food by that means is not in itself to offer any sufficient explanation; the story needs to be filled in by further details and what is important is the direction in which we must look to provide such details. As I have indicated, to look to the possible mechanism underlying the responses is to concern oneself with necessary conditions only, not sufficient conditions. The details that one needs when concerned with sufficient conditions are such facts as that the animal sees that pressing the lever produces food, that this is what it wants, and so on. If it does not see that pressing the lever produces what it wants why should it go on pressing the lever? This consideration is very relevant to other examples such as Guthrie's cats (Guthrie & Horton, 1945), which in moving a lever in order to open a cage from which they wanted to escape, tended to do so in a stereotyped way; if they originally moved the lever, so opening the cage, by backing into it rather than by moving it with a paw, they continued to open the cage in this stereotyped and uneconomical way. But is this surprising? If it is the case that the animal's action is explained by saying, amongst other things, that it sees that moving the lever causes the cage to open, we are demanding of it some kind of insight into the connection between the lever and the door. Why should one suppose that the cat must see the principle of this connection? To find the cat's stereotyped action surprising is really to suppose that intelligent action here is to be expected; and intelligent action would involve seeing the connection as it really is, with the result that moving the lever with the paw would be the rational form of action. But the real situation is quite different. We have no grounds for expecting the cat to see the principle of the connection between lever and door. Anything that happens as a result of the movement of the lever must in a sense be to the cat a form of magic, and in a magical situation of this kind the obvious behaviour to which to have recourse is that which originally produced the effect. We have no reason to expect in a cat behaviour which looks more rational. The same applies to the rat in the Skinner box; what it has to see is that pressing the lever produces the food. (It does not matter *precisely* how it happens, except that here efficiency will produce more food, while with the cat in the cage efficiency has the same effect as before.)

Without such considerations the actions of these animals remain unintelligible *as actions*. And even if one could explain the physical movements which take place in terms of physical and physiological mechanisms, it would still not be *actions*, what the animal *does*, that is so explained. What I have been trying to make clear, and the reference to the type of explanation which is relevant reinforces the point, is that with so-called instrumental or operant conditioning one is certainly concerned with what the animal does, not what happens to it, as may be the case in classical conditioning. There is all the difference in the world between the salivation of a dog in response to the smell of food, or the patellar jerk in response to a tap on the knee, and the pressing of a lever by an animal in the cage. The difference is not that between two sets of phenomena which remain nevertheless instances of the same genus of cases; it is a more fundamental difference, one between radically different phenomena, and it is this which makes any suggestion that they are all cases of conditioning so misleading. What I have said about the relevant explanations reinforces this point, since the explanation of what the animal does makes reference to what it sees to be the case, etc.—something quite foreign to the concept of conditioning in its classical form.

It is possible to make a larger claim still in this connection—that it is only in this second kind of case that it is feasible to speak of learning, and that for this reason classical conditioning is irrelevant to learning, while learning proper has nothing to do with conditioning in any intelligible sense of that term.[1] By 'learning proper' I mean the following. Not all modifications of behaviour can be taken as the result of learning, not even if they lead to a form of success in behaviour that has not previously been the case with the person or animal concerned. Analogously, not all cases of the inducement in a person or animal of such modifications of behaviour can be taken as teaching or training.[2] Certainly, not all processes which lead to the increased probability of a given response can be called 'learning processes'. What then is the proper characteristic of learning? It seems to me that what is essential is the use of experience, however that is to be analysed; a person or animal which has learned to do something does it by making use of past experience. It does not have to do so consciously; it does not have to be clear to it at any one time that it can do the thing in question because of the experience that it has had. Thus the possibility of learning while asleep or under drugs, etc., does not invalidate my claim, although there are obvious difficulties involved here. In order to make use of experience it is necessary at least that one's consciousness be modified as a result of something that happens to us which is in principle an object of experience.[3] I use the term 'in principle' so that learning while asleep is not ruled out, but in order to rule out the effects of processes which cannot be called learning processes.

Given all this, it should be clear why I say that the rat in the Skinner box can be said to be learning to press the lever in order to obtain food; it is doing so because it is making use of the experience of the effect of pressing the lever, even if its idea of the connection between cause and effect is rudimentary. On the other hand, Pavlov's dogs were not making use of experience in salivating to the sound of a bell, or at any rate were not doing so if the situation was as Pavlov supposed. If, on the other hand, the situation was really different and the animals were salivating because they

---

[1] Cf. Vesey, 1967. I read Vesey's paper after writing this one, but I am in general very much in agreement with his line of argument.

[2] This point was clear to Aristotle, *De Anima* 417$^b$ 2 ff.

[3] I should be more sceptical about the acquisition during sleep of skills which involve the *organization* of experience, e.g. playing the piano.

took the bell as a sign of food, they were indubitably making use of their experience, even though they were not properly speaking *doing* anything. They were not then learning to salivate in response to a bell; they *had* learned that a bell means food and salivated as a result. The salivation is in some way a causal result of the belief that they had come to hold. In just what way this can be so is a further matter.

(What I have referred to as the use of experience could be described as the use of *information*. To describe the matter in this way would be in line with the approach followed by J. J. Gibson (1966) in his *The Senses Considered as Perceptual Systems*. Gibson, indeed, says explicitly (p. 272) that sign-learning at least can be subsumed under the theory of information pick-up, and he speaks of both Pavlov's dogs and Skinner's rats as detecting invariant associations in the situation. Learning that something is the case thus involves the recognition of connections of one kind or another. Implicitly if not explicitly, Gibson's approach entails a rejection of stimulus–response theory, as at any rate a significant theory of our perception of the environment and, *a fortiori*, of our behaviour in accordance with our perception of the environment. And how is the behaviour of men and animals to be explained without reference to how they see the world?)

The situation which we have reached, then, is that there are the following alternatives. (1) There is such a thing as conditioning, and its prototype is the Pavlovian or classical case. Being restricted to the production of muscle-movements, glandular reactions, etc., in the animal, it cannot serve to explain behaviour, if this is something that the animal does. The explanation of behaviour, even simple and relatively stereotyped behaviour, must involve reference to other factors, how the animal sees things, what it takes to be the case, etc. (2) There is no room for a concept of conditioning even in the Pavlovian case, since even there, for the phenomenon to appear, the animal has to take something as a sign of something else. Why an autonomic reaction should then follow may then seem a problem, but it is no more of a problem than why it follows the smelling of food in general. It might be held that even though the animal *could* be taken to be treating the sound of the bell as a sign of food, a more plausible hypothesis in view of the automatic character of the reaction is that a connection is set up between the effects on the nervous system of the conjunction of two kinds of stimulation. As explained earlier, the fact that a reaction *appears* automatic may be very misleading, and this would be true even of a subject's introspective reports on his own reactions. The fact that a person is not explicitly aware of taking something as a sign of another thing is not in itself clear evidence that he is not doing this. In saying that a person does take one thing as a sign of another we are saying that he has certain beliefs about the situation, and people can certainly believe things without being aware of the fact that they are doing so. The most persuasive factors in this connection are the facts that the salivation is a *reflex* reaction, and that the transference of a reflex reaction to a belief raises some difficulties. But this in turn may simply raise doubts about the validity of the concept of a reflex in this connection. For, as suggested earlier, the reflex is itself something of an artificial abstraction, since the notion presupposes the isolation of one neurological connection from the rest of the nervous system. Moreover, that reflexes can be inhibited is a well-known fact; if this kind of modification is possible, why should not other modifications be possible without changing the apparent nature of the phenomenon? The matter must be left with a question-mark hanging over it.

Nevertheless, what remains clear about alternative (1), if this be accepted, is that conditioning is not a concept that can be applied to what animals and people *do*, as opposed to what happens to them. It thus has no proper role in the explanation of behaviour, if that notion be applied to what they do. *A fortiori*, it has no proper role in theories of learning. A final question that might now be raised is whether the notion can be applied to such phenomena as imagery. Can one speak of conditioned imagery? After all, is it not the case that images sometimes simply come into our mind, so that there is no question of our *doing* anything? They are then something that happens to us. The difficulty here is that if the image which is to be conditioned is to be something that occurs to us it must be *made* to occur to us in the first place; it must, that is, be produced causally in the way that a knee-jerk is produced by a tap on the knee. The causal connection must then be transferred to another cause. But how can the initial image be set up, except by suggestion or a process like the association of ideas, unless what the subject is being asked is to *do* something, e.g. to think of something. I mention suggestion and the association of ideas, because these might be represented as possible ways in which someone might be got to have a certain image, without being told to think of something by its means. But they are, of course, notoriously unreliable ways and it could not be claimed that any means whereby someone could be got to have an image in this way has the same kind of putative intelligibility that belongs to the reflex. It may be the case that when someone generally or characteristically has a certain image in certain circumstances he may be got to have that image in response to some further occurrence like the sound of a bell, if that sound is correlated sufficiently often with the situation in question. But the ordinary explanation of this would be that the sound of the bell conjures up the image or suggests it to the man because of its association with the original situation. This is not an association between stimuli, since it was not postulated that the original situation acted in any way as a stimulus to produce the image. There is *no* way of producing an image in someone's mind in the first place which has the reliability and explicability of the reflex. We may have devices for making people think of certain things, but their intelligibility is as *devices* not as mechanisms, and we have no idea of what underlying mechanisms may be involved. Thus once again, there are important differences between this kind of case and the conditioned reflex proper.

I began this essay by expressing doubt whether it was better to say that the concept of conditioning has outlived its usefulness or that it never had any utility in the first place. That doubt has not really been resolved, since it remains unclear whether conditioning in the sense explained in Pavlov's official theory ever occurs. What *is* clear is (*a*) that if it occurs the number of cases in which it does so is very restricted and (*b*) that most of the instances which have come to be subsumed under the concept of conditioning should not be so subsumed. There are great and substantial differences between these cases, and for the most part they imply some recognition of or belief about what the situation is on the part of the animal or person concerned. They involve, that is, the animal or person taking something, e.g. as a sign of something else; they involve the idea that the animal or person must derive information from its environment and put it to use. This cannot be expressed in stimulus–response terms. But only a story of this kind will be relevant to behaviour; conditioning applies at the most to what happens to an animal, not to what it does. Hence, if nothing else is true, it is certainly true that conditioning is not a notion that can have a place in a science of behaviour. It is time, therefore, that the notion was dropped from psychology and with it all theories of behaviour that depend on it.

### REFERENCES

Gibson, J. J. 1966. *The Senses considered as Perceptual Systems.* Boston.
Guthrie, E. R. & Horton, G. P. 1945. *Cats in a Puzzle Box.* New York.
Skinner, B. F. 1938. *The Behaviour of Organisms.* New York.
Taylor, C. 1964. *The Explanation of Behaviour.* London.
Vesey, G. N. A. 1967. Conditioning and learning. *The Concept of Education.* Ed. R. S. Peters. London.
Watson, J. B. 1925. *Behaviourism.* New York.

# COMMENT

## *by* A. J. WATSON

Professor Hamlyn's essay is deceptive; at least, I was misled. Upon first or second reading, I believed my task would be easy. I had only to lend support to the conclusions and arguments of his essay, to say that the difficulties and idiosyncrasies in the nature of conditioning to which he draws attention cannot be too widely appreciated, especially by physiologists, and to add a few further morose remarks upon the state of the subject in general. But, upon reflection, I am less content to follow this course, despite the satisfaction it would afford. I now find myself still largely in agreement with the spirit of his essay and I would support a part of his conclusion, at least if this is taken in a certain sense. But I believe there may be a second conclusion closely intertwined with the one I accept, and this seems to me of far more doubtful validity. In addition, I am not confident that these conclusions, correct or erroneous, are really justified by the arguments from which they stem. My task is therefore to attempt to disentangle these different themes and, in so far as I can do so, to assess them separately. I do not regard this prospect with confidence, for it is all too clear that, even if I am right in believing that there are two distinct themes in the essay, they are inherently closely related.

I shall attempt first to state what I take to be Professor Hamlyn's main conclusions; second, to consider several peripheral points which, although relatively isolated, seem to require comment; third, to mention the arguments by which I believe one of the conclusions in the essay should be supported; and finally, to indicate the grounds of my disagreement with the second conclusion.

Professor Hamlyn concludes that classical (Pavlovian) conditioning is a phenomenon which is at best rare and possibly fictitious. Since, therefore, most instances of behaviour cannot be considered as examples of conditioning, psychology should abandon the belief, apparent in so much literature, that a reference to conditioning affords a full and convincing explanation of the behaviour under examination. Subject to minor qualifications which will be mentioned below, this conclusion I too accept; as Professor Hamlyn says, between the phenomenon of classical conditioning and the bulk of animal and human behaviour 'there are great and substantial differences'. I am, however, less content with his account of these differences. The second conclusion, which I wish to distinguish from the above, is that conditioning, or an explanation in terms of conditioned reflexes, could not in principle be applicable to behaviour. Behaviour consists of things which the organism does, not of things which happen to it. Professor Hamlyn argues that conditioning could in principle provide a sufficient explanation of the latter only (granting, as above, that it does not in fact succeed even to this extent) and hence *must* be irrelevant to the explanation of behaviour. For such an explanation involves the use of concepts which 'cannot be expressed in stimulus–response terms', nor indeed, in any 'mechanical' terminology. In summary, then, the first conclusion is that most behaviour is different in kind from the phenomenon of classical conditioning in such a way that reference to conditioned reflexes does not afford an adequate explanation; and the second conclusion is that anything which we count as behaviour has, logically, a characteristic which precludes not only explanation within the specific framework of conditioned reflexes but also within any context which, so to speak, shows the 'mechanical' character of such theories. As I have said, the former I accept, but the latter appears dubious.

I turn now to some preliminary minor points. Professor Hamlyn stresses the theoretical status of the 'connections' to which reference is often made in accounts of conditioning and in learning theory in general. 'All that can be observed is that the response, or whatever it is, takes place and becomes more probable in the given situation.' It should, however, be noted that these connections are not theoretical in the sense that the term refers to entities essentially unobservable. Rather they are to be compared to, say, the transistors in a computer where a man has no means of removing the casing in order to examine the internal mechanism. In fact, some progress has been made towards the physiological identification of these connections; although, at least in my view, ignorance rather than knowledge is more conspicuous in this field of research. It is to be doubted, however, that Professor Hamlyn would agree that this is the kind of theoretical status pertaining to the connections of learning theory. But, if so, a subsequent part of the argument seems to be overstated; or, rather, to be too cursory. It is noted that not all those who employ the vocabulary of conditioning refer to connections and that, for them, the term 'conditioned' may refer only to the observed association between stimulus and response: '. . . it is merely a loose term used to cover any situation in which the probability of a response increases due to its association in some way in the experience of the organism with some other factor in the situation'. Use of the term in this way thus involves some statement concerning the history of the development of the responses but implies no theoretical explanation of the kind outlined above. Professor Hamlyn doubts, however, whether so empirical a use of the term can really have been intended. 'Surely it is not just *this* that is meant by "conditioning".' Yet many psychologists would insist that it is just this that they mean when they speak of conditioning and would claim that the elimination of theoretical implications is to be regarded not as an impoverishment, but as a virtuous abstention from pernicious, idle and confusing dispute.

I am concerned, however, not to enter into this long-standing controversy, but to qualify Professor Hamlyn's conclusion. For he continues that if 'conditioning' is used in this way '. . . then the term has certainly no explanatory value; and psychologists have certainly behaved as if it did have such value'. Many authorities differ upon the conditions which must be satisfied if a statement, or set of statements, is to have 'explanatory value'. In particular, it is not clear how far the use of a classificatory term may provide some kind of explanation. If I ask why a dog has ridiculously short legs and am informed that it is because it is a Sealyham, I may gain considerable illumination concerning the creature. First, the dog becomes one of a class with which I am familiar; it is no longer an oddity. This, if explanation, is so by virtue of generalization. I may also infer several other characteristics of the animal with which I am confronted—that its preposterous shape has been produced by deliberate selective breeding, that it is probably affectionate, loyal, possesses a will of iron, is in no circumstances to be trusted with other dogs, and so on. I do not wish to argue here that I have been given an *explanation*; I leave the question open. My purpose is to emphasize the utility of the name of the breed. It seems to me that 'conditioning', when so used, has a similar value. If a response has been conditioned, I know that it did not appear as the result of maturational processes, is not dependent upon previous or present interference with the nervous system of the animal, may be expected to be related to stimuli in one of the several known ways, and so on. I also know that the behaviour will, or will not, be permanently modified by the application of defined experimental procedures—for instance, that it can be extinguished. It appears, therefore, that whether or not reference to conditioning in this non-theoretical way may be said to afford an 'explanation' for a response or feature of behaviour, those who so employ the term cannot be condemned for a useless, or uninformative, analysis of the subject of study.

Professor Hamlyn later argues that an account of classical conditioning involves reference to an 'understood connection', the physiological value of which must be known in general, though not precise, terms. It follows, he claims, that a particular 'sense has to be given to the terms "stimulus" and "reaction" in these circumstances', a sense in which the stimulus is specified in terms of physical energy at the sense organs and reaction in terms of particular muscular contractions. I am uncertain of the scope of this argument. Certainly, with some reservations, Pavlovian conditioning has often been discussed in physiological terms and the sense of stimulus and reaction which is here regarded as an important feature of such accounts has been employed. Certainly, the physiological implications of this kind of behaviour were of major interest to Pavlov himself. Yet it seems perfectly clear that all the features of such accounts could be translated into non-physiological terminology; all, that is, save those which make reference to *specific* and particular physiological processes. But since, as Professor Hamlyn implies, our knowledge is not yet such as to support particular 'physiologizing' to any great extent, it follows that nearly all the important, explanatory (if such they be) aspects of Pavlov's story can be restated without reference to physiology. It is, indeed, a feature common to theoretical approaches to behaviour which are couched in physiological terms that these may be discarded or replaced by others without prejudice to the fundamental structure of the theory, except perhaps for a certain embarrassment arising from the revelation of the puny skeleton within. In these circumstances it is not clear to what extent, and in what way, the sense to be given to the terms 'stimulus' and 'reaction' is *necessarily* determined. The 'understood connection' may be comprehended in terms of its theoretical properties and there does not appear to be any logical step from such a concept to the form which those of 'stimulus' and 'reaction' must take; and, if there is, it cannot stem from any physiological considerations, which are relatively irrelevant to the basic nature of conditioning.

In this paper, two distinctions are employed which seem to me of dubious utility. And, if I have understood him correctly, Professor Hamlyn himself finds difficulty with the first. This is the distinction between 'reflex' and other behaviour drawn on the basis of the overt, observable characteristics of the behaviour. Professor Hamlyn points out that classical conditioning may, perhaps, more correctly be presented in other terms. 'One could always construe Pavlov's dogs as having responded to the sound of the bell in the light of their perception of it . . .', and 'Phenomenology is certainly unreliable in such cases, there are many instances of movements which are automatic and habitual, without being in any way like a reflex reaction . . .'. But I am not sure that Professor Hamlyn would wish to discard the distinction to the extent of saying that the behaviour of the intact animal is *never* an example of purely reflex processes. The point is not argued in the paper and I must here be content with an affirmation that, in my view, no such distinction, both workable and useful, can be drawn on the basis of observation of the characteristics of behaviour alone.

The dubious second distinction, related to the first, is that between 'natural', and 'intelligent', and other behaviour. I do not, of course, wish to contend that behaviour or responses cannot be characterized with profit as more or less intelligent or natural. Rather, I wish first to stress that this is always a quantitative matter. There is not one kind of response or behaviour which is 'blind', 'automatic', 'just a reaction' and another which is 'insightful', 'deliberate', 'planned' and so on. I am not clear what Professor Hamlyn's view on this point may be. In his examination of the operant escape situation originally employed by Thorndike, Professor Hamlyn points out that the sequence of events must have a certain magical quality for the cat, and seems to imply that therefore here intelligent action is not to be expected; '. . . and intelligent action would involve seeing the connection (between the lever and the door) as it really is, with the result that moving the lever with the paw would be the rational form of action.' However, 'the same applies to the rat in the Skinner box; what it has to see is that pressing the lever produces food'. Reference is then made to the relative efficiency of response with the paw in the two tasks. I am unsure whether on this view, an action is natural, or intelligent, (a) when the processes whereby the consequences of the action take place are comprehended by the agent; or (b) when that action is selected which is most efficient (presumably in terms of effort and reward) for the achievement of some prescribed good. Without further argument, I would accept that either of these qualities contributes to the naturalness of an action. But I wish to claim that the first, certainly, is not a necessary condition for the ascription of 'intelligence' to behaviour; and, further, that much, perhaps most, behaviour of both men and animals is almost entirely devoid of this quality.

Professor Hamlyn raises the difficulty always encountered in attempts to define 'learning' and, accepting these, suggests that an essential feature is 'the use of experience'. This seems to me to be a little obscure and not to take sufficient account of the distinction between learning and performance. For I believe that not only some psychologists would wish to leave open the possibility that an organism may learn something, or acquire some information, which in fact it never has occasion to use. (Other psychologists would raise issues of principle here, but I do not think they are germane to this discussion.) This may, however, in itself be no more than a verbal misunderstanding. Its importance arises as the first step in the argument which leads to the conclusion. 'In order to make use of experience it is necessary at least that one's consciousness be modified as a result of something that happens to us which is in principle an object of experience.' The introduction of consciousness at this point raises all manner of questions. Discussion of many, no doubt, would be irrelevant; the examination of others would prove lengthy; and, in any case, I should not know the answers. Nevertheless, from my point of view, this is strong language. First, let us separate this claim from another important point which the paper seeks to make, namely that perception is an important ingredient in action, or in the processes from which actions stem. With this further point I shall be concerned below. Setting aside, however, this further argument, I do not entirely understand the conclusion, or perhaps, the reasoning upon which it is based. *Why* is it necessary, in order to make use of experience, that consciousness should be modified? I can find no grounds for this assumption, unless 'experience' is here employed to mean something which, inherently, can have no effect upon anything except consciousness. But if 'experience' in some such sense as this is intended, I find myself in a similar difficulty, for then I cannot find the basis for the claim that learning is necessarily based upon experience. Again, I can only affirm that there seem to me many examples of learning, in addition to the special cases such as learning during sleep, in relation to which there is no evidence of any modification of consciousness. The organism can do what hitherto it has been unable to do. But what evidence is there of any change in conscious processes, other than this demonstration of learning itself? I should perhaps here emphasize that I am not at this point disputing the occurrence of conscious processes, or reflecting on the relation between perception and behaviour. My claim is simply that quite different organisms show evidence of learning where either there cannot be any independent evidence of changes in consciousness or, where this is possible, the evidence is obscure or clearly negative.

Finally, I wish to consider that aspect of Professor Hamlyn's main conclusion with which I disagree. The difference between most examples of ordinary behaviour and the phenomena of classical conditioning is such that a far more sophisticated mechanism must underlie the former. The extension of investigation to operant learning and more complex behaviour 'affects drastically our conception of what mechanisms may conceivably be involved'. Professor Hamlyn regards our comprehension of such a mechanism as a possibility, though perhaps a little remote. 'What I have said implies that it should be possible in principle to set out the kind of mechanism which makes possible the tendency of the animal to press the lever when hungry.' It is, however, 'quite clear that its complexity makes it very unlike the mechanism implied by the concept of classical conditioning'. I agree that this is so. But I believe that the reasons which lead Professor Hamlyn to this conclusion are more relevant to another point; and that to demonstrate clearly and unequivocally that the mechanism proposed in the theory of classical conditioning is not involved in the control of most behaviour, it is necessary and sufficient to show that this theory makes *wrong* predictions. There seems little difficulty in this demonstration. It is clear that such a device would, for the most part, produce behaviour different from that actually observed and would not arouse organisms, even at a relatively primitive level, to behave as in fact they do. I shall assume that such arguments, now well known, are valid and accepted. The arguments employed by Professor Hamlyn seem to me to bear principally on the following further point.

It is argued that the theory of classical conditioning could provide sufficient conditions for the phenomena concerned. But that a more complex and sophisticated mechanism, related to operant behaviour, could at most only provide necessary conditions for the actions to be explained: '. . . the reference to the mechanism is reference to the *necessary conditions* of action, not the sufficient conditions. This is not the case in the classical conditioned reflex, where the reaction is produced by the conditioned stimulus, so that the story about the stimulus and the system which it interacts with is a story about the *sufficient conditions* of the resulting movements.' I am not clear whether Professor Hamlyn would allow that a story concerned with interaction of sensory events and more complex mechanisms could provide sufficient conditions for the movements which take place when people and animals act. But I can find nothing in his essay which refutes this view. And it appears that, since in both cases we are dealing with the interaction of inputs with a mechanism resulting in movements, if the story could provide sufficient conditions in the case of classical conditioning it could similarly afford them in relation to the movements involved in other instances of behaviour. If this point is allowed, the argument hinges on the difference between actions and movements, sufficient conditions for the former necessarily requiring the use of concepts outside the range of 'mechanical', 'causal' theories. Professor Hamlyn perhaps expresses this point when he gives as the reason for saying that reference to a mechanism could not provide sufficient conditions for action, 'for pressing the lever is something that the animal *does*'. And, again, 'If one wishes to give the sufficient conditions of the *action* of the animal . . . one must, that is, say such things as "The animal continues to press the lever because the effect of doing so is a state of affairs which is desirable to it".' And it will similarly be necessary to refer to how the animal 'sees' the situation, the probable nature of a contemplated action, and so on.

I believe this argument to be in error, at least as a contention based on points of principle. Presumably, we are able to identify and classify actions with a degree of reliability sufficient to justify attempts at explanation of *some* kind. It may very well be that to identify an action as of a certain kind we depend upon knowledge, or beliefs, about the context of the action, the intentions of the agent, and so on. But still we identify it. If, then, with or without the aid of a theory we were to isolate conditions, prevailing antecedent to the action in question, which were immediately followed by the action such circumstances would presumably be sufficient conditions for the action concerned. And there appears to be no necessary reason why such conditions should not be entirely of a kind which could form part of a 'mechanical' or 'theoretical' story. Let us take an example of complex behaviour, since presumably the force of Professor Hamlyn's argument should be most apparent in such cases. Generally speaking, we seem to be able to identify cases of suicide; not infallibly, of course, but sufficient for our purpose. Suppose, now, that we believe that we have identified conditions so unpleasant that any adult person, given certain facilities, will kill himself rather than endure them. We could be wrong. But, if we are not, have we not established sufficient conditions, in a causal sense, for the act of suicide?

It might be argued, in reply, that much is concealed in this example. Even if we have succeeded in establishing sufficient conditions for the movements exhibited by our unfortunate victims, we have not thereby identified sufficient conditions for suicide. To know, or believe, that someone has committed suicide, we must believe that he thought that his behaviour would have fatal consequences. Thus, someone might take poison in the mistaken belief that the substance would merely alleviate his suffering. And this would not be suicide, though the movements involved might be identical whatever his beliefs or intentions. Hence the example contains implicit reference to the agent's beliefs, to how he sees the situation, to his desires and so on. I accept that reference to such considerations may be essential in the process of identifying any particular action as an instance of suicide. But it does not seem to follow that therefore similar references must appear in a sufficient explanation of the act. For suppose that it could be shown that in all those cases where the movements made are held to be movements made whilst committing suicide, the movements are preceded by one specified condition of a part of a hypothetical mechanism; while the same movements made in the course of other actions are preceded by different states of the mechanism. There is nothing impossible about this suggestion (and nothing very plausible about it either, given our present knowledge); and if such things could be demonstrated, it is difficult to see the grounds upon which it could be held that, in a causal sense, our mechanism provides sufficient conditions for movements but not for actions. There would be no lack of sufficiency in the explanation; and reference to, for instance, how the agent perceived the situation would be replaced by reference to states of the mechanism. It is not necessary to argue that the reference to a state of the mechanism is equivalent to the reference to the agent's perceptions, beliefs, and so on. It is necessary only to point to the possibility of an alternative, sufficient explanation. I conclude that although there is little immediate prospect of the provision of sufficient explanations of behaviour in terms of the general kind of theory mentioned above, and still less prospect of such an account in physiological terms, such explanations are 'in principle' possible.

There is little that can be said to draw these disparate points together. I hope, however, that I may have given some vague indication of the reasons which seem to me to support Professor Hamlyn's rejection of the theory of classical conditioning as an explanation of behaviour, but to argue against the limitations he places upon the use of more sophisticated efforts of the same kind.

A few years ago, I had many interesting discussions in this general field with Mr P. Hirschmann. I cannot now say how far the views I hold at present stem in general or in particular from suggestions which Mr Hirschmann then made. But, where this may be the case, I am anxious to acknowledge my debt. It may safely be assumed that I am entirely responsible for errors and infelicities in the argument.

# REPLY

## *by* D. W. HAMLYN

Alan Watson agrees with much of my paper, and it is gratifying to discover such broad agreement with a psychologist. Since he has shown so much willingness to agree with me, let me begin by showing an equal willingness to accept one of his major criticisms. This is his comment on what he calls my 'second conclusion'. He is undoubtedly right in pointing out that what I have said about the idea that the sufficient conditions of action must make reference to such things as how the animal sees the situation is quite wrong. He is equally right in pointing out that in the admittedly improbable event of a given circumstance (whether external or one relating to the neuro-physiological set-up of the animal or person) always being followed by a given action, we should say that the circumstance is the sufficient condition of the action's taking place. This is indeed the logic of the case. That the animal should see the situation in a given way and so on is a condition of what is taking place counting as an action, and it might for that reason be considered as another necessary condition of the action; but there may well be other things which can count as sufficient conditions. In all this he is undoubtedly right, and I should have known better than to claim otherwise.

But it seems to me that he takes this point to have wider implications than I should be prepared to give it. Let us suppose that there is some circumstance given which a certain kind of action always follows, so that we can say that the obtaining of the circumstance is a sufficient condition of this kind of action. In that case, as he rightly says in commenting on the extent to which mere generalizations 'explain' the events which they cover, we can give some sort of explanation of this kind of action in terms of the antecedent circumstance. But this would be a minimal form of explanation and no scientist would have the right to be content with it (a comment which I should equally make on Mr Watson's attempt to defend those psychologists who treat the notion of conditioning in a non-theoretical way). Given this, it is worthwhile asking why there is a certain improbability and implausibility about the attempt to produce sufficient conditions of action in this way. (Mr Watson's example of suicide seems to me rather more implausible than some other possible ones, but let that pass.) The implausibility is not one which is relative merely to our present state of knowledge; it has a more intrinsic character. It may be admitted that certain circumstances are sufficient for the production of gross and general forms of behaviour. It is clear, for example, that taking certain drugs may always produce certain forms of behaviour. I have no wish to dispute as inconceivable suggestions of this kind. It is, however, another matter to suppose that a certain definite action, as opposed to a *form* of behaviour, always ensues in a given circumstance.

If I think that there is a kind of absurdity about this last suggestion it is because whether or not an action does ensue in a given circumstance raises just that question which Mr Watson wishes in effect to put on one side— the question of the criteria of identity for the action in question. As Mr Watson says, we do identify actions, even if in doing so we presuppose 'knowledge, or beliefs, about the content of the action, the intentions of the agent, and so on'. But this implies that even if it is true that in certain circumstances a certain general form of behaviour always ensues, what precise action is performed will be dependent on how the person sees the situation, what his beliefs about it and perhaps about other things are, what his wants and intentions are, and so on. Given the possibility of variation in these factors, there is great implausibility in the suggestion that an action of a *precise* kind may have as a sufficient condition a circumstance of the kind that Mr Watson has in mind in his suicide example. If it were the case that attempts at suicide always took place in certain unpleasant circumstances this would be in a curious sense a fact about human nature; it would be a supposed fact that conflicts with everything that we know about human beings, but that does not matter for present purposes. What I mean by this is that what would make the fact intelligible would be some kind of understanding about the kinds of being that men are. Hence, if we are to look for further explanations of the fact it is in this direction that we must look. Analogous, though in many ways different, considerations would apply to the fact about the effects of maternal deprivation. But Mr Watson's argument involves another step—that if there could be facts of this kind it could equally be a fact that an attempt at suicide always takes place when people's nervous systems are in a certain state. This step requires justification, and we have to ask what, if it were a fact, would make it intelligible.

The intelligibility of the supposed fact that human beings cannot stand situations of an extreme degree of unpleasantness and hence always try to commit suicide in these conditions is one of a quite different kind from that of the supposed fact that they always act similarly when their nervous systems are in a certain state. We might even say that while the first claim is in the running for truth, even if it lacks all plausibility, the second is a truism of an unilluminating kind, if it is anything. It will be a truism to the extent that it may be supposed that there must be some neurological state that goes with the intention to commit suicide (or at least this will be so on a certain view of the body–mind relationship, a view which, it will be evident from my original paper, I accept). But in that case the intelligibility of the claim that an action has as a sufficient condition a certain state of the nervous system comes only through the understanding that the action is due to a certain intention (whatever the status of that 'due to'); the identification of the neurological state is dependent on a prior recognition of the intention. It might indeed be said that the neurological state is identifiable only via the intention, and it is a piece of pious optimism that it can ever be identifiable separately. Mr Watson seems to believe that this is merely a function of our present lack of knowledge, but in effect his own words belie him.

He says that it might be the case that the movements involved in an action are always 'preceded by one specified condition of a part of the hypothetical mechanism, while the same movements made in the course of other actions are preceded by different states of the mechanism'. And he adds that if this is true of the movements it must also be true of the action itself. The fact that the movements involved in an action are always preceded by a condition of the mechanism does nothing to show that this condition is a *sufficient* condition of the movements. It shows only that the condition is a necessary condition of the movements. When he adds that the same movements made in the course of other actions are not preceded by the same state of the mechanism, the force of his remark is to stipulate that the condition of the mechanism is not even a necessary condition of the movements as such, but only of the movements involved in the action in question. To grant that the condition of the mechanism is a sufficient condition of either movements or action we need to be told that whenever the condition of the mechanism obtains the movements or action always follow. As far as his words are concerned, we are told nothing to this effect. But it is also clear that we could not come to knowledge of sufficient conditions of action of this kind in the way contemplated. Whether or not there can be sufficient conditions of action, whether of the form represented by a specific situation in which men always do one sort of thing or of the form represented by the state of the nervous system in which that specific action takes place, one cannot investigate the sufficient conditions of an action by trying to find out what mechanisms are responsible for it; the most that one can find out *in that way* are the necessary conditions of the action. Hence if a given neurological state *is* a sufficient condition of a certain action, its intelligibility as such is not as the mechanism responsible for that action. If it is intelligible at all, and if it is not, so to speak, a brute fact that the action takes place in this circumstance, then the intelligibility must come, as I indicated above, from the fact that the neurological state in question is one that goes with something that really does make the action intelligible—that is to say an intention to perform that action.

What I have said could be put by saying that a story about the state of the nervous system does not provide a sufficient explanation of action. It would be wrong of course to say that reference to the necessary conditions of action or to what may in certain circumstances be its sufficient condition does nothing by way of explanation of the action. But it should be clear that reference to physiological conditions cannot afford a complete explanation of action; for the point of a complete explanation is to make the explicandum intelligible. Thus a complete explanation of action must make it intelligible as action, and this presupposes reference to the family of concepts from which that of action is drawn. What I would say to the psychologist here is: 'By all means go on investigating the mechanism of action. It is extremely important that you should. But do not think that you are thereby making actions more intelligible as actions.' Mr Watson seems at one point to want to make the burden of decision between a theory of behaviour based on conditioning and one based on other facts depend on the validity of the predictions that each theory makes. I do not think that this can be so. The issue is over which theory makes behaviour intelligible. My thesis is that a theory based on conditioning *cannot* make behaviour (if by this is meant anything more than a few very limited bodily movements) intelligible; and I think that my thesis still stands, despite the error of what I said about the sufficient conditions of action, the point on which Mr Watson rightly fastened. Mr Watson himself agrees with my conclusions about conditioning but hopes that a more sophisticated approach to mechanisms may be more promising. Whether or not this is so depends on what the sophistication consists in, of course; but it will certainly not do to suppose that the proper approach can consist of a mere extension of what is supposed to be the case with conditioning. Not only should the notion of conditioning be dropped from psychology, but anything that resembles it, however effective may be the procedures presupposed in what is often called conditioning for producing certain results in animals and human beings.[1]

There are a number of other points which are raised by Mr Watson which are more incidental to the central issue. I agree with his comments

[1] I mention this point because it is sometimes said that all that matters for e.g. Skinner is that he is able to arrive at predictions about behaviour in given circumstances, on the basis of which it may be possible to arrive at procedures for bringing about desired behaviour. I cannot think that this *is* all that matters. The apparent efficacy of a procedure may say nothing for our understanding of what makes it efficacious. In any case, the use of such procedures may raise moral issues, as may their justification in the name of science. But that is a matter for a separate discussion.

on my use of the word 'intelligent'. I should have made my point about Thorndike's or Guthrie's cats in terms of the notion of rationality rather than that of intelligence. I shall mention only one further point—what he says about learning. I agree that my interpretation of that notion in terms of the use of experience is vague, but in reply to Mr Watson's question why I say that this involves a modification of consciousness I can ask only in turn: 'What else could be meant by "experience"?' Experience is not anything that happens to an animal, and not anything that happens to an animal falls within its experience. Mr Watson asks: 'What evidence is there of any change in conscious processes, other than this demonstration of learning itself?' My answer can only be 'None', for my point is meant to be a conceptual one and hence not one to be confirmed or refuted by empirical evidence; it is not just a contingent matter that there is a connection between learning, experience and consciousness. Mr Watson claims in return that there is evidence of learning where there is not or cannot be evidence of changes in consciousness. I am not fully aware of the basis on which his claim rests. Perhaps I may content myself with yet another question. 'What is it that constitutes "evidence of learning"?' After all, not all cases in which an 'organism can do what hitherto it has been unable to do' are cases of learning. There is such a thing as maturation. What marks off learning from this and from other phenomena such as the imprinting that ethologists speak of? I think that this is a difficult question, to which I am not sure what answer is to be given other than the one which I have given already. And on this Mr Watson has not convinced me that I am wholly wrong. Nevertheless, Mr Watson appears to agree with me that conditioning, if it is useful to speak of it at all, has nothing to do with learning; and if that is so, it constitutes a point of agreement which is of some substance.

# IMPERFECT RATIONALITY

## *by* JOHN WATKINS

### I. INTRODUCTION

Historians are not much concerned with also-rans and drop-outs. They have a bias towards success. I do not intend to preach against this bias (which in any case has some self-correcting tendencies: a success-minded historian of Russian Bolshevism has to pay *some* attention to the Mensheviks). But I will argue against certain methodological ideas in recent philosophy of history which tend to encourage it. The most extreme of these could be summarized thus:

(1) Only *successful* actions are open to historical explanation.

But there are more modest and plausible views which tend in the same direction. Consider the following two propositions:

(2) A past action is open to historical explanation only if the rationality principle can be applied to it.

(3) The rationality principle can be applied only to *rational* actions.

The conjunction of (2) and (3) will come the closer to yielding (1) the more closely is successful action correlated with rational action.

My own position, very briefly, is this. I agree with (2) but I disagree, despite its blandly plausible air, with (3)—unless the term 'rational' is used in such a relaxed sense that an action is trivially 'rational' merely in virtue of being an action. I hold that a rationality principle is likely to be at its most heuristically fruitful when it is applied determinedly to recalcitrant cases—to actions that are manifestly unsuccessful and seem more or less irrational or even downright crazy. The darker the region, the more light may the principle spread. Of course, it applies to rational and successful actions too. My point is that it is not restricted to these but can be used to rationalize irrationality and failure.

My plan is this. In §2 I will consider Collingwood's arguments for proposition (1) above. In §3 I will defend the rationality principle against some of the charges that have been levelled against it. We shall have to distinguish a strong and a weak version of the principle. At its strongest, it becomes an *optimality* principle. This will be discussed in §4 and again, in its game theoretical interpretation, in §5. I shall contend that in this strong form it breaks down even as a normative principle: there are problem-situations for which there can be no determinate optimal solution;

and in any case a considerably weaker kind of rationality principle is required for historical explanation. An 'imperfect rationality principle' will be adumbrated in §6. Finally, in §6.3, I will consider a veritable paradigm of a botched, crazy-seeming and disastrous action for which no explanation was given by the agent (who died as a consequence of it); yet even this action can be explained and quasi-rationalized by the application to it of an imperfect rationality principle.

## 2. COLLINGWOOD'S THESIS

I am using Collingwood, here, chiefly as a stalking-horse. It would be silly to take his thesis about the unexplainability of unsuccessful actions too seriously when he only flirted with it. Alan Donagan has pointed out that Collingwood himself implicitly repudiated this thesis in the very book—his *Autobiography*—in which he advanced it, by offering an explanation for Caesar's invasion of Britain in terms of a purpose which Caesar *failed to achieve*.[1]

The thesis is assuredly false—but interestingly false, a bold exaggeration of something rather plausible. Collingwood's argument for it may be condensed into the following three propositions (the wording is mine):

(1) Every action is an attempt to solve a definite problem that has arisen in a definite situation.[2]

(2) A historian can explain a past action only if he knows what problem it was intended to solve.

(3) A historian can come to know what problem a past action was intended to solve only (in Collingwood's words) 'by arguing back from the solution' (Collingwood, 1944, p. 8).

It follows from (3) that, in the case of an unsuccessful action, an action which *failed* to yield a solution for the problem it was intended to solve, the historian cannot discover the problem and so, by (2), cannot explain the action. In a well-known passage Collingwood wrote:

How can we discover what the tactical problem was that Nelson set himself at Trafalgar? Only by studying the tactics he pursued in the battle. We argue back from the solution to the problem. What else could we do? Even if we had the original typescript of the coded orders issued by wireless to his captains a few

---

[1] Donagan, 1962, p. 268. Donagan adds: 'No historian could long maintain that history can make sense of success and victory, but not of defeat and disappointment.'

[2] The word 'action' is used here in a rather extended sense; thus painting a picture would count as an action. Collingwood said that while still a boy he learnt from watching and imitating his 'father and mother, and the other professional painters who frequented their house . . . to think of a picture . . . as the visible record . . . of an attempt to solve a definite problem in painting' (Collingwood, 1944, p. 8).

hours before the battle began, this would not tell us that he had not changed his mind at the last moment, extemporised a new plan on seeing some new factor in the situation. Naval historians think it worth while to argue about Nelson's tactical plan at Trafalgar because he won the battle. It is not worth while arguing about Villeneuve's plan. He did not succeed in carrying it out, and therefore no one will ever know what it was (Collingwood, 1944, pp. 50–1).

My main disagreement is, of course, with (3). Why on earth did Collingwood suppose that the *only* way to recover a problem is via its solution? Why did he exclude the possibility of independent evidence about what the problem was? There are, I think, some partial answers at least to these questions.

In the first place it is relevant that the two main fields in which Collingwood himself worked as an historian were archæology and the history of ideas.

Now an archæologist usually has no documentary evidence. If he is trying to interpret the remains of some physical artefact (such as the Roman Wall, on which Collingwood himself worked), he is likely to have to conjecture what its purpose was, what problem it was intended to solve, primarily on the basis of evidence supplied *by the thing itself.* And this in turn is likely to mean that he will have some hope of making a realistic conjecture *only if the thing was reasonably well designed for its purpose.* For suppose that it had been a botched job. Suppose (to give a fanciful example) that the ancient Egyptians had invented a diving-bell which, however, had the disadvantage that anyone in it would suffocate when it was submerged. An archæologist, we now imagine, comes across the well-preserved remains of this inefficient device. His assistant ventures the conjecture that it was some sort of diving-bell. Our archæologist examines it, and then replies: 'No, it can't possibly have been a diving-bell—there's no air-supply. Perhaps it's a coffin for burial at sea.'

In other words, archæologists in their interpretations of the remains of artefacts have to assume that these (to put it in Collingwood's question-and-answer terms) constituted tolerably good answers to the problems they were intended to meet. And this may have inclined Collingwood towards proposition (3) above: in the absence of independent evidence, the archæologist has to 'argue back' to the problem from its (more or less) successful solution as embodied in the artefact (or the remains thereof).

However, the assumption that an artefact was tolerably well designed and constructed does not, of course, imply that it was *successful* in a larger sense. The Maginot Line was, I presume, well enough designed and constructed, engineering-wise, to enable an archæologist of the future to guess its purpose. Collingwood's understanding of the Roman Wall was no doubt shrewd and realistic. But this, of course, does not imply that the Roman Wall successfully fulfilled its larger purpose: perhaps it was as unsuccessful as the Maginot Line.

A word, now, about the bearing on proposition (3) above of Collingwood's work on the history of ideas. Here, the evidence does indeed consist of documents. But Collingwood claimed that these will have significant gaps, at least in the case of past philosophical writings. For a philosopher of the past was writing for people who were already asking the question that he was trying to answer; consequently, *the philosopher did not spell out that question.* Thus there is, after all, an important parallelism between philosophical historiography and archæology. In both fields the student has to *infer* what the problem was from the solution of it that he is now studying, whether this is Plato's *Republic* or Roman remains. Moreover (Collingwood insisted) he will be able to do this, in the case of a philosophical work, only if the solution was *successful.* Had a philosopher made 'a complete mess of the job of solving his problem, he was bound at the same time to mix up his tracks so completely that no reader could see quite clearly what his problem had been . . . The fact that we can identify his problem is proof that he has solved it; for we only know what the problem was by arguing back from the solution' (Collingwood, 1944, p. 50).

Collingwood's argument here can be strengthened a little. He usually spoke as though a philosopher addresses himself to one definite question at a time;[1] and this rather suggests that the difficulties of retrieving past philosophical problems would never have arisen if only philosophers had taken the trouble to state, wherever appropriate, just what their question was.

Now it is certainly a great help if a philosopher or a scientist tries to explain his problems to the reader (as did Descartes and Kepler, for example, and also Collingwood himself). A lot of unnecessary perplexity and incomprehension has been caused by the absurd convention that a 'scientific' writer should adopt an austerely impersonal style,[2] eschewing all intellectual autobiography. However, it is *not* normally the case that a philosopher faces one well-defined problem which he could easily spell out

---

[1] He almost invariably spoke of *the* question or *the* problem that a certain philosopher was dealing with on a certain occasion. Indeed, his 'logic of question and answer' obliged him to speak in this way: according to it, the answer to a foregoing question turns into a presupposition which causes the next question 'to arise'—at any stage in the process just one question is being asked. See Collingwood, 1940, pp. 23 f. For a criticism of his 'Logic of question and answer', see Donagan, 1962, pp. 58 f.

[2] 'Didactic dead-pan', as I called it in a broadcast sermon, reprinted (1964) in *Experiment*, David Edge (ed.), B.B.C. This booklet also contains a famous broadcast on the same theme by Sir Peter Medawar.

if only he would take the trouble to do so. I incline to Popper's view that problems are objective at least in the sense that they may be seriously misappraised and underestimated, though we may come to appreciate them better when we find what went wrong with our earlier attempts to deal with them (Popper, 1966, p. 26). This consideration deepens the problem of recovering past philosophical problems; it also tells against Collingwood's solution of it. (Indeed, his solution is, strictly speaking, incoherent: if one cannot understand a philosophical work unless one knows of what problem it was a solution, and if one could come to know this only *from* an understanding of that solution, then one could never get started.)

Let us rather think of a past philosopher as having worked, not at one clearly perceived problem at a time, but within a developing problem-situation. At any one time he is likely to be sharply aware of some of its salient features, dimly aware of others, and unaware of yet others (though some of these will, no doubt, subsequently obtrude themselves upon him). To gain a historical understanding of this philosopher's thought we shall have to try to reconstruct his problem-situation; and we can do this only by a process of successive approximations; in this process the man's own writings will, of course, provide invaluable evidence. But there will be much evidence outside his writings.

Nor is it necessary that the philosopher should have *solved* his problem for us to be able to reconstruct it. A historian may have certain *advantages* over the man whose problem-situation he is trying to reconstruct. His distance from the intellectual scene may lead him to miss things; on the other hand, with the advantage of hindsight, the historian may recognize difficulties in his hero's problem-situation which the latter did not see. And his ability to bring out discrepancies between the objective problem-situation and his hero's subjective appraisal of it may enable the historian to explain how his hero came to accept what was actually an inadequate solution which had to be discarded later.

With this last point we have made a start on the problem of explaining unsuccessful actions. A historian may explain both why someone did what he did, and why what he did was unsuccessful, by first hypothetically reconstructing the agent's own estimate of his problem-situation, and then pinpointing certain significant discrepancies between this and the objective problem-situation (as hypothetically reconstructed by the historian). I will return to this in §6.2.

## 3. A FIRST LOOK AT THE RATIONALITY PRINCIPLE

### 3.1 *Introduction*

It would seem that the first thing to do, in a discussion of the rationality principle, is to formulate it so that we know what we are talking about. The trouble is that we can hope to formulate it at all adequately only *after* we have seen what serious difficulties beset it. There is a pure or extreme version of the principle that identifies rational decision-making with *optimal* decision-making. (An optim*al* decision is one that could not be bettered, though it might be equalled. An optim*um* decision is one such that any alternative to it would be less good.) At first sight this identification seems plausible: if there are two outcomes open to a decision-maker, one good and the other even better, then it would hardly be rational for him to choose the first. However, we shall find in §4 that the idea of optimal decision-making involves such wildly optimistic assumptions about our capacity for self-knowledge, especially when risks and uncertainties are introduced, that it would not serve even as a normative principle. Moreover, it breaks down theoretically at some points. Thus we are *driven* to accept some less extreme principle for the purpose of historical explanation. But in the meanwhile we cannot yet tell in what ways the pure principle of optimization should be weakened and modified. Like an expectant parent I will content myself, at present, with finding a name for the unborn baby. I will call it the 'imperfect rationality principle'.

So at this stage our formulation of 'the' rationality principle can only be rough and provisional. An individual is placed in a certain objective problem-situation.[1] He has certain aims (wants, preferences) or perhaps a single aim, and he makes a factual appraisal (which may be a misappraisal) of his problem-situation. The rationality principle says that he will act in a way that is 'appropriate' to his aim(s) and situational appraisal.[2]

Various charges have been levelled against the rationality principle: for instance, that it is trivial,[3] that it is false,[4] and that it is redundant.[5] I will consider these charges in turn.

---

[1] For this concept, see p. 171 above.

[2] The appropriately vague term 'appropriate' is due to Popper. By incorporating the agent's aim(s) into his *situation*, Popper can formulate the principle even more shortly: 'Agents always act in a manner appropriate to the situation in which they find themselves.' (See Popper, 1967.)

[3] This charge was forcefully argued by H. F. Spinner in a discussion with Hans Albert and myself in Heidelberg in 1967.

[4] Popper (1967) caused a certain amount of consternation recently by holding that the rationality principle is (*a*) integral to nearly all social theories, and (*b*) false.

[5] George C. Homans (1967, pp. 38–43) has claimed that the 'rational choice model' is an (innocuous but) redundant duplication of a small part of behavioural psychology. This is presumably connected with his deterministic conviction that we have only 'the illusion of choice' (1967, p. 103).

## 3.2 *The charge of triviality*

Triviality may not be such a bad thing. A rule of inference may be trivial in the sense that it has no content, but this does not mean that we should throw it out. I do not claim that the rationality principle, as vaguely formulated above, has the perfect triviality of a logical rule of inference, but I do claim that it has the sort of 'triviality' that goes with indispensability.

I drive a foreign friend to London Airport. He has a seat reserved on a flight to his home city. He is looking forward to getting back to his wife, and he has a lecture to give in his own university tomorrow. I watch him check in, we say good-bye, and I drive off, never doubting that he will in fact leave on that flight.

Clearly, my prediction does not follow logically just from my knowledge of his situation. The flight may be cancelled, or he may have a heart-attack. So I add the assumption that there will be no such physical interruption. But this is not enough: he may be handed a telegram telling him that a sister in Edinburgh is dangerously ill. So I add the assumption that there will be no such change in his situational information. *Still* my prediction does not follow: it is *logically* possible that he will change his mind for no reason. If my prediction is to follow logically I must add the further assumption that he will act rationally. It looks trivial; perhaps it *is* trivial; but we need it if we are logically to derive predictions about actions from premises about the agents' situations.

The charge of triviality is often linked with the charge of unfalsifiability: we predict that a man will act in a certain way from our understanding of his aims and situational beliefs, together with the rationality principle; and when he fails so to act, *we cling to the rationality principle* and revise our hypotheses about his aims and beliefs. I agree that this is what we usually do, and I would add that it is what we *ought* to do. For a policy of clinging to the rationality principle in the face of falsified predictions can be justified even *from a falsificationist point of view.*

In any science a considerable body of premises is usually necessary before falsifiable predictions become logically derivable. It will usually not be too difficult to replace an existing premiss without diminishing the system's empirical falsifiability. However, there may also be premisses which it seems practically impossible to dispense with without seriously diminishing the system's falsifiability, or even rendering it untestable. Such a premiss may be called a *principle*, that is, a privileged component that is treated as *unfalsifiable* in the interest of the *falsifiability* of the whole system. (The conservation laws of physics have sometimes been treated as principles in this sense.) I am not saying that such principles should be sacrosanct: they *may* eventually get replaced by something better, but this is likely to be a major operation; and in the meantime we need them too badly to be free to discard them in response to a falsified prediction.

If the rationality principle is needed to derive action-predictions from situational descriptions, then it seems clearly to fall within this category.

### 3.3 *The charge of falsity*

I will not argue for the truth of our rationality principle, but will try to show that it is not hit by the main kinds of evidence that have frequently been cited against it.

(i) The rationality principle is not hit by the existence of perverse or wicked *aims* (or preferences). An old lady is coshed and her handbag containing a few shillings is snatched. Sickening; but is it evidence against the rationality principle? It would be if the person who did it finds coshing old ladies a distasteful business, has a horror of prison, and in any case knew that there were only a few shillings in her handbag. But it is not by the rationality principle that coshing should be condemned. This principle is morally neutral.

(ii) The rationality principle is not hit by the existence of perverse social *situations*—situations which the people involved deplore, but wherein each of them has no rational alternative but to act in ways that collectively cause the situation to persist (or even worsen). Thus a war may have reached a stage where both sides (*a*) curse the day on which it started and (*b*) believe that the only hope of ending it is to prosecute it more violently. In such cases, instead of lamenting the collective irrationality of nations, it might be more helpful to try to find a way out of the impasse by situational analysis.

(iii) The rationality principle is not hit by institutional 'irrationality'. (This is only a development of the previous point.) The actions of individual people are the proper domain of the principle, and one should be cautious about extending it to institutional acts. It may be safe to do so in cases where the decision (taken by, say, a board of directors) was made in the light of common information and a common set of institutional values. But very often, of course, institutional 'decisions' are the result of various pressures and interests. Ideally, such 'decisions' should be explained by applying the rationality principle to the moves of the various people involved, and then tracing how these led to the institutional 'decision' (which may be a poor compromise which pleases nobody).

(iv) The rationality principle is not hit by the fact that most human conduct is not of the 'coldly calculating' variety. Every experienced motorist has known it happen that his foot was pressing the brake-pedal *before* he asked himself, 'What is that cyclist going to do?' One can act appropriately to one's situation without explicitly formulating one's aims or explicitly describing one's situation.

(v) The rationality principle is not hit by the existence of perverse *beliefs*. The principle takes not only an agent's aims but also his situational appraisal as given, even although this last may be badly wrong. (I have little doubt that what is typically condemned as irrational behaviour is usually behaviour that is in fact more or less rational relative to a perverse situational appraisal. A paranoic may repel friendly advances because he believes that people are feigning friendliness in order to trick him.) It follows that the rationality principle is not hit by ignorance, stupidity, or a low level of problem-solving ability; a man's appraisal of his problem-situation includes his appraisal of the various possibilities of dealing with it. If a landlubber, out for the day in a leaky boat, does not realize what the bailer is for, and decides as the water rises that his only chance is to swim for it (although he is a poor swimmer and wind and tide are against him), then the coroner would be wrong to certify him as insane.

A rationality principle might be extended to the ways in which a person modifies his beliefs. But then everything would turn on the *end* (which need not be consciously entertained) that the changes are to serve. And the end may be, not the acquisition of truth, but the reduction of mental discomfort. Members of a religious sect sit waiting for the great flood which, according to their leader and prophet, will at midnight destroy their town, believers (only) being rescued by flying saucer. The flood does not come. Result: enhanced faith in their leader.[1]

You smile. But consider their situation. To accept the refutation of the prophecy at its face value would have led to the disintegration of the sect, which would have been painful, given their emotional (and financial) investment in it. The alternative was for the leader to explain it away and even to elevate it into a major new confirmation of the sect's tenets—and for the followers to swallow this enthusiastically; which is what happened.

According to the theory of cognitive dissonance, due to Festinger and his associates (Festinger, 1957; Festinger and others, 1964), a person who has made a difficult decision, and one that he cannot easily reverse, will tend afterwards to strengthen what I call his 'decision-scheme'[2] in such a way that his decision will come to seem better justified than it seemed to him

---

[1] I am relying on the summary in Roger Brown (1965), pp. 590–3, of the original account in Festinger, Rieken & Schacter (1956). Miss Margaret Boden drew my attention to this odd, yet in some ways very human, story.

[2] This term is explained in §6.1 below.

when he made it. Again, this seems rather rational—if the aim is peace of mind. If, before he finally decided to marry Ann (who is pretty but not wealthy), it had been touch and go whether Jack would not marry Jane (who is wealthy but not pretty) instead, then it is quite understandable if Jack subsequently becomes scornful of the idea that money can buy happiness.

### 3.4 The charge of redundancy

The redundancy of any rationality principle within the social sciences is implied at one extreme by *physicalism* (according to which social and psychological processes are reducible to physical processes), and at the other by *social organicism* (according to which people are steered, in their social behaviour, not by *their* aims, etc., but by the needs of the social wholes to which they belong). I shall not argue with either of these (to my mind, pretty outlandish) positions here.[1] I want, rather, to consider a view held by some people *within* the individualist camp (namely, certain behavioural psychologists) which goes something like this. Social science should consist of a system of empirically testable generalizations about human behaviour. The behaviour recorded in such generalizations may be rational or it may be irrational. The social scientist ought not to prejudge this question by laying down a rationality principle *a priori*. And if he does find that behaviour in a certain sphere is generally rather rational, there is no need for him to introduce a rationality principle *a posteriori*. His behavioural generalizations by themselves will suffice. If one of these states that a certain stimulus, say $S_1$, will lead to a certain response, say $R_1$, then *it does not matter*, for scientific purposes, whether $R_1$ is linked in some rational way to $S_1$: it suffices to know that it *is* linked. From the point of view of *S–R* psychology (as it is usually called), an appeal to a rationality principle smacks too much of the unsavoury idea that people have minds. We should rather adopt a 'black box' approach to what goes on in people's heads,[2] and content ourselves with correlating observable stimuli and observable responses.

The standard paradigm of a stimulus–response relation is an experimental set-up in which a rat can secure a pellet of food by depressing a bar with, say, his left forepaw. If the rat learns the trick, the set-up is called the 'stimulus' and the rat's behaviour the 'response'. *S–R* psychology is often

---

[1] I have offered an argument, derived from Popper's conception of a scientific reduction, for the impossibility of a *scientific* (as opposed to a merely philosophical) reduction of psychology to physics, in my contribution to the forthcoming volume on Popper's philosophy in *The Library of Living Philosophers* edited by P. A. Schilpp.

[2] For a critical discussion of the idea of 'black box' theories, see M. Bunge, 'Phenomenological Theories' in Bunge, 1964, Chap. 16. Bunge instances *S–R* psychology (p. 234).

accused, understandably, of trying to reduce men to the level of rats.[1] I regard it as *unfair to rats*. The rat's experimental set-up should not be called a 'stimulus': *it* does not stimulate the rat to depress the bar. The stimulus to action *is within the rat*: he wants food. What the set-up constitutes is rather his *problem-situation*: how to extract food from a seemingly foodless cage. A tough problem; three cheers for the rat who solves it! No doubt he engaged in some random trial-and-error movements before hitting on the solution—rather as an anagram-solver engages in random letter juggling. But the decisive point is that the rat, like the anagram-solver, *retains* the solution that clicks.

In fact, the *S–R* psychologist can only *pretend* to treat our heads as black boxes, because a psychological 'stimulus' cannot be understood in a purely objective way (see Hayek, 1952, Chap. 5). Alf and Bert are at a crowded, excited meeting. Their external circumstances are similar: they are being jostled, the room is noisy and hot. But Alf is experiencing incipient claustrophobia, whereas Bert is a pickpocket. Their *problem-situations* are very dissimilar: to Alf, an 'Exit' sign would be a 'stimulus'; to Bert, a wallet peeping out of a hip-pocket.

Someone who uses his understanding of what goes on in people's heads can easily construct counter-examples to *S–R* generalizations. (The rationality principle revenges itself on those who would kick it out, by showing the way to refutations.) Take a principle that is fundamental to *S–R* psychology, the 'reinforcement' principle. This says that the more a certain response—it may originally have been a random response—to a given stimulus is rewarded, the more probably will this response be made on future occurrences of that stimulus; and vice versa. Obviously, there is something in this; if we hit upon a successful solution to a practical problem, we are likely, other things being equal, to use it again if the problem recurs. But with the help of the rationality principle we can easily think of counter-examples. For instance, many people in Britain have been filling in football coupons for years, often at considerable cost, without any return so far. Why have they not desisted long ago, as the 'reinforcement' principle suggests? Because they know that failure to win so far does *not* reduce the probability of their winning in the future. (Some of them may indulge in the 'gambler's fallacy' and believe that their past failure actually *increases* the probability of their winning in the future. These people have an additional reason—admittedly an unsound one—for defying the 'reinforcement' principle.)

No doubt someone who wins a modest sum on the pools is likely to be encouraged thereby, as the 'reinforcement' principle requires. On the

[1] See the opening chapter of Arthur Koestler's admirable *The Ghost in the Machine*, 1967.

other hand, someone who wins a *huge* sum may quite rationally give up football pools altogether. Why should he continue his laborious coupon-filling when he could be away on his yacht?

### 3.5 *The rationality principle and social organicity*

The rationality principle is tied up with an individualist methodology which I shall not defend here, having spilt a good deal of ink over that issue on previous occasions. But it may be instructive to consider briefly a few points of similarity and difference between our rationalist–individualist approach and what I regard as the best of the collectivist methodologies, namely functionalism.

An individualist, although he denies that a human society *is* an organism in any serious sense, can very well allow that social institutions may show something of an *organic-like* character, developing in unplanned ways and coming to serve social functions for which they were not intended. The organic-like character of such institutions as money and the common law was stressed by Carl Menger, one of the main spokesmen for this approach in the nineteenth century;[1] while in our own day one of the staunchest of individualists, F. A. Hayek, has also been a powerful advocate of the idea that 'the independent action of many men can produce coherent wholes, persistent structures of relationships which serve important human purposes without having been designed for that end' (Hayek, 1952, p. 80).

Functionalism likewise discounts the idea that social institutions are, typically, deliberately planned (even those that are, are likely to develop in unforeseen ways); and it likewise emphasizes the large disparities that may exist between the *intentions* of people engaged in a certain social practice, and the latter's social *effects*.[2] But here the two approaches begin to diverge. The rationalist–individualist holds that a practice exists because people have their individual reasons for engaging in it. The functionalist holds that it exists because of its socially beneficial consequences. Yet, as its critics have often pointed out,[3] functionalism leaves us in the dark about the mechanism by which a society is supposed to nurture and preserve practices that are beneficial to itself, and to suppress ones that are dysfunctional; for there is no analogue in human societies to the kind of evolutionary process that justifies functionalism in biology. And since the functionalist holds that social practices often serve *unobvious and*

---

[1] See Book Three, 'The Organic Understanding of Social Phenomena', of Menger, 1963.
[2] Jarvie (1964) has drawn attention to this affinity; see his *The Revolution in Anthropology*, Chap. 6, §3.
[3] See, for example, Harsanyi, 1968, p. 307.

*unsuspected* functions, he cannot suppose that they are sifted by some sort of deliberate political process.

For an individualist to find that a certain social institution or process has an organic-like character is always a *discovery*: he finds that he can show *how* it is that the unintended effects of the activities of individuals who are pursuing (more or less rationally) their own aims, conspire to, say, produce a system with homeostatic tendencies (see Menger, 1963, p. 133). The full-blooded functionalist, by contrast, dispenses with a rationality-principle and *postulates* a kind of social organicity. The individualist, on the other hand, is (or should be) as prepared for the discovery of unintended consequences that are socially injurious[1] as for ones that are socially beneficial. Social optimism is not built into his methodological approach.

## 4. OPTIMIZATION

### 4.1 *Introduction*

On its strongest interpretation, the rationality principle becomes an optimality principle: to decide rationally is to make a decision that could not be bettered, given one's present situational information (which may, however, be more or less incomplete or incorrect).

It is agreed on all sides that people's decisions are, in fact, very often sub-optimal. But it is widely held that theories of optimal decision-making may at least play a *normative* role.[2] However, my thesis will be that they break down even as normative theories; so we are *driven* to adopt some weaker principle. But a critical consideration of the extreme principle will help us to decide how its extremism might be moderated to yield a principle that it is realistic, non-vacuous, and heuristically fruitful.[3]

In decision theory decisions are usually classified under three main heads according to whether they are made: (1) 'under conditions of certainty', (2) 'under conditions of risk', or (3) 'under conditions of uncertainty'. The difficulties that beset the idea of optimal decision-making increase as we proceed from (1) to (3). I will begin with (1); even here the difficulties are by no means negligible.

Under 'conditions of certainty' the decision-maker proceeds as if he knew (*a*) all the alternative decisions open to him in the given choice-

---

[1] See Keynes, 1936, pp. 358 f., for his famous survey of anticipations of the thesis that thrift can have bad economic consequences.

[2] See, for example, Savage, 1954, p. 97, and Popper, 1957*a*, p. 141.

[3] It was not without trepidation that I ventured into this field, with its vast literature and technical intricacies. I am particularly grateful to my colleague, Mr Lucien Foldes, for saving me from some bad mistakes. I should also like to thank Dr L. D. Bernard for helpful comments.

situation, and (b) the outcome for him that would infallibly follow from each of these possible decisions. (An 'outcome' should be understood as the net return to him up to a certain more or less arbitrary cut-off point. There is no end-means dichotomy. If you are wondering whether to knock up an omelette or slip out to the local restaurant, cooking and washing up will be a—presumably negative—part of one outcome, as is paying the bill part of the other outcome.)

Conditions of certainty have one big advantage over conditions of risk or uncertainty: for optimal decision-making under conditions of certainty it is enough if the decision-maker can order the possible outcomes available to him in a merely linear way. (Two outcomes are ordered linearly by him if he either prefers the first to the second, or is indifferent between them, or prefers the second to the first.)

But even the requirement of a merely linear ordering (or ordinal comparison) raises difficulties. For the ordering to be consistent not only the relation of preference but also the relation of *indifference* must be *transitive*: if someone is indifferent between $A$ and $B$ and indifferent between $B$ and $C$, then he should, furthermore, be indifferent between $A$ and $C$.

But suppose that he values them according to some physical property such as their size, and that $A$ is in fact slightly larger than $B$ and $B$ than $C$, although these differences are, singly, too small for him to notice; and suppose that he can just notice the combined difference between $A$ and $C$. Then should he not prefer $A$ to $C$ and yet be indifferent between $A$ and $B$ and between $B$ and $C$?

According to Ward Edwards, this difficulty is spurious because the notions of preference and indifference should be understood in a *statistical* sense: the statement that someone is indifferent between $A$ and $B$ should be taken to mean that, in a long series of choices between $A$ and $B$, he would choose the one as often as the other; moreover, if $A$ is imperceptibly better than $B$, then he will (according to Edwards) choose $A$ slightly more often than $B$, and so in a statistical sense will not be perfectly indifferent between them 'even though $A$ and $B$ are less than one j.n.d. [just noticeable difference] apart in utility' (Edwards & Tversky, 1967, p. 24).

I do not know how much warrant there is for this assumption that the unnoticeable difference between $A$ and $B$ will have a sort of subliminal influence on the decision-maker which will show itself statistically in his choices. It looks rather fishy to me. After all, it would be perfectly in order for him *always to choose $A$*, say (rather as some people always call 'Heads'), *just because $A$* seems to him to be just as good as $B$.

Another familiar difficulty that arises already under conditions of certainty is sometimes called the problem of multi-dimensionality. Suppose that I evaluate $A$ and $B$ along two dimensions, or in two respects, and that $A$ is better than $B$ in one respect while $B$ is better than $A$ in the other. Can I rationally decide between them if I can make only ordinal comparisons along these two dimensions?

A function of two (or more) variables becomes indeterminate when one (or more) of the variables is only ordinally measurable; for if one of them alters in a way that would, by itself, decrease the value of the function while the other alters in a way that would increase it, then we cannot calculate the net effect unless we know by *how much* they respectively altered. So the answer seems to be that in this case I cannot *rationally* decide between $A$ and $B$: I must just *decide*.

This problem becomes more serious under conditions of risk, to which I now turn.

### 4.2 Risk

Under conditions of risk the decision-maker is again assumed to know all the alternative decisions that are open to him in the given choice-situation. And while he does not know what outcome a decision would have, he does know, in the case of each possible decision, each of the possible outcomes it might have; and he is able to assign a non-arbitrary numerical probability to every possible outcome of every possible decision. (I shall assume throughout this sub-section that the probabilities are calculated *correctly*.) Casinos and lotteries may provide conditions of risk in this sense, but horse-racing does not.

We seem straightaway to need some quantitative and non-arbitrary measure of utility; for the attractiveness of a possible decision is now some function (*what* sort of function we will consider later) of two variables, the respective probabilities and utilities of its possible outcomes.

Admittedly, there is a way of circumventing this need.[1] A possible risky decision, with its set of attendant possible outcomes with their respective probabilities, may be regarded as a lottery; and a choice-situation in which the decision-maker must make one or other of, say, six such decisions may be regarded as a situation in which there is a choice between six lotteries. Now instead of trying to *calculate* the relative values of the lotteries in terms of their component prizes and associated probabilities, the decision-maker might treat each lottery as *one* unit, *decide* (somehow) how to rank them, and then proceed as under conditions of certainty.

But this would result in a seriously weakened theory of optimal decision-making. For it allows the decision-maker to rank lotteries in ways that could be shown to be sub-optimal *if* we were allowed to analyse the

[1] As Mr Foldes pointed out to me.

lotteries instead of treating them as unanalysed units. If lotteries are treated as unanalysed wholes we could no more criticize someone for preferring a lottery offering a 1% chance of getting £1,000 and a 99% chance of getting nothing to a lottery offering a 2% chance of getting £2$^{10}$ and a 98% chance of getting nothing, than we could criticize him for preferring a small, dried-up lemon to a big, juicy grapefruit. But if we are allowed to analyse the lotteries we can point out to him that the second lottery offers him twice as good a chance of getting a *larger* prize (namely, £1,024). It is easy to construct more sophisticated examples in which the sub-optimality is less obvious.[1]

It would seem then, that a theory of optimal decision-making under conditions of risk must, if it is to avoid serious inadequacy, treat the utility of a lottery as a function of the probability-cum-utility components of the lottery; and this in turn seems to call for some sort of cardinal utility-measure.

*The metric of von Neumann and Morgenstern.* There is a famous method for measuring utility-values on an interval-scale,[2] which is due to von Neumann and Morgenstern (1953: see §3, and also the Appendix).

---

[1] A tedious example. Three lotteries, $L_1$, $L_2$, and $L_3$, offer respectively: (i) a quarter chance of prize $A$ and a three-quarters chance of nothing; (ii) a half chance of prize $B$ and a half chance of nothing; and (iii) a quarter chance of prize $C$ and a three-quarters chance of nothing. They may be represented thus:

$$L_1: \tfrac{1}{4}A$$
$$L_2: \tfrac{1}{2}B$$
$$L_3: \tfrac{1}{4}C$$

Someone says that he prefers $L_1$ to $L_2$ and is indifferent between $L_2$ and $L_3$. We may represent this by: $\qquad u(L_1) > u(L_2) = u(L_3)$.

We now present him with two further lotteries, $L_4$ and $L_5$, which offer respectively: (iv) a quarter chance of $B$ and a quarter chance of $C$ and a half chance of nothing; and (v) an eighth chance of $C$ and a quarter chance of $A$ and a five-eighths chance of nothing; or

$$L_4: \tfrac{1}{4}B, \ \tfrac{1}{4}C$$
$$L_5: \tfrac{1}{8}C, \ \tfrac{1}{4}A.$$

He says that he prefers $L_4$ to $L_5$ (perhaps because $L_4$ offers him a better chance of getting a prize). But if we were allowed to analyse the lotteries we could criticize this as inconsistent with his previous statement of preference. In view of his '$u(L_2) = u(L_3)$' he should be indifferent between the first disjunct of $L_4$ and the first disjunct of $L_5$ (namely, $\tfrac{1}{4}B$ and $\tfrac{1}{8}C$ respectively); and in view of his '$u(L_1) > u(L_3)$' he should prefer the second disjunct of $L_5$ (namely, $\tfrac{1}{4}A$) to the second disjunct of $L_4$ (namely, $\tfrac{1}{4}C$). Thus he should prefer $L_5$ to $L_4$.

[2] Temperature is measured on an *interval*-scale, whereas length, weight, and money are measured on a *ratio*-scale. On an interval-scale the ratios of the *intervals* between pairs of values is constant, wherever the zero is located and whatever unit is used. Thus on a Centigrade scale the ratio of the interval between 0° and 15° to the interval between 15° and 20° is 3 to 1; on a Fahrenheit scale the corresponding positions are 32°, 59° and 68°, and the ratio of the two intervals is again 3 to 1.

Their method is ingenious in that it derives numerical values from merely ordinal judgments of preference and indifference. Numerical values are got by finding the probability-value at which a subject becomes indifferent between on the one hand the certainty of a middling outcome, and on the other hand a lottery offering either an inferior outcome or a superior outcome. Suppose that someone prefers $C$ to $B$ and $B$ to $A$, or:

(1) $$u(A) < u(B) < u(C).$$

We now devise a sort of variable lottery, $[p.A, (1-p)C]$, where $p$ can take values between 0 and 1, '$p.A$' stands for the chance of getting $A$, and '$(1-p)C$' stands for the complementary chance of getting $C$. That there should be a value of $p$ such that our subject will be indifferent between $B$ and the $A$-or-$C$ lottery, or such that:

$$u(B) = u[p.A, (1-p)C],$$

is justified by the following consideration. We know that if we put $p = 0$, he should prefer the lottery to $B$, since the lottery now offers the certainty of $C$, and he prefers $C$ to $B$. As we increase the value of $p$, so the value to him of the lottery will decrease until, if we put $p = 1$, he should prefer $B$ to the lottery, since the lottery now offers the certainty of $A$, and he prefers $B$ to $A$. Hence at some intermediate point there should come a value of $p$ ($0 < p < 1$) at which he is indifferent between $B$ and the $A$-or-$C$ lottery.

Suppose that, as a first step, we put $p$ at $\frac{1}{2}$, and our subject reports that, at this value, he still prefers the lottery to $B$, or:

(2) $$u(B) < u(\tfrac{1}{2}A, \tfrac{1}{2}C).$$

This tells us that his preference for $C$ over $B$ *exceeds* his preference for $B$ over $A$: a half chance of winning the superior prize $C$ more than compensates for the half chance of getting the inferior prize $A$. We can now say that the *interval* between $u(B)$ and $u(C)$ is greater than that between $u(A)$ and $u(B)$.

Suppose, next, that our subject becomes indifferent between $B$ and the lottery when we raise $p$ to $\frac{3}{4}$, or:

(3) $$u(B) = u(\tfrac{3}{4}A, \tfrac{1}{4}C).$$

This is taken by von Neumann and Morgenstern as telling us by *how much* his preference for $C$ over $B$ exceeds his preference for $B$ over $A$: the interval between $u(B)$ and $u(C)$ is *three times* greater than that between $u(A)$ and $u(B)$. If we arbitrarily put $u(A) = 0$ and $u(C) = 1$, then we would be obliged by (3) to put $u(B) = \frac{1}{4}$. (It should be stressed that the numerical values 0 and 1 are here quite arbitrary. These numerical values could be transformed by any linear transformation, for that would leave the ratio of the *intervals* unchanged. For instance, we could equally have put $u(A) = 100$, $u(C) = 200$, and $u(B) = 125$.)

This method of eliciting numerical utilities involves a *gamble*. And this raises the question whether the idiosyncrasies of different people's *gambling temperaments* might not distort or invalidate applications of the method. This question was touched on by von Neumann and Morgenstern (1953, §3.7.1), but in a rather offhand way. They say that their postulates 'do not attempt to avoid' the [awkward] possibility that there may 'exist in an individual a (positive or negative) utility of the mere act of "taking a chance" of gambling . . .' But the reader who takes this to mean that they do not *shirk* the difficulty but meet it is in for a disappointment: he will find that they 'do not attempt to avoid it' in the sense that they steamroller right over it. For one of their postulates—namely (3 : 1 : b)—requires numerical utilities to satisfy the calculus of mathematical expectations; and this latter is incompatible with the idea of idiosyncratic gambling temperaments, since it prescribes that the value of a-specified-probability-of-winning-$X$ shall equal the-probability-of-winning-$X$ multiplied by the-value-of-$X$, or:

$$u(p.X) = p.u(X).$$

Another way of putting this is to say that the utility of a given chance of obtaining a certain outcome is its *expected utility*, that is, the utility of the outcome multiplied by its probability. According to this doctrine, everyone should evaluate a numerically calculable risk in the same way, no matter whether or not he is cautious and averse to risk-taking, and whether or not the risk is to be taken once only or repeatedly.

Von Neumann and Morgenstern's postulate (3:1:b) in effect says this: given an *either-or* lottery of the form $[p.A, (1-p)C]$, then the utility of the lottery equals the sum of the expected utilities of its two disjuncts, or

$$u[p.A, (1-p)C] = p.u(A)+(1-p).u(C).$$

Clearly, this excludes the possibility that one's evaluation of a lottery may be affected by one's gambling temperament: 'We have practically defined numerical utility as being that thing for which the calculus of mathematical expectations is legitimate', they wrote; and they added: 'concepts like a "specific utility of gambling" cannot be formulated free of contradiction on this level' (1953, p. 28).

Since that was written it has become almost routine, among utility-theorists, first to make a stiff little bow towards the awkward fact that people have variable gambling attitudes, and then to dismiss it and reaffirm their loyalty to the doctrine of expected utility.[1]

---

[1] According to Keynes, this doctrine originated with Leibniz (Keynes, 1921, p. 311*n*).

That von Neumann and Morgenstern's method involves the doctrine of expected utility can be indicated without reference to their postulate (3:1:b) in the following way. In an $A$-or-$C$ lottery let $A$ be a null prize: to 'win' $A$ is to win nothing. And let us assign zero utility to winning nothing. Suppose that our subject again reports

$$u(B) = u(\tfrac{3}{4}A, \tfrac{1}{4}C),$$

from which we again extract the numerical utilities,

$$u(A) = 0,\ u(B) = \tfrac{1}{4},\ u(C) = 1.$$

Since $A$ is now a null prize, this means that he is indifferent between the certainty of getting $B$ and a quarter chance of getting $C$, or

$$u(B) = u(\tfrac{1}{4}C).$$

But since $\qquad\qquad u(B) = \tfrac{1}{4} = \tfrac{1}{4}u(C),$

it follows that $\qquad\qquad u(\tfrac{1}{4}C) = \tfrac{1}{4}u(C),$

as the doctrine of expected utility requires.

*If* the doctrine of expected utility were satisfactory, optimal decision-making under 'conditions of risk' would be no more problematic than under 'conditions of certainty'. The von Neumann and Morgenstern method for ascertaining numerical utilities would be available; numerical utilities could be multiplied by probabilities to yield expected utilities; and the decision-maker could proceed as if under conditions of certainty, but with *expected* utilities in place of utilities.

But this doctrine is far from satisfactory. It has had a curious career. From the outset, more or less serious objections were raised against it; some of these were answered; but the answers tended to aggravate other objections. (For instance, the objection that people are willing to insure their houses at statistically 'unfair' rates that allow insurance companies to make a profit was answered by invoking the 'law' of diminishing marginal utility; but this answer aggravated the difficulty posed by the fact that some people are prepared to buy 'unfair' national lottery tickets.) Yet the doctrine seems to be more firmly entrenched today than ever. For it is widely regarded as *indispensable* for a theory of rational decision-making 'under conditions of risk'; it has a built-in defence-mechanism against counter-evidence; and seemingly divergent behaviour can in any case be dismissed as irrational.

For my part, I believe the doctrine to be false *qua* psychological hypothesis and unreasonable *qua* rational prescription; and I feel no uneasiness about pressing home objections to it since I regard it as replaceable by Shackle's idea of a 'gambler-indifference map' (about which more later).

Everyone admits that different people may evaluate a given numerically calculable risk very differently. However, the doctrine of expected utility can always be defended against such variations by postulating appropriate variations in the marginal utility of money (or of whatever is involved in the gamble). If someone has such a horror of risks that he would become willing to venture £5 on a half chance of winning a prize only if the prize were raised to £100 or more, a defender of expected utility would no doubt reply that for this person the marginal utility of money diminishes so rapidly that the utility of an additional £100 is only twice that of £5. Thus Mosteller and Nogee, in the course of describing an experimental attempt to ascertain people's marginal utilities, mention the case of a 'conservative' person who consistently requires 'more than a mathematically fair offer before taking a risk'; they take this to mean, *not* that he has a cautious gambling temperament, but that for him the utility of money diminishes significantly; yet they were considering bets involving *only a few cents* (Mosteller & Nogee, 1967, p. 129).

My contention will be that, although a defender of the doctrine of expected utility may cope well enough with typical quirks of people's gambling temperaments *one at a time,* he cannot defend it against them *collectively* without distorting the classical idea of diminishing marginal utility into a ridiculous caricature of its old self.

A gamble 'under conditions of risk' typically involves paying a *stake* in return for a specified *probability* of winning a specified *prize*. I will make a few remarks about the psychological significance of each of these variables. My remarks will be unsystematic, and negative in intent: my aim is to rebut the doctrine of expected utility rather than to establish anything positive about the shape of people's gambler-indifference maps.

Consider, first, the question of the size of the stake. It is a commonplace that most people tend (other things being equal) to become increasingly reluctant to wager the larger is the proportion of their total fortune that would be at stake.[1] The defender of expected utility explains this in terms of diminishing marginal utility. But this explanation has an awkward implication. The theory of diminishing marginal utility implies that it should in principle be possible for a man to name the sum that would be worth twice as much to him as his present fortune. Suppose that someone worth £10,000 names £100,000 as such a sum. We now invite him—this piece of research would have to be financed by one of the bigger Foundations—to stake his whole fortune on a half chance of winning £101,000.

---

[1] In Kipling's poem *If—*, it is not your *fortune* that you are expected to risk 'on one turn of pitch-and-toss' but only 'all your *winnings*' that you lose 'and never breathe a word about your loss'.

According to the doctrine of expected utility it would be irrational for him to decline this wager. But most of us would sympathize with him if he replied that he would be crazy to accept a half chance of total destitution. According to the doctrine of expected utility (in conjunction with the 'law' of diminishing marginal utility), for a given probability—say, $0 \cdot 5$—there corresponds to each stake that a man could pay out, up to and including his whole fortune, a prize large enough to entice him to venture that stake for that probability of winning that prize. The fact seems rather to be that most people have some (ill-defined) cut-off point: when the stake has increased beyond a certain magnitude, relative to one's total fortune, one becomes essentially unwilling to wager however great the expected utility of the uncertain prize. Thus our imaginary subject might be unwilling, in the case of a once-only wager, to venture even half his fortune on a half chance of even £1m.

Consider now the second variable mentioned above, namely probability. Suppose we have a series of lottery tickets all offering the same prize but with diminishing probability. Thus the first ticket might give one a half chance of winning £100, the second a quarter chance of winning the same amount, the third an eighth chance, and so on. Keynes raised the question (although he did not put it in quite this form) whether the value of the tickets may not diminish more than proportionately to the reduction in the probability, contrary to the idea of mathematical expectation (Keynes, 1921, p. 313). It seems to me that there is no general answer to this. Some people (myself included) are little interested in betting unless they believe themselves to have a better than even chance of winning: we have a rather high cut-off point such that, if the probability in a once-only wager lies below it, we are unlikely to be interested in the wager. On the other hand, the existence of sweepstakes and national lotteries shows that there are people willing to venture a small stake on a miniscule probability of winning a large prize, even though the gamble is far from being mathematically fair, and still further from being fair in terms of expected utility. In short, different people's evaluations of wagers involving low probabilities seriously diverge in opposite directions from what they should be according to the doctrine of expected utility.

As to the magnitude of the prize: few people are much interested in a once-only wager for a very trifling sum. Maybe I am too affluent, but I would not stake a penny on a 60% chance of winning tuppence; and I am ready to swear that this is *not* because the utility to me of an extra penny would be significantly diminished.

Collectively, these various cut-off points—or, if that sounds too sharp and definite, these rather vague, flexible and inter-dependent maxima and minima—tend to confine a person's gambling activities (if any) within a fairly narrow compass. For instance, a man may enjoy putting an occasional fiver on a horse who would not think of putting five shillings or five hundred pounds on it; and so on.

*Gambler-indifference maps.* I will now introduce Shackle's idea of a gambler-indifference map, somewhat modifying it as I go along (partly because Shackle did not operate with probabilities and partly because I can no longer accept his idea of 'standardized' focus-gains and losses).[1] I assume that stakes and prizes are either in monetary terms or can be represented by monetary equivalents. Stakes are measured along an $x$-axis and prizes along a $y$-axis. The map consists of a family of gambler-indifference curves each associated with a given probability. To draw such a curve we mark off points at short intervals along the $x$-axis, select some probability-value, and then seek to discover, for each stake, the prize such that our subject would be indifferent between gambling and not gambling with that stake for that probability of winning that prize. If we call the given probability $r$ we can say that a point on our gambler-indifference curve represents a stake $X$ and a prize $Y$ such that our subject is indifferent between $X$ and $r . Y$, or

$$u(X) = u(r . Y).$$

By repeating this process for a range of probability-values between 0 and 1 we shall generate a gambler-indifference map representing our subject's gambling temperament.

Let us now, for each probability-value for which we have a gambler-indifference curve, draw the straight line passing through the origin that represents numerically 'fair' combinations of stake and prize. (Thus for $r = \frac{1}{2}$, the numerically 'fair' prize is always twice the stake.)

Now according to the doctrine of expected utility, in conjunction with the classical theory of diminishing marginal utility, we should expect each gambler-indifference curve to start at the origin and to draw gently away from and above the corresponding numerically 'fair' straight line as the stake increases, but without becoming quite vertical since the marginal utility of money approaches zero asymptotically.

However, according to our own cursory review of typical gambling idiosyncrasies, we should *not* expect this. If a person has low maxima for the tolerable magnitude of the stake, his gambler-indifference curves will become vertical quite soon. If he is willing to buy a ticket in an unfair national lottery, then some of his gambler-indifference curves (associated

---

[1] See Shackle, 1952, pp. 29 f. His 'standardized focus-gain/loss' will be criticized below, pp. 194–5.

with very small probabilities) will dip *below*, or to the right of, the corresponding straight line. If, on the other hand, he is not interested in wagers involving low probabilities, then the straight lines for low probabilities will in his case have *no* associated gambler-indifference curves. And if he is not interested in gambling for trifling sums, his gambler-indifference curves will meet the *y*-axis *above* the origin.

In short, a man's gambler-indifference map is likely to be, if not exactly kinky in the mathematical sense, at any rate perversely distorted compared with what it should be according to the prevalent doctrine. I will introduce the term *gambler-effect* to refer to divergences between someone's actual evaluations of numerically calculable risks, as represented by gambler-indifference curves, and the curves as they should be according to the doctrines of expected utility and diminishing marginal utility.

Of course, the doctrine of expected utility *can* be preserved against gambler-effect very easily by postulating appropriately peculiar marginal utility curves. Indeed, this is just what utility-theorists have got into the habit of doing. Mosteller and Nogee, who follow von Neumann and Morgenstern in raising the question of what I call gambler-effect only to drop it,[1] approvingly report that Friedman and Savage, in a famous paper (Friedman & Savage, 1948, pp. 279–304),

also admit (indeed, require) the possibility of *increasing*, as well as diminishing, marginal utility, allowing the introduction of *inflection points* in utility curves ... Friedman and Savage argue that, *owing to the peculiar shape of their utility curves*, individuals [who simultaneously buy insurance and lottery tickets] *may be maximising their expected utilities* (Mosteller & Nogee, 1967, p. 125, my italics).

A *classical* marginal utility curve looked elegant and seemed plausible; a Friedman–Savage marginal utility curve, distorted *ad hoc* in response to each discovery of an idiosyncrasy in a man's gambling temperament, would —if it were drawn and held out for inspection—look ridiculous. Yet it is easy to see why utility-theorists have felt constrained to subject the classical idea to this maltreatment. They supposed that the theory of decision-making under conditions of risk requires at least an interval-scale for utilities, since the attractiveness of a risky decision is a function of the probability and utility values of its possible outcomes; and they further supposed that for the construction of such an interval-scale the doctrine of expected utility is indispensable. Thus this doctrine had to be upheld at any cost. But they were wrong. For a theory of decision-making under conditions of risk can be constructed with the help of a *monetary* scale—a

---

[1] 'Nevertheless, this criticism has a tendency to be overworked' (Mosteller & Nogee, 1967, p. 167).

ratio-scale rather than an interval-scale—and gambler-indifference maps. Such a theory at once replaces elusive utility-values with hard cash-values *and* releases us from the pretence that there are no significant variations among different people's gambling temperaments.

*Expected utility and the law of large numbers.* So much concerning the falsity of the doctrine of expected utility considered as a general psychological hypothesis. But it may alternately be considered as a normative doctrine that prescribes how rational people *should* evaluate numerically calculable risks. The evaluation of a once-only risk in terms of expected utility has often been justified by appealing to what *would* happen if the risk were accepted many times over. Thus Jevons once wrote: 'When it is as likely as not that I shall receive £100, the chance is worth but £50, because if, for a great many times in succession, I purchase the chance at this rate I shall almost certainly neither lose nor gain.'[1]

I agree with Shackle's forcible protests against such inferences from the statistically fair rate to the value of a single chance.[2] For one thing, there are some kinds of chance (as in playing Russian roulette, for instance) that may not allow of being taken again and again by the same person. But let us confine ourselves to chances that *are* repeatable. Why should someone who is contemplating a *single* purchase of a half chance of winning £100 and an equal chance of *losing his entire stake*—why should he evaluate *that* risk at a rate that would make it statistically probable that he would (in Jevons' words) 'almost certainly *neither lose nor gain*' if he purchased the same chance 'a great many times in succession'? Why should he evaluate a single *win-or-lose* gamble at the rate that would *minimize the risk* of loss or gain in another situation? Utility-theory does not tell someone contemplating buying a bottle of champagne that he should first imagine himself buying bottle after bottle of the stuff and then proceed from this to the calculation of the utility to him of one bottle. It leaves him free to evaluate it as he will. I hold that utility-theory should likewise leave us free to evaluate risks as we will. Some people have a horror of risks rather as some people have a horror of alcohol. If utility-theory has no business to lay down that a

---

[1] Jevons, 1957, p. 36. According to Keynes (1921, pp. 313–14), Condorcet had already defended the idea of mathematical expectation by appealing to Bernoulli's law of large numbers.

[2] See Shackle, 1955, Ch. 1, for example. But I disagree with his inference that numerical probabilities are therefore inapplicable to once-only cases. Shackle considers probability only in its *frequency* interpretation, an interpretation which does indeed essentially involve the idea of a long—actually, an infinitely long—hypothetical sequence of repetitions. But there are other interpretations which do not do so: for instance, the *classical* interpretation (see, e.g., Laplace, *A Philosophical Essay on Probabilities*, Ch. 3), and the *propensity* interpretation (see the papers of Popper (1957*b*) in *Observation and Interpretation*, pp. 65–70, and in *Brit. Jour. Phil. Sc.* **10**, 1959, 25–42).

rational man ought to prefer champagne to lemonade, why should it lay down that a rational man ought to prefer a 99% chance of £100 to a 100% chance of £98? Yet this is what the theory of expected utility would imply (unless, of course, the implication were nullified by recourse to alleged variations in marginal utility).

*A farewell look at von Neumann and Morgenstern's method.* From now on I shall assume that people do have distinctive gambling temperaments and that there is such a thing as gambler-effect. I shall now consider whether, on this assumption, anything can be salvaged from von Neumann and Morgenstern's apparatus for measuring utilities.

At the outset we shall have to detach their lottery-method from the axiom system in which they embedded it (since the latter excludes the possibility of gambler-effect) and treat it as an independent device. This means that there is no longer any guarantee that it will yield consistent results. I shall argue that we must expect it to yield inconsistent results.

I suggested earlier that an important variable in the appraisal of wagers is the magnitude, relative to the gambler's personal fortune, of the *stake*. What is at stake when someone faces a choice between a middling outcome $B$ and a lottery offering either an inferior outcome $A$ or a superior outcome $C$? The answer, I suggest, is: $B$ minus $A$; for if he opts for the lottery and loses, he forgoes $B$ but is compensated by $A$. (Of course, the compensation will be negative if $A$ is a negative prize.) Thus gambler-effect is likely to vary with, among other things, the relative difference in value between $A$ and $B$.

Suppose that we get a wine-lover to indicate the value of $p$ such that he is indifferent between $B$ and $[p.A, (1-p)C]$, where $A$ is a bottle of hock, $B$ a bottle of sparkling burgundy, and $C$ a bottle of champagne. If he enjoys sparkling burgundy only *slightly* more than hock, then whether he is a cautious or a daring gambler is unlikely to affect his answer significantly. Suppose now that we replace, in the lottery, the bottle of hock by a bottle of lemonade, a beverage which our wine-lover finds unsatisfying. If he is a daring gambler, he may—lured by the glittering prospect of the champagne—put quite a high value on $p$. If he is a cautious gambler the dread prospect of getting only the lemonade may constrain him to put a significantly lower value on $p$.

Suppose that he marginally prefers a case containing $2n$ bottles of hock to one containing $n$ bottles of champagne. Then we can confront him with an array of outcomes, $A, B, C, \ldots, J, K, L$ (where $A = 1$ bottle of hock, $B = 1$ bottle of sparkling burgundy, $C = 1$ bottle of champagne, $D = 2$ bottles of hock, $E = 2$ bottles of sparkling burgundy, $\ldots J = 8$ bottles of hock, $K = 8$ bottles of sparkling burgundy, and $L = 8$ bottles of champagne) such that he marginally prefers $L$ to $K$, $K$ to $J$, ... and $B$ to $A$. We now seek to locate the utility for him of each of these outcomes on one interval-scale.

A method of measurement ought to be invariant with respect to the order in which the measurements are made. But if gambler-effect varies according to the magnitude of the stake, then we cannot expect such invariance from the von Neumann and Morgenstern method. Suppose that we begin by applying it to successive triplets of neighbouring outcomes— first $A$, $B$, $C$, then $B$, $C$, $D$, ... and finally $J$, $K$, $L$. This will determine a ratio of the interval $AB$ to $BC$, of $BC$ to $CD$, hence of $AB$ to $CD$, and so on. Every outcome should thereby get located on the scale; and in each individual application, gambler-effect should be small since the middling outcome is preferred only marginally to the inferior outcome.

But suppose that we now work through in a different order taking successively the triplets $A$, $K$, $L$, then $A$, $J$, $K$, then $A$, $I$, $J$, ..., and finally $A$, $B$, $C$. In the first 'measurement' of this second round the 'stake' —$K$ minus $A$—will be at a maximum (for the prize of winning eight bottles of champagne he must risk forgoing eight bottles of sparkling burgundy and getting only one bottle of hock). Gambler-effect should be strongest at this first 'measurement', slightly less strong at the second, and so on until, at the last 'measurement', it is as weak as it was in the first of the first round of measurements.

Suppose now—I trust I am not wearying the reader—that we make a third round of 'measurements' taking triplets consisting always of the extreme outcomes $A$ and $L$ and of each of the intermediate outcomes in turn. Suppose that after the first round $G$ was located midway between $A$ and $L$. Throughout that first round, gambler-effect was small. In the course of our third round we invite our wine-lover to give the value of $p$ such that

$$u(G) = u[p.A, (1-p)L].$$

For this 'measurement' to be consistent with the earlier one he must put $p = \frac{1}{2}$. But there is no reason to expect this, because gambler-effect will now be rather strong. If he is a rash gambler, the lure of $L$ may tempt him to put $p > \frac{1}{2}$; if he is cautious, fear of getting merely $A$ may constrain him to put $p < \frac{1}{2}$.

If numerical measurement of utilities presupposes the doctrine of expected utility, then we cannot have numerical utilities. But this does not matter. For the theory of optimal decision-making can treat the present value of a certain chance of some future outcome as a function of its probability and its *monetary* value, and this function may be characterized by a gambler-indifference map.

But conditions of risk, where non-arbitrary numerical probabilities can be ascribed, are rarely met with. We no longer meet with them when we venture out of the casino on to the race-course; still less do we meet with them in the worlds of business, politics and war. In these arenas we may be justified in asserting that one outcome is more or less *likely* than another, but we can seldom if ever ascribe a non-arbitrary numerical value to such comparative likelihoods. Let us now see how the theory of optimal decision-making fares when we turn to conditions of uncertainty.

### 4.3 *Uncertainty*

Under 'conditions of uncertainty' the decision-maker is again assumed to know all the alternative decisions open to him in the given choice-situation and, for each possible decision, all the alternative outcomes it might have; but though he may judge one outcome to be more, or less, or equally likely than another, he is *not* able to assign to each outcome a non-arbitrary numerical probability.

Obviously, the problem arises again, here, of the indeterminacy of a function of two variables, one of which is only ordinally measurable. Suppose that a punter contemplates two alternatives: either to put £10 on horse $A$ to win at 10–1, or to put £10 on horse $B$ to win at 20–1, horse $A$ being more likely to win than horse $B$. Which horse should he back? We would like to answer that if it were only a *little* more likely that horse $A$ will win, then he should back horse $B$, whereas if it were much more likely, then he should back horse $A$. But such quasi-quantitative comparisons cannot be made on a merely ordinal scale of likelihood.

*Potential surprise.* Shackle proposed that likelihood-estimates should be plotted along a 'potential surprise' axis: if I regard outcome $A$ as less likely than outcome $B$, then I would be more surprised by the occurrence of $A$ than by the occurrence of $B$.[1] And he has subsequently claimed to have put potential surprise on a cardinal footing (Shackle, 1961, Chaps. 16, 17). (His claim is not restricted to potential surprise: it implies that *any* kind of feeling experienced with varying degrees of intensity can be ordered on an interval-scale.) He claims that, if we can experience two different intensities of a feeling, then we can intuitively discern the *mid-point* between them (Shackle, 1961, p. 131).

This is a psychological hypothesis which might be tested by asking subjects to identify such mid-points, and then testing their answers (if answers were forthcoming) for consistency. (For instance, if a subject

---

[1] See Shackle, 1952, Chap. 2, and 1955, Part 1.

located $C$ midway between $A$ and $E$, $B$ midway between $A$ and $C$, and $D$ midway between $C$ and $E$, then he should go on to locate $C$ midway between $B$ and $D$.) But I fear that if I were selected as an experimental subject, I would be unable to vouchsafe such answers. I would be surprised if it snowed in London on 31 May, and more surprised if it snowed on 14 June. But I cannot name a date snowing in London on which would cause me a degree of surprise just midway between those two degrees of surprise. If Shackle's hypothesis gets confirmed by experiments on subjects other than myself, I shall have to admit that this is just my personal deficiency. In the meanwhile I shall prefer to suspect that Shackle, not getting the answer he wanted in other ways, did what several celebrated intuitionists from Descartes to G. E. Moore have done: invoked an intuitive faculty to deliver it.

In any case, Shackle's concept of zero potential surprise cannot fulfil its intended function. His idea was that a number of possible outcomes $A$, $B$, . . ., each associated with some degree of potential surprise and some (positive or negative) utility or monetary value, could be *standardized* and rendered comparable with one another by finding outcomes $A'$, $B'$, . . ., each having *zero potential surprise* and such that the decision-maker would be indifferent between $A$ and $A'$ and between $B$ and $B'$. I suggested in 1955, and I believe that I was interpreting Shackle correctly, that a standardized outcome could be regarded, if positive, as the price at which one would just be prepared to sell the uncertain prospect of a larger gain (and, if negative, as the price one would just be prepared to pay to get rid of the uncertain prospect of a larger loss).[1]

Now Shackle insisted that potential surprise does not behave like a probability (or rather, like an improbability): for a possible outcome may have a probability of one half, say, or one third, or even less, *and still carry zero potential surprise*. I will not be in the least *surprised* if the coin I am about to spin lands heads up; and if Brown, Jones, and Smith are the three candidates for a post and all seem to me equally suitable for it, I will not be *surprised* if Brown gets it.

It seems obvious to me now (though I overlooked this in 1955) that this is ruinous for Shackle's idea of standardizing potentially surprising outcomes by reducing them, indifference-wise, to outcomes with zero potential surprise. For standardized outcomes should be ordered, preference-wise, in a determinate way (see Shackle, 1961, p. 191). For instance, an outcome consisting of a gain of £150 and associated with zero potential surprise should be unequivocally superior to one of £100 likewise associ-

---

[1] Watkins, 1955, p. 70. (Reprinted in *Uncertainty and Business Decisions*, Carter, Meredith & Shackle (eds.).

ated with zero potential surprise. *But there is no assurance that this is so.* The gain of £100 may be potentially unsurprising because it is *certain,* whereas the gain of £150 may be potentially unsurprising because it is *as likely as not*—in which case, of course, most people would prefer the former.

*A step-wise likelihood scale.* I now wish to resurrect an idea that offers a way out of the 'strictly numerical' or 'merely ordinal' dilemma. This idea was suggested independently by Carter (1957, pp. 54 f.) and by Popper.[1] It consists of a somewhat idealized elaboration of a realistic hypothesis, namely, that we register changes of feeling in ourselves (e.g. feelings of warmth) induced by some continuously changing external factor (e.g. room temperature) in a discontinuous, step-wise manner. For instance, if the temperature in my study creeps up, then after an hour or so I may notice that I am no longer equably but rather uncomfortably warm.

The elaboration of this hypothesis consists in supposing that a person might learn to identify *and re-identify* the discrete *levels* that he intermittently experiences of a certain feeling. Suppose that this person was in the habit of registering his reactions to temperatures in his diary. After a time he goes through his diary collecting the various expressions he has been using and ordering them by the 'hotter than' relation. He may sometimes come across two or more expressions that he cannot so order. In such cases he infers that he has been using them interchangeably to denote the same level of temperature-sensation, and selects one of them to be the mark for that level henceforth.

He has now constructed a subjective warmth-scale for himself. It is a step-wise scale. It might read something like this: 'Icy', 'Shivery', . . ., 'Cool', 'Equable', . . ., 'Sweltering' 'Oven-like'. There may be gaps in it; but these may be detected and filled in later. Also, he may become more discriminating later and be able to replace one band by two or more bands. When he eventually feels that he has got his scale into something like final shape he may number off the bands and henceforth use the numbers as labels in place of the old descriptive expressions.

Now it may be psychologically far-fetched but it is not, I think, in principle absurd to suppose that someone might undertake an analogous exercise for the construction of a likelihood-scale. Indeed, in one respect this would be easier, since the two ends of the scale, 'Certain' and 'Impossible', are fixed. The object of the exercise would be to determine what discrete levels of likelihood an individual can distinguish and register. People do seem to operate, in a rough-and-ready way, with discrete

---

[1] In a conversation with myself. I presented this solution with acknowledgements, in my 1955 paper (reprinted in the above symposium).

likelihood-classifications. A person may say, 'But surely that's *extremely* unlikely', and get the reply: 'I admit that it is rather unlikely, but I would not call it extremely unlikely.' We might begin by presenting the subject with a wide variety of meteorological possibilities ('Snowing in London on 31 May', etc.) and ask him to sort them into groups of essentially equal likelihood. Then we could present him with a variety of political possibilities (e.g. that war will have broken out in ten years' time between: U.S.A. and Canada, Britain and France, China and Russia, Israel and Syria), and ask him to allocate these among those groups. I do not say that the exercise would succeed but I do say that it would not be bound to fail. Suppose that it did succeed. Then the separate steps or bands might again be numbered off. It might be convenient to label the lowest (viz. 'impossible') by 'o' and the highest (viz. 'certain') by '1', and to use fractions or decimals as labels for the intermediate ones. But, of course, we cannot assume that these will behave like numerical probabilities. For instance, if $A$ and $B$ are distinct (non-overlapping) possible outcomes, we cannot assume that the likelihood of $A$-or-$B$ equals the likelihood of $A$ plus the likelihood of $B$.

I assume as before that the values of the possible outcomes can be expressed in monetary terms. Taking whatever is convenient as a smallest monetary unit, we can form a step-wise monetary scale. We can then put this together with a step-wise likelihood-scale to form a grid, each cell of which represents a *determinate* intersection of a value-band and a (re-identifiable) likelihood-band. The decisive point is that gambler-indifference curves—or rather, gambler-indifference *bands*—might be applied to such cells. And this in turn means that something like *standardization*, in Shackle's intended sense, should be possible: the author of such a grid should be able, with respect to any cell above the level of certainty, to name a cell at the level of certainty such that he is (approximately) indifferent between them.

One last point: a likelihood-scale of the above kind may be regarded as a weakened or relaxed or 'degenerate' version of a numerical probability scale (as opposed to Shackle's potential surprise scale, which is essentially irreconcilable with a probability scale). Thus there is no major discontinuity between 'conditions of risk' and 'conditions of uncertainty'; and the theory of optimal decision-making, though under strain, has not yet broken down. I will now turn to a case where it *does* break down.

## 4.4 *A breakdown in optimization*

Even under 'conditions of uncertainty' the decision-maker is required to know a great deal more than an actual decision-maker usually knows about his situation. I imagine that it is rare for an entrepreneur or a politician to be able to list all the possible outcomes of all the possible decisions open to him in a given problem-situation. Most of us have to make most of our important decisions under conditions that are far more un-certain than the so-called 'conditions of uncertainty' of decision theory. A decision-maker may be so under-informed about his situation that he cannot calculate an optimal way of dealing with the uncertainties that beset him.

However, the breakdown of the idea of a determinable optimal solution that I shall now consider does not occur because the decision-makers are under-informed. On the contrary, it occurs in spite of their being *maximumly* informed (in the sense that *all* the relevant evidence is available to them).

Consider a 'game' with the following simple structure:[1] Alf and Bert are each to choose either heads or tails. If both choose heads, Alf gets £3 and Bert £2. If both choose tails, Alf gets £2 and Bert £3. If one chooses heads and the other tails, each gets nothing. No negotiation is allowed and the game is to be played once only. The outcomes, or 'pay-offs', can be represented by the following matrix:

|  |  | Bert $B_h$ | Bert $B_t$ |
|---|---|---|---|
| Alf | $A_h$ | 3, 2 | 0, 0 |
| | $A_t$ | 0, 0 | 2, 3 |

The symmetry of the game seems to preclude the possibility of a rational decision by either player about which choice to make. However, if Alf and Bert are *under*-informed, this may allow certain asymmetries to enter into their calculations; and these asymmetries *may* make possible a rational choice. Perhaps Alf is inclined to conjecture: (*a*) that Bert is more grasping than Alf; (*b*) that Bert recognizes this; and (*c*) that Bert believes that Alf also recognizes it. Alf would then be able to make a rational choice (namely, tails). (Whether it would also turn out to be a *successful* choice is another matter.)

[1] I take it from Schelling, 1963, p. 60.

Now let the situation be changed by a gift to the two players of two magnificent computers, Alpha for Alf and Beta for Bert. Alf is told that Alpha continuously collects *all* the evidential data provided by its near-environment, is stored with a complete set of all the known laws of nature, and can compute the answer to any prediction-problem that is in principle soluble on the basis of those data and those laws. Bert is told the same about Beta. Alf concludes that it would be looking a gift-computer in the mouth to rely on his hunches and conjectures about how Bert will decide when he could get Alpha to make a scientific prediction for him on the basis of all the extant evidence. And Bert concludes likewise. Alf sets Alpha the task of predicting how Bert will decide: his intention is, of course, that if the prediction turns out to be that Bert will choose heads then he too will choose heads, if tails then tails. Bert also sets Beta the corresponding task, with the corresponding intention.

But their computers let them down—and not because they are not good enough computers. Each computer now constitutes a major part of the other's environment, so that they have become what Popper called *strongly interacting* predictors, which in turn means that they *cannot* complete their respective prediction-tasks.[1] Another way of putting it is that, with the introduction of the computers as physical representations of the idea of maximum information, the original symmetry of the game has been restored.

A game theorist may object that an optimal decision was unforthcoming because it was being sought in the wrong way. The two players each hoped to be able to decide how to act on the basis of a *prediction* about how the other will act. But the general idea of game theory is that a player should try to counter his opponent's *best possible* strategy irrespective of any predictions he may make about how his opponent will actually play the game.[2] He can ignore the question of how his opponent will actually play by selecting a strategy such that, should his opponent depart from *his* best strategy, then that can only redound to the other's benefit. Game theory offers a method for coping with uncertainties which dispenses with the need for predictions, probability-estimates, or likelihood-appraisals concerning the possible outcomes. The agent considers all the possibilities impartially. His aim is to ensure that the worst possibility that can befall him, irrespective of its likelihood, is as little bad as possible (or, if that sounds gloomy, to ensure that the least good possibility is as good as possible).

[1] See Popper, 1950, esp. pp. 128 f.
[2] The game theorist would have to admit, however, that he has no optimal solution for the above non-zero-sum game.

Thus the game theoretical approach to optimal decision-making can dispense with the difficulties (which have preoccupied us in this section) involved in comparing one possible outcome with another that is at once more desirable and less likely than it. (However, game theory does require numerical 'pay-offs'.)

Let us now consider game theoretical rationality.

## 5. GAME THEORETICAL RATIONALITY

Game theory has been strikingly successful in extending the theory of optimal decision-making to problem-situations that one might have intuitively supposed to allow of no optimal solution. Its original programme, one might say, was to establish a general method for solving all (suitably formulated) rationality-problems. But one of the most philosophically interesting results of this programme was, I believe, precisely the discovery of rationality-problems that defy game theoretical methods. I shall consider this negative result in the second part of this section. We will, I think, appreciate it better if we first consider the positive achievement of game theory.

### 5.1 *Introduction*

In the world of big business, of politics, of military contest, the decision-maker seldom responds simply to the situation as he now finds it: he must usually consider also the possible counter-moves which a move of his may provoke from his competitors, rivals, or opponents. It is with the formal structure of these kinds of decision-problem that game theory is concerned.

For an '*n*-person game' to be a genuine game in the game theoretical sense, the number *n* of players (or sides[1], or teams, or firms, or governments, or armies) should be quite small, but greater than one (see von Neumann & Morgenstern, 1953, §2.4). If it sinks to one, as in patience, the game becomes a 'one-person game against nature'. Game theory extends to these special cases; but it treats them as degenerate 'games' because the other 'player' ('nature') is not trying to win and employs no strategies. On the other hand, if *n* becomes large, so that a condition of near-perfect competition obtains, then each player may rightly assume that the other players will pay comparatively little attention just to *his* strategies and decisions; in which case we really have, not an *n*-person game, but *n* one-person games against 'the rest' or 'against nature'.

The ideal number of players, for game theory, is *two*. Indeed, it is only within the domain of two-person games that classical game theory has

[1] Bridge is a *two*-person game for game theory.

achieved definitive results.[1] For one thing, a three-person game may be rendered indeterminate by *shifting* coalitions. (Alf and Bert gang up against Charlie and agree to divide the swag between the two of them. Then Charlie wins Bert over by offering to divide it unequally between the two of them with Bert getting a more than half share. Then Alf wins Charlie over by offering him a full half share. But now they are back where they started except that Charlie has taken Bert's place.)[2]

In game theory there are no tactics, only strategies. A 'strategy' in the game theoretical sense specifies a move for every possible situation that may arise in the course of play (a 'strategy' is at once perfectly flexible and perfectly rigid: it allows *and prescribes* for every eventuality). To play a game in a game theoretical manner, a player sets out, normally in the form of a matrix, all the possible strategies available to him and to his opponent. Then he calculates the outcomes or 'pay-offs' of each intersection of a strategy of his with a strategy of his opponent. The game will then be laid out in 'normal form' as a disjunction of all its possible outcomes. The player pits himself against what his opponent *could* do rather than *will* do. And he really makes only *one* decision: having decided upon an optimal strategy (whether 'pure' or 'mixed'), his subsequent moves are made in an essentially mechanical way. (If the strategy is 'mixed', i.e. composed of pure strategies in certain specified ratios, then *which* pure strategy should control a particular move must be decided by some randomizing device.)

Game theory may seem unrealistic in precluding the idea of *learning* and improving one's performance as play proceeds. But this idea has its dangers which game theory avoids. For your opponent may, in his early moves, try to deceive you into 'learning' the wrong lessons. He will not succeed if you abide by game theoretical canons.

How does one determine an optimal strategy for oneself? To take the easiest case first, one may find that a 'dominating' strategy is available. A dominating strategy is one that will always give a better pay-off than would any alternative strategy *whichever* strategy one's opponent may play. In game theory such a strategy is always regarded as optimal. (But this can lead to a strange result, as we shall see in §5.2 below.)

If no dominating strategy is available, then one should try to determine an optimal strategy by the following rule: select a strategy whose worst (or least good) possible pay-off is better than (or at least as good as) that of any alternative strategy. In other words: minimize your maximum loss (or maximize your minimum gain). This is the famous Minimax (or Maximin)

---

[1] It should be mentioned, however, that John C. Harsanyi claims (1968) to have 'succeeded in developing a new approach to game theory which does yield determinate solutions for *all* classes of games' (p. 309; see also pp. 338 f.). I regret that time did not allow me to investigate Harsanyi's claim.　　[2] See Rapoport, 1967, p. 92.

principle. It can be justified in this way. Suppose that Alf is hesitating between two strategies, $A_1$ and $A_2$. $A_1$ is endorsed and $A_2$ is ruled out by the minimax rule; but $A_2$ is tempting because many of its best pay-offs are much higher than the best that $A_1$ can offer. Now game theory lays down the equalitarian postulate that the players are equally knowledgeable and rational. Thus Alf should reckon that his opponent Bert will defend himself against $A_2$ by playing *his* minimax strategy (or one of them if he has more than one); in which case Alf will be *worse* off if he plays $A_2$. So Alf should eschew the illusory prospect of greater gain and the real prospect of greater loss held out by $A_2$, and stick to a minimax principle which guarantees him a best minimum. Should Bert fail to play *his* minimax strategy, then Alf will get a windfall benefit.

It may be (though the chances rapidly diminish with games of increasing complexity) that both players have a (pure) minimax strategy; in which case, if it is a 'zero-sum' game (i.e. a game where one player's gain is the other player's loss), then their two minimax strategies will jointly determine a unique pay-off, positive for one and negative for the other. Such an outcome is called a 'saddle-point' (so called because the *lowest* point of the upper profile of a saddle is also the *highest* point of a cross-section through it).[1]

If the game has no saddle-point, then one can do better (at least if the game is going to be repeated a number of times) minimaxwise by selecting a *mixed* rather than a pure strategy. In this case, each of one's possible strategies is given a probability (which may be zero), the probabilities summing to one, and one plays according to the strategies as selected at random in ratios corresponding to the probabilities.

In a zero-sum two-person game where a single pay-off (positive for one player, negative for the other) is jointly determined by their respective minimax strategies, that pay-off is called the *value* of the game. A fundamental result got by von Neumann in 1928 was that *every* zero-sum two-person game has a value. (In the interest of generality one may consider a pure strategy as a limiting case of a mixed strategy, namely a 'mixed' strategy where all but one of the component strategists gets zero probability. Then a game with a saddle-point determined by the intersection of two pure minimax strategies becomes a special case of a game with a value determined by the intersection of two mixed minimax strategies.) This result may be received with mixed feelings by people who actually enjoy *playing* games; for there is no need for rational people to play out a game with a value; they can simply calculate where they would end up.

Admittedly, the idea of an optimal mixed strategy becomes a bit dubious

---

[1] The concept of a saddle-point is explained particularly clearly in Williams, 1966, pp. 24–8.

if the game is to be played *once only*.[1] Suppose that Alf's optimal strategy is a mixed strategy $A_M$ consisting of five parts of $A_1$ to one part of $A_2$. Suppose, further, that considered as a pure strategy $A_1$ is superior to $A_2$. The game is to be played once only. His way of playing $A_M$ is to throw a die and play $A_1$ if it does not land six up and $A_2$ if it does. He throws and it lands six up. Should he now, in accordance with $A_M$ which is superior to $A_1$, play $A_2$ which is inferior to $A_1$? What is the optimal strategy here seems to be indeterminate. However, I will not press this. Instead, I will turn to a famous (non-zero-sum) game widely discussed in game theory literature under the title 'The Prisoner's Dilemma'. I shall take the liberty of re-vamping this game in the hope of endowing it with some interest for students of political philosophy.

### 5.2 *The 'State of Nature' Game*

A much-discussed question among Hobbes's critics is this: could Hobbesian men ever get themselves out of the 'state of nature'? They all *want* to do so; and Hobbes assumed that they *could* do so. But they are so divided and unorganized that it is hard to see how they could come together and covenant with one another.[2]

An encounter between just two individuals in a Hobbesian state of nature could very well exemplify the structure of the Prisoner's Dilemma game, whose detailed circumstances (involving police bribery, etc.) are rather sordid and without political interest. So I will transpose it into a Hobbesian setting. The game now goes like this.

Alf and Bert are two Hobbesian men in a Hobbesian state of nature. Both carry a murderous weapon. One afternoon, out grubbing for acorns, they stumble upon one another in a small clearing in the woods. The undergrowth makes flight impracticable. (Fortunately, though owing no allegiance to any monarch, both speak the Queen's English.) Alf cries: 'Wait! Let us not hack each other to pieces. Let us rather submit ourselves to some common power who will protect us from each other.' Bert replies: 'I share your sentiments. I will start counting. When I reach "ten" each of us will hurl his weapon into the trees behind us.' Each man now starts furiously to think: should he or should he not throw away his weapon when Bert reaches 'ten'?

Each assumes, first, that if they both double-cross each other a fight to

---

[1] I am indebted, here, to the lucid discussion in Luce & Raiffa, 1957, §4.10.

[2] Even if it is just I do not consider this a very serious criticism. Hobbes intended his picture of men's natural condition as a *warning*: it showed how bad things would get if men had no sovereign power over them. If the critics are right, that merely *intensifies* the warning: not only would conditions be very bad but there would be *no escape* from them.

the death will result from which the survivor, if there is one, will probably emerge badly wounded; and second, that if one throws away his weapon and the other keeps his, the latter will triumphantly kill the other.

According to Hobbesian psychology, they will be motivated by two main passions: the desire to triumph and the fear of being killed. But the latter is the stronger passion: given a choice between, on the one hand, submitting to a common power and, on the other hand, an equal chance of either triumphing or being killed, they will choose the former. Let '$S$' stand for submission, '$T$' for triumphing, and '$K$' for being killed. Then we can express, in the notation used previously, this preference thus:

(1) $$u(S) > u(\tfrac{1}{2}T, \tfrac{1}{2}K).$$

Let '$F$' stand for fighting it out. How will they evaluate $F$? Fighting involves at least a half chance of being killed (Hobbesian men are supposed to be, and to realize that they are, equally good at killing each other), and at most a half chance of triumphing. So they will prefer $S$ to $F$ and $F$ to $K$. The order of their preferences will be:

(2) $$u(T) > u(S) > u(F) > u(K).$$

Let us now, in order to have a matrix with numerical pay-offs, choose arbitrary numerical values that satisfy (1) and (2). Beginning at the two ends of (2), let us put $u(T) = 10$ and $u(K) = -10$. These values, in conjunction with (1), require us to assign a value to $u(S)$ such that

$$10 > u(S) > 0$$

so let us put $u(S) = 5$. It seems reasonable to put $u(F)$ closer to $u(K)$ than to $u(S)$ since $F$ involves at least a half chance of being killed *and* a high probability of being wounded. So let us put $u(F) = -5$. (In view of the Hobbesian principle of the uniformity of men, we can give the same utility-values to Alf and Bert.)

The 'State of Nature' game can now be presented in a matrix. Alf has two strategies, $A_1$ (= discard weapon) and $A_2$ (= retain weapon); and Bert has two corresponding strategies, $B_1$ and $B_2$. Alf's pay-offs are represented by the bracketed numeral on the left in each cell, Bert's by the one on the right.

|  |  | Bert | |
|---|---|---|---|
|  |  | $B_1$ | $B_2$ |
| Alf | $A_1$ | (+5)  (+5) | (−10)  (+10) |
|  | $A_2$ | (+10)  (−10) | (−5)  (−5) |

From an orthodox game theoretical point of view, it looks as though Hobbes's critics are right. Alf and Bert seem to be trapped in the state of nature. Alf's $A_2$ and Bert's $B_2$ dominate their $A_1$ and $B_1$ respectively. When Bert (who must surely have nearly finished counting by now) reaches 'ten', both should retain their weapons. Here, as in the original Prisoner's Dilemma game, if both players are 'rational' in the game theoretical sense, they each get a pay-off of $-5$. Had they both been 'irrational' they would have got $+5$.

Is there any way by which game theory could, in such a game, avoid the $A_2B_2$ outcome? In the original Prisoner's Dilemma game, the two prisoners are in separate cells and cannot communicate, which may tend to suggest that, if only they could communicate, they could *negotiate* their way to the $A_1B_1$ outcome. But our State of Nature game shows that negotiation leaves the situation unaltered. Alf and Bert *did* negotiate: they 'covenanted' (to use Hobbes's term) to play $A_1$ and $B_1$. But that left them with the problem of whether or not they dare to *keep* the covenant. Likewise, if the two prisoners had managed to telephone each other, and had agreed on $A_1B_1$, as soon as they put the telephone down they would have been back in their old dilemma. In game theory the idea of a contract is essentially dispensable. A contract is either enforceable (there are penalties for breaking it) or unenforceable. If enforceable, then the penalties can be incorporated in the pay-offs and the game governed by these revised pay-offs. If unenforceable, then the game is governed by unrevised pay-offs. Game theory imputes a kind of ruthless self-interestedness to players very like that which Hobbes imputed to men in general; and game theory endorses the spirit of Hobbes's saying: 'Covenants, without the sword, are but words, and of no strength to secure a man at all.'

We may feel shocked that Alf and Bert should no sooner have reached an agreement than they start wondering whether to keep it. However, in extenuation it should be emphasized that what drives each of them to consider breaking it is not only, and perhaps not primarily, the hope of getting $+10$ rather than $+5$, but the fear of getting $-10$ rather than $-5$ if he abides by his co-operative strategy.

It has sometimes been suggested that game theory could avoid the $A_2B_2$ outcome in games of the Prisoner's Dilemma variety if the game were to be repeated many times; for then a player could afford to signal, in the opening plays, his preference for the $A_1B_1$ outcome, and he could stand by his co-operative strategy so long as the other did not let him down.

Here as elsewhere I am against coming to the rescue of theories that are in trouble over once-only situations by appealing to a hypothetical sequence of repetitions of it. Our State of Nature game could only be played once by Alf and Bert (they either get themselves out of the state of nature or at least one of them gets killed). And even if, *per impossibile*, they could play it again and again there is a well known objection to the claim that they could thereby avoid the $A_2B_2$ outcome.[1] Suppose the game is to be played 100 times. When the 100th play is reached, *that* will be a once-only game on which they will revert to $A_2B_2$. Which leaves 99. But then they will revert to $A_2B_2$ on the 99th game. Which leaves 98 . . .

Most game theorists agree that there is no escape, so long as we remain within game theoretical assumptions, from the conclusion that, in a game of this structure, $A_2B_2$ is the 'equilibrium' outcome for 'rational' players. But game theorists vary considerably in their interpretation of this result. Luce and Raiffa regard it as but an illustration of the fact that people sometimes find themselves in painful situations where everyone would *like* everyone (himself included) to act differently from the way in which the situation obliges each of them to act. Luce and Raiffa suggest that games of this sort show up the anti-social structure of certain situations, and typically reveal a need for State intervention to alter the situation so as to allow people to act as they would like to (Luce & Raiffa, pp. 96–7).

I am not entirely persuaded by this. If a cry of 'Fire!' in a cinema starts a stampede, everyone may wish that everyone would leave in an orderly way. But *one* person *cannot* leave in an orderly way when everyone else is jostling and pushing. By contrast, Alf *can* throw away his weapon. So can Bert. The $A_1B_1$ outcome *is* a possible outcome.[2]

My own view is indebted to that of Anatol Rapoport: surely there *is* something fishy and paradoxical-seeming about $A_2B_2$ being the required outcome for rational players (Rapoport, 1966, pp. 128 f.). I do not say that there is an optimal solution for the problem confronting Alf and Bert, and that game theory provides the *wrong* solution. I say that there is *no* optimal solution, that the idea of optimality breaks down here. Let us look at the problem through Alf's eyes. I suggest that his reasoning will lead, step by step, to *oscillating* conclusions, as follows:

(1) 'My strategy $A_2$ dominates $A_1$; so I must obviously choose $A_2$.'
(2) 'But wait. If Bert reasons as I am doing we now both realize that we are both heading for $-5$. We would both prefer $+5$. So I should choose $A_1$, confident that he, by similar reasoning, will choose $B_1$.'

---

[1] Its reasoning will be familiar to philosophers acquainted with the 'prediction paradox' (see O'Connor, 1948).
[2] When games of this structure have actually been played a surprisingly large number of pairs of players actually *achieved* this outcome. See the summary by Rapoport, 1960, of experiments carried out by Morton Deutsch.

(3) 'But wait. He *may* choose $B_2$, in which case $A_1$ will give me a terrible pay-off. I must guard against this. Moreover, if he does choose $B_1$, as I was supposing just now, then I will get $+10$. So I should choose $A_2$.'

(4) 'But wait. If he reasons similarly we are both heading for $-5$ again . . .'

And so on. It looks as though which way his decision will go depends on the point at which he breaks off his reasoning.

## 6. IMPERFECT RATIONALITY

### 6.1 *Decision-schemes*

I call the set of all those considerations, whether consciously formulated or not, that enter into a particular piece of decision-making a 'decision-scheme'. According to normative decision theory, a decision-scheme should consist of a complete specification of the possible outcomes, a complete preference-map or a complete allocation of pay-off values to the outcomes, and (where appropriate) a comprehensive apparatus for dealing with risks and uncertainties.

Judged by this, an actual decision-scheme is usually very imperfect indeed. An ideal decision-scheme is pictured as being present to the agent's mind in its entirety, a completed whole in which the several components simultaneously play their due role. An actual decision-scheme is usually built up bit by bit, so that the arrival of an isolated bit of situational information may have a quite disproportionate influence.[1] And even when all the evidence is in, the practical significance of different parts of it may wax and wane as the decision-maker attends now to this factor, now to that.

Not only is an actual decision-scheme more or less vague and fragmentary compared with the ideal, but the agent will usually *reduce and simplify it further* as he proceeds towards a decision. Instead of the complete enumeration of possibilities demanded by normative theory, we usually seize upon a few features and pick out a few interesting possibilities in the given problem-situation. This selectivity has its dangers, but it is practically essential in any but the simplest problem-situations. A well known obstacle to computerizing chess is the lack of any known way to program a computer to concentrate upon *interesting* developments: like the ideal decision-maker of normative theory, the computer prefers to survey the entire board and take every possibility into account.

---

[1] See Audley, 1967, p. 47. Any skilled committee-man knows how important can be the *order* in which evidence is presented.

Schelling has argued against the traditional game theoretical approach, with its impartial enumerations of strategies and pay-offs, that the need often arises to single out and endow with special significance some little asymmetry or discontinuity: for instance, in a 'game' where both sides badly need to co-ordinate their strategies but cannot achieve this by negotiation, they will both need to pick out some feature of the strategic landscape which the other can also be expected to pick out, in the hope that this will serve as a co-ordinating mark (Schelling, 1963, *passim*).

In 'one-person games' also there is often a compelling need to *narrow* one's attention, to reduce very drastically the possibilities to be taken seriously. When you *start* planning your summer holiday, you may cheerfully list all sorts of places; later, you will welcome any excuse to cross a name off the list.

Much the same can be said of the other side of a decision-scheme: we tend to simplify and exaggerate our *preferences* much as we do our situational assessments. Instead of the unified preference-map of normative decision theory, most of us operate with one or other of a number of preference-systems, according to the 'hat' we are currently wearing. When we come home from the office or go out to a party, we may switch easily and unnoticingly from one system to another. (A holiday-maker lazes on the beach in hot sun and hedonic mood. Then he jumps up: a swimmer is in difficulties. His hedonic calculus switches off, moral concern switches on.) But sometimes, of course, the other preference-system stays switched on. If they indicate conflicting decisions the agent will have to make some sort of meta-decision about which system shall now predominate before he can decide what to do.

Most decision theorists (game theorists are an exception) like smooth curves and continuous functions. Decision-*makers* prefer discontinuities; they like desiderata of the kind that either are satisfied or are not satisfied, rather than values that may be satisfied marginally better or marginally worse. Where the situation offers a fairly continuous range of possibilities, the decision-maker will often, in the interests of psychological determinacy, *create* discontinuities by drawing more or less arbitrary lines. It is Schelling, again, who has emphasized the importance of such line-drawing in certain political and strategic conflict-situations where it may be the only answer to creeping encroachments ('salami-tactics') and continuous escalation (Schelling, 1963, pp. 75–7).

Much the same can again be said of our estimates of risks and uncertainties. Again, we tend to split these up into a few discrete categories. If a stockbroker advises a client that he expects share $A$ to do better than share $B$ in the longer term, but share $B$ looks more interesting as a short-term speculation, it is likely that his client, far from asking him to place numerical values on these expressions, will treat them as if they stood for important qualitative differences. He may regard quick profits as sinful and opt for $A$; or he may enjoy a flutter and opt for $B$.

We dislike having things hovering around our threshold of practical significance. Faced by a small probability of a large calamity, the average decision-maker is likely *either* to treat it as if its probability were zero and to exclude it from his decision-scheme, *or* to treat it as if its probability were significantly higher than it is and to include it in his decision-scheme, where its presence may lead him to take out an insurance policy, re-route his journey, etc. Marginal utility theory endows economic man with a refined discrimination apropos border-line cases. Actually, border-line cases make us nervous and we deal with them by forcing them into one category or the other.

*If* people ever operate with anything at all resembling the grid mentioned on p. 196 above, I am confident that, far from having a fine mesh, it would be crudely divided into a few main boxes.

An actual decision-scheme, then, is something like a crude caricature, drawn with a few bold strokes, of the complete situational picture and preference map of normative decision-theory.

Some decisions are made so rapidly that there is hardly time for the decision-maker to articulate to himself any part of his decision-scheme, though afterwards he may formulate some of his reasons for deciding as he did. In the case of a decision made with due deliberation and after much discussion, the decision-scheme behind it may be well articulated. It can now be regarded as a body of propositions, some in the descriptive and some in the imperative mood, issuing in a practical conclusion. However, it is logically impossible that a decision-scheme should be *fully* articulated; for every proposition has infinitely many logical consequences. Thus a decision-scheme is bound to have *unnoticed implications*. With luck, these will be without practical significance. But they may not be: a decision-maker may have overlooked a serious implication, an implication such that, had he attended to it, he would have drawn a very different practical conclusion from his decision-scheme. This point will be important when we come to the explanation of seemingly irrational action.

### 6.2 *Explaining bungled actions*

An (imperfect) rationality principle is needed to bridge the gap between a decision-scheme and a consequential action. That there is at least a logical gap was claimed earlier (p. 173 above). That claim may be reinforced by pointing out that the logical gap becomes an actual gap when we turn from waking life to things that happen when we are asleep. Bradley once asked, 'Why, when we strive to move in dream, do we not always move?' (1935, p. 303). The short answer is that no rationality principle applies to *dreamt* decision-schemes. Nor does one apply to somnambulist behaviour like sleep-walking. But in waking life the situation is different. A rationality principle effects a two-way connection. It says that a man who has a decision-scheme issuing in a practical conclusion will try to act in accordance with that practical conclusion. (My friend *did* leave on that flight from London Airport.) It also says that, given that a man was not drugged, hypnotized, sleep-walking, etc. but *acting*, then his action was the acting out of the practical conclusion of a decision-scheme. In this form the principle amounts to a factual, metaphysical assertion:[1] to every practical conclusion drawn from a decision-scheme there corresponds, physical circumstances permitting, an appropriate action, and behind every action there is an appropriate practical conclusion drawn from a decision-scheme.

But the principle can also be cast in the form of a methodological rule that enjoins historians and other investigators of human behaviour, not necessarily to accept the principle *qua* factual postulate as true, but to proceed on the supposition that it is true.

In this last form (which is the form in which I shall henceforth consider it) it says, first of all, that to provide a conjectural explanation for a past action is to postulate a decision-scheme which has a practical conclusion of which that action could be the natural outcome.

Of course, we want such a conjectural explanation to be as little *ad hoc*, as plausible, and as well supported by evidence (especially evidence other than that provided by the action itself) as we can make it. But I shall not discuss these desiderata further. Rather, I shall concentrate on the application of this rule to actions that went more or less badly wrong. I claimed at the outset of this paper that the rationality principle may lead to more revealing results the more recalcitrant is the conduct to which it is applied, and I shall now try to vindicate that claim.

We are at liberty to seek an explanation for *any* action, successful or otherwise. But I happen to believe that, as a matter of psychological fact, we are often more curious about unsuccessful than successful actions. 'Why did you do that?' is often a polite way of asking 'Why on earth did you do *that*?' In seeking an explanation we are often seeking a rationalization of a seemingly irrational action.

Philosophically speaking, the easiest kind of historical explanation of an

---

[1] 'Metaphysical' in a sense indicated in my 'Confirmable and Influential Metaphysics' (July, 1958).

action that ended in failure is this. The main components of the agent's decision-scheme have been ascertained to the historian's satisfaction; these point pretty unambiguously to a certain practical conclusion; the action in question was in line with that conclusion; *but* there is a significant discrepancy between the situational appraisal contained in the agent's decision-scheme (as reconstructed by the historian) and the agent's objective problem-situation (as reconstructed by the historian); and the failure of the action can be explained in terms of this discrepancy.

No doubt this is over simple. For one thing, it treats the agent's decision-scheme as static; but he presumably began to realize, as things went increasingly wrong, that he had misjudged his situation and must revise his plans. But a simple statement of the idea may be sufficient for our purpose.

The more difficult case is one where, not only did the action end in failure, but the main components of the agent's decision-scheme, as presently understood by the historian, do *not* point to a practical conclusion corresponding to the action in question. In such cases the action may seem not just unsuccessful but downright irrational or even crazy. However, our imperfect rationality principle (methodological version) enjoins the historian not to throw up his hands and declare that the agent must have taken leave of his senses. Rather, his task is to look for new evidence (or for a reinterpretation of existing evidence) that allows him to revise his reconstruction of the agent's decision-scheme so that it does now yield a practical conclusion appropriate to the peculiar action in question. (Of course, the revision should not be merely *ad hoc*: it should be arguable on grounds other than that it yields the desired conclusion; but I am not here discussing this desideratum for the revised explanation.) In other words, the historian should try to transform a case of this sort into a case of the previous sort where the agent, though frustrated or defeated by objective factors, acted reasonably enough given his misappraisal of his situation.

When a historian seeks to revise his reconstruction of an agent's decision-scheme, one possibility is that he will discover evidence indicating that the agent had overlooked a serious implication of the information at his disposal, an implication which became so painfully obvious later that it is hard, now, to imagine that it could have been overlooked at the time. Consider the example of a foolish move in chess. White moves his queen to an attacking position. But no sooner has he removed his hand than he realizes, to his horror, that Black can now fork his queen and his king with a knight. The situational information available to him before he made this fatal move did carry the implication that he should lose his queen if he made it. But he overlooked that implication. If he is now to account for that move he must, as it were, block out that unnoticed implication (which now stands out like a sore thumb) and show how those implications of his decision-scheme that he did notice suggested such an unfortunate practical conclusion.

I now turn to a more dramatic example of a bungled action.

### 6.3 *Rationalizing the irrational: a case-study*[1]

If Collingwood's thesis concerning the unexplainability of unsuccessful actions applies anywhere, it should surely apply to the disastrous naval manoeuvre I am about to describe. For a long time I, along with most other people interested in it, regarded this as an explanation-defying enigma. It seemed to be an all too good approximation to a gratuitous act. But my view of it was transformed by Richard Hough's *Admirals in Collision* (1959). Mr Hough practised the kind of methodological rule I was adumbrating in the previous sub-section; and he succeeded in rationalizing this irrational-seeming affair. In what follows I shall rely on, and make free use of, his historical material.

I begin with an 'objective' summary of the gist of the story.

On the morning of 22 June 1893, the Mediterranean Fleet left Beirut for Tripoli. It consisted of thirteen ships, eight of them battleships, divided into two divisions. The C.-in-C. was Vice-Admiral Tryon in *H.M.S. Victoria*. Rear-Admiral Markham, in *H.M.S. Camperdown*, was in command of the Second Division. The ships were in line abreast. By 2.20 the Fleet was nearing Tripoli. Tryon now formed it into two columns behind *H.M.S. Victoria* and *H.M.S. Camperdown*, *H.M.S. Camperdown* being *six* cables on *H.M.S. Victoria*'s port beam. At 3.25 Tryon hoisted a signal ordering the two divisions to turn *inward* 180° in succession. This signal caused utmost alarm: the turning-circles of the flagships were both about *four* cables. Markham kept his repeat-signal at the dip, indicating that he had read but not understood the Admiral's signal, and ordered it to be queried by semaphore. But before his query could be transmitted Tryon semaphored to him: 'What are you waiting for?' and hoisted the *Camperdown*'s pendants as a rebuke for her slowness. The chastened Markham now hoisted the repeat-signal close up, whereupon Tryon immediately executed the signal. The *Camperdown* went hard-a-starboard, the *Victoria* hard-a-port, and the two ships swung towards each other. In those days battleships were built with powerful rams. I have never heard of an enemy ship being rammed. But now the *Camperdown*'s ram knifed through the *Victoria*'s side. The following ships sheered clear, and began lowering boats. However, the Admiral signalled 'Annul sending boats'. Shortly

[1] I discussed this case on a previous occasion (*The Listener*, 10 January 1963, pp. 69–70).

afterwards the *Victoria* lurched over and sank bows first, her still turning screws massacring many of the men in the water: 356 officers and men were drowned, including Admiral Tryon. He was reported to have said, before he went under, 'It was all my fault.'

The picture of Tryon given in his official biography (Fitzgerald, 1897) is that of a brilliant officer, energetic, resourceful, imaginative, and destined for the top. Then comes the last chapter, and the astonishing, ridiculous, incredible end of it all. A veil descends: the extraverted commander turns into an enigma, leaving only question-marks dancing on the waves as he goes under.

Let us leave Tryon, for the time being, and look at the situation through the eyes of his second-in-command. Markham's first reaction to the incredible signal was to exclaim, 'It's impossible.' But the signal was flying; and Markham had entire faith in his commander. So he groped for some reasonable interpretation of the signal. It actually consisted of two signals, one ordering the First Division to turn in succession to port, the other ordering the Second Division to turn in succession to starboard. Markham's first hypothesis was that the signals would be executed separately. Then it occurred to him that, if they were executed together, it was Tryon's intention to turn the *Victoria* slowly on a wide turning-circle that would keep her outside the *Camperdown*'s turning-circle. The rule of the road requires approaching ships to pass port side to port side, which in this case meant that it would be incumbent on the *Victoria* to pass outside the *Camperdown*. Stung by Tryon's semaphored rebuke and the hoisting of *Camperdown*'s pendants, Markham muttered, 'We shall have to do it.' He was confident, by now, that he had made sense of the signal that had at first struck him as 'impossible', and he ordered the repeat-signal to be hoisted close up.

The twofold signal on the flagship was then executed simultaneously, thereby refuting Markham's first hypothesis. In accordance with his second hypothesis, the Rear-Admiral now ordered the *Camperdown*'s helm to be put hard-a-starboard. But he was then shocked to see that the *Victoria*'s rudder-signals indicated that *her* helm was hard-a-port. With hindsight we may judge that he should have taken this as a refutation of his second hypothesis and should now have eased off the *Camperdown*'s helm in order to pass outside the *Victoria*. But rather as a motorist who gets into a skid by braking too hard tends to brake still harder, so Markham responded to the new information by making the *Camperdown*'s turning-circle still smaller: he ordered the starboard engine full astern; and as the ships drew closer he ordered both engines full astern.

Now let us return to Tryon. But before we enquire into the manoeuvre itself, let us first look into his signal to Markham before the manoeuvre was executed ('What are you waiting for?') and his signal after the collision ('Annul sending boats'). Unexplained, these two signals reinforce an impression of an impatient megalomaniac, too sure of his crazy plan to brook any delay and too proud, after it had failed, to accept any aid. However, both signals have straightforward situational explanations. As to the first: before the turn the Fleet was heading towards the shore and would soon have been in shallow water: a turn was becoming urgent. As to the second: after the collision it was not at first believed by the officers on the *Victoria*'s bridge that she was in any immediate danger of sinking: it looked as though it should be possible to beach her, which is what Tryon decided to try. (This is why her screws were still turning when she sank.) If the *Victoria* was going to head for the beach at the best speed she could make, it would be silly to let the other ships lower their boats *now*. After 'Annul sending boats' Tryon ordered a further signal to be sent instructing the other ships to follow the flagship with boats slung out at the ready. But this signal had not been made when the *Victoria* sank.

Now to the manoeuvre itself. Tryon had planned it some hours before it was carried out. He outlined it to his Flag-Captain and Staff-Commander. The latter remarked: 'It will require at least eight cables for that, Sir.' With hindsight we can see that this was an unfortunate comment, for it failed to indicate that the Staff-Commander thought that the ships were to turn inwards upon each other. He should have demanded at least twelve cables: at *eight* cables there would still have been every danger of collisions among the eight battleships with turning-circles of four cables. At the time, Tryon assented to eight cables, which was also unfortunate since it had the effect of allaying his staff-officers' anxieties sufficiently to deter them from further remonstration with their formidable commander. But when he came to prepare the Fleet for the manoeuvre Tryon told his Flag-Lieutenant to signal the Fleet to form two columns *six* cables apart; and for emphasis he wrote the figure '6' on a piece of paper. When the Staff-Commander saw this signal flying he sent the Flag-Lieutenant to the Admiral to query the distance. (One wishes that he had sent a stronger message or gone himself to protest that this would lead to a collision.) When the Flag-Lieutenant got there he found that the Flag-Captain was already with the Admiral and querying the distance. Tryon brusquely replied, 'Leave it at six cables.' He seems to have had a remarkable ascendancy over his officers, for after that they desisted from further protest.

Various non-rational pseudo-explanations for Tryon's manoeuvre were offered. It was rumoured that Tryon was drunk, and also that he was suffering from fever. There is no evidence for either allegation. His official biographer, Admiral Fitzgerald. toyed with the idea (mentioned at the court-martial) that by a mental miscalculation he had halved the distance needed. A ship's turning-circle is *twice* its radius. One cable is *two* hundred yards. *Two* ships' turning-circles were involved. Perhaps Tryon left a two out somewhere? A decisive objection to this is that, independent of mental arithmetic, officers could see that the *Victoria* and the *Camperdown* would collide if they turned inwards together with full helm. Indeed, it was the *obviousness* of the danger which persuaded people that Tryon must have taken leave of his senses. 'His brain must have failed him', exclaimed a fellow Admiral afterwards.

Now I come to Richard Hough's rational reconstruction of Tryon's intended manoeuvre. Not that he claims originality for it. He points out that the gist of it had already appeared in a letter to *The Times* one month after the collision, only to be largely ignored and later forgotten. The writer of the letter was Sir William Laird Clowes, a naval historian who had known Tryon personally. Ten years later, in a footnote in the concluding volume of his great naval *History* he ruefully remarked that he was not aware that his arguments about the manoeuvre had ever been adequately discussed.

One point made by him in his letter to *The Times* was that Tryon's signal did not merely order the two columns to turn inwards in succession: it added that *the order of the fleet was to be preserved*. In other words, on the completion of the manoeuvre the *Camperdown* was still to be on the port beam of the *Victoria*; whereas if, *per impossibile*, the two ships had turned inwards 180° with turning-circles of less than three cables, the *Camperdown* would then have been on the *starboard* beam of the *Victoria*. To preserve the order of the fleet one column would have to wheel round outside the other.[1]

So the question becomes *which* column was to pass outside the other. We know that Markham concluded that Tryon would lead his column round outside, because this was what the rule of the road required. But Clowes drew attention to a Queen's Regulation then in force which ran: 'If two ships under steam are crossing so as to involve risk of collision, the ship which has the other on her starboard side shall keep out of the way of the other.' Now as the two ships began to turn towards each other it was

---

[1] Two correspondents (Mr P. W. Bowyer-Smyth in *The Listener*, 24 January 1963, and Mr A. House in a private communication) have questioned whether the two signals to the two columns were each separately governed by 'order of the fleet to be preserved' (in which case this could mean merely 'order of the *column* to be preserved'), or whether this important qualification was made in a separate signal to the whole Fleet. I do not know the answer to this. I can only say that 'order of the fleet to be preserved' would have been *pointless* if addressed just to the columns, since the order of the *columns* would in any case have been preserved.

the *Camperdown* which had the *Victoria* on her starboard side and therefore, by this Regulation, it was for Markham to keep out of Tryon's way. And this Regulation would override the rule of the road.

If we accept the arguments of Clowes and of Hough, our hypothetical reconstruction of Tryon's decision-scheme prior to the manoeuvre would go something like this: 'The Fleet's course will have to be reversed as it approaches the shallow water off Tripoli. I will do this with one of those daring manoeuvres which help to keep my captains on their toes. I will bring the two divisions fairly close together in line ahead, after which they will wheel round, the *Camperdown* leading the port column outside the starboard column. That one column is to pass *outside* the other will be made clear by adding "order of the fleet to be preserved" to the signal. It will be unnecessary to include in the signal that it is the *Camperdown* who is to pass outside since that is implied by a Standing Regulation. Moreover, when Markham sees that my helm is hard-a-port he will be left in no doubt that it is I who am turning inside and that he must go outside me. How close should the two columns be? The Staff-Commander suggested eight rather than six cables and I tentatively agreed to that; but on reflection, I think that six would be better: this will help to make it perfectly clear that there is no question of the columns turning inwards on each other.'

Supposing something like this to have been Tryon's decision-scheme, the ground on which it should be criticized is, of course, his assumption that his signal would be read as he intended it. We know that his staff-officers, the Rear-Admiral, and the captains of the other ships all read it differently. In partial extenuation of Tryon, however, it can be mentioned that no one came right out and told him that it looked as though his manoeuvre would lead to a collision. If someone had, Tryon would presumably have realized to his horror that his signal was being read by at least some officers in a disastrously wrong way. As it was, Markham started semaphoring when it was too late, and Tryon's staff-officers demurred only at the *distance*. Tryon was not the sort of man who would be sensitive to perplexed looks, head-shakings, and mutterings on the part of his subordinates. His alleged exclamation, 'It was all my fault', could be taken as his admission that he should have perceived beforehand the dangerous possibility of his signal being misread.

It is not my intention to defend Tryon: look at it how you will, it was not a good manoeuvre. I have been arguing that it is, however, *explainable*— by reproducing a plausible explanation for it, and one that was got by systematic adherence to something like our imperfect rationality principle.

If Collingwood's thesis that unsuccessful actions are unexplainable were right, or even if some more restricted thesis were right to the effect that actions that are both unsuccessful and *unreasonable* are unexplainable, two things would be lost that we do not want to lose. One is the historian's rightful assurance that he may try to penetrate any human enigma. The other is the right of someone who makes a real mess of things, to continue to be regarded by other people as a fellow human being. To regard the doings of another as essentially unexplainable is to regard him as not quite human. Before Mr Hough's book appeared, this had happened to Tryon: he had become, in the public imagination, a comi-tragic unperson, the perpetrator of an unfunny joke, an eminently lampoonable Victorian.[1] Indeed, the process of dehumanization began immediately: 'Though bodily he was present on the afternoon of June 22,' a naval officer declared, 'the guiding brain that made him so dear to us was absent.'

I said that I would not preach against the historiographical bias towards success. But I may conclude this essay by remarking that the explanation of failure may be not only historically enlightening but humanizing as well.

## REFERENCES

Audley, R. J. 1967. What makes up a mind? *Decision Making*. London.
Bradley, F. H. 1935. *Collected Essays*. Oxford.
Brown, R. 1965. *Social Psychology*. New York.
Bunge, M. (ed.). 1964. *The Critical Approach to Science and Philosophy*. New York.
Carter, C. F. 1957. A revised theory of expectations. *Uncertainty and Business Decisions*. Eds. C. F. Carter, G. P. Meredith & G. L. S. Shackle. Second Edition. Liverpool.
Collingwood, R. G. 1940. *An Essay on Metaphysics*. Oxford.
Collingwood, R. G. 1944. *An Autobiography*. Harmondsworth.
Donagan, A. 1962. *The Later Philosophy of R. G. Collingwood*. Oxford.
Edwards, W. & Tversky, A. (eds.). 1967. *Decision Making*. Harmondsworth.
Festinger, L., Rieken, H. W. & Schacter, S. 1956. *When Prophecy Fails*. Minnesota.
Festinger, L. 1957. *A Theory of Cognitive Dissonance*. Stanford.
Festinger, L. and others. 1964. *Conflict, Decision, and Dissonance*. Stanford.
Fitzgerald, C. C. P. 1897. *The Life of Vice-Admiral Sir George Tryon*. London.
Friedman, M. & Savage, L. J. 1948. The utility analysis of choices involving risk. *Jour. Pol. Econ.* **56**.
Harsanyi, J. C. 1968. Individualistic and functionalistic explanations in the light of game theory. *Problems in the Philosophy of Science*. Eds. I. Lakatos & A. Musgrave. Amsterdam.
Hayek, F. A. 1952. *The Counter-Revolution of Science*. London.
Homans, G. C. 1967. *The Nature of Social Science*. New York.
Hough, Richard. 1959. *Admirals in Collision*. London.

[1] He was lampooned, for instance, in the film *Kind Hearts and Coronets*.

Jarvie, I. C. 1964. *The Revolution in Anthropology*. London.

Jevons, W. S. 1957. *The Theory of Political Economy*. Fifth edition. New York.

Keynes, J. M. 1921. *A Treatise on Probability*. London.

Keynes, J. M. 1936. *The General Theory of Employment, Interest and Money*. London.

Koestler, A. 1967. *The Ghost in the Machine*. London.

Luce, R. D. & Raiffa, H. 1957. *Games and Decisions*. New York.

Menger, C. 1963. *Problems of Economics and Sociology*. Trans. F. J. Nock. Urbana.

Mosteller, F. & Nogee, P. 1967. *Decision Making*. Eds. W. Edwards & A. Tversky. Harmondsworth.

O'Connor, D. J. 1948. *Mind*, **57**, 227, July.

Popper, K. R. 1950. Indeterminism in quantum physics and in classical physics. *Br. J. Phil. Sci.* **1**.

Popper, K. R. 1957 *a*. *The Poverty of Historicism*. London.

Popper, K. R. 1957 *b*. *Observation and Interpretation*. Ed. S. Körner. London.

Popper, K. R. 1966. *Of Clouds and Clocks*. Washington.

Popper, K. R. 1967. La rationalité et le statut du principe de rationalité. *Les Fondements Philosophiques des Systèmes Economiques*. Paris.

Rapoport, A. 1960. *Fights, Games and Debates*. Michigan.

Rapoport, A. 1966. *Two-Person Game Theory*. Michigan.

Rapoport, A. 1967. International relations and game theory. *Decision Making*. London.

Savage, L. J. 1954. *The Foundations of Statistics*. New York.

Schelling, T. C. 1963. *The Strategy of Conflict*. New York.

Shackle, G. L. S. 1952. *Expectation in Economics*. Cambridge.

Shackle, G. L. S. 1955. *Uncertainty in Economics*. Cambridge.

Shackle, G. L. S. 1961. *Decision, Order and Time*. Cambridge.

von Neumann, J. & Morgenstern, O. 1953. *Theory of Games and Economic Behaviour*. Third edition. Princeton.

Watkins, J. W. N. 1955. Decisions and uncertainty. *Br. J. Phil. Sci.* **6**, 21.

Watkins, J. W. N. 1958. Confirmable and influential metaphysics. *Mind* **67**, 267.

Williams, J. D. 1966. *The Compleat Strategyst*. Revised edition. New York.

# COMMENT

## by ALAN DONAGAN

Most of Professor Watkins' conclusions are so congenial to me that what I have to say may afford him too meagre an opportunity to confirm and develop them. He rightly reminds us of two reasons why a failure may call for study.

First, one man's failure is not always another's success. It may be conceded that Villeneuve's failure at Trafalgar is to be understood less by considering what he did, than by considering what the successful Nelson and his fleet did. Yet Nelson's grand tactics at Trafalgar are intelligible only in the light of his and others' reflections on previous engagements that were, from the British point of view, unsuccessful: specifically, inconclusive actions between fleets both drawn up in orthodox line of battle. When a failure occurs because of a blunder, or the malfunctioning of some institution, men of affairs need to know what went wrong, and why.

A second, and perhaps a more generous reason for investigating failure is that, even when neither blunder nor malfunction occurs, the race is not always to the swift, nor the battle to the strong. In his last campaign, Lee's generalship is not less worth attention than Grant's. Sometimes this is obvious. In studying recent politics in the Soviet Union and its satellites, it is difficult not to occupy ourselves wholly with the defeated.

Hence I do not contest Watkins' censure of Collingwood for maintaining in his *Autobiography* that

Naval historians think it worth while to argue about Nelson's tactical plan at Trafalgar because he won the battle. It is not worth while arguing about Villeneuve's plan. He did not succeed in carrying it out, and therefore no one will ever know what it was (Collingwood, 1939, p. 70).

Watkins acknowledges that Collingwood did not advance this objectionable view in his other writings. Elsewhere I have explained his brief aberration as an intelligible if inadequate attempt to forestall a difficulty in his theory of metaphysics as an historical science; a theory which, in turn, was an intelligible if inadequate attempt to square his conception of metaphysics with developments in his theory of conceptual thinking after the publication of *An Essay on Philosophical Method* (see Donagan, 1962, pp. 267–8, 282–4).

Writing directly about history and anthropology, Collingwood himself

218

exposed the worst errors entailed by the conception of history as the history of success: the dehumanization of cultures and societies deemed backward by the investigator ('primitive' and 'underdeveloped' are sometimes, but not always, used with such an implication); and the disparagement of whole periods in which a given culture or society is held to have remained in backwardness, or to have relapsed into it. His methodological diagnosis of the notion that some historical periods are 'Dark Ages' is fairly well known:

Certain historians, sometimes whole generations of historians, find in certain periods of history nothing intelligible and call them dark ages; but such phrases tell us nothing about those ages themselves, though they tell us a great deal about the persons who use them, namely that they are unable to rethink the thoughts which were fundamental to their life (Collingwood, 1946, pp. 218–19).

But Collingwood's attack upon a related view, which is even more dangerous in practice, because it is sanctioned by the great name of Freud, appears to have been overlooked by many philosophers of history. In *Totem and Taboo* and other writings, Freud set a fashion not yet dead, for treating primitive societies and cultures as calling, not for historical enquiry, but for psychopathological diagnosis.

... Freud's programme is to reduce the differences between non-European and European civilizations to differences between mental disease and mental health ... This is not the place to lay bare in detail the quibbles and sophistries by which Freud persuades himself (and others too, apparently) that his programme has been carried out. My purpose ... is to remark that a person who can attempt to equate the difference between civilizations with the difference between mental disease and mental health ... is a person whose views on all problems connected with the nature of civilization will be false in proportion as he sticks honestly to his attempt, and dangerously false in proportion as his prestige in his own field stands high.[1]

In passages like these, Collingwood sober made amends for the few misdeeds of Collingwood drunk.

As an alternative to the pessimistic view of Collingwood's *Autobiography*, Watkins advocates what he calls an 'Imperfect Rationality' Principle. This

---

[1] Collingwood, 1938, p. 77 *n*. The fashion of reducing the differences between non-European and European civilizations (substituting the name of whatever civilization you please for 'European') to differences between mental disease and mental health still reigns. It is rarely decried. Hence it was encouraging to encounter Martin Bernal's Collingwoodian criticism of Lucian Pye's recent book, *The Spirit of Chinese Politics* (1968): 'Its subtitle *A Psychocultural Study of the Authority Crisis in Political Development* makes it explicit that, by analogy at least, the author believes that China is suffering from a neurosis, if not a deeper mental illness ... The analogy also suggests that while China is adolescent and irrational, America is adult and responsible' (*New York Review of Books*, 11, 7 November 1968, p. 29).

Principle, as he employs it in the closing section of his paper, has the form of a methodological rule concerning historical explanation. In that form, it enjoins two things:

(1) 'that to provide a conjectural explanation for a past action is to postulate a decision-scheme which has a practical conclusion of which that action could be the natural outcome' (p. 209);

and,

(2) that 'such a conjectural explanation . . . be as little *ad hoc*, as plausible, and as well supported by evidence (especially evidence other than that provided by the action itself) as we can make it' (p. 209).

A 'decision-scheme' is Watkins' term for 'the set of all those considerations, whether consciously formulated or not, that enter into a particular piece of decision-making' (p. 206).

The elements of a set of considerations that constitute a decision-scheme are divided by Watkins into three sub-sets, the third of which may be empty. They are: (1) those considerations that compose the agent's 'complete situational picture'; (2) those that compose his 'preference-map'; and (3) those that, when appropriate, constitute his 'apparatus for dealing with risks and uncertainties' (pp. 206, 208). An agent does not necessarily articulate his decision-scheme to himself. It may be unconscious; but, even when it is not, he may act too rapidly to be able to formulate it before acting. Yet a decision-scheme may be articulated, and, when it is, Watkins declares that it 'can . . . be regarded as a body of propositions, some in the descriptive and some in the imperative mood, issuing in a practical conclusion' (p. 208). I take this to be elliptical for something like the following: an articulated decision-scheme can be regarded as a body of propositions expressible in sentences of which some are in the indicative mood, some in the imperative, issuing in a practical conclusion. The agent's situational picture will of course be expressible in sentences in the indicative mood; his preference-map and his apparatus for dealing with risks and uncertainties will presumably be expressible in the imperative. Even so, questions remain. What does Watkins mean by 'propositions [expressible by sentences] in the imperative mood'? In what sense do the propositions that constitute an articulated decision-scheme 'issue' in a practical conclusion? And what is a 'practical conclusion'?

The third question is perhaps the easiest to begin with. Watkins employs several phrases that presuppose a distinction between an agent's practical conclusion and his action: for example, 'action . . . in line with [his practical] conclusion' (p. 210), and 'practical conclusion appropriate to the peculiar action in question' (p. 210). Moreover, in order to show that, while lacking 'the perfect triviality of a logical rule of inference', the Rationality Principle nevertheless has 'the sort of "triviality" that goes with indispensability' (p. 173), Watkins offers the example of a man who down to the moment of action, has a preference-map showing certain ends as strongly preferred to all others, and a situational picture showing that those ends can be achieved, and only achieved, by a course of action in which he takes a certain air-line flight; who, in addition, admits neither uncertainty whether the results he prefers will be achieved by his proposed course of action, nor fear that some undesired result may follow; and who, finally, remains physically able to take the air-line flight in question. That a man in such a condition may yet not take that flight is, Watkins contends, logically possible. In terms of explaining actions by decision-schemes, this contention may be elaborated as follows: it is logically possible that a man's decision-scheme may issue in a certain practical conclusion, and yet that he will not act in line with that practical conclusion. Watkins' rationality-principle in §3.2 may then be rendered as the rule: in providing conjectural explanations of past actions, to disregard the logical possibility that an agent may not act in line with the practical conclusions he holds at the time of acting.

Further light on the question, 'What is a practical conclusion?' is cast by answers to our two other questions. First, in what sense do the propositions that constitute an articulated decision-scheme 'issue' in a practical conclusion? Watkins' remarks about unarticulated decision-schemes are crucial here. He appears to conceive an unarticulated decision-scheme as a set of propositions such that all the possible implications of any proposition in the set, or of any conjunction of them, are also included in the set. The inference follows that

it is logically impossible that a decision-scheme should be fully articulated; for every proposition has infinitely many logical consequences. Thus a decision-scheme is bound to have *unnoticed implications* (p. 208).

Hence Watkins concludes that the practical conclusion that an agent draws from his decision-scheme may not be what his decision-scheme logically implies; and that an agent may change his practical conclusion, not because he has changed his decision-scheme, but merely before his awareness of its implications has changed.

This conclusion forestalls one answer to the question, what is it for a decision-scheme to 'issue' in a practical conclusion? If a man may draw a practical conclusion (e.g. that a move in chess is right, though in fact it is fatal) which does not logically follow from his decision-scheme (cf. p. 210), then for a decision-scheme to issue in a practical conclusion is not the same as for it logically to imply that conclusion. What other answers are possible? There seems to be only one. It is that for a decision-scheme to issue in a practical conclusion is for that conclusion to be drawn from it as an inference. The drawing of such inferences, in Watkins' view, may be either conscious or unconscious. In the former case, the decision-scheme from which the conclusion is drawn will presumably be articulated by the agent.

It follows from these results that no practical conclusion is generated by an agent's entire decision-scheme. Whether his practical conclusion is arrived at consciously or unconsciously, it issues from only that part of his decision-scheme which he has consciously or unconsciously noticed. It is only that part that a conjectural explanation is required to postulate.

Yet there is a sense in which conjectural explanations necessarily exhibit an agent's practical conclusion as logically following from his decision-scheme. In explaining why Mr Smith decided not to buy himself a new camera it suffices to state that, in casting his monthly accounts, Mr Smith found that, from the balance he had allowed for (£125), an unexpected expense of £70 had to be deducted, leaving him only £55, which was not enough. It is not necessary to explain that Mr Smith thought that $125 - 70 = 55$. Suppose, however, Mr Smith was trying to discover why he disastrously overdrew his account. He had been certain that his balance covered the £85 cheque he wrote for his new camera. He discovers, to his horror, that the counterfoil of the cheque for his unexpected £70 expenditure reads: 'Bt. fwd. £125; this ch. £70; Balance, £105.' In explaining what happened, he—and we—must mention that he had taken 125 less 70 to be 105. Presumably there will be an explanation for this, for example, that he had mistaken the digit '7' for the digit '2'; and for that there may be a further explanation. Yet there is no need to explain why a man who took 125 less 70 to be 105 arrived at the balance Mr Smith arrived at. On his utterly wrong rule of inference, his operations were impeccable. In general, a mistake in reasoning is equivalent to adopting, however transitorily, a wrong rule of inference; and only by so treating them, can such mistakes be represented in conjectural explanations. The logical smoothness of a conjectural explanation does not rest on any foolish assumption of the logical competence of human agents; rather, it is a necessary condition of revealing the precise points at which their thinking was illogical.

Further light on Watkins' conception of a practical conclusion is thrown by his description of an articulated decision-scheme as in part (presumably the preference-map and the apparatus for dealing with risks and uncertainties) composed of 'propositions . . . in the imperative mood' (p. 208). Sentences in the imperative mood are not usually described as expressing propositions. R. M. Hare has erected a theory of moral reasoning in part on the principle that sentences in the imperative mood, which are used to tell somebody *to* do something, are not used to tell somebody *that* something is the case; that is, are not propositional. Sentences used to tell somebody *to* do something, Hare calls 'prescriptive'. He goes on to assert that a man who wholeheartedly accepts what is expressed by a prescriptive utterance is likely to live, not merely talk, differently from one who does not (see Hare, 1963, p. 23). Watkins has shown that this also holds of a man who wholeheartedly accepts certain situational pictures (which are expressed in indicative sentences): for example, that crime does not pay. The difference Hare presumably had in mind was that, while a man who believes that crime does not pay is likely not to take to a life of crime, if he does, his doing so, although not tending to show that he does not wholeheartedly accept the proposition that crime does not pay, on the other hand does tend to show that he does not wholeheartedly accept his poor but honest father's admonition, 'whatever you do, keep straight!'

Up to a point, Hare's view is undeniable. A man who wittingly takes to a life of crime necessarily disobeys his father's prescription to go straight; but he does not necessarily contradict his father's supporting statement that crime does not pay. Watkins' practical conclusions are prescriptions an agent can obey or disobey; that is the force of saying that actions may be 'in line' with them or not, 'appropriate to' them or not. In terms of Watkins' theory, what is correct in Hare's view can be stated in two propositions.

(1) From a situational picture alone, however elaborate, no practical conclusion follows.
(2) It is because practical conclusions are prescriptions that actions can be 'in line with' them or not, 'appropriate to' them or not.

I do not see any reason why Watkins should deny either of these propositions.

Yet Hare and Watkins do not wholly agree. For Hare maintains that if a man wittingly acts out of line with a certain prescription, then he cannot wholeheartedly accept it. Watkins, however, appears to allow that such a man may accept it. His example in §3.2, as we have seen, appears to imply that a man may wholeheartedly accept a practical conclusion prescribing that he take a certain air-line flight, and yet *irrationally* not take that flight. Misleadingly, Watkins describes such a man as 'chang[ing] his mind for no reason' (p. 173). His point, however, is that the man in question does not cease to think that the practical conclusion really does follow from his situational picture and his preference-map, according to his apparatus for dealing with risks and uncertainties. Nor does he change his decision-scheme. He continues to accept his practical conclusion, but he does not act on it. In not acting on it, he is irrational. And that is why Watkins holds that the Imperfect Rationality Principle, in its form as a methodological rule, is not purely logical. It rests on a postulate about human conduct: namely, that unless prevented by physical causes, people act in accordance with the practical conclusions of their decision-schemes (p. 209). If I understand him, Hare regards this postulate as analytic, and therefore logically superfluous; Watkins considers it synthetic.

That may be why Watkins described as 'propositions' even those parts of decision-schemes that are expressed in sentences in the imperative mood (p. 208). It is not self-contradictory to assert of somebody that he holds a certain proposition, and yet does not act on it. To act contrary to one's beliefs may be either irrational or immoral, or both; but a man may be either or both. If accepting a preference-map, or an apparatus for dealing with risks and uncertainties, were no more than accepting a proposition, then Watkins would be right and Hare wrong.

Is a preference-map, or an apparatus for dealing with risks and uncertainties, no more than a set of propositions? There appears to be a distinction between adopting a preference-map and accepting such propositions as that it is a good one, is the best one available, and the like. It is reflected in the distinction between accepting a command and accepting propositions about it, for example, that it is legitimately given, or intelligent, or the like. Both adopting a decision-scheme and accepting a command appear to be quite distinct from accepting a proposition. This is obscured by the intimate connections between them. If I think a command addressed to me to be legitimate, I have a reason for obeying it, which, with most men, in most institutions, with respect to most commands, is decisive. Hence the temptation is strong to confound commands with certain propositions, and to overlook the fact that those propositions are about the commands. In describing an articulated preference-map as a set of propositions, I think Watkins has succumbed to such a temptation. Hare, I think, has succumbed to the converse temptation of confounding certain propositions about preference-maps (e.g. that acting according to a certain preference-map would be right) with the preference-maps they are about. Both temptations must be resisted.

It might be objected that nevertheless some distinction must be drawn between accepting a preference-map and acting on it, or trying to act on it. Can a man not accept the command to perform some task in the afternoon, and yet, when the afternoon comes, not perform it? May he not then accept a preference-map which, together with a situational picture and an apparatus for dealing with risks and uncertainties, both of which he also accepts, yields the practical conclusion to perform a certain action at a certain time, and yet, when that time comes, not perform it?

Two issues here must be separated. (1) A man may accept a certain preference-map until he discovers that, together with a situational picture and apparatus for dealing with risks and uncertainties that he also accepts, it yields a certain practical conclusion. In refusing to act upon that conclusion, while retaining his situational picture and apparatus for dealing with risks and uncertainties, he rejects his preference-map. This case I take to be unproblematic. Nobody taking Hare's position that preference-maps are non-propositional denies that a man may reject both a preference-map, and a practical conclusion to which it leads, when he discovers what that conclusion is, even though he accepted that preference-map before he made that discovery. (2) A man may accept, about an action to be performed in the future, a practical conclusion yielded by his decision-scheme; and yet, when the time for that action comes, he may not perform it. This case presents a genuine difficulty; for it presupposes that the agent both genuinely did accept a practical conclusion, and did not act on it. The difficulty may, however, be resolved by distinguishing the time at which the agent accepted the practical conclusion, from the time at which he did not act on it. An historian constructing a conjectural explanation of an action is concerned with the agent's decision-scheme *at the time of acting*. Evidence of the agent's decision-scheme before that time may be pertinent; for, since every change in an agent's decision-scheme calls for explanation, historians do not postulate such changes unless they are obliged by evidence to do so. This may obscure (as I think it does in §3.2 of Watkins' paper) what on reflection is obvious, that to explain an action at time $t$ by a practical conclusion accepted by an agent prior to $t$ will not suffice, because it will not exclude the possibility that the agent changed his mind.

Some difficult examples remain. Throughout a period of time an agent may intend within that period to perform a task that will itself take time, and yet may not perform it. Thus, you may intend to mow the lawn during the afternoon, and may hold to your intention without actually beginning, until you find it is too late to begin. The point here is that although throughout the period from $t$ to $t_n$ you intend during that period to carry out a complex task $k$, which would consist in performing the acts $a_1$, $a_2 \ldots a_m$, at no time between $t_1$ and $t_n$ do you intend then to do $a_1$, or any other of the acts that constitute a performance of $k$.[1] Throughout the period $t_1$ to $t_n$ your state is that of intending at some future time within that period to begin carrying out $k$. Hence such cases are not exceptions to the

---

[1] I owe both point and example to Professor Judith Jarvis Thomson.

position that an agent cannot knowingly accept at $t$ the practical conclusion to perform a certain *act* at $t$, and yet not perform it or attempt to perform it.

I conclude, then, that for an agent at any given time knowingly to accept a practical conclusion that calls for a certain action at that time, logically implies that he perform or attempt to perform that act. Hence an historian, explaining an action by showing that it has in accordance with a practical conclusion yielded at the time of action by the agent's decision-scheme, neither requires a methodological Principle of Imperfect Rationality, nor presupposes a synthetic proposition to the effect that, unless prevented by physical causes, agents act, or attempt to act, in accordance with their decision-schemes. The Principle of Imperfect Rationality presupposes nothing that is not analytic. It is, therefore, both true and logically redundant; but it is not trivial in an important sense that Watkins does not mention: it is not obvious.

It should now be evident that I endorse what, from the point of view of historians and social scientists, is Watkins' major thesis: that unsuccessful actions are to be explained in exactly the same way as successful ones, by reconstructing the decision-schemes of their agents.[1] My principal disagreement with Watkins is over what most historians and social scientists would dismiss as a tiresome philosophical technicality: whether such explanations presuppose a non-analytic Imperfect Rationality Principle. I have one further, and less technical, difficulty. Granting Watkins' major thesis, that unsuccessful actions are to be explained by reconstructing the decision-schemes of their agents, does it follow that the agents whose actions are so explained will not appear as comi-tragic unpersons, or as dehumanized in some other way?

The explanation of Admiral Tryon's disastrous manoeuvre advanced by Richard Hough and accepted by Watkins indeed rehumanized the Admiral, but why? Because it provided his action with a decision-scheme? But people do all kinds of grotesque and monstrous things that succeed, in the sense of accomplishing their purpose; and presumably the decision-schemes of such actions will contain grotesque and monstrous elements. Hough's explanation rehumanizes Tryon, not because it provides his action with a decision-scheme, but because the decision-scheme it provides for it is a humanizing one.

Even when tragi-comedies like the Profumo case, or crimes like Christie's murder of Timothy Evans by procuring a miscarriage of justice, have been explained by reconstructing the decision-schemes of the principal agents, those agents do not, as a rule, cease to be figures of mockery or horror. It

---

[1] Watkins' theory of historical explanation closely resembles Collingwood's. Compare, for example, Donagan, 1962, pp. 192–6.

may be that mockery and horror of those who caricature or exaggerate our absurdities and vices is a means of self-control. Yet, however much we understand the decision-scheme by which Christie's crime may be explained, to most of us Christie will remain something of an unperson. We may understand, but there are limits to what we can bear.

## REFERENCES

Collingwood, R. G. 1938. *The Principles of Art.* Oxford.
Collingwood, R. G. 1939. *An Autobiography.* Oxford.
Collingwood, R. G. 1946. *The Idea of History.* Oxford.
Donagan, A. 1962. *The Later Philosophy of R. G. Collingwood.* Oxford.
Hare, R. M. 1963. *Freedom and Reason.* Oxford.
Pye, L. 1968. *The Spirit of Chinese Politics.* Cambridge, Mass.

# REPLY

## *by* JOHN WATKINS

I have gained a good deal from Professor Donagan's careful and courteous examination of my (all too) rough sketch of the idea of a decision-scheme. But on the main issue between us I am obdurate. He writes (p. 226, my italics):

> I conclude, then, that for an agent at any given time knowingly to accept a practical conclusion that calls for a certain action at that time, *logically implies* that he perform or attempt to perform that act ... The Principle of Imperfect Rationality presupposes nothing that is not analytic. It is, therefore, both true and logically redundant ...

Against this I still hold that there is a logical gap between decisions and their execution and that to bridge it a synthetic principle is needed.

A behaviourist may say that deciding to do something and starting to do it are one and the same thing. But Donagan does not accept this behaviouristic identification (nor do I): he emphasizes that a man may decide now to do a certain thing later, and he allows that in cases of this kind it does not follow logically that the man will actually do it; it is only where a man decides to do a certain thing here and now, that it follows logically, according to Donagan, that he does it or at least attempts to do it.

Let us discuss this in relation to a simple example. Jones and his friend, Smith, reach a bus-stop in time to see their bus drawing away. Smith (who has near-telepathic insight into Jones' thought-processes) says to himself:

(1) Jones is thinking that it will be some time before the next bus comes; that if he waits for it he will get very chilly since it is a wintry evening; that if he walks home briskly he will warm up and will get home not much later than if he waits for the bus.

(2) Jones would rather get home a bit later but warmed by the walk.

(3) So Jones' practical conclusion is that he should start walking now.

(4) So Jones will now start walking.

Does (3) logically imply (4)? No; for (4) may be false when (3) is true. It is not merely that Jones may be struck by a thunderbolt, etc. We might try to exclude that sort of thing by weakening (4) to:

(4') So Jones will now start walking, physical circumstances permitting.

More importantly, Jones may be struck by a new thought: he may say to himself:

> But wait! Smith will have to stay for the next bus—it's too far for *him* to walk. And he'll be disappointed if I leave him now. I don't want that. So I'll wait for the bus.

We might try to exclude this sort of thing by strengthening (3) to:

> (3') So Jones' practical conclusion is that he should start walking now, and no further consideration will occur to him that would disturb that conclusion.

Does (3') entail (4')? Well, if we go on appropriately strengthening (3) and weakening (4) we shall eventually render the latter logically derivable from the former—but not so long as our revised (3) continues to record only what is going on in Jones' head while our revised (4) predicts something (however hedged) about the movement of his limbs.

Donagan says 'that for an agent knowingly to accept a practical conclusion that calls for a certain action at that time, logically implies that he perform or *attempt to perform* that act' (my italics). This would be true if 'attempt to perform' means no more than 'decide now to do'. Let us test this against an amusing passage in Iris Murdoch's novel, *The Bell*. Dora, a young woman who has decided to rejoin her husband, gets to Paddington Station early and finds a seat on the train. It is a very hot afternoon and the train becomes very crowded, with many people standing. An elderly lady halts in the doorway of Dora's compartment: it seems that she is a friend of the person sitting next to Dora and that they had hoped to travel together. It occurs to Dora that she ought, perhaps, to give up her seat. She silently argues against this:

> She had taken the trouble to arrive early, and surely ought to be rewarded for this ... There was an elementary justice in the first comers having the seats ... The corridor was full of old ladies anyway, and no one else seemed bothered by this, least of all the old ladies themselves! Dora hated pointless sacrifices. She was tired after her recent emotions and deserved a rest. Besides, it would never do to arrive at her destination exhausted ... She decided not to give up her seat.
> She got up and said to the standing lady 'Do sit down ...'.

Donagan could hardly accept this as it stands. He might roundly declare that 'She got up' is inconsistent with 'She decided not to give up her seat'; or he might say that Dora, although *attempting* to remain sitting, was raised to her feet by some mysterious force; or perhaps the author has suppressed the fact that Dora *did* remain sitting for a time in accordance with her decision, and then re-thought the matter and arrived at the conclusion that she should give up her seat after all. But if one regards the rationality principle as synthetic and non-necessary one can accept this passage just as it stands, as a humorous portrayal of human inconstancy.

Apropos Donagan's last point: in my slang there are *unpersons* and *anti-persons*. The former seem less than human to us because we cannot discern any decision-scheme behind their actions; these people are grey enigmas. The latter seem less than human to us because we discern in their decision-schemes certain horrifying elements.

# UNDERSTANDING AND EXPLANATION IN SOCIOLOGY AND SOCIAL ANTHROPOLOGY

## *by* I. C. JARVIE

### I. INTRODUCTION

It is easy to conflate problems of explanation and of understanding, perhaps because the ordinary words 'explanation' and 'understanding' are used almost interchangeably.[1] However, in the philosophy of science we can, by adopting a deductive explication of the former, serve clarity. Whatever understanding is, explanation is the process of deducing one statement from others in accordance with some formal and also some material requirements.[2] So, while what one man understands another can be puzzled by, it cannot be that what obeys the rules of valid deduction to one man does not to another. Understanding *might* seem amenable to analysis—or, rather, is it misunderstanding that is so amenable? Understanding or making sense of, like perceiving, would seem to be an interplay between expectation and feedback; the attempt to 'impose' a mental set and at the same time to correct that set. Our preconceptions are thus essentially involved in the process of understanding: and among these preconceptions are what appear to resemble standards: we find unsatisfactory those explanations of human behaviour which do not render it 'reasonable'; we are doubtful of theories attributing to atoms asymmetric properties. The analogy between rationality and symmetry in nature may not be close, but their roles as standards of understandability are: understanding a human being and understanding an atom are not different processes.

All this will seem like uncontroversial platitudes until it is put against the standard view that we should approach the study of alien societies without prejudice and preconception. It becomes my counter-contention that we *cannot but* attempt to translate alien societies into terms of ours, and only where there occur breakdowns and inconsistencies in our translation do we scrutinize our preconceptions and change them. The problem raised by my contention is how to study other societies objectively; that is,

---

[1] Robert Brown (1963, p. 41) defines the one in terms of the other: 'Explaining away, then, is the removing of an impediment, an impediment either to someone's relationships with other people or to his intellectual understanding.'

[2] For the requirements see Popper, 1959, pp. 59–62, and 1957, pp. 24 ff.; and Bartley, 1962, pp. 15–33.

231

how to be critical of our preconceptions in order to test for their truth-values. In the heyday of intellectualist anthropology the scholar took it for granted that he and his society were models of rationality and understand-ability. He was thus led quickly into misunderstanding and misjudging alien societies simply because they were different. In the more liberal climate of this century it has been argued that we can avoid misjudgment by making no judgments at all—since moral value systems are incommensurable—and this will possibly help us to have fewer misunderstandings.

Like the liberal critics of intellectualism, I find myself against both misunderstandings and misjudgments. However, I do not regard relativism as either the only respectable or as a satisfactory alternative. Indeed my contention above leads me to conclude that *all* our efforts to understand will be misunderstandings, misjudgments and oversimplifications. All we can do about it is face the fact and be as critical as possible of our efforts.

Recently there has appeared a new and plausible doctrine which con-cedes many of the relativist points but seemingly avoids total commitment to it. Its author is Peter Winch, and he has published it in a paper, 'Under-standing a Primitive Society' (Winch, 1964, pp. 307–24), which is a follow-up to his intriguing book *The Idea of a Social Science* (Winch, 1958). Winch argues that there are certain necessary conditions, social and moral, for there to be any social life at all, thus he is no relativist. He also argues, if I may indulge a crude first approximation to his position, that cross-cultural value-judgments will always be misjudgments—and therefore should be avoided—because there is no language game in which cross-cultural value-judgments could be legitimate moves. The conflict between his position and my previous contention about the whole nature of under-standing itself is obvious. I therefore wish to argue that Winch yields too much to relativism, that there are language games in which cross-cultural value-judgments are legitimate moves, that these moves are played in all cultures as well as the sociologists' sub-culture. Indeed, such cross-cultural value-judgments, such use of one's own society as a measuring instrument or a sounding board, is the principal way sociological under-standing of an alien society is reached.

## 2. WINCH'S POSITION IN OUTLINE

Winch's concern in 'Understanding a Primitive Society' is the broad problem of how we can understand the customs of an alien primitive society. He focusses his discussion not on understanding primitive customs in general, but on understanding those special cases of primitive customs which involve beliefs—like beliefs about how reproduction occurs—which conflict with beliefs taken for granted in our society. Thus we have a narrower problem of understanding and a broader one. The broader one is of understanding primitive customs in general; the narrower one is of understanding those customs involving ideas which conflict with our ideas. Winch apparently regards this narrower problem of understanding as crucial to the broader problem of understanding, and concentrates on it presumably in the belief that if we can understand the special cases, understanding primitive customs in general will be easy.[1]

Winch comes at his problem by discussing a remark in Evans-Pritchard's book on Azande magic (Evans-Pritchard, 1937) to the effect that Azande magical beliefs not only run counter to beliefs of ours, but do not themselves 'accord with objective reality'. In dealing with Azande witchcraft, Evans-Pritchard asserts that not only do some of its component beliefs contradict some of ours (which is not so serious, since both could be false), but also that ours are true and theirs are false.

Winch's central argument is that in saying that the beliefs of primitives do not accord with objective reality we are talking in terms of reality seemingly conceived of outside language and culture. This, however, Winch believes we cannot legitimately do. Even if there is such a thing as objective reality outside language, to conceive it outside language and culture is impossible. In saying that primitive beliefs are false we are saying they do not accord with *our* conceptions of reality; with reality as *conceived* by us, if you prefer. But imposing our conceptions in this way is no way to come to understand primitive society. It is no way to understand it because magical beliefs, for example, may well be tied into a whole system of other beliefs, a veritable world-view in which the disputed beliefs play an essential social and theoretical role. This system constitutes an on-going way of life and it shows grave lack of understanding (in several senses) to single out elements of it and attempt to adjudge them not in accord with objective reality. They may not accord with reality as seen from the standpoint of our language and culture, but they can hardly be said not to accord with the same reality they themselves conceive—even the reality, linguistic and cultural, they themselves embody.

It would thus seem that in certain ways our society's beliefs and primitive beliefs are incommensurable, they cannot be compared. The road to understanding is nevertheless to seek out the universal problems of human life; but not to give them universal solutions, since these would be outside language and culture, which is impossible. Rather, we have to try to show

---

[1] Since Winch does not set out these relations between the problems explicitly I will not take up this latter contention. But that the solution of the general problem will flow from a solution of the narrower problem does not seem obviously true.

how the primitive society solves them in its own way. Here Winch has skilfully pulled back from the relativist precipice by asserting that all societies do have something in common, the problem of human life in general and its general root in universal facts, i.e. birth, sex and death.[1] So if we begin social studies with these universal facts and their ensuing universal problems, we have some bases of comparison, some universals in terms of which we can begin to make sense of any society: namely, how each society handles the universal facts of birth, sex and death.

### 3. AN ALTERNATIVE TO WINCH

Before expanding and discussing what I take Winch's position to be,[2] I should like to contrast it with the admittedly not very sympathetic position from which my criticisms will come.[3]

I have no quarrel with the suggestion that in the social sciences we aim to understand other societies. But I would go further: social science also aims to *explain* and *appraise* societies, their institutions, and beliefs. Social science, like any other science, uses language in attempts to make true universal statements about the world, which will explain the world. We have these aims because of something like an innate curiosity which drives us to ask questions, formulate problems, seek answers which will improve our knowledge and understanding. The phylogenetic explanation of our curiosity and concern with problems sees it as an adaptive mechanism. The ontogenesis of problems seems to be something like disappointed expectations, i.e. refuted beliefs and theories.[4] The (conscious and unconscious) expectations and theories we hold are clearly deeply influenced by the upbringing and education—the culture, in fact—of which we are products.

[1] In another paper, Winch argues that there are also moral universals, especially truth-telling, integrity and justice. See Winch, 1959–60, pp. 231–52. Winch's universals bear comparison with Malinowski's theory of basic needs (see Ralph Piddington, 1957), which were mainly biological, and with Goldschmidt's recent theory that they are socio-cultural, see Goldschmidt, 1966.

[2] Winch's ideas have already been discussed in the following: Brodbeck, 1963, pp. 309–24; Gellner, 1962, pp. 153–83; Louch, 1963, pp. 273–86; Winch, 1964, pp. 202–8; Louch, 1965, pp. 212–16; Saran, 1964–5, pp. 195–200; Martin, 1965, pp. 29–41. There are also three papers on Winch in the forthcoming volume *Problems in the Philosophy of Science, Proceedings of the International Colloquium, London, 1965*, vol. 3, eds. I. Lakatos & A. Musgrave (Amsterdam 1968). They are: 'The new idealism' by Gellner, 'The very idea of a social science' by Cohen, and 'Anthropomorphism in social science', by Watkins.

[3] There are some criticisms of Winch's book in my review in *Brit. J. Philos. of Sci.* 12, 1961, 73–7. J. Agassi and I have developed our own view of how to understand magic in 'The Problem of the Rationality of Magic', *Brit. J. Sociol.* 18, 1967, 55–74. The two extremely important papers of Gellner listed in the previous note have influenced me considerably, and a certain amount of overlap has been unavoidable.

[4] This is the general tenor of the philosophy of K. R. Popper.

It follows from all this that the entire activity of social science involves, and necessarily involves, bringing the ideas, standards and concepts of our language and culture to bear on those of other cultures, seeking to find correspondences and trying them out, then following through on the feedback and modifying both our concepts and the correspondences. What we find problematic will at first be a result of our culture and expectations (knowledge). What we find good and bad, true and false, will depend at first on our culture and knowledge. In trying to understand other peoples we resort to something rather resembling the act of translation from one language to another (see Gellner, 1965). We do our best to find equivalents in translating, but must also rely on analogy, similarity, metaphor and the occasional directly appropriated, then explicated, untranslatable concept —which litter the anthropological literature. This act of translation fails if it leaves us unable to appraise the truth of the beliefs of the other culture, or to discuss the values of that culture as we can our own. It would fail in the same way as a translation from a French or Russian science or religion or ethics text would fail if it left us in no position to appraise the finished product.

(One curious phenomenon which I will return to below in §6 is how constant acquaintance with alternative sets of concepts and ways of classifying the world tends to shatter a lot of one's own preconceptions. The bizarre ceases to be bizarre, and a degree of self-awareness about one's own culture and society is gained which only the greatest of imaginative leaps can otherwise obtain.)

## 4. WINCH'S POSITION DEVELOPED

Winch has produced a plausible, even a beguiling, argument. I am uneasy about more than I can coherently criticize; I can criticize more than I have space for. The general Wittgensteinian underlay has been discussed elsewhere.[1] My aim will be to come to a fuller understanding of how we understand alien societies, and I will proceed via a detour through Winch's pretty flower-bed of argument.

My entire account of understanding would seem to beg the questions Winch wants to raise. It seems that in his view whether a statement is true will depend upon what it means. What it means, in his view, will depend upon how it is used, how it functions as part of the form of life it belongs to. The notion, then, of translating one form of life into terms of another doesn't make much sense. The way beliefs operate in a form of life is

---

[1] Criticism of the meaning/use thesis can be found in Gellner, 1959; and Wisdom, 1963, pp. 335–47.

peculiar to that form of life. In particular, there is no reason to suppose that a statement true-to-them is translatable into a statement true-to-us; but if it is translatable into a statement false-to-us that does not show that it is false-to-them or false *tout court*.

Winch would I think agree that anthropologists have always struggled to make beliefs we cannot share, held by others, e.g. magical beliefs, understandable. This involves giving an account of them which satisfies our 'scientific' culture's criteria of rationality[1] or intelligibility. In some cases we find ourselves studying beliefs that are not true, yet they are believed to be true by the culture we study. Winch would contend that the last sentence is illegitimate: that it cannot be uttered intelligibly because the words 'not true' and 'true' belong to two quite separate universes of discourse (ours and theirs), or they both belong to ours and are not found used, of the beliefs in question, in theirs. But our universe of discourse cannot appraise other universes of discourse, or appraise itself as the only true universe of discourse. Reality is built into a universe of discourse, and the outcome of a question about it in another cannot be appraised in the first. Winch gives an example from religion: 'What [God's] reality amounts to can only be seen from the religious tradition in which the concept of God is used . . .' (p. 309). Winch is not here quite repeating the hoary old chestnut that only the religious can handle and discuss religious concepts. Unlike those religious people who claim this, Winch nowhere says there is anything to stop the non-believer coming fully to grasp and being able to play the religious language game. (It can hardly be a rule of the game that sincerity of belief is required, even though display of sincerity may be required in ritual fashion.) But the ability properly to play it, and only this, shows true understanding of it. 'Reality is not what gives language sense. What is real and what is unreal shows itself *in* the sense that language has' (p. 309).

There is then for Winch no external or objective reality against which to measure a universe of discourse; or, assuming that Winch thinks there is an external reality, discussion about it cannot take place within that universe of discourse. (The opacity of Winch's argument at this point may be a consequence of the classical difficulty of trying to express the ineffable (see Russell, 1922, p. 18).) This is true of the universe of discourse of science, apparently, as well as that of magic. Science exists in a cultural context outside of which the actions and statements which constitute it would not make sense. Those who do not play its game can hardly appraise,

---

[1] Although, incidentally, it would be a mistake to think that our scientific culture admits as rational only those views which agree with current orthodoxy. No one calls Newton's optics, or Bohr's model of the atom irrational, though they are by now superseded.

e.g. its theories or experiments; those who can play that game cannot use the game to appraise itself.

The question then about Azande magic is simply whether it is a coherent game, universe of discourse, world-view, form of life. Yet Winch denies that a 'yes' in answer to this question commits him to endorsing all magic. It only commits him to that which is 'one of the principal foundations of their whole social life' (p. 310). Magic in our culture is *not* foundational, it is parasitic as the black mass is on the mass. The former is not understandable without reference to the latter; but not vice versa. This aside, then, is the answer 'yes': are Azande beliefs a coherent system?

To begin with, not only do Azande beliefs lead to inconsistencies, but the Azande use *ad hoc* devices to evade these. Azande, however, do not themselves press their beliefs until they yield contradictions; and they showed no theoretical interest if these were pressed on them by the outsider. In their game so to speak they do not press; if we press them we are trying to force them to play our (scientific) culture's game. Azande do not regard the revelations of the oracle as hypotheses open to discussion and criticism: 'Oracular revelations are not treated as hypotheses and, since their sense derives from the way they are treated in their context, they therefore *are not* hypotheses' (p. 312). Indeed, 'to say of a society that it has a language is also to say that it has a concept of rationality' (or intelligibility? p. 318) and that standard may involve not bothering about *ad hoc*-ness, or about checking inconsistency; indeed, what is to *count* as consistency depends on the wider context of life (p. 318).

Are there then any universals on which the social sciences can get a purchase so that other forms of life can be understood without living them through? Winch isolates T. S. Eliot's trio of 'Birth, copulation, and death' as points every society has to be organized around. From motherhood, marriage and funerals, social science can start.

## 5. CRITICAL DISCUSSION OF WINCH

Many kinds of criticism have been levelled against Winch's ideas. One is that his conclusions fly in the face of what social scientists constantly and successfully do (cf. Gellner, 1968). Another is to show that his methodological conclusions do not follow from his philosophical premises (cf. Cohen, 1968). Convincing as these are, they by no means exhaust what is to be said. For my part, I shall criticize certain philosophical ideas Winch holds which seem to me to lead to totally mistaken views about what can and cannot be truly said about societies other than our own. Let me start where Winch is surely in the right: there is no point in defending Evans-

Pritchard's statement that Azande magical beliefs do not accord with objective reality. On Evans-Pritchard's own evidence, Azande magic is irrefutable, quite metaphysical. It asserts that there is a presence in the world causing things to happen. Whatever happens can be made to fit this theory. For this reason it is a question whether we can claim that our (culture's) physiological explanation of a man's death is nearer to the truth than the Zande magical one. Certainly any conceivable crucial experiment could be interpreted to fit the magical theory.

What can be said—and surely Winch will agree (he even makes use of it) —is that whereas on the one hand, our explanation comes from an open and critical intellectual system in which ideas about the world are constantly scrutinized, criticized, and revised to meet these criticisms, on the other hand, Zande magic is part of a closed and unrevisable system of beliefs which may have been that way for a long time. The issue then is not whether magic or physiology truly explain the death, but whether the choice between an open and a closed world-view can be rationally argued. And here Winch's view seems to be, roughly, that the question cannot be rationally argued because there are different standards of rationality embedded in each position, and we could only decide the issue if one side was allowed to impose its standards on the other. If this happened it would indeed settle the argument before it started—as Winch himself tries to do by suggesting the outcome will be inconclusive.

It is at this point that my agreement with Winch stops. I would hold that there is something like a community of rationality shared by all men, but recognized or fostered by different societies in varying degrees (none being perfect). This rationality consists at the very least in learning from experience, and especially from mistakes. All the ethnography I have ever read turns up no people, however primitive, who do not in some matters learn from their mistakes. The minimum standard is thus ever present even though seldom acknowledged. Those unaware of having it could have it disclosed to them that they possessed it by outsiders, like white men. But self-discovery is also possible. This could come about in the following way. Their closed system of ideas comes into contact with another closed system of different ideas. Suddenly for the first time there is the shock of choice: another view of the world is possible. Once this is realized, to accept either becomes difficult, since what if there is a third, fourth, etc.? Of course I am idealizing and simplifying, but all I seek to establish is the possibility. It would not then be an unreasonable conclusion to decide that whatever world-view is chosen it had better be held tentatively in case something better comes along. The first culture we know to have broken through to this stage is that of Ionia and its intellectual heirs, including ourselves (Popper, 1963, Chap. 5). That it is something of a feat to get so far should not be forgotten, since other reactions to culture clash are possible, especially dogmatism (we just know we are right); mysticism (truth surpasses understanding); and scepticism (all world-views are equally doubtful and thus equally arbitrary).

Having suggested there is a community of rationality recognizing at least learning from mistakes; and having also shown how awareness of choice of world-views could lead to a more tentative attitude, a more rational attitude, one that recognizes and seeks to avoid the possibility of making mistakes; I shall now discuss certain *consequences* of Winch's view that different societies have different standards of rationality and that therefore there can be no mutual appraisal between those standards. I shall argue that, in contradiction to Winch's view that standards of rationality are incomparable, rational argument is possible between open and closed systems: indeed the way we discover intellectual problems and make intellectual progress (both ontogenetically and phylogenetically, so to speak) depends on such argument being carried out (not only between primitives and primitives, primitives and westerners, but even between Einsteins and Bohrs).[1] Winch's view seems to: (i) deny standards to other societies which they in fact possess; (ii) suggest that we are our own best interpreters, that a society's own view of itself is the best possible; (iii) postulate a metaphysical distinction between the core of a culture and what is parasitic on it; (iv) presuppose that societies are demarcatable and seamless; (v) deny empirical content to the beliefs of primitive peoples.

(i) If, when we say that there are inconsistencies within Azande ideas we are imposing our standards on them, then a curious situation is created. Do the Azande recognize this inconsistency as a problem? If they get contradictory advice from their oracles what do they do? They can't do both things; they must choose. How do Azande choose? They certainly do not allow that their oracles are in any way discredited by the contradictions. What they do is what we could call resorting to *ad hoc* reinterpretations which resolve the inconsistencies; they show no theoretical interest in exposing and extirpating these inconsistencies. But the problem confronting them—inconsistency of prescriptions—is there and cannot be evaded. In whatever way Azande reach decisions in such cases we can report truthfully that Azande ideas sometimes yield contradictions which they resolve by *ad hoc* measures. A contradiction exists whether or not someone faced with it fails to see it, refuses to see it, or is indifferent to whether he

---

[1] I am referring to that part of the Einstein–Bohr debate which concerns standards of rational appraisal, see Niels Bohr, 'Discussion with Einstein on Epistemological problems in Atomic Physics', in Schilpp (ed.), 1949.

sees it or not. Its objectivity—if you like its reality—is affected by none of these attitudes. In fact with Azande it is not that they can't see any problem when they get contradictory advice, but that they adopt a poor strategy for dealing with it. Yet one should also acknowledge that *ad hoc* defence is common in our culture. This only goes to show that the standard which invokes consistency is by no means universally accepted in our society: but at least our culture has argued the issue.

Thus the Azande do possess some standard of rationality recognizable in the west as rational, though as rather poor. This contradicts Winch's claim that 'the context from which the suggestion about the contradiction is made, the context of our scientific culture, is not on the same level as the context in which the beliefs about witchcraft operate' (p. 315).

(ii) The above quotation from Winch already indicates what his probable reply would be: not that Zande have no standards, but that they have different and non-comparable standards. Even if I have shown that there is overlap between our standards and theirs, I have not shown that the non-overlap makes no decisive difference. Winch insists that it doesn't make sense to say that Azande magic is inconsistent, only: by our standards of consistency Azande magic is inconsistent. My claim is that overlap of standards is more important because in overlap may lie the seeds of appraisal and change. Winch's view somehow freezes them within their contemporary standards, unable to appraise and revise them. Most cultures will take their current beliefs as obviously true; in freezing current views Winch accepts the claim that they are (obviously) true. Assuming the alternative view for a moment, namely that a culture does not have to accept its own appraisal of itself, we would wonder why Winch thinks social anthropologists have to. Even in closed systems the *possibility* of rational discussion of current beliefs exists. Winch's view is based on a picture he seems to have of primitive society as clearly bordered, internally coherent and interlocking, all the parts reinforcing each other; he seems to think that Azande live by their magic and never sit around saying 'this is what we believe and it is true', for to say this they would have to have some conception of what it would mean if it were false. Winch claims their beliefs are not hypotheses and not, therefore, doubtable in this way. This amounts to the claim that a culture accepts its own appraisal of its own beliefs.[1] It is a false claim, and empirically refuted every time any society reappraises itself—whether from the inside or as the result of outside intrusions. To this Winch may say, from the inside never and from the outside only accompanied by violence of sorts which thus doesn't count.

[1] Around this point the argument draws very heavily on Gellner, 'The New Idealism', see note 2, p. 234.

This answer, be it good or bad, cannot be seriously considered except on the view that societies are seamless and coherent, and especially that their systems of ideas are closed and cut off in some clear-cut sense. But consider the following argument.

An idealized model of Azande magic would be one of an isolated group of people bowling happily on their way in a social system heavily dependent on magic. Not being aware of alternative views they never consider the truth or falsity of their magical beliefs. They are just a part of their way of life and it makes no sense for us to declare some arbitrary part of that way of life false. But what then do they make of their own member who one day announces: it says in your book of laws . . . but verily I say unto you . . .? Reforming kings and priests and prophets can change societies. They sometimes condemn previous beliefs not just as wicked, but as false, and previous deities and demons as false gods. So even within closed societies a monolithic world-view can come to be undermined.

Then, of course, if we look at culture clash we see other ways in which the Azande might come to doubt their own appraisal. Those Azande who have been educated abroad and have come back, what shall they believe? They can live as Azande half-believing in the old ways. Trying not to press the issue, under pressure they will usually hold to what their education has taught them. Consider the case of a member of a primitive society who has become a western doctor and returned to his society. He might turn a blind eye to local remedies for minor disorders; he will be more concerned to interfere in serious cases, or where the local remedy is harmful; when the case is hopeless he will again turn a blind eye to traditional practices.[1] What could be more reasonable? For minor matters and for hopeless matters it medically doesn't much matter what you do, so long as it is not positively harmful. Whereas in terms of public relations for when western medicine *can* do something, or culturally where the practices are intricately interlaced into the social system, it may matter a great deal not to interfere with custom. Yet there is a large class of serious but curable cases where traditional practices and modern medicine come into direct conflict and the former is to be rejected as false. It is a matter of empirical fact that in most parts of the world when this is put to practical test the population opts for the western magic (cf. Gellner, 1965). This may be because the westerners can often propose and then carry out crucial experiments. These have great power to impress—which also suggests that standards of rationality have universal components. Chinese in Hong Kong know that western medicine,

---

[1] I owe this rational reconstruction to a lecture given by Professor J. Agassi. He was discussing the uneasy coexistence of Chinese traditional medicine and western medicine in Hong Kong.

given a chance, can cure most cases of TB: traditionally to catch it was to be doomed, as the folk-tradition and the many fictional characters who die of it testifies. Of course, viewing western medicine as magic is not in conflict with Winch's view. But the preference for western magic by local standards is a refutation of Winch's thesis that a culture accepts its own ideas as true; cannot but operate with them; would be 'utterly lost and bewildered' (p. 311) without them; and therefore can't judge them against others. Whenever a culture can adapt itself to new ideas, and implement them into its way of life, Winch's view doesn't hold.

(iii) In the light of Winch's assertion, quoted immediately above, and his citing with approval Wittgenstein's saying '*The limits of my language mean the limits of my world*', and his citing of Evans-Pritchard's view that, 'In this web of belief every strand depends on every other strand, and a Zande cannot get outside its meshes because this is the only world he knows. The web . . . is the texture of his thought and he cannot think that his thought is wrong' (both quotes in Winch at p. 313), it is not easy to reconstruct what Winch's reply to point (ii) would be. Perhaps, that culture clash is a special case unrelated to the case of widely separated societies trying to come to terms with one another. This raises the question of whether there is a Zande 'web' of thought in the required sense: whether there is a Zande culture utterly different from those it comes into contact with, and having at its core certain fundamental meanings or conventions in terms of which it organizes experience. Or whether it isn't rather typically the case that societies interact constantly both with their neighbours and so to speak with themselves, with their history, and that no demarcatable system of ideas or meanings ever is fixed and permanent.

Winch says he avoids relativism and is able to say magic in our society is false because it is parasitic on the main ideas of our society—i.e. religion—and therefore can rightly be judged false in terms of, and by the standards of, the dominant beliefs. This cannot be done with Azande magic because it is not parasitic but fundamental to their way of life. In other words, he is unwilling to separate Azande magic from Azande culture, but insists on separating Azande culture from English culture. Is this not arbitrary? Let us go slowly here. Not separating Azande magic from Azande culture involves not separating it from any other Azande aspect. Now the Azande have a 'scientific' (in our sense) technology as well: can they then judge the one in terms, and by the standards, of the other? If Zande technology is parasitic on Zande magic, then it should be judged in terms of Zande magic. If Zande magic is parasitic on Zande technology, then it should be judged by the standards of Zande technology. Should Zande technology be judged in terms of Zande magic or vice versa? Or both in terms of both?

What is parasitic exactly? Winch's claim that magic in our society is parasitic is based on his view that it is a parody: the black mass parodies and distorts the idea of the mass. The mass, however, is a parody of Jewish synagogue practices. These in their turn may be parodies of Egyptian, Babylonian and other ceremonies. Here internal reform makes Winch's idea of a parody a parody of the real situation. What goes on in present-day witches' covens probably antedates the mass (see Lethbridge, 1962; Hughes, 1952). What is parasitic on what? When does the parasitic parody become the mainstay? Moreover, our society contains science and religion, and these two clash constantly, or at least they did—rightly or wrongly—a few times in the past. That is, they did try to judge each other false, despite frequent attempts to avoid these clashes by *ad hoc* redefinitions of the scope and nature of religion (tending on the whole to make it ever more vacuous and metaphysical).[1] Which of these is parasitic? Or is neither? Is it not pure metaphysics to seek a core of a culture which makes it what it is?[2] Are not now most societies pluralistic, and does not that empirical fact refute Winch's attempt to escape relativism?

(iv) Winch's response to this might be that we are trying to impose our pluralistic standard of rationality on the Azande. My reply is the much more radical view that we and the Azande are all parts of one pluralistic world culture. Winch's whole view presupposes separate ways of life which enable their members to live untroubledly within their existing system of ideas and brush off all contact with the rest of the world. Once upon a time, in small isolated Pacific island societies this might have been the case (although Kon Tiki and later investigations of Thor Heyerdahl and others showed that even that isolation was not complete) (see Heyerdahl, 1950, 1958). Now it is not. No culture really is a closed system, and any that fancies it is tends to get awakened rudely. In other words, Popper's model of the closed society probably never was actualized, and certainly is not now.[3] There are only different degrees of openness. And since they are not seamless the possibility for contrast and self-appraisal is always there. Azande differ from our society mainly in not having a tradition of self-reappraisal, and institutions to carry that tradition. They are *capable* of both things. And they certainly don't have our metatheory that self-appraisal and criticism is good. The differences between us are ones of

---

[1] Clashes over policy are still frequent nowadays, e.g. birth control. Logically, ideas which lead on to absurd or false conclusions have doubt cast on them. It seems *false* to say the world will be a better place without birth control.

[2] This kind of metaphysics is what Popper calls 'essentialism', see his *The Open Society and Its Enemies*, 1962, pp. 9–21.

[3] Popper nowhere suggests that it ever was actualized, or even could be.

empirical fact. But still there is a break-point. And that is when the conscious idea of critical appraisal and discussion is introduced.

This break-point may have much to do with the acquisition of literacy. For until there is literacy the past is in no way 'fixed' as a standard of comparison, the notion of learning is thus much less rich and definite. Indeed a very great strengthening of rationality in our sense of learning from experience seems to turn on literacy (see the important article by Goody & Watt, 1962, pp. 304–45). Moreover, our world-view seems more powerful because literacy enables us to incorporate and discuss other world-views.[1]

Winch's thesis that 'The concepts used by primitive peoples can only be interpreted in the context of the way of life of those peoples' (p. 315) founders because no such utterly cut-off and circumscribed 'way of life' exists.

(v) My last critical point is that Winch's curious view that primitive beliefs cannot be critically discussed along western lines empties them of empirical content. Western science is realistic. We believe it makes assertions about the world which are true or false. The reality of that world is extra-linguistic. Winch denies this last (p. 315) and thus denies realism. For Winch there is no linguistically external reality against which to match a dispute between our beliefs and primitive beliefs. However, this realism is presupposed by our science, to which Winch confesses no hostility. Science, in other words, sees itself as exploring reality, not *a* reality. Winch would presumably hold this is a misconception. Above I have argued that primitive peoples can come to appraise their own beliefs about why things happen against our competing suggestions. What is happening here? Are the primitives revising their conception of reality? How do they do that? For Winch reality is not what gives sense or meaning to language, but shows itself in the sense that language has. But how can one adopt another's conception of reality without judging it better, without claiming it is a truer conception of what is? If reality (or the world) shows itself *in the sense that language has* then there is no such thing as a truth independent of the ideas and wishes of man. Provided a culture is coherent, and one works within its well-entrenched beliefs, then there is no way of saying that these are false. They must be true. Only beliefs which run against the general grain of the culture, while accepting its terms of reference, can be false.

Does Winch want to deny that there exist primitive people who have beliefs, some of which make factual assertions about the world of existing things? Does he believe that factual assertions about the world can in any extra-linguistic sense be true or false? If he does not, then he is the sort of

---

[1] For actual demonstrations of how alien world-views can be handled see Lawrence, 1964; and Horton, 1964, pp. 85–103.

Protagorean relativist he strives to convince us he is not. A non-relativist would hold that the world is not altered by the language in which it is being discussed. Therefore it is in some sense extra-linguistic. Truth and consistency are qualities we attribute to statements *à propos* their relationship to this 'external world'. Inconsistent statements cannot possibly be true together of any world; true statements are true of this world; false statements are false of this world. That the Azande do not have explicit notions corresponding to these, show little interest in them, etc., is simply an empirical fact. These ideas are great discoveries in the history of mankind. They are accepted in a wide diversity of cultures, from Ancient Greek and Jewish, to modern European and American. They are not some special and weird peculiarity. They are at the core of what I earlier referred to as a universal standard of rationality. Diffusion isn't perfect and so they are not to be found everywhere yet. But, like industrialization, they will be.

This fact in itself must give Winch some discomfort. For how and why can two incommensurable belief-systems like that of the west and that of the Azande lead to one completely ousting and replacing the other? Did it just happen, or was there some rational discussion?

## 6. SOCIOLOGICAL PROBLEMS AND THE INESCAPABILITY OF MUTUAL AND SELF-APPRAISAL

Because of the existence of the universal standards of truth and consistency it seems to me that no two belief systems are incommensurable. My conviction is reinforced by Bartley's argument (Bartley, 1963, pp. 134–75) to the effect that there are no logical difficulties in maintaining that all elements of a belief system can be held tentatively, open to criticism. What he calls the checks of the problem, logic, and the facts are available, as well as the check of other components of the belief system. Since criticism is always possible, it cannot be dismissed as peculiar to one system and out of place in others. Those others may suppress it or ignore it, but it can erupt and since it can it 'belongs' there as much as anywhere. Thus it is my conclusion that there is nothing extraneous about our standards of truth and consistency and that provided they are manipulated with sympathy, and provided we do not make them the excuse for condescension, we are at liberty to discuss whether other people's beliefs have these properties.

However, I wish to claim more in this final section of the paper. It seems to me that if evaluative standards of different societies were incommensurable, there would be and could be no social science, indeed no history. After all, history is the attempt to explain the past in terms of the present— what we nowadays regard as satisfactory explanation is very much a product of current ideas. We have to explain historical events like why a man did something apparently stupid. One explanation would be that he happened to hold a false belief. The falseness cannot be ignored, since from it may stem the consequences we are trying to explain. We cannot explain medieval plagues without reference to medieval ideas of disease and hygiene *and their falseness*. We no longer accept that race explains behaviour: few will get away with explaining wars by an aggressive instinct; theories that say societies degenerate are not much used. But in history we are in a way looking at, explaining, and evaluating the beliefs of, another society. As one explores further into the past of one's culture, or country, the gap becomes larger between them and us. Do we stop ourselves at some point and say we can't criticize, e.g. witch- and heretic-burning because 'what a witch's or a heretic's reality amounts to can be seen only from the (religious) tradition in which the concepts of witch and heretic are used . . .'? There is something absurd about this. We do it, and there is nothing wrong philosophically or otherwise with it. Similarly, in our contemporary society, we may not be able to explain actions without pointing out the falsity of the beliefs on which they were based. If we can appraise a past state of affairs of our own society, and segments of our own society today, then we can do it to others.

As to social science in general: my thesis would be that there is very unlikely to be any social science in a smoothly working closed society; that social science in fact is a product of other cultures' impact on one's own. What happens is that other possible ways of ordering social arrangements are seen, one's own ways come into question, if only in the sense that they have to be explained and defended to oneself. If our system of monogamy is better than say polygyny, what sociological and other arguments can we produce that it is? First we question our own society. Then others raise puzzles. It is intriguing how many social scientists are themselves marginal to their society. They are refugees, foreigners, from minority groups, or otherwise 'loners'— perhaps they are trying to explain to themselves what others unthinkingly belong to. Thus my remark about ontogeny and phylogeny. Both curiosity about sociological questions as a matter of history, and its growth in individual men seems to me to be connected with contact, culture shock, disruption, which lead one into evaluative and comparative questions.

All this seems obvious enough, yet Winch flies in the face of it. He maintains in his book that understanding a society is a kind of conceptual empathy which imprisons you in a universe of discourse that cannot evaluate itself. He thus attacks Popper for saying that sociological concepts are explanatory models.[1] Since Winch holds sociological concepts play a role

---

[1] Martin's and Cohen's criticisms (see note 2, p. 234) are especially pertinent here.

in the actions themselves, they somehow cannot explain it. There is some vacillation here in the use of 'understanding' and 'explanation' (cf. Rudner, 1966, pp. 81–3). True, participants in a war hold a concept or model of war. But this is purely conventional—it is *their* explanatory model of what they are doing. Whether it is a war or not is not decided by them or explained by their use of the concept. They may understand that they are in a war but be mistaken. War is a label that cannot be strictly applied and which is useful only if it satisfies someone asking for an explanation of what is going on.

Among the main problems of economics, sociology and social anthropology are the explanation of certain events, especially large-scale events like depressions, suicide of democracy, disintegration in circumstances of contact. Problems of this kind are not easy to handle. Unlike smaller-scale problems they cannot be put down to someone or some people aiming to bring them about (as can orderly traffic flow, or social security payments). On the contrary, these are events which everyone wants not to happen, they are unintended and unwanted. They are also inter-societal, not confined to one social system either in their occurrence or in the scope of the explanation. Social scientists can neither come face to face with them nor solve them without indulging in some evaluation of beliefs. However long an economy has been depressed (or underdeveloped), the economist can say that this needs explanation. He may find himself saying: 'the British have to learn that their twin desires to support an international reserve currency, and have a high growth rate, are inconsistent with each other'. Or that 'the Indians have to see that their desire for economic development is being vitiated partly by their religious love of cows'. Sociologists may in all conscience have to point out that 'total freedom for everyone and the *maintenance* of democratic systems are not simultaneously realizable in this unhappy world'. Anthropologists may say to missionaries: 'your declared love of this people and their way of life makes you blind to the fact that you will destroy it by your proselytizing and interference'. All these could be made logical inconsistencies by different formulations. But also, the British are being told their values are in conflict, so are the Indians, lovers of freedom, and missionaries. Such evaluations of evaluative systems are not somehow illegitimate straying by the social scientists, but what the social sciences are all about.

REFERENCES

Bartley, W. W. 1962. Achilles, the tortoise, and explanations in science and history. *Br. J. Phil. Sci.* **13**.
Bartley, W. W. 1963. *The Retreat to Commitment*. London.

Brodbeck, M. 1963. Meaning and action. *Philosophy of Sci.* **30.**

Brown, R. 1963. *Explanation in Social Science.* London.

Cohen, P. 1968. The very idea of a social science. *Problems in the Philosophy of Science, Proceedings of the International Colloquium, London, 1965,* vol. 3. Eds. I. Lakatos & A. Musgrave. Amsterdam.

Evans-Pritchard, E. E. 1937. *Witchcraft, Oracles and Magic Among the Azande,* London.

Gellner, E. 1959. *Words and Things.* London.

Gellner, E. 1962. Concepts and society. *Transactions of the 5th World Congress of Sociology,* vol. 1. Louvain.

Gellner, E. 1965. *Thought and Change.* London.

Gellner, E. 1968. The new idealism. *Problems in the Philosophy of Science, Proceedings of the International Colloquium, London, 1965,* vol. 3. Eds. I. Lakatos & A. Musgrave. Amsterdam.

Goldschmidt, W. 1966. *Comparative Functionalism.* Berkeley and Los Angeles.

Goody, J. & Watt, I. 1962. The consequences of literacy. *Comparative Studies in Society and History,* vol. 5.

Heyerdahl, T. 1950. *The Kon-Tiki Expedition.* London.

Heyerdahl, T. 1958. *Aku-Aku.* London.

Horton, W. R. G. 1964. Ritual man in Africa. *Africa* **34.**

Hughes, P. 1952. *Witchcraft.* London.

Lawrence, P. 1964. *Road Belong Cargo.* Manchester.

Lethbridge, T. C. 1962. *Witches.* London.

Louch, A. R. 1963. The very idea of a social science. *Inquiry* **6.**

Louch, A. R. 1965. On misunderstanding Mr Winch. *Inquiry* **8.**

Martin, Michael. 1965. Winch on philosophy, social science and explanation. *Phil. Forum* **23.**

Piddington, R. 1957. Malinowski's theory of needs. *Man and Culture.* Ed. R. Firth. London.

Popper, K. R. 1957. The aim of science. *Ratio* **1.**

Popper, K. R. 1959. *The Logic of Scientific Discovery.* London.

Popper, K. R. (ed.) 1962. *The Open Society and Its Enemies,* vol. 2. London.

Popper, K. R. 1963. *Conjectures and Refutations.* London.

Rudner, R. 1966. *Philosophy of Social Science.* Englewood Cliffs.

Russell, B. 1922. 'Introduction' to L. Wittgenstein's *Tractatus Logico-Philosophicus.* London.

Saran, A. K. 1964. A Wittgensteinian sociology? *Ethics* **75.**

Schilpp, P. (ed.) 1949. *Albert Einstein: Philosopher-Scientist.* New York.

Watkins, J. W. N. 1968. Anthropomorphism in social science. *Problems in the Philosophy of Science, Proceedings of the International Colloquium, London, 1965,* vol. 3. Eds. I. Lakatos & A. Musgrave. Amsterdam.

Winch, P. 1958. *The Idea of a Social Science.* London.

Winch, P. 1959–60. Nature and convention. *Proc. Arist. Soc.* **60.**

Winch, P. 1964. Mr Louch's idea of a social science. *Inquiry* **7.**

Winch, P. 1964. Understanding a primitive society. *Am. Phil. Qu.* **1.**

Wisdom, J. O. 1963. Metamorphoses of the verifiability theory of meaning. *Mind* **72.**

# COMMENT

## *by* PETER WINCH

Mr Jarvie says that 'a crude first approximation' to my position is 'that cross-cultural value-judgments will always be misjudgments—and therefore should be avoided—because there is no language game in which cross-cultural value-judgments could be legitimate moves' (p. 232). One relatively minor point to be made here is that, in the main body of the article to which Jarvie is chiefly referring (Winch, 1964; reprinted in *Religion and Understanding*, Ed. D. Z. Phillips, Blackwell, 1967) I had nothing much to say specifically about '*value*-judgments' at all. In the groping and tentative remarks which I made towards the end of that article (to the precise form of which it seems to me Jarvie pays insufficient attention in his references to them) I did, indeed, make reference to the importance which conceptions of good and evil have in our understanding of our own and other forms of social life. And in 'Nature and convention' (Winch, 1959) I was specifically concerned with questions of morality. It is remarkable, though, that in both these places I was explicitly arguing for a position contrary to the one Jarvie ascribes to me here. This makes it difficult for me to see how his formulation can appear to him to be even a 'crude approximation' to my position. It is true that, in these places and elsewhere, I was exercised to bring out some of the peculiar *difficulties* which face anyone who wishes to make moral judgments about actions belonging to a culture different from his own. But it does not seem to me from the way Jarvie writes that he would want to deny that such difficulties exist; and, as I have said, one of my main concerns was to see how some of those difficulties might be overcome. If I devoted a lot of attention to trying to make explicit precisely what the difficulties are, well this seems to me a not unreasonable preliminary to any serious attempt to overcome them.

On the wider question of cross-cultural comparisons involving conceptions of rationality, Jarvie repeatedly, in the body of his article, states it to be my view that 'standards of rationality are incomparable' (cf., for example, p. 239). I do not quite understand what sense is to be attached to 'incomparable' here, but I have certainly never held any view of which I would be willing to accept this as a proper paraphrase. What, indeed, was I doing in the article about the Azande if not 'comparing' their conceptions of rationality with some of those with which we are familiar in our society?

In the course of this article I criticized certain views expressed in Alasdair MacIntyre's 'Is Understanding Religion Compatible with Believing?' (MacIntyre, 1964). A point which I emphasized there, to which I think Jarvie gives insufficient attention, is that the very asking of the kind of question Jarvie is raising presupposes that we have already been able to identify certain features of the life of the society we are studying as involving something we are prepared to call 'an appeal to standards of rationality'. Our willingness to speak in such a way is obviously based on the way we already speak about such standards in our own social life; so in a sense a comparison between our society and theirs is already involved and we may speak, with Jarvie (p. 232), of using our own society 'as a measuring instrument or a sounding board'. The question is not *whether* we can do this, but *what sort* of comparison is involved.

There is also the question of what we are comparing Zande (or whatever) standards with. Jarvie shares with MacIntyre a tendency which I criticized in the latter: to speak of the standards of 'our scientific culture' in this connection in a way which suggests that the only standards available to us against which to compare Zande standards are the standards involved in the practice of *scientific* work. Now it is of course true that the role played by such work in the culture of western industrialized societies is an enormously important one and that it has had a very far-reaching influence on what we are and what we are not prepared to call instances of 'rational thought'. But it was an essential part of my argument, both in *The Idea of a Social Science* and in 'Understanding a primitive society' (1964), to urge that our own conception of what it is to be rational is certainly not exhausted by the practices of science; and in the latter essay one of the main thrusts of my argument was an attempt to show (*a*) that misunderstandings of the sense and purport of an institution like Zande magic arose from insisting on just *this* comparison; and (*b*) that an understanding of such an institution may be furthered by comparing it with quite other sectors of the kind of life we are familiar with. This is very far removed from the claim which Jarvie wishes to foster on me that we cannot make any comparisons at all between our institutions and theirs. I must also remark that Jarvie's holistic way of talking about 'our scientific culture' sits very ill together with his allegation (itself unfounded) that I think of the life of a culture as somehow 'seamless': this, I suggest, is precisely his way of looking at our own culture.

Jarvie says (on p. 240): 'Winch claims their (sc. the Azande's) beliefs are not hypotheses and are not therefore doubtable in this way. This amounts to the claim that a culture accepts its own appraisal of its own beliefs.' Leaving aside the obscurity I find in the assertion that the first of these claims 'amounts to' the second, there are several comments I should like to make on what Mr Jarvie says about 'appraisal' which are connected with the point I have just been making. I am confused—and I think Jarvie is too—by his way of talking about 'societies' and 'cultures' as having beliefs and appraising them. Beliefs may be more or less widely current amongst the members of a society and there may be stronger or weaker tendencies to criticize and discuss such beliefs amongst those members. Some societies have, to a greater or lesser degree than others, institutions and traditions of a predominantly critical nature. Where such traditions do exist, naturally individuals are more likely to be stimulated to think critically about the ideas and practices current in their society than where they do not. Jarvie quite rightly thinks it an important feature of Western culture that there are such critical traditions in it. He emphasizes the importance of science in this connection; I should be more inclined to emphasize the importance of philosophy and I did in fact so emphasize it in *The Idea of a Social Science* (*passim*; but cf. especially pp. 102–3, where special importance is attached to the recurrence within philosophy of discussions about the nature and possibility of philosophy itself). I do not know how Jarvie reconciles this passage with his claim that '[Winch] maintains in his book that understanding a society is a kind of conceptual empathy which imprisons you in a universe of discourse that cannot evaluate itself' (p. 246).

However, even leaving aside the existence in a society of institutions of a self-consciously and explicitly critical nature, it is a mistake to say that the view I developed about the relation between forms of social life and the standards according to which people act within these forms rules out the possibility of such standards being changed as a result of criticism. Indeed, one of my fundamental lines of argument was that the way in which characteristically human behaviour does involve the possibility of discussion and criticism itself shows the intimate connection between men's actions and the social context of rules and standards within which they perform those actions. This point is made most explicitly in Chapter 2 of *The Idea of a Social Science*; I draw Jarvie's attention especially to Section 5 of that chapter, where, in opposition to Michael Oakeshott, I argued that the concept of social life cannot be understood unless we give due prominence to what I there called 'reflectiveness'.

But more important than Jarvie's mis-statement of my position is the confusion which he himself betrays when he speaks of being 'imprisoned' in a universe of discourse. There are, of course, situations where we might quite properly speak of someone in this way. Consider, for example, a man, who, in the pursuit of business success, does something morally unjustifiable; and who, when we try to remonstrate with him on moral grounds, fails completely to respond to the moral categories involved in our arguments, but continues to think about his actions solely in terms of criteria of business efficiency. We might perhaps speak of such a man as being 'imprisoned' in the categories of the world of business. Now on my view, we, who oppose such a man on moral grounds, can think and argue as we do only in so far as we are masters of certain moral concepts which are intelligible to us because of the sort of life we lead. This does *not* mean, however, that we are 'imprisoned' in this life or by these concepts.

Why the difference? One important consideration is that, in the case of the man in my example, we can make clear sense of an *alternative* way of thinking about his situation to that which he in fact follows. His situation *has* an ethical dimension whether he recognizes it or not; it has this dimension because of its place in the life of the society in which it arises. In not seeing this his eyes are closed to a possibility which in fact exists for him— compare the sense in which the fly is 'imprisoned' in the fly-bottle. Someone who wanted to say that we are equally 'imprisoned' in the universe of the moral concepts with which we criticize this man would have to show that we, too, are being blind to a certain possibility. But then he must *specify* this possibility; and how is he to do this? It may be, of course, that in some particular circumstances he will be able to show further moral possibilities to which we are blind; or he may be able to argue that there are considerations—perhaps religious—which 'transcend morality'. But to do this he will, naturally, have to appeal to moral, or religious, considerations; and if we ask about the conditions under which it is possible for us to understand *these*, we are right back where we started.

There are two moves which might be made at this point towards specifying an alternative in terms which are not related as I have argued to the forms of life which our social environment makes available to us. On the one hand it might be argued that social anthropology, for example, presents us with forms of life very different from our own which we can come to understand; and that thereby we can both come to see possibilities to which our involvement in the life of our own society blinds us and also see more clearly certain features of that life to which for one reason or another we have been insensitive. On the other hand it might be held that our escape route lies through some direct confrontation with an 'objective, external reality'—'direct' in the sense that it is not mediated by concepts which are in any way culture-dependent.

Now the first of these moves is essentially the one I was making in 'Understanding a primitive society'. As I have argued, Jarvie does not see this because of his idea that I deny the possibility of ever understanding a form of social life very different from one's own. The second move is the one I was mainly opposing in my criticisms of Evans-Pritchard and MacIntyre in that article. Jarvie's position seems to be roughly as follows: We *are* able to understand forms of social life different from our own by something like a translation from one language into another. (There are some difficulties about this analogy which I will not pursue here.) Of course we are always liable to misinterpretations and 'mistranslations', but the critical apparatus provided by western science gives us the means of learning from our mistakes and so getting nearer the truth. I think probably the fundamental point of contention between Jarvie and myself concerns the relation between these methods of getting nearer the truth and the truth which we thus get nearer. He is not very explicit on this point, but I read him as thinking that the relation is an 'external' one in the sense that the methods of science are not essential to our understanding of the meaning of the truths which they enable us to discover; rather, it has just been found, by trial and error, that these methods do as a matter of fact constitute the best way of checking on the truth or falsity (the correspondence with some 'external', 'objective' reality) of thoughts which we can have quite apart from our participation in our 'scientific culture'. If something like this is not Jarvie's view, then I altogether fail to understand the nature of the disagreement between us. I think it is because he feels that the ultimate appeal is to this 'objective external reality', which is there and given quite independently of our methods of conceiving and discovering it, that he thinks, in opposition to me, that he has a way of comparing scientific modes of thinking with other modes of thinking and pronouncing it better; the point being that they are all, science, religion, magic or what have you, thought of as ultimately aiming at the same thing (perhaps 'understanding how things really are'), and science is plainly the best way of achieving this aim.

In opposition to this, I was arguing against the idea that all forms of human intellectual activity are comparable in *this* way. Even where we are dealing with modes of investigation, where it is appropriate to speak of 'understanding how things really are', it is a mistake to suppose that such modes of investigation are necessarily in competition with each other, or that their results, when different, are necessarily in conflict with each other. I do not, of course, rule out the possibility that there *may* sometimes be such conflict: whether this is really so or not in a given case will only be determinable by detailed examination of the particular case.

Thus, Jarvie discusses the Zande strategy for dealing with apparently conflicting oracular advice and comments that this shows that they 'do possess some standard of rationality recognizable in the west as rational, though as rather poor' (p. 240), claiming that this contradicts a remark of mine about the different levels on which witchcraft beliefs and the western scientific approach operate. Once again he here identifies 'western scientific culture' with 'the west' *tout court*. But, apart from that, I never of course denied that Zande witchcraft practices involve appeals to what we can understand as standards of rationality. Such appeals also involve behaviour which we can identify as 'the recognition of a contradiction'. What I was urging, though, was that we should be cautious in how *we* identify the contradiction, which may not be what it would appear to be if we approach it with 'scientific' preconceptions. Against the background of such preconceptions Zande standards might indeed seem 'rather poor', but whether they really are rather poor or not depends on the point of the activity within which the contradictions crop up. My claim was that this point is in fact very different from the point of scientific investigations.

At this point I must tackle head-on the remarks Jarvie makes towards the end of §5 of his paper concerning the relation between such notions as language, reality, truth and consistency. As he rightly sees, it is here that the crux lies. The aim of his remarks is to represent me as 'the sort of Protagorean relativist he strives to convince us he is not'. Do I, he asks, 'want to deny that there exist primitive people who have beliefs, some of which make factual assertions about the world of existing things'? Answer: of course not. Do I believe 'that factual assertions about the world can in any extra-linguistic sense be true or false'? I have a difficulty in answering this question, since I do not immediately understand the force of the phrase 'in any extra-linguistic sense'. Jarvie explains as follows: 'A non-relativist would hold that the world is not altered by the language in which it is being discussed. Therefore it is in some sense extra-linguistic.' Well, if this is all he means, I can happily answer 'yes' to his question; but obviously the matter cannot be left there.

Ben Nevis is 4,406 feet high. Let us suppose that there were no relevant geological changes between a time $t_{-n}$ at which there did not exist men having a language and techniques of measurement which would have enabled them to say, either truly or falsely, that Ben Nevis had a certain height, and a subsequent time, $t$, at which there did exist men having such language and techniques. Now this supposition is of course completely irrelevant to the truth or falsity of the statement: 'At time $t_{-n}$, the height of Ben Nevis was 4,406 feet.' What we need, in order to determine the truth or falsity of this statement is geological, not anthropological or linguistic, information. Using Jarvie's terminology, I might express this by saying 'the reality of Ben Nevis' height is extra-linguistic'.

Does this mean, though, that the reality of *height* is extra-linguistic?

Well, we might speak in this way if we wanted to express the point that the kind of remark I have made about the sort of information which is relevant to questions about Ben Nevis' height would apply equally to *any* questions about a mountain's height (or, with suitable modifications, about the height of anything else).

Now Jarvie refers to a remark of mine to the effect that 'reality is not what gives sense or meaning to language, but shows itself in the sense that language has' and comments that, if this is so, 'then there is no such thing as a truth independent of the ideas and wishes of man'. I do not know about 'the ideas and wishes of *man*', but certainly it is a corollary of what I have said in the last two paragraphs that what makes true any statement you like to mention about the height of a mountain (or anything else) has nothing to do with the ideas or wishes of any man or group of men you like to mention. But when I said that reality shows itself in the sense that language has, I was not denying any of this.

What I was denying was that considerations of this sort are sufficient to deal with the kind of questions which philosophers have asked of the form 'What is the reality of $x$?' A certain sort of logical positivist might at one time have said that statements about the heights of mountains are analysable into statements about the results obtained when surveying instruments are applied in a certain way. I should want to resist such a claim and I might express my resistance by saying that the logical positivist analysis fails to capture or express the reality of height. Alternatively, I could say that the logical positivist fails to bring out the point of using surveying instruments in the way he describes. I should have to amplify my objection by describing some of the more important of the countless ways in which we make use of the results of such measurements and calculations in connection with activities very fundamental to our whole way of living. That is, I should have to describe what Wittgenstein called certain language games. Having done this I might comment: 'the reality of height shows itself in the sense that these language games have'; and I should still not be contradicting any of what I said at the beginning of this discussion of height.

Let us now try to imagine a people amongst whom none of these language games is played. (I do not think we can get very far with this and the difficulty is a measure of the very fundamental role they play in our life and thought.) We shall have to postulate perhaps physiological peculiarities which mean that it makes no difference to them whether they are moving up, down, or on a level; and we shall have to postulate the absence of any interests or activities in the context of which it is important to them to know whether one thing is obscured from their view by another. (As I said, it is difficult.) We might say of such a people: 'For them height has no reality'; and *this* would be a comment on the kinds of language game they play. Suppose that a visitor from Britain wanted to teach them about height. I do not see what possibility there would be of their 'translating' statements about the heights of things into any statements which could be made in their own language, as it then existed. What would be necessary would be for the Briton to teach them new techniques and new ways of speaking. How successful he was in this would depend on a great many diverse factors: whether, for example, their culture were familiar with any other technique of measurement, or something like it; how physiologically different they were from us; and so on. I can see no reason *a priori* to say that it *must* be possible eventually to teach them this or not.

Of course, nothing in what I have said implies that the actual height of anything in the environment of this people depends on the way they live or the language they use. When I talk about the height of things in their environment I am using *our* language of measurement and, as we have seen, the truth or falsity of statements made in that language depends on, e.g., geological information and not anthropological or linguistic information.

Imagine now a member of the 'heightless' tribe, who has been taught western techniques of measuring height, so that he is in a position to give answers which we should call correct to questions about the heights of things. Such a being might still utterly fail to see the point of all this activity; for him it is all meaningless mumbo-jumbo. He might even come to think of 'Height' as the name of a tribal deity and of our complex measuring behaviour as forms of religious worship; and he might regard this all as mere superstition. We might say of such a being: 'For him height has no reality'; or: 'He has not grasped the proper reality of height'.

What I am suggesting is that the westerner who feels there is no reality in Zande magic *may* be in a position very analogous to that of the 'height-less' stranger *vis-à-vis* our society. He has grasped most of the rules of the game, in a sense, but is unable to grasp the spirit which informs those rules.

The heightless stranger might express his bewilderment about our measuring activities by saying 'There is no height.' He would not of course mean by that that he has found everything to be on a level: for that, if it were a possible proposition at all, would have sense only within measuring institutions. Rather, he is saying something like 'The whole idea of height as something to be measured is an illusion.' In a similar way, most philosophers who have said that 'Time is unreal' have not wished to deny the truth of propositions such as 'Before I had breakfast I washed myself.' The proper role of drawing attention to such propositions as the latter in a philosophical discussion about the nature of time would be as reminders about the language games into which the concept of time enters, not, *pace* G. E. Moore, as straightforward 'refutations'.

A western philosopher or anthropologist who says of Zande institutions: 'There are no witches' is not reporting the outcome of some empirical investigation he has conducted; he is expressing doubts he feels about the whole concept of a witch. He may continue to have these doubts even though he is perfectly well aware of the particular techniques by means of which Azande 'identify witches' and the kinds of thing they say about 'witches'. But he can see no sense in this whole complex of procedures; for him, witches have no reality.

When I, in 'Understanding a primitive society' (1964), objected to someone's saying, in this kind of context, 'there are no witches', I was not *contradicting* this claim. I was not saying 'Oh yes, there certainly are witches'. That is, I was not, as Jarvie puts it, 'endorsing' Zande beliefs, or saying that they are 'obviously true'. That is something I should find as hard to understand as I should someone's saying that they are 'obviously false'. The heightless stranger who says 'There is no height' is not, whatever he may think he is doing, denying the truth of certain propositions which we affirm. He is betraying his lack of comprehension of the institution in the context of which we do affirm those propositions.

Confusion on this matter is encouraged by talk about Zande 'beliefs'. If we think about the Zande institution of witchcraft in terms of what the Azande 'believe', we seem to invite the question: 'Well, is what they believe true or false?' I say we 'seem' to invite this question, because I should as a matter of fact be willing to argue that it is *not* always possible to ask this question with regard to what it is perfectly proper to call 'beliefs' (cf. the 'belief in other minds'). But it might be better in this context to say that the subject of discussion is not Zande beliefs but Zande *concepts*. Zande concepts are, of course, exercised or applied in the beliefs which Azande hold; and we can learn what Zande concepts are by studying the various beliefs they hold in particular contexts and by enquiring into what those beliefs mean to them in those contexts, into the importance those beliefs have for them in the lives they lead.

There is a passage in Wittgenstein's *Philosophical Investigations* which is fundamental here:

241 'So you are saying that human agreement decides what is true and what is false?'—It is what human beings *say* that is true and false; and they agree in the *language* they use. That is not agreement in opinions but in form of life.

242 If language is to be a form of communication there must be agreement not only in definitions but also (queer as this may sound) in judgements. This seems

to abolish logic, but does not do so.—It is one thing to describe methods of measurement, and another to obtain and state results of measurement. But what we call 'measuring' is partly determined by a certain constancy in results of measurement.

Let us apply this to the way Azande think about witches. The Azande 'agree in the language they use' about witches. It is a feature of this agreement that, in certain circumstances, certain procedures having been carried out with due precautions, there would be overwhelming agreement to the effect that such and such a person is a witch. This does *not* mean that anyone is made into a witch simply by being thought (even unanimously thought) to be a witch. But to suppose that the judgment of the totality is wrong on a particular occasion is to invoke the standards appropriate for judging such matters and the existence of such standards does depend on the fact that in general, when certain criteria are applied, agreement is usually reached.

Of course, what I have just said applies to cases where we take the institution of witchcraft for granted and ask about the position of particular judgments made within the context of that institution; somebody may want to raise doubts about the whole institution; but these will have to be of quite a different sort. Such a person would not be straightforwardly contradicting any particular claim but would be arguing against a whole way of speaking.

Now I do *not* deny the possibility of ever mounting such an argument. I have criticized a certain way of trying to mount it, i.e. by appealing to western 'scientific' conceptions, in order to show Zande beliefs false. But this of course is not the only possibility. And I agree with Jarvie here about the importance of the fact that an institution like that of Zande witchcraft does not exist in a vacuum, but that Zande life contains other important elements which may develop in such a way as to make witchcraft practices lose their foothold. This process will involve the making of certain sorts of criticism, but here it is of the utmost importance to be clear about what kinds of criticism they are. It may be that Azande come to think of certain new ways of living (perhaps introduced from the West) 'better' than their traditional ways. I certainly do not wish to argue that they must inevitably be mistaken if they take such a view and it is easy to point out, as Jarvie does, obvious advantages in tackling, e.g. disease by the methods of western medicine rather than by the methods of Zande magic. It is also easy to overlook the good things that may be lost in such a transition; though again, of course, I do not claim that the losses must outweigh the gains. Such questions can only be settled, if at all, by patient and sensitive attention to particular cases. I do, however, want to protest very strongly against the sort of claim that creeps into Jarvie's paper towards the end to the effect that the almost universal success of western ways of life in ousting other 'more primitive' ways shows anything about the superior rationality (or superior anything else, except persuasiveness) of western institutions. In this connection I should like to remind Mr Jarvie both of Plato's remarks about the difference between persuasion and instruction in his *Gorgias* (as well as what he says about 'The Great Beast' in *The Republic*) and also of Sir Karl Popper's criticisms of moral and other forms of historicism.

### REFERENCES

MacIntyre, A. 1964. *Faith and the Philosophers*. Ed. J. Hick.
Winch, P. 1959. Nature and convention. *Proc. Arist. Soc.* **60**.
Winch, P. 1964. Understanding a primitive society. *Am. Phil. Qu.* **1**.
Wittgenstein, L. 1954. *Philosophical Investigations*. Blackwell.

# REPLY

## *by* I. C. JARVIE

Consider the following claim:

Certain social institutions (including ways of conducting argument and of using language) impede attempts to discover what the world is like.

An unexceptionable enough piece of sociology, one would think. Yet, unless I continue to get Professor Winch wrong, his position should lead him to discover grave difficulties in saying just what sense this claim makes. Winch has argued that what the world is like (what is real and unreal) is a distinction of language, i.e. a social usage. The activity of discovering what there is would seem to go together with the question of what is real and unreal and become a linguistic matter. My counter to Winch is to ask whether it is a social usage to say the sun rises and sets every day. Perhaps it is; it is also a grave error. Now maybe geocentrism stands at the centre of a complex, satisfying, even beautiful social system. Well and good, but nothing to get sentimental about: that social system is predicated on a mistake. The question is, do its members know it, and—where they don't—if it were brought home to them would it make any difference?

Is this 'scientific' example fair? Aren't the cases of witchcraft or of God different? In what ways? Well, unlike geocentrism, they are allegedly non-empirical, metaphysical.[1] Even if that is true, have they not also been used to account for empirical phenomena? Yes. In so far as witches and God are used to account for disease, disaster, crop failure, success in life, health, etc., we are entitled to appraise them as explanations, perhaps note the logical lacunae in the deductions and in that case the incorrectness of any claim that they have explanatory power. Like atomism, they may one day graduate to empirical testability but critical appraisal of them need not stop in the meantime. Especially as some who accept witch-beliefs and God-beliefs indicate how they can be efficacious, this opens up the possibility of comparison with the more efficacious western explanations. (Efficacious both in the practical and in the intellectual sense.)

I contend, then, that my opening claim *makes sense*. It says that, e.g. magical and tabooistic attitudes to things are inimical to intellectual

---

[1] Since Winch traces the pedigree of his ideas to Wittgenstein, I must be equally frank and trace mine to Popper. Some Popperians, e.g. Agassi (1964), might want to exclude God and witches from metaphysics. Professor Agassi has helped me very much with this reply, for which I thank him.

260

progress; that oracular and authoritarian traditions of education inhibit learning; that entrenched metaphysical systems are inimical to progress; that witchcraft is wasteful and redundant as a treatment of disease though it may be useful when combined with western remedies—not useful medically, but as a means to overcome prejudice, induce calm, etc. And so on.

If the claim I have made is *true*, then it is reasonable to remove or alter those social institutions which impede discovery. It seems to me that Professor Winch's views on languages and their limits, and on the institutions of rationality, are inimical to my initial claim, and therefore inasmuch as they are proposals for institutional reform of the social sciences are themselves to be rejected or altered because they impede attempts to discover what the world is like. In what follows I should like to look only at a handful of the points in his 'Comment' which are relevant to the aforementioned claim.

Winch considers the very interesting case of a tribe with no concept of height, and none of the ancillary concepts that might be used to describe it indirectly. Winch holds that to say that Azande magic has no reality may be 'very analogous' to saying height has no reality which in turn is like 'time has no reality'. To argue against the latter, 'before I had breakfast I washed myself' is a reminder about the language game into which the concept of time enters, not a straightforward refutation. The problem may be in talking of beliefs, with the invited question, 'are they true or false?', rather than of concepts. Winch is not saying 'there are witches' but rather that if I say there are not, he holds I shall be 'betraying . . . lack of comprehension of the institution in the context of which we do affirm those propositions' in question (about witches).

Before coming to beliefs, some remarks on concepts. Here I find myself in very strong disagreement with Winch—I think *his* confusion is connected with his misplaced interest in *concepts* rather than theories. Little purpose is served, I would say, by discussing concepts—what is interesting is what the concepts are used to *say*. Indeed I incline to think that any interesting doctrine can be reformulated so as not to use a disputed word or concept. This view may have grave difficulties, but Winch's view that concepts, like institutions, depend on agreement in their use, has equally grave ones. Winch quotes Wittgenstein (1954) as saying that for statements to communicate, there must be agreement in definitions *and in judgments* (*Philosophical Investigations*, 242). Not, of course, in any particular judgment but, at least, in standards for judging such judgments. This is a key passage in Wittgenstein; I for one am grateful to Winch for drawing attention to it. The idea it expresses might be called Wittgenstein's theory of language as a social institution, and in language he includes both concepts and standards of judgment. It is a theory which deserves close critical scrutiny. I consider it open to several serious objections.

For one thing, when using concepts and statements which employ them, we attempt to do with them what we do not attempt to do with most institutions, i.e. to *say* something about the world. Some are used to describe, some to explain, some to evaluate, some to perform and so on. Concepts are simply tools for performing these specific acts; and they are in this sense unique as well. We use institutions for various purposes, but these are seldom substitutable within the same social context. Rituals like applause cannot be altered (e.g. to hissing as in Japan) without losing their significance altogether. Even magic words are not allowed much variation ('abracadabra', or 'open Sesame' must be pronounced just so). But ordinary concepts and words are all too easily substitutable.

Some of the statements we make make truth-claims. Whether or not any particular statement *is true* does not depend only on an agreement in the use of the concepts. Even whether or not we can *agree* on whether what is said is true does not depend only on an agreement in the use of concepts. Such agreement depends on whether we can agree on the truth or falsity of other statements which they contradict. How can we agree on the truth or falsity of statements which contradict them: if we can falsify them by agreeing on the non-falsity of statements which contradict them; or, if we can't falsify them, then by continuing the discussion. And so on. Agreement, in a strict or tight or wide or comprehensive sense is *never* reached: all we have is partial and temporary; tentative acceptance with the possibility of revision any time. This is as true of the standards of judgment and the rules of the discussion, as well as of truth-claims and meanings. Disputes over all these are regular, inescapable, fruitful and conducive to progress. So if we must 'agree in the language we use', that is, in the form of life, we never do.

Revisability is a matter of correction cum expansion. Correction often involves changes in the language, and thus of shifts of concepts; these can be easily explained in specific cases. Now when one grasps a new use of an old concept—say the new hippie use of 'bag'—one sometimes does it *before* general agreement has been given to the concept. And to say of someone, 'man he can't get out of that folk–rock bag' can be true despite the fact that most people in the society would not be able to face the question of whether it is true or not because they can't handle the concept. And hence, Wittgenstein seems to be in error. Wittgenstein himself perhaps noticed this and tried to cover it by his idea of *loose*, expandable conventions, of language games involving family resemblances. So let us pursue this idea.

Consider, then, the doctrine that the one 'given' in language is not words, or rules, or meanings, but the form of life of which the language is a part. The language games of the building trade are adapted to the building trade form of life. Now, the building trade is a role, or a complex of roles we call an institution. A person often finds that his different roles conflict. Is there an analogue where we can see persons involved in different language games that conflict with each other? People who have changed their social class sometimes can play the language games both of their class of procreation and of their class of orientation. Which form of life does their double language game belong to? Or are there two? Bilinguals are an extreme case of this—is their form of life and two-language game *sui generis*? How to demarcate forms of life that contain persons who play different language games simultaneously? Ah, but they never do it at the same time! Double-spies? They never *reveal* the other game they play, how do we know they play it correctly—either one? Not by asking! What is the problem? The double-spy has no 'given' form of life—so what is his language-game?[1]

As there is violation of game rules, so there is the question of misuse of concepts; but if the rules are loose, is there violation, and how? Wittgenstein thinks that philosophers violate the rules. There are non-philosophical violations too and I think Wittgenstein cannot account for them because of his loose conventions. Compare the two cases, first the permissible use of 'rises' in 'the sun rises' or 'bag' in 'man he can't get out of that folk–rock bag'; second, a vicious misuse. Sir Karl Popper, in *The Open Society and Its Enemies*, has a passage where he imagines someone reversing the concepts of democracy and tyranny: Popper's strategy—not to be deflected from the point by playing with concepts—seems to me sensible and sound. Wittgenstein's theory leaves us helpless. Within communist language games 'democracy', 'formal democracy', 'dictatorship of the proletariat', all have well-defined usages which are viciously parasitic on their usages in the ordinary language. The pro-feelings aroused by their ordinary use are transferred to their perverted use, which certainly bears a family resemblance to their normal use. Let me come back sharply now to the issue of whether or not there are witches. Winch grants that it is possible to raise doubts about the whole institution, to argue against a whole way of speaking. But when I say there are no witches I am doing much more than attacking a way of speaking. I am also attacking an Azande claim about the furniture of the universe. Winch agrees that people like the Azande make factual claims (e.g. witchcraft-substance exists and can be detected in autopsies, and by administering medicines to chickens), he also

---

[1] Interesting recent criticism of Winch is to be found in Nielson, 1967, pp. 191–209.

agrees that the world is not changed by the claims made about it. So: we cannot conclude that there are witches *because* Azande believe and act as though there are; the Azande happen to be mistaken, just as the followers of Galen—i.e. almost all medieval and renaissance doctors—who for so long believed there was a hole in the wall between the two sides of the heart were mistaken. To be mistaken is not to be stupid, unthinking, primitive, immoral or anything else. It is to be honestly, decently, and yet hopelessly wrong. I cannot show that the Azande are wrong directly by a refutation because not only is their doctrine of *benge* hedged around so much as to make it virtually irrefutable, but also, when the *post mortem* that is supposed to discover it fails to, they have a barrage of *ad hoc* devices for handling this, and one can always defend any claim infinitely long if *ad hoc* devices are freely to hand. Rather would I argue with them that if, as Evans-Pritchard suggests, the problem their magic comes to solve is mainly the explanation of disaster, there exists a much more powerful explanatory scheme in my culture. For example, we can wipe out termites so that they do not pick on anyone's granary; we can immunize people from diseases witches cause altogether. We cannot do everything and we cannot convince all doubters, and we cannot answer the allegation that what we are offering is merely a more powerful magic. But we can in all integrity peddle our notions in the belief that they accomplish aims the Azande already have, better than the means indigenous to them.

I leave aside Winch's concluding hint that my endorsement of all this is to affect superiority or to be historicist. I do think that Azande adoption of western ways might be an improvement and also that it need not involve much changing of their ways of speaking. We in our culture still speak, with complete justice, of the sun rising and setting, of the land around airports as 'flat', of the sky being 'up', the earth 'down', etc. Even though these ways of speaking force our educators to spend time confuting them, they are still in order. But using the language of medicine or of science as a sort of mumbo-jumbo is highly reprehensible. If this is condescending to magic, so be it; if viewing science as a higher order activity then magic is historicism, I concede the charge.

Let me return to Winch's problems with concepts, namely, their close connection with the institution in the context of which the propositions employing them are asserted. We remember it came out sharply in his discussion of witches, time and height. I maintain that witchcraft is, aside from all else, certain beliefs about the world, and that whether or not there are witches is a fact outside of language. This leads Winch to pose the very curious question: is whether or not there is height a fact outside of language, or is the reality of height extra-linguistic?

# REPLY BY I. C. JARVIE

I don't know what 'the reality of height' can mean. Winch is surely right that operationist logical positivists could not possibly reduce 'height' to measuring operations language games. Then he says 'the reality of height shows itself in the sense that these language games have'. What can this mean? Can a language game have a sense? What kind of sense? A move in a game can make sense; but chess, for example, does not possess 'sense'. To a non-player it may seem a meaningless waste of time—nonsense. Does Winch mean that the reality of height shows itself in the sense that language games involving the word 'height' make when they are played? But games do not 'make sense' when they are played. What would it mean to say this game of soccer does not make sense when played (or when not played)? When one says 'there are no witches' one is *not* saying that witch-talk makes no sense, any more than the man without a concept of height who has learned our concept could think of it as mumbo-jumbo (*pace* Winch). We can, since Evans-Pritchard, very well distinguish 'making sense of' and 'understanding'. The point being that 'there are no witches' says not that 'witches' has no *sense*, but that it has no *reference*, in the same way as 'phlogiston' has no reference or perhaps 'god'. Of course, to claim that phlogiston exists is to make a different *kind* of claim than that God exists. However, when I deny them both there is a strong similarity: I am denying existence, on one level or another, to putative entities, of one kind or another (hitherto believed in by many).[1] But I cannot really say how great or small this similarity is because, after all, the difference between a difference in kind and a difference in degree is itself a difference of degree. This applies even to the very abstract difference referred to by 'height'.

Winch thinks we can't analyse very far the case of a people with no concept of height because height is so fundamental. He says a visitor from Britain would have great trouble because he would have to teach the heightless society new techniques and new ways of speaking. For my part, I think this is a typical, not a special, case. I think that Newton and Faraday and Einstein and Planck and Bohr (and Jesus, and Buddha, and Luther) all taught the human race new techniques and new ways of speaking. Einstein argued against certain views or beliefs, not against certain ways of speaking or a certain form of life. Indeed, Koestler has said rightly that Newtonian ways of speaking still dominate our form of life. And our forms of life still use Newtonian concepts—even centrally, e.g. The National Aeronautics and Space Administration. Winch is drawing attention to a genuine problem, but I think his views make it insoluble: the whole problem when someone like these great figures, the Einsteins,

---

[1] Time and height involve no entities, putative or otherwise, only relations. For this reason I think Winch's introduction of them confuses both the discussion and himself.

Buddhas, etc., come along is just 'is he a genius or a crank?', very much the reaction of the heightless ones to Winch's Briton. And indeed the solution is very difficult, for reasons noticed by Winch. But that we do, regularly, solve these problems is undeniable. Winch's theory, however, is this: when we speak or accept the genius's language we have decided he is a genius, and vice versa. This is true, but irrelevant: the problem is by what criteria *do* we decide to accept or speak? On Winch's theory this is inexplicable: as long as we use an older language game that game *a priori* excludes all other language games, old or new. This corollary of Winch comes out most clearly when Winch argues that a businessman who fails to grasp a moral criticism can be said to be 'imprisoned' within his viewpoint. Winch allows that we can criticize the businessman on moral grounds because we possess moral concepts he does not. The question is, does Winch think the only way out of the limitations of the businessman's moral conceptual scheme is through being taught by those possessing concepts he does not? Then, who teaches the teachers? Are we not all imprisoned? Winch agrees that to say *we* were imprisoned someone would have to show that we, too, are being blind to a certain possibility, as the businessman was blind to our different moral concepts. But then, Winch argues, we must specify this possibility. This is an odd claim. Why can it not be established in principle that all conceptual schemes are limited—how else could they and their forms of life grow and develop? To refuse to concede this, as Winch does, until the areas they can grow into can be specified—before they have grown there—is having it both ways. 'We won't agree we are limited until you show us how, if you can show us how we are no longer so limited.' Few of us doubt that our scientific knowledge is limited, but not only because we can see where, but also because of a general metaphysics which holds open the possibility that our horizons could be radically expanded tomorrow. But now, here is precisely the metaphysics we do not share with the Azande. And I would not baulk at the idea that this metaphysics makes sense because of its place in the life of the society it and we inhabit. The Azande do not live it, and to grasp it they may have to; but the way the Azande live is captured and made intelligible by *our* conceptual apparatus as well as theirs.

Thus Winch's argument about the people with no conception of height is a beautiful one, but it turns the tables nicely: *we*, who lack witchcraft, are, when we discuss witchcraft, like the heightless people trying to discuss height with us. Those without height, witches, or God, inhabit so to say the same world, but they 'see' it in different ways: the reality of it is different for each of them since the reality is constituted by their conception of it, their concepts. The radical conventionalism of this view is striking, for it might lead one to conclude that any possible conceptual scheme is viable. This, however, is not Winch's view. He believes that all viable conceptual schemes must have the central events of forms of *life*—without which life still does not go on—central to *them*. I don't follow this myself, since a radical materialistic conceptual scheme without the concept of life would seem possible and even viable—because if forms of life with no conceptual scheme at all are possible as in the animal, fish, insect and plant kingdom, then can we possibly say things must be different for man?

So far I have tried to show where there is a clash between the claim I made at the beginning and Winch's views on language and its limits. Now I turn to the problem of rationality and consistency where I think both Azande views (*ad hoc* evasions) and Winchian views (the Azande know what they are doing) disastrous. But first there is a minor misunderstanding to be cleared up. Winch contends that my crude first approximation to his position is the contrary of what he holds. I wrote in paraphrase of him 'cross-cultural value-judgments will always be mis-judgments—and therefore should be avoided—because there is no language game in which cross-cultural value-judgments could be legitimate moves'. Winch claims that in 'Understanding a Primitive Society' he *was making* cross-cultural comparisons—although not value-judgments, which he treats elsewhere. I confess to using 'value-judgment' to include cross-cultural comparisons (evaluations?) of the rationality of actions and the truth of claims. However, even if we now write 'comparisons' for 'value-judgments', I cannot grasp what the contrary position is that Winch was arguing for, since I am sure he would not want to assert either of these contraries: 'cross-cultural comparisons will *never* be mis-comparisons—and therefore need not be avoided—because there is a language game in which cross-cultural comparisons could be legitimate moves', or 'cross-cultural comparisons will not be mis-comparisons—and therefore need not be avoided—because in *every* language game cross-cultural comparisons can be legitimate moves'. That aside it is unclear in what language game Winch permits or forbids comparisons. He is ostensibly using our language to explain that it is not possible for us to pass judgment on whether Azande witch-talk corresponds to reality. Holding that what is real and unreal is a distinction belonging to each language or form of life, and that in the Azande form of life where the witch-talk occurs witchcraft is not treated as a hypothesis subject to such a question, the question makes no sense. Here then is our language being used by Winch to teach us about another language and its contrasts with our own. Winch's discourse, then, is at least second-order, or meta-linguistic. (He is not speaking the English of the English magician, priest, or scientist, but of an English anthropologist or philosopher.) We do then have a way—anthropological or philosophical—of discussing the adequacy of Azande language—such as their real/unreal distinction—the *adequacy of their form of life* (as compared and contrasted with western ways of life or not). To some extent the whole point of the attack in my paper is that the form of life is not a 'given'—unquestionable and basic. Not only is it that our form of life permits consideration of other forms of life and our own, but that the whole 'justificationist' belief that there can be a 'given', is a fundamental mistake. A 'given' form of life is open to precisely the same objections as a 'given' sense-datum, a 'given' protocol-sentence, or whatever.[1]

What does Winch say about rationality? It seemed to me he said that 'standards of rationality are incomparable'. Yet he writes he has 'never held any view of which' he would be willing to regard this as 'a proper paraphrase'. But he writes, 'MacIntyre seems to be saying that certain standards are taken as criteria of rationality because they are criteria of rationality. But whose?' (p. 317). In his unexplained question Winch here seemed to me to be denying that criteria of rationality can be compared to the extent that they are criteria of rationality. Q.E.D. To ask 'whose?' is to relativize the whole question and beg it. If he was not denying it, then Winch and I are agreed, since I hold that standards of rationality *are* comparable and western ones are *better* than Zande. Does Winch agree? If yes, no quarrel; if no, my paraphrase is quite in order.

As an example of why I rate western standards of rationality as better, I used Winch's own example of the Azande attitude to contradictions. Zande standards of rationality which are indifferent to certain contradictions can only seem 'rather poor', Winch thinks, depending on the point of the activity in which the contradictions grow up. But why can't standards of rationality seem rather poor as standards of rationality? What purpose do standards of rationality serve? As a minimum requirement they must serve the acquisition of knowledge. Now if our knowledge is greater than Azande knowledge (as it is in at least one way: we know how to go to them and make sense of their way of life, they don't know how to reciprocate) we would seem to have a *prima facie* argument that our standards do the job their standards try to do, and do it better. Therefore might they not well come to agree that ours should be theirs, and that theirs are rather poor?

Claiming that western standards of rationality are better, I instanced science. Winch then says, 'Our own conception of what it is to be rational is certainly not exhausted by the practices of science'. Quite. I should only claim (see notes 1, p. 236 and 1, p. 239 to my paper) that science is a

---

[1] For the term 'justification' and the main philosophical criticism of it see Bartley 3, 1962. Bartley considers his work an extension of Popper's.

paradigm of rationality—for us and *tout court*. I should go further than Winch and say that the *rationality of science* is not exhausted by the *practices of science*—why else is there continuing methodological debate? Yet the paradigm of rationality is action taken with full knowledge; the paradigm of full knowledge is scientific knowledge; therefore action taken to gain scientific knowledge is at the heart of any idea of rationality.[1]

## REFERENCES

Agassi, J. 1964. The nature of scientific problems and their roots in meta-physics. *The Critical Approach*. Ed. M. Bunge.

Agassi, J. & Jarvie, I. C. 1967. The problem of the rationality of magic. *Br. J. Sociol.* **18**.

Bartley, W. W. 1962. *The Retreat to Commitment*. New York.

Nielson, Kai. 1967. Wittgensteinian fideism. *Philosophy* **42**.

Popper, K. R. 1945, 1962. *The Open Society and Its Enemies*. London.

Wittgenstein, L. 1954. *Philosophical Investigations*. Blackwell.

[1] See my paper with Agassi, 1967.

# SITUATIONAL INDIVIDUALISM AND THE EMERGENT GROUP-PROPERTIES

## *by* J. O. WISDOM

One of the fundamental problems in the philosophy of the social sciences is to locate the source of power possessed by groups, if it is in any measure independent of individuals. If a group is no more than the sum of the individuals composing it, there is no problem: the power of the group is only some compound of the power of its members. But, assuming that a group is more than this, we seem forced to suppose that not all its power can derive from them. Is a dual control understandable? If not, are individuals merely the pawns of group-power? Or is there some way, after all, of seeing group-power as an outcrop of the power of individuals?

This last is the answer given by Popper (1964) through his thesis of methodological individualism. His thesis goes a long way towards solving the problem, but seems to contain a certain gap. My aim is to show that in filling it we are led to a framework of dual control. (The view of group-power as wholly overriding individuals lies outside the scope of this discussion.)

### METHODOLOGICAL INDIVIDUALISM AND ITS DUAL CLAIMS

However wrong-headed, the most natural presumption to adopt to begin with is that individual human beings began as individuals related to one another only as individuals, and that in the course of time they banded themselves together into groups or communities. Society would then appear as a product arising from individuals. This presumption probably stems, at least in part, from the atomistic approach of western thought.

In the course of time, unacceptable developments ensue, such as severe taxation or the impossibility of disburdening oneself of a nagging wife. Individuals feel helpless before these states of affairs. The source of some of them may be, for a time, attributed to a god, but the source of others—and later on almost all others—becomes attributed to society itself. It is noticed that institutions (in a broad sense) have great effects on people which they are powerless to alter: we all have to wear some clothes; we all have to go to school; many have to follow religious observances, or to follow some social observances, such as lifting one's cap or driving on the left side of the road; all may even have to vote at an election. Whence do institutions/society derive their power? If to begin with the power is supposed to reside in a god, such a belief becomes attenuated with the (gradual) growth of secularism. It is a natural development for the power to be located firmly in society itself. Thus Marx (1906) regarded the way large numbers of individual workers were exploited as a manifestation of the institution of exploitation and explained such exploitation as a consequence of the institution of capitalism.

Thus the locus of power, as we see it, shifts from individuals to society. In extreme forms it is not shared: in the extreme form of individualism, society is a mode of individual operations and partakes of no independent power; in the extreme form of what, following Gellner, we may best call holism, societal power overrides individuals who have no share of it (beyond being for Marx midwives (catalysts) or brakes).

Thus we reach the idea of an institution or of society as an *independent* source of power. Doctrines of holism, historicism, group-minds, a collective unconscious are (nearly always) versions of the idea of society with wholly independent power. Institutions seem to have a life of their own.

Then holism itself suffers a twist of the screw. Secularism, which has given up a belief in transcendent gods above, begins to doubt that any form of god is even immanent in society. A dash of empiricism and liberalism salt the scepticism. And Popper, spokesman for English distrust of the doctrinaire, for English liberalism and minimum interference with *laissez-faire*, reacts against holism[1] with the assertion that institutions do not have aims—only individuals have aims, interests, needs, intentions, or take decisions.

This thesis is one of Popper's most important contributions to the philosophy of the social sciences. But unlike nearly all his other ideas in print, his account of it is sketchy, and the Popperians have not sufficiently filled it out. Agassi's (1960) masterly exegesis almost completely succeeded, but still, I think, left a basic point obscure. This would be of no moment if the intuitive idea were wholly clear; but the very significance of it lies in the combination of two factors that, on the face of it, are incompatible.

For methodological individualism looks like a 'reductionist' theory, for it holds that institutions (or social wholes) consist of individual aims, intentions, interests, and power; but this does not quite correctly represent Popper, for he holds that institutions can never be expressed wholly in terms of individuals. So he is saying that there is nothing in society but individuals and their motivations, and yet that institutional activities have a life of their own. If this position is not explicated fully in such a way as to

---

[1] Popper (1963, 1964) has made a sustained attack on holism. For me also it is untenable, but it cannot be examined here.

dispel any appearance of contradiction, the likelihood is that an account of it will achieve coherence at the expense of leaning towards one pole of the thesis or the other. Thus Watkins' discussions (notably Watkins, 1957) have left the impression with some readers that Popper's theory is fully reductionist.[1] This is the interpretation to be expected, not the opposite one, because Popper so stresses the role of individuals that no one would interpret him as reducing the activity of individuals to that of institutions. In fact, Popper's view allows a place to both poles, and (assuming that this is *not* an inconsistency, and I hold it is not) the first task is to dissect the thesis and present it free of apparent contradiction (cf. Scott, 1961).

### FORMS OF REDUCTIONISM

Popper himself recognizes that institutions have in some sense a life of their own. He is fully alive to such facts as that successive British Governments in recent times intended to protect sterling currency. The problem that opens up is that of squaring such facts with individualism. And the most obvious way of trying to do so is by means of the device of 'reduction', by which an institutional aim is 'reducible' to that of individuals.

It is commonly supposed that there is just one process of 'reduction'; but this is not so—there are at least two. In the usual version, the broad procedure is to interpret a statement about institutional intentions as a *shorthand* for a statement about individual intentions, convenient in practice because of brevity but expendable in principle.[2]

---

[1] Popper himself, though only very slightly, opens the door to this misinterpretation. The following passages convey it. 'We must try to understand all collective phenomena as due to actions, interactions, aims, hopes, and thoughts of individual men, and as due to traditions created and preserved by individual men' (Popper, 1964, 157–8); '... the "behaviour" and the "actions" of collectives...must be reduced to the behaviour and to the actions of human individuals' (Popper, 1963, 2, 91).

[2] This method of reduction has a noble history containing some exquisite examples. Dedekind (1963) 'reduced' irrational numbers to rationals, thus solving the two-thousand-year-old Pythagorean problem of the incommensurability of the two sorts of numbers. Whitehead (1920) 'reduced' points and lines to Chinese nests of boxes, thus liquidating the two-thousand-year-old contradiction in the Euclidean concepts. Broad (1933) 'reduced' propositions to co-referential judgments in an attempt to eliminate from propositional logic any shadow of a Platonic idea. The method is excellent in mathematics and logic. One of the great failures was in philosophy, in the phenomenalist theory of perception, according to which physical objects are 'reducible' to families of sense-data. The most influential use of the idea in philosophy was Russell's (1920) theory use of descriptions ('The author of *Waverley* exists' 'reduces' to 'One and only one man wrote *Waverley*'), in which a description denotes no Platonic idea or object but is 'reduced to', or disappears on translation into, phrases denoting two objects (man and book) and one relation (writing). In relation to the present theme of 'reducing' institutions, Russell's theory of descriptions ('logical constructions' and 'incomplete symbols') was adapted by John Wisdom (1933, 1934) before he entered the Wittgenstein orbit, when he worked along the lines of Russell and Moore. He (1933) 'reduced'

Agassi appears to underwrite this method of 'reduction' as what Popper meant, for he speaks of the 'aims' of an institution as being a shorthand. I think, however, there is reason to doubt whether this is what Popper meant. This is a tricky point. As Agassi himself well points out, Popper allows of attributing an aim to an institution only in so far as *individuals* give it an aim. Thus, parliament legislates because we give it the aim of legislating; and, of course, we could take away this aim, as happens when a dictator abolishes parliament (or suffers it to remain, obliged to rubber stamp his measures). Now this construction does not 'reduce' parliament to the activities of ministers and members of parliament, judges and police; for it involves the activities of individuals outside the functioning of parliament, namely the members of the electorate. However, the intentions of these individuals should figure in the 'reductionist' analysis if this is to be at all adequate. So the 'reductionist' programme would misrepresent Popper only if it were too restricted as regards the range of the individuals included in it. Still it does apparently represent him correctly, that is that 'institutional aims' is a shorthand, provided the longhand is full enough. Is this really so?

The 'reductionist' programme aims at dispensing with *all* institutional wholes after the 'reduction' is carried out. Now Popper is fully aware, I think, that this is impossible, for you can dispense with one or even more institutional wholes, but *only in an institutional setting* (Popper, 1963, 2, 90). This is one of the factors unstressed in existing accounts of his methodological individualism.

For ease of reference I propose to reserve the expression 'reductionist' or 'reductionist individualism' for the extreme form of individualism, according to which all institutional wholes are 'reducible' without remainder to terms concerning the purposes of individuals, i.e. all institutions are epiphenomena of individual purposes. 'Reductionist individualism' is a position sometimes attributed to Popper, the position sometimes put upon Watkins' accounts. In my view, although this is a misinterpretation of Popper's position, his view involves a partial reduction, in which any one institutional whole is 'reducible' though not all such wholes at once. That is to say, when any given institution is 'reduced' to the aims of individuals, this is effected only at the cost of introducing some other whole,

'England in a monarchy' by translating it into 'Englishmen acknowledge a monarch', and (1934) 'reduced' 'Some nations invaded France and some did not' by translating it into 'There were groups of people each with common ancestors, traditions and governors such that the members of each group selected from among themselves soldiers and those soldiers forcibly entered the land owned by Frenchmen.' ('Groups' he eliminates similarly.) These 'reduce' the institutions, England, monarchy, nations, and France, to individuals, Englishmen, monarch, and Frenchmen.

which in turn can be reduced but only at a similar price. Thus, the social whole, the Government, can be replaced by an individual, Mr Gladstone, when 'the Government decided . . .' is replaced by 'Mr Gladstone signed on order . . .'; but this can be done only because Mr Gladstone acted in his institutional capacity of Prime Minister. The reference to a further whole may be covert, but it is present. Thus, with 'reduction' in this form, whatever whole is 'reduced', some whole is always left over 'unreduced'. This I take to be Popper's position. And I propose to call it 'situational individualism', to bring out that for him individual aims exist only in an institutional situation, i.e. that there are two poles in his thesis.

### UNINTENDED CONSEQUENCES

I wish now to turn to a different facet. Popper (1963, 2, 324; 1964, 65) has laid great emphasis on the significance of *unintended consequences*. By this he refers both to individuals and to institutions; he is more concerned with the unintended consequences of our individual actions than with those of institutions, but both occur and are important. We all know they occur, but few may agree that they have theoretical importance. For Popper they are basic and are inevitable. As an example, suppose you want to buy two similar houses; you begin by buying one of them and this in some markets will put up the price of the other, although no one intended this to happen.

I will first argue for a conclusion of Popper's that there must always be unintended consequences of our actions, which he did not trouble much about, leaving it as empirically obvious; but it is significantly more than that.

Unintended consequences commonly take the form of what I would describe as stable side-effects. Now supposing that the number of consequences were definite, they might or might not be all foreseeable. But an institution has an indefinite number of effects. Hence, even if *each* consequence were foreseeable, *all* would not be—just as *every* whole number is countable yet *all* whole numbers are not countable. This might be expressed by saying that unintended consequences may be *distributively* predictable but are not *collectively* predictable.[1] All which is just a logical way of bringing out the point that whatever effort we make to foresee unintended consequences and however successful we are, there must logically always be some we shall have failed to foresee.

The reason why I have introduced the logical argument is that it enables a decisive point to be made. The unintended consequences that are not

---

[1] At the risk of over-sophistication, it may clarify for a few readers to say that unintended consequences are 'simply predictable' but not 'omega-predictable'.

foreseen are of the same kind as those that are actually foreseen. So totally unknown consequences no longer have the aura of stemming from a mysterious origin. Their origin is the same as the origin of foreseen consequences, and these may not seem so mysterious. In fact, we may consider regarding an institution as a concatenation of individual intentions plus unintended consequences of individual intentions that are unforeseen. Then an institution would be made up of intentions and their consequences only and yet be more than actual intentions. This, I think, explains the point made by Popper, that only intentions and their consequences produce a certain institution, yet no one may actually have intended to produce it. Otherwise expressed, an institution is not a complex of goal-directed activities but results solely from goal-directed activities.

It is now easy to see that this construction of unintended consequences is identical with the construction put upon 'partial reductionism'; and we could say that 'situational individualism', which I attribute to Popper, is 'distributively reductionist', while 'reductionist individualism', which I claim Popper rejects, is 'collectively reductionist'.[1] For the unforeseen consequences of our actions are severally though not collectively 'reducible' to individual intentions and their consequences.[2] The sense of these rather abstract terms may be more easily retained in mind if we replace the two kinds of 'reductionism' by 'piecemeal reductionism' for Popper's view and 'global reductionism' for the misinterpretation of his view.

Thus the independent power of every institution lies in the capacity to produce unforeseen unintended consequences; but these are the results only of individual intentions, so that nothing over and above the individual intentions and their consequences is needed for building up the content of institutions. Thus Popper can maintain a position with two poles, which are apparently incompatible, namely individualism and institutionalism.[3]

I shall, however, introduce a far-reaching modification later.

[1] Or respectively 'simply reductionist' and 'omega-reductionist'.

[2] Popper's criticisms of 'reductionist individualism' are that it has difficulty in accounting for the relations of institutions to one another and to individuals, for the independent power of institutions, for their unintended consequences, and also for the facts that some institutions originated unintentionally, and that some persist against individual intentions; and it would have to operate with that dubious concept, the beginning of society. 'Situational individualism', on the other hand, has no such difficulties.

[3] It is clear from the above discussion that the objects scheduled for 'reduction' in Popper's theory are primarily existents or facts. Nonetheless natural science is concerned to explain natural phenomena; social science to explain institutions. And Popper's theory of 'situational individualism' is a *metatheory* stating the kind of theory an explanation of social institutions ought to be, namely that laws about institutions are 'distributively reducible' or 'partially reducible' to individuals. Thus his theory applies both to existents or facts and also to laws.

## PSYCHOLOGISM AND SOCIAL GROUPS

Contrasting with 'situational individualism' is psychologism, which is a form of individualism that Popper rejects.

Psychologism (or psychologistic individualism) is the thesis that an institution (or society) is a purely psychological product, manifestation, expression, or symbol of human nature, which, moreover, has to assume a life of its own even though composed of nothing else; or, in short, that society depends on the 'human nature' of its members (Popper, 1963, 1, 83), or more explicitly is the product of interacting minds (Popper, 1963, 2, 90).

Psychologism differs from 'methodological individualism'. Some critics have understandably been unable to see the difference. Psychologism refers to human nature, i.e. the nature of a person or a mind, mental processes, or to the psychological theory of what these are. 'Methodological individualism' refers to none of this structure or explanatory factors but only to human purposes, etc., which are attributes of human activity—the facts, if we like, about human beings (for which the factors coming under psychologism consist of law and theory or explanation). Thus for Popper, it is not psychologistic that parliament depends upon the intentions of many people to have the country governed; it would be psychologistic to interpret parliament as a societal manifestation of individual intentions to run their lives in an orderly fashion.

Since the issue of psychologism lies on the fringe of the present investigation, it will suffice to say here that Popper gives, in my opinion, a valid criticism of the doctrine (though he leaves a loophole, which I believe could be closed). Nonetheless, although institutions are not to be interpreted purely psychologistically, there is a social whole, which is just as fundamental as the social institution, that may have to be interpreted thus— namely the social group.

One example of a group is a set of people who go to the same college. Another is a set of doctors. A group may, but need not, meet. We all have great experience of groups, but hardly any articulate knowledge. Fairly recently, however, a great advance has come from Bion (1961), whose work, *Experience in Groups*, is perhaps the most important yet written on group psychology.

It is noteworthy that Bion repudiates all attempt at constructing a psychology of the group upon that of the individual. Just as for Popper, an individual is always an individual in a setting of institutions, so for Bion an individual is always a member of some group (usually several). Group activity is not 'reducible piecemeal' to individual activity; group psychology is not 'reducible piecemeal' to individual psychology. We shall have to consider whether group activity is even distributively reducible to individual activity.

Analogous to the situation described by Popper, though not discussed by Bion, there are unintended consequences of group activity and of individual actions in groups. But again analogously, a group has no source other than individual activity; its origin, persistence, and development involve no other ingredients. Thus far a group has the same sort of structure as an institution.

But there is a difference. An institution does not have intentions; a group, according to Bion, does. It also has a catalogue of things like desires, fears, hopes, anxiety, guilt feelings, self-preservation tendencies, and so on. Whether a group intention is formed only out of individual intentions, or whether a group intention is not reducible, even 'piecemeal', to individual activity, which would violate Popper's presupposition that only individuals have intentions, is an important question to which I shall return at the end. All I need for the moment is the existence in a significant sense of a group intention, and that there is a theory of its action, for Bion gives not just a concept, but a theory of group intention, which is in some measure testable.

## METHODOLOGICAL PROBLEMS

The problems I wish to highlight are now beginning to emerge.

(1) The foregoing exegesis of Popper's 'situational individualism' aims at removing a misunderstanding of his position by distinguishing (a) a faulty interpretation of it, namely, 'global reductionism' or 'reductionist individualism', which allows no independence to the power or aims of institutions over and above what is possessed by individuals, from what he really seems to hold, namely, (b) 'piecemeal reductionism' or 'situational individualism', which does admit of such an independence in some measure. This distinction enables us to square Popper's individualism and reductionist tendency with the fact that he undoubtedly recognizes that institutions have in some sense a life of their own. Whether or not his thesis, understood in this way for institutions, is satisfactory, I suspect that for groups the thesis is not fully adequate.

(2) There is, however, a closely connected problem to do with unobservables. It is not certain that these entities exist in the social sciences; when reviewing Brown (1963), I was under the impression (Wisdom, 1964) that he had found an example in Simon (1957), but it turned out not to be one.

It seems evident that Popper's metatheory of 'situational individualism' precludes the existence of unobservables in the social sciences. Now it is undesirable that a metatheory should impose restrictions on the kinds of entity that shall appear in a hypothesis. Indeed the policy I nail my flag to is that certain sorts of problems will not give up their secret without them, any more than in the natural sciences.

In order to discuss these problems, I need an example, and, since there is none that I know of readily available in the literature that is clear-cut outside economics, apart from Bion's and the evidence relating to his is difficult to specify, I have had to put one together. It is in the area of social psychology.

The problem I propose to consider concerns the social pathology of Great Britain.[1]

### GREAT BRITAIN'S ILLS

There is a problem about the ills from which Great Britain is suffering. The stimulus for tackling it was a special number of *Encounter* edited by Arthur Koestler in 1963. Distinguished contributors made some distinguished contributions: their aim was a diagnosis of Britain's condition. They did not reach a diagnosis, but I think that one can be made, and I shall propose a group hypothesis. Provided this should prove testable, irrespective of whether it proves false or not, we shall be in a position of having a genuine example with which to discuss the interlocking problems of this paper.

There are numerous types of complaint the nation seems to be suffering from. These are factual, though a certain amount of interpretation creeps in and they are to some extent controversial; I will cluster them under two complexes.

### The Leo–Struthonian Complex

The first complex is epitomized strikingly by Koestler under the heading of the lion and the ostrich. The nation was like a lion over the Battle of Britain, but like an ostrich over the Nazi threat. Koestler has coined a nice new expression 'struthonian'[2] from the Latin for an ostrich. I would prefer to add to this, to bring out the bipolar nature of the attitude, and speak of Leo–Struthonianism—after all, those who wear the Old Struthonian tie

---

[1] I put this forward first as two talks on 8 and 13 August 1966, on the B.B.C. Third Programme, and they appeared in *The Listener* for 18 and 25 August 1966, under the title 'The Social Pathology of Great Britain'. They are rearranged, somewhat rewritten, with one part telescoped and with a small addition, but there is no basic alteration.

[2] One correspondent in *The Listener* has criticized Koestler for using this form of adjective, and me for following him, on the grounds that the normal way of coining a word from Latin would require the form 'struthionine'. I suspect the era is over when English remains rigid in such matters. And 'struthonian' is smoother. But I would readily acquiesce if 'struthionine' were adopted.

might agree that one can afford to be struthonian if, when real need arises, one is really leonine. This attitude, however, is not well adapted to the battle of living during times of peace.

A whole crop of manifestations of Leo–Struthonianism can be found.

There are post-war cases of ineptitude on a large scale, concerning, for example, the railways, roads, housing, hotels, sidetracking recommendations of royal commissions, unrealistic approaches in business firms, an out-of-date structure of trade unions, the lack of initiation of managements, absenteeism, unofficial strikes; sometimes justified on the grounds that efficiency is chilling. Then self-damaging actions carefully contrived by various governments, concerning, for example, the failure to take the initiative over the Common Market, dividing children into sheep and goats according to their performances at the age of eleven, the heavy restraints placed on university development, a fine disregard for research, a self-effacing economic policy pursued complacently for some fifteen years through the fifties and early sixties.

It is worth looking at an interesting economic example of 'safe' action, promoting incompetence, and highly self-damaging—the devaluation of 1949. What it did was to enable incompetent firms to export more easily without recourse to more economical production. Thus, there was no incentive towards improvement, only a respite for the lazy. I do not say that no symptomatic relief should have been given in an emergency; but some temporary expedient should have been adopted, not a permanent debilitating drug. Instead, for longer term purposes, a reasonable export coefficient could have been devised, which firms might be expected to reach, and a tax on failure to achieve it could have been introduced as a spur to efficiency. The Government directly protected inefficiency and indirectly promoted national damage.

This list of deficiencies forms a symptom characterized like the previous one by ineptitude, but it is more directly self-damaging.

Great Britain's isolation from the Common Market was also a sign, the most tangible one, of deliberately abdicating from her role of international leadership, disparaging her own prestige, achievements, influence, contributions, and developments as being of no importance. Further, there is the callousness and dishonesty to be found in political life. Ministers in the House of Commons, for political purposes, have lied or given information so misleading as to be morally and psychologically equivalent to a lie, yet as individuals you could trust them with your safe. Thus the political dishonesty does not match the personal but is split off therefrom. It is not political dishonesty that is an ill, but the peculiarity that ministers of unimpeachable honesty as individuals become dishonest as a group.

There is the rise of crime, the backward state of prisons, and the gross inadequacy of psychiatric services. And a fine piece of struthonianism is the failure to accept as a reality that teenagers are now, in practice, sexually active.

In short we can find widespread leo–struthonian ineptitude, self-damaging national actions, abdication from giving a lead, shifty group-behaviour, and ignoring the seamy side.

### The Gentleman–Amateur Complex

We come now to a different type of phenomenon.

(a) Great Britain is a country consisting of first-class citizens and second-class citizens. The fact is less marked than it used to be, but it remains to a significant extent a fact.

There is a parallel attitude about foreigners. Individuals are of course charming, but, as a group, foreigners are regarded as slightly quaint; they therefore do not even approach resemblance to first- or second-class citizens.

These attitudes may be seen to be a particular expression of a very special dichotomy, namely that between amateurs and gentlemen on the one hand, and, on the other, professionals and players. It has been vividly portrayed by Austen Albu (1963). To my mind it is of fundamental importance.

The concept of a 'player' refers to a professional performer with expertise, and a player has no other virtues. There is something machine-like about him and he has a one-track mind. An engineer or any technologist (with the possible exception of doctors or lawyers) is a player; he is an expert at managing gadgets and thus ministering to part of the material comforts of his 'betters'.

A gentleman on the other hand is held to have the gift of flexibility and improvisation, which is highly prized; expertise is not needed by the gentleman because his powers of improvisation will cope with any problem that turns up. But, more than that, expertise is despised.

Parallel with this we find the dichotomy between character and brains: brains are unbecoming to a gentleman because they smack of expertise; character, on the other hand, is an absolute value, but it has, I think, a special meaning over and above dependability, loyalty, etc., namely, the ability to cope flexibly with the unexpected. In other words it is the improvisation attribute of the gentleman over again.

Expertise-like drill is for 'other ranks'. Apprenticeship is a form of drill and leads to expertise. Expertise is thus a drill-skill, and leaves no room for improvisation.

Further, contempt for expertise leads to the opinion that inefficiency is satisfactory, and Goronwy Rees (1963) is right to call this the cult of incompetence.

(b) The way Great Britain wastes its most outstanding public men is startling. Although it is to go back further than the post-war period I am concerned with, the tendency is worth tracing back. After World War I, no use was found for Lloyd George; in the thirties Churchill was kept on one side; in the fifties and sixties Butler was warded off; Grimond, though not actively barred, was not bothered about. Does the country enjoy such a plethora of great men that it can afford to do without the distinctive contributions such men could offer? This is hardly the implication of the long line, with hardly an exception, of second-rate rulers, irrespective of political persuasion, that have held power for generations. The explanations for these cases are of course easy to find in the situations of the times; but nothing in the group structure has opposed the tendency—quite the contrary, the group structure has confirmed the tendency. For the country does not like really able men. One or two of those cited have been written off as 'too clever by half'. Certainly the country's distrust of the expert who has no vision is sound. But no one could maintain such a charge against these men. The country distrusts not only the 'mere expert', but distrusts expertise altogether, however leavened by the imagination of the amateur.

This extraordinary national misuse of talent could have been classified under struthonian incompetence, or self-damaging procedures. But I think it belongs more closely to the Gentleman–Amateur Complex, because most of these men were shut out for being too able and they combined the gifts of the amateur with expertise.

We have now a formidable list of failings, falling broadly under the headings of Leo–Struthonianism and the cult of the Gentleman–Amateur.

### Diagnosis

Now, if I met an individual with this set of disabilities, I would regard them as a syndrome, and I would make a diagnosis of *depression*, though I would qualify this as *sub-acute* (that the depression is sub-acute is because the intensity of the symptoms, or more correctly signs, is not great, and virtues exist that could not be present if the depression were deep). My thesis is that Great Britain is a group suffering from sub-acute depression —although there may be no more depressives among the population now than at any other time. (And I mean this in the psychiatric and not in the economic sense.) I am putting it forward not as an individual but as a group hypothesis.

I am not arguing by analogy. It is invalid to argue from the individual to the group; argument by analogy does not hold. I have simply made a hypothesis about a group, which I have derived from a psychiatric diagnosis about individuals. The success or failure of the hypothesis will depend on whether it can be used satisfactorily or not, but it does not depend upon argument by analogy; that is to say, it will depend on whether it can be applied and tested.

I have thus put the numerous ills from which Great Britain is suffering under two complexes, Leo–Struthonianism and the Gentleman–Amateur Complex; and these are interpretations of the facts as Koestler and others have found them. Koestler and Albu have virtually given these interpretations. I have gone further in suggesting that these two complexes are basic to the syndrome. And I have added a diagnosis of this syndrome and the complexes (namely, sub-acute depression).

We have now to examine depression as a group-psychiatric phenomenon.

### The diagnosis applied to Leo–Struthonianism

The core of the structure of depression in a person consists of ambivalence, i.e. hostility to what he values most highly, and he becomes weighed down by a sense of guilt at his unkindness to the object of his greatest concern. Most characteristically, the weight of guilt removes the sunshine, makes nothing seem worthwhile, and leads to self-reproach and even self-punishment and self-inflicted damage. Already you will see a couple of the symptoms I have described falling into place, the muted *élan* and inefficiency, and the list of policies carried out against the national interest.

A person has various means of defence, all of which must be based on detaching the hostility from the object of value. In an extreme case, a person might blind himself altogether to the hostility and indulge in a burst of elation. But milder alternatives are open, for instance the denial that the hostility does any real harm, which is likely to be accompanied by a damping down of hostility generally, even where it is needed, and thus lead to giving in all round, however unsuitably. How does this apply to Chief Enahoro, who was deported from the United Kingdom at the request of the Nigerian Government? Individuals thought he was wrongly treated—and indeed that a British value was set at naught—but the Nation, or at least the Government of the country, as a group, acted on the claim that he would receive a fair trial when he reached home, though it was pretty obvious he would not. The British Government appeared unable to give offence and show ordinary firmness to the Nigerian Government. The parallel with an individual depressive is very close. The depressive is not a liar, but he is sometimes apt to use various kinds of rationalizations, prevarications, excuses, and justifications, to defend himself against reproach (which is not incompatible with self-reproach), and to him they may be convincing, while to the outsider they may appear rather thin. Despite what appeared to be the fatuousness of the Government's statements about the Chief's freedom to have counsel of his choice, although the one he was known to prefer was a lawyer who would not be permitted to land in Nigeria, it would seem quite probable that the Government regarded their action as morally justified.

Thus the hypothesis of sub-acute depression can account for a further symptom in the presumed syndrome.

Further it explains why the ostrich becomes lion-hearted when physical danger actually develops; for *hostility can focus on a real enemy without ambivalence, and once the hostility is thus withdrawn from cherished values ambivalence is dispelled for the duration.*

What of crime, prisons, and psychiatric services? I will mention only one small point: the depressive is not in a position to cope with the seamier aspects of his being; if he were able to, he would not be depressed. So far from coping, he even tries to deny their existence; and this is just what the Nation does about these matters—it averts its gaze.

Split attitudes and inhibition about sex again are usual in depression.

### *The diagnosis applied to the Gentleman–Amateur Complex*

The gentleman used to be one of the chief values of Great Britain. Now a depressive structure could not permit its values to be sullied. Hence, to prevent overt depression, it was vital to keep a sharp split between the gentleman and the player.

In the past it was easy to maintain this split. But after the war this was no longer so, and we have to enquire into the change.

There came the not-at-all surprising developments of the welfare state, increased literacy among other ranks, increased authority among other ranks, depletion of the gentleman's means—there were plenty of changes that helped to put the gentleman off his pedestal. Hence the split between gentleman and player could no longer be maintained.

Connected with this is a factor of a different sort. One of Britain's basic characteristics, which was pointed out by Renier (1933), was that it was a ritualistic society. The idea of taking pleasures sadly means that *playing* the game, important though that may be, was more important than *enjoying* the game. That is to say, playing the game was not fun but ritual. Now the effect of the democratizations resulting from the war, in dethroning the gentleman, was to break down his ritual. And this must lead to an upsurge of the state of affairs controlled by the ritual. We are thus faced with the question of the functions of ritual. Put very generally, what ritual controls involves aggressiveness; so, if ritual breaks down, the problem is how to contain it. For the Nation, neither open aggressiveness nor paranoia are in character, and depression seems to be the only alternative open for containing aggressiveness.

Further, a contributory factor lies in an unintended consequence of the welfare state. For the care he bestowed on his inferiors (some inferiors, e.g., servants and subjects of his charity) was part of the role of the gentleman, and you can hardly deprive a man of his altruistic activities without making him depressed. The transformation of a colony into a dominion would have in some degree the same effect. The outcome was that the gentleman lost his function and lost therefore his confidence.

Indeed, these days it is sometimes hinted that gentlemen are disappearing. Now this is physically impossible in the short time that has elapsed. So what must be meant is that the gentleman is no longer prominent, no longer counts for much. And, indeed, it is sometimes felt that to be a gentleman is 'square', and there are some who try to efface the manifestations of their type. Thus, there is ambivalance towards one of the country's most prized values—and the ambivalence emanates not just from the envy of others, which is not new, but emanates from gentlemen themselves, which is new. There is, in short, a national ambivalence—hence, the depression.

### The gentleman's reaction to ambivalence

Now, the gentleman's loss of prestige does not mean that the split between citizens first-class and second-class is overcome, for the gentleman has lost value to gentlemen without his being reconciled to players.

There seem to me to be interesting confusions underlying this phenomenon.[1]

With the growth of the self-reproach of being a gentleman, it has not occurred to him that a gentleman is a complex of several different kinds of qualities, and if he now feels ashamed of snobbery, for instance, he need not overlook the fact that some characteristics of the gentleman can function well. Indeed there are several of these. The gentleman had a code of decency in day-to-day behaviour (subject to limitations and exceptions, but it existed), a code of manners, a standard of taste. Naturally these would all be bogus if they owed their existence to the need, for example, to maintain snobbery; but has the gentleman questioned whether they may not have

---

[1] Psychiatrically they involve faulty schizoid identifications and splits.

another source? Does he regard it as impossible to value these qualities while jettisoning the attitude to citizens second-class, to foreigners, and to professionals and players?

### A valid conception of the amateur

It seems to me that the reason why he cannot do this is because he clings to the idea of the amateur in a faulty way. The expert, as I shall call the professional and player, he conceives of narrowly as drilled to carry out one task with precision. No doubt this is what a number of experts are now actually like. But the conception overlooks an important methodological point: that *where a new problem is involved the expert is always an amateur*— it is precisely because a problem is new that the expert has not got a drill ready to hand for dealing with it. So the idea of the amateur is not wholly misguided. In fact, all original scientists have to have this character of the amateur in them. Einstein was very much the amateur. Engineers developing atomic energy, with nothing in previous experience to guide them, can only be amateurs in their new work. But this does not mean that they can afford to neglect expertise.

Expertise without the leavening of the amateur cannot cope with anything off the beaten track. By contrast, the amateur without expertise may range from being incompetent to being irrelevant. The one can perform, but only on tram lines; the other is off the rails. Our question is not whether to go on the rails or off them, but whether we can dispense with them. The question for the gentleman, then, is whether he can revalue the amateur, not as a hot-house plant, but as a rose grown upon the briar of expertise. If he could, this would overcome the ambivalence, the split with citizens second-class, and the depression.

The problem is a crucial one. Many other countries do not have it because they either never had, or they have got rid of, that group allied to the upper middle classes and merging into the upper classes. But Britain has to handle the problem. Elimination of the gentlemen, even if practicable, say by educating them not to be gentlemen, would not eliminate the national depression, because it would not eliminate the ambivalance in any reasonably short run. But the idea of the Expert–Amateur (not the amateur-expert, i.e. specialist who does not know his job) could transform this state quite quickly.

## *Prognosis*

Let us now turn to the question of *prognosis*.

There are several possibilities and the prognosis is a conditional prediction. It is, however, possible to weigh the likelihood of the several possible conditions and therefore to select the most likely outcome.

Broad possibilities for the depressive are: getting worse, becoming chronic, and getting better. If he gets worse, he may do so by adopting schizoid defences; if he gets better, he may do so by effecting neurotic adjustments of a more or less workable kind; if he becomes chronically depressed, this may be because his adjustments work badly.

What about getting worse? This would be indicated if the signs were that the overt disorder was not the central disorder but a defence against one; in that case I would look for activities designed, not simply to prop up the patient, but to protect him from something that is not actually present. So I would look to see whether such protective measures were attended by anxiety. And with depression what would have to be protected would be the core of the personality against being broken up. Arresting the nuclear disarmament squatters is, it is true, a minute sign of such protective measure, aimed at preventing the group of squatters from developing into a fragment broken away from the rest of society. But we should have to see repressive measures against riots and the like before the depression would get worse.

Now there are other possibilities that, on the surface, look more satisfactory. Thus the Nation might become more struthonian, to the degree of believing that all is right with its world. But this would produce an irresponsibly elated national life, which might feel good but which would be unstable and a fool's paradise. However, such a development would seem to be wholly out of character. Another possibility is that confidence might be recovered by imagining grave national threats to be emanating from many quarters of the world.[1] This would undoubtedly set the country on its feet; but it is not a serious alternative for a genuine depressive, and also it would seem to be out of character. Great Britain has displayed nothing like the anxiety of some other countries, e.g. France, Germany, Russia, and the U.S.A., at threats, real or imagined, from abroad. Indeed she has often underrated these; and, although this might conceivably be disguised indication of a paranoid trait, it would be very much a paranoid trait under control. An interesting possibility, which does seem to be real, would be an access of obsessionality, in an attempt to control aggressiveness caused by the disruption of ritual; for ritual is a controlled form of

---

[1] This mechanism was used by certain totalitarian countries.

obsessionality. Nonetheless, I do not rate the chance of such an outcome very high, because it would depend on national aggressiveness threatening to get out of control; and, while the Suez episode is evidence of this, Britain does not look like repeating such a thing.[1] The protectorate may have been an obsessional interregnum, but look what it took to produce it; and as a whole it seems somewhat out of character. All these alternatives are notable for affording ways of producing an outward show of recovery. But, in fact, they would be unstable psychotic or neurotic defences against getting worse—the first manic, the second paranoid, the third obsessional. Although unstable, they could work reasonably well in the short or moderate run. Still they are not a likely choice for Britain.

If the depression is to be chronic, I would expect it to be an acceptable, though not welcome, settlement with fate on the (national) assumption that any move to improve things might make them worse and dare not be risked. There has been some evidence of this shown by several examples of refusal to take a bold line, e.g. over the Common Market, town planning, etc., but some against it as shown by examples of enterprises, e.g. the welfare state, atomic power stations, new universities, etc. This depressed attitude is reflected in the deeply ingrained conservativism of the Nation and of all sections of it, and it is a danger, but I do not think it wholly characteristic, because offset by some refreshing activities.

What about recovery? The prime condition to be satisfied for genuine recovery, and not the façades referred to, would be to come to terms with the national ambivalence. The national ambivalence has two facets, the disparagement of its own values—the gentleman—and the split in society between gentlemen and players. How might this be tolerated or alleviated?

The national ambivalence would be controlled by the growth of the Expert–Amateur. For such an image would undermine the distinction between citizens first-class and second-class, promoting economic efficiency, removing the tendency to act nationally against the national interest, acting with more forthrightness politically, and allowing Old Struthonians to fade away or remain as eccentrics; it would undermine Leo–Struthonianism. The mind should not look down its nose at its hand. The main manifestations of ambivalence would be removed or smoothed.

To satisfy this condition is difficult. But certain things suggest that it is a reasonable possibility. Gentlemen in the past have shown their own characteristic of the amateur in changing their own way of life. Thus they moved into some forms of commerce—no doubt with heart-searching but the upper lip stiffened, they did it, and got over it. Let us not underrate

---

[1] And in fact has not repeated it in a case (Rhodesia) where it could be claimed to have some justification.

that step; it meant accepting the humiliation of earning money instead of having it as a natural endowment. It should be no worse a step to accept the degradation of becoming versed in expertise. What may help to bring about such a change of attitude most effectively and most quickly is perhaps the new universities—provided the new enterprise they display is not smothered at birth by getting under the control of Old Struthonians. This is a great challenge to the new universities, and it would reflect a great change, in view of the country's reprehensible history over education, if the new universities should save it from itself. If this eventuality should make you too optimistic, you could enquire whether, and how far, standard vintages are being given new labels.

Since getting worse and pseudo-recovery seem to be out of character and to depend on conditions that do not seem to hold, I would say the choice lies between remaining chronically depressed and recovery. There are some developments that would militate against chronicity. On the other hand, the condition for coming to terms with ambivalence, the growth of the Expert–Amateur, is fraught with difficulty, that is to say, national resistance to it is bound to be very strong. I would put the odds on recovery slightly higher than on chronic depression; but more likely still, I would think, would be a partial recovery, a compromise formation between the two possibilities, quite good enough to make most people proud, but disappointing in comparison with what might have happened.

What would favour chronicity is the conservative ingrained fear of taking enterprising and bold measures; against it there is the fact that some initiative is left. What would favour recovery is the flexibility of the gentleman, who might be able to evolve a new conception of his type, together with the institutional development of the new universities, which might foster a new development of his type; against it there is the distrust by gentlemen of both gentleman and player. Since the conditions for all four possibilities are all present, I would expect a measure of each to ensue; and hence, say over the next fifteen years, I would expect to find a mixture of chronicity and recovery. [Written not later than the summer of 1964.]

A favour prognosis lies neither in Kipling nor Cousins, but in Culture grafted upon Craft. I would expect the graft to take in some measure; but that we shall find also the conservative spirit of Kipling and of Cousins in some measure still present.

### The problems further specified

That completes my example, of a social psychological kind, with which to make certain points. (i) The diagnosis of a community depression seems to attribute an 'irreducible' purpose to the Nation, in a sense running counter to 'situational individualism'. (ii) The diagnosis makes use (at least ostensibly) of a concept of an unobservable. The hypothesis is testable, for it yields a prognosis asserting that (within a limited span) there will be a mixture of some chronic depression with some recovery, subject to the conditions (*a*) that class distinctions retain some of their sharpness, to provide for the retention of social conservativism, and (*b*) that education undergoes radical alteration, to provide relief from national ambivalence.

The general relationship between (i) and (ii) is more complicated than it looks. For though 'reductionist individualism' is obviously incompatible with the use of unobservables (since a 'global reduction' removes everything except observables), it is not so obvious that 'situational individualism' is incompatible also (cf. Morgenbesser, 1967).

We shall have to consider whether the example of a social depression requires us to modify or go beyond 'situational individualism' and involves an unobservable.

We may now try to come to grips more closely with Popper's thesis of 'situational individualism'.

Above I have claimed that he does not mean that social wholes are 'globally or collectively reducible', i.e. that in the reduction all mention of a social whole would disappear; such a version would render social wholes mere *epiphenomena* of individual activities. What I have claimed is that for him, social wholes are 'distributively reducible' in a 'piecemeal' way, i.e. that any and every social whole is 'reducible' but that in the reduction there will always be some other unreduced social whole. I wish now to present this in somewhat different terms.

It is well known that in the philosophy of sense-perception, phenomenalism, from its birth in Berkeley and Mill to its death-throes in Ayer, provided a reductionist account of physical objects in terms of sense-data, in which all reference to all physical objects would disappear and the reduction would refer solely to sense-data (many of the early criticisms were essentially of the form that attempted reductions, eliminating reference to one physical object, covertly brought in reference to another). All versions of phenomenalism attempting this could be regarded as 'globally reductionist'. In the twilight of its life-span, Ayer (1954) gave a new twist to the theory, in which he admitted openly that such a reduction is impossible in principle, and substituted a weaker form of phenomenalism in which all he claimed was that, although the reduction could not be carried out, a reference to a physical object involves nothing beyond a reference to sense-data (and their mutual relations). Thus a physical object was no longer, on this construction, 'globally reducible' to sense-data, in the sense that all of it was reducible, but was only 'reducible' piecemeal, i.e. we could go on 'reducing' endlessly and so could not complete the reduction, and yet what was left would refer to nothing other than sense-data—to no other category of entity. Ayer thus replaced 'globally reductionist' phenomenalism by 'piecemeal reductionist' phenomenalism.[1]

The point I wish to bring out is that the 'global reducibility' of social wholes is analogous to classical phenomenalism,[2] and 'situational individualism' or the 'piecemeal reducibility' of social wholes is analogous to Ayer's 'piecemeal phenomenalism'. This point (a) throws doubt on whether social wholes, even in the milder form, can have the independent role we require them to have, for this is a doubt phenomenalism evokes. On the other hand, (b) if social wholes do have some source of power or function independent of individuals, the question arises, where does it come from; for phenomenalism stimulates the question, if there is more to physical objects than sense-data, what is the nature of the additional factor? These questions arise on their own, but the parallel with phenomenalism underlines them.

A further significant point is that a physical object, not only for classical phenomenalism but even for piecemeal phenomenalism, would be an observable in that it would contain nothing that could not be 'reduced' (even though 'piecemeal') to observables; likewise on Popper's theory, social wholes, which are 'piecemeal' though not 'globally' 'reducible' to individuals' purposes, etc., would be an observable in that it would contain nothing 'irreducible' (even though 'piecemeal') to observables. In other words, Popper's 'situational individualism' leaves no room for unobservables.

Let us turn at last to the question whether 'situational individualism' allows sufficient weight to the independent power of wholes. This theory asserts not that social wholes are 'distributively reducible' purely to individuals' purposes, but to these plus their unintended consequences. At first sight one may wonder whether this reduction does not include every feature that could be ascribed to a social whole. We may, however, find that the conception of 'unintended consequences' is narrower than might be expected.

Popper's conception of unintended consequence seems pretty definitely

---

[1] Ayer gives me the impression, though I cannot vouch for its correctness, that he was reluctantly conceding the failure of classical phenomenalism, and that he was desperately trying to save the theory somehow and so introduced a twist he did not quite like. He does not seem to have realized he was giving an interesting new theory, not an *ad hoc* twist to wriggle round a jam. The classical form made physical objects epiphenomena of sense-data; his new form apparently allowed them an independent role.

[2] This parallel has also been noticed by other writers.

to refer to consequences that, though unintended, could have been intended. Popper is concerned with the way our *calculations* go wrong. For example, measures to end the slump in the 1930s led to the unintended consequence of exacerbating it (which could have been, though it was not, intended by arch-conspirators). Thus, what is denoted consists of consequences of a familiar sort. There may be plenty of surprises, but a surprise in this context is a surprise in virtue of its *occurrence* and not of the *nature* of the thing that occurs.

However, there exists a possibility of a different type of phenomenon, where there is total surprise, i.e. the *nature* of the consequence is unexpected rather than, or as well as, its occurrence. Popper gives the impression of being concerned—very reasonably—with what might be described as 'things that might have turned out differently'. But it is another order of eventualities if a new structure comes along. Examples may prove treacherous but must be tried. Consider a situation in which an experimenter is trying out combinations of substances. He gets some surprises, such as that when he sparks hydrogen and oxygen he gets water (an 'expected surprise') or when he exhausts the supposed constituents from air and finds a residue (argon, an unexpected surprise). But then he puts together copper and zinc rods in sulphuric acid, and finds that a wire joining the rods deflects a magnet. The *nature* of the phenomenon, electricity, is not just a surprise in the sense of one thing rather than another, but a new order of things where nothing of that sort would hitherto have been expected at all.

Here I am, of course, discussing 'emergent properties'.[1] In this context there are to be considered 'emergent concepts' (standing for 'emergent facts') and 'emergent laws', both of which may be candidates for 'irreducibility' to the concepts and laws of the ingredients from which the emergent phenomena emerged. The supposition of emergence is that neither concepts nor laws are *deducible from* the given constituents (i.e. of course without empirically established discoveries consisting of generalizations bridging the given constituents and the emergent properties).

Now the group-structure I have conjectured, of a psycho-social depression, is an example of such an emergent phenomenon. It is not at all just a social reflection of widespread individual depression. It is not just a surprising occurrence that might have been otherwise; it is not the kind of thing at all that might or might not be expected. It is a different order of eventuality. And not only is it unforeseeable but it may even be unrecognized (like a psychological depression) after it has arisen. (And it may exercise some control over our behaviour without our being aware of it.)

---

[1] Mandelbaum (1955, 312n) has referred briefly to 'existential emergents'.

Popper's conception of unintended consequence seems to be confined to consequences that are unintended but could have been intended, that might be foreseen but are not foreseeable. Here I am considering a conception of 'emergent consequence', which lies outside his conception of unintended consequence.

## TRANSINDIVIDUALISM

We are now in a position to resolve the difficulty arising from Popper's theory. Naturally, it seems obvious or virtually tautologous that there could be nothing in a social whole beyond individual purposes and their unintended consequences. And in the literal sense of consequence, this is so, for an emergent property is still a consequence. But it is so different in kind from what Popper seems to have had in mind as to fall outside his domain of unintended consequence.[1]

I would reformulate: that a social whole may consist of individuals' purposes, their unintended consequences, and their 'emergent consequences'.

I would suggest that this last is what gives a separate character in some measure to certain social wholes, i.e. social groups: it is this that endows them with the independent power I have been trying to isolate.

I could accordingly describe the position, just reformulated, as 'transindividualism'.

In describing the position thus, I do not mean the extreme collectivist or historicist doctrine Popper has so successfully attacked. But I do mean more than is ascribed to 'situational individualism': The thesis it refers to gives specificity to the claim made by several critics of 'methodological individualism', e.g. Mandelbaum (1955), Gellner (1956), Goldstein (1956), Scott (1961), who contend in one way or another for something 'irreducible', e.g. a 'diachronous' framework (Goldstein) or 'societal facts' (Mandelbaum). Their criticisms, however incisive and significant in many ways, lacked thrust, occasionally by minor misunderstandings but mainly by being unable to show that the supposedly 'irreducible' fact really is 'irreducible'; for Popper has never denied that some such phenomena as these exist, he contended only that they are (distributively) 'reducible'. So these critics were making no case against him merely by pointing to such facts; what they would have had to do would have been to give some argument against their 'piecemeal reducibility'. It is this that I have attempted to do. I have not tried to show that the social pathology of

---

[1] My case has been presented as holding for social groups. In this paper I have tended to assume it is not needed for institutions, i.e. to assume that Popper's theory is fully adequate that far; but this needs examination, as it may be conceding too much.

depression is 'irreducible' by any form of argument to the effect that we cannot see how the reduction would run; I have aimed at showing that such a depressive constellation is of a different order from the straight-forward meaning of 'unintended consequence' and therefore outside the 'piecemeal reducibility' of Popper's theory.

It is true that such separate constellations are further consequences of the unintended consequences and therefore a sense of 'reducibility' could be found for them; even so, a radical innovation in Popper's theory would have been made. This comes out additionally by reverting to the question of unobservables; for, while 'situational individualism' excludes unobservables, 'transindividualism' requires them.[1]

It seems clear that 'situational individualism' imposes upon Popper's theory of scientific method a restriction (noted by Scott, 1961) in addition to testability, for it constitutes a framework for theories in the social sciences which imposes a 'reduction' and precludes unobservables. In other words the general prescription, that a hypothesis needs only to be testable, would allow the hypothesis of a societal depression to be part of social science, but the additional restriction, that it has to be 'reducible piecemeal', would debar such a hypothesis from a place in social science.

Neither Popper nor Agassi have discriminated the extreme form of holism, according to which social wholes *dominate* individuals absolutely so that no one can resist them, from a moderate idea, e.g. 'emergent power', according to which social wholes *influence* individuals so that individual action is determined by a combination of two factors, social wholes and individual purposes, either of which may happen to predominate on a given occasion. Thus, if we revert to the three possibilities mentioned at the beginning, (i) of social wholes necessarily dominating individuals (ii) of social wholes necessarily deriving their power from individuals, and (iii) of dual control, the first would be held by Marx, the second by Popper, while the third is the view put forward here.

This point could equally be put in terms of the idea of 'organism'. The view that society is an organism is usually associated with historicism. Therefore, the idea of an organism is excluded from 'situational individualism'. But it need not be associated with historicism. And it is appropriate for describing the thesis of 'emergent power' I have been developing. I

---

[1] The critics of methodological individualism have dissociated themselves from all forms of holism, that is to say, while contending for 'irreducible' societal facts they do not wish to classify these as 'holistic'. They probably object to the conception of 'holism' because of its ideological overtones—the overwhelming domination associated with it coupled with the aura of religious or political veneration with which it is often endowed. The conception of 'emergent power' adopted here is not necessarily overwhelming or venerable.

do not mean that a society or an institution is an organism but that it is only in *part* an organism. (Even the human body is not *in toto* an organism, for the hair and nails are not organic. I would think that a much larger part of society, a very large part, is non-organic.) The idea that society is in part organic does not have any 'ghostly dictatorship' or 'closed society' overtones; for inorganic substances can and do have great effects upon organisms, and, if society is in part organic, this does not preclude individuals from influencing it. The holistic or organic idea of society may have come in the first case from oracular, romantic, or obscurantist philosophers. But the idea may be salvaged. And I would claim to have given it a place, by means of the example of the social pathology of depression, in Popper's methodology of testability or refutability.

Historicist theories are monistic; so is Popper's. The modification I have introduced constitutes a bipolar theory (some would call it dualistic). It allows complete flexibility of explanation, diagnosis, and prescription, in that both individuals and the independent power of a group are factors in explaining a social phenomenon; in some cases one factor will be the greater influence, in other cases the other; in certain cases one factor may dominate and even overwhelm the other, and there is no *a priori* way of knowing without investigation which factor is the more significant in any given case. If you think that women have not yet achieved emancipation or that men are still very largely adolescent, you might seek to explain such a situation (or remedy it) by investigating their individual psychologies or you might try looking for something in the group constellation.

'Transindividualism' allows equal weight in principle, or any proportion of weights, to individuals or the societal facts, in governing the course of individual life in a group or society.

## REFERENCES

Ayer, A. J. 1954. Phenomenalism. *Philosophical Essays*. London.
Agassi, Joseph. 1960. Methodological individualism. *Brit. J. Sociol.* **11**, 244–70.
Albu, Austen. 1963. Taboo on expertise. *Encounter* **21**, 45–50. Ed. Arthur Koestler.
Bion, W. R. 1961. *Experience in Groups*. London.
Broad, C. D. 1933. *Examination of McTaggart's Philosophy*, vol. 1, Chap. 4.
Brown, Robert. 1963. *Explanation in Social Science*. London.
Campbell, N. R. 1921. *What is Physics?* Ref. to Dover ed., 1952, 85–6.
Dedekind, Richard. 1963. Continuity and irrational numbers. *Essays on the Theory of Numbers*.
Gellner, Ernest. 1956. Explanation in history. *Proc. Arist. Soc.* (Suppl. vol.) **30**, 157–76.

Goldstein, L. J. 1956. The inadequacy of the principle of methodological individualism. *J. Phil.* **53**, 801–13.

Koestler, Arthur (ed.). 1963*a*. *Encounter* **21**.

Koestler, Arthur 1963*b*. The lion and the ostrich. *Encounter* **21**, 5–8. Ed. Arthur Koestler.

Mandelbaum, Maurice. 1955. Societal facts. *Br. J. Sociol.* **6**, 305–17.

Marx, Karl. 1906. *Capital*, Chap. 24. New York.

Morgenbesser, Sidney (Ed.). 1967. Psychologism and methodological individualism. *Philosophy of Science Today*, 163. New York.

Popper, K. R. 1963. *The Open Society and its Enemies*, vols. 1 and 2. New York.

Popper, K. R. 1964. *The Poverty of Historicism*. London.

Renier, G. J. 1933. *The English: are they Human?* Chap. 9. London.

Rees, Geronwy. 1963. Amateurs and gentlemen. *Encounter* **21**, 20–5. Ed. Arthur Koestler.

Russell, Bertrand. 1920. *Introduction to Mathematical Philosophy*, 167 ff., esp. 177. London.

Scott, K. J. 1961. Methodological and epistomological individualism. *Br. J. Philos. Sc.* **11**, 331–6.

Simon, H. A. 1957. Mechanisms involved in pressures toward uniformity in groups. Mechanisms involved in group pressures on deviate-members. *Models of Man*, Chaps. 7 and 8 with H. Guetzkow. New York.

Watkins, J. W. N. 1957. Historical explanation in the social sciences. *Br. J. Philos. Sc.* **8**, 104–17.

Whitehead, A. N. 1920. *The Concept of Nature*, Chap. 4. Cambridge.

Wisdom, John. 1933. Ostentation. *Psyche*, **13**, 175–76.

Wisdom, John. 1934. Is analysis a useful method in philosophy? *Proc. Arist. Soc.* (Suppl. vol.) **13**, 96.

Wisdom, J. O. 1964. Review of Brown (1963). *Economica* **31**, 219–20.

# COMMENT

## *by* ROBERT BROWN

Do human groups have powers that are distinct from, and independent of, the powers possessed by the individual members of those groups? Dr Wisdom believes that the correct answer to this ancient question is 'yes', and that social groups have emergent properties which 'are the further consequences of the unintended consequences' of individual actions. As an example of a set of such properties he gives the sub-acute depression that he takes contemporary Britain to display. This set, while yielding a testable prognosis, is not observable. Nor is it reducible to any set of observable actions and their observable but unintended consequences. Wisdom concludes that social groups possessed of emergent properties have an independent power to influence the actions of individuals, that this power is sometimes exercised, and that it is a matter for empirical investigation to determine, in any given case requiring explanation, the relative influence of social groups and individual agents.

These conclusions will be acceptable to many philosophers, myself among them, who are dubious about the supporting argument, and dubious, also, about some of that argument's deeply rooted assumptions. Both sorts of doubt are generated early by Wisdom's treatment of the relation between emergent properties and their unpredictability. The problem under discussion there is how to characterize the unforeseeability of emergent properties. In what sense are such properties unforeseeable?

In order to answer this question we must begin by distinguishing between two other questions: (*a*) Are emergent properties the further consequences of the *unintended* consequences of actions? (*b*) Are emergent properties the further consequences of the *unforeseeable* consequences of actions? It makes a difference which question is at issue. Many foreseeable (predictable) consequences are unintended and conversely. For example, if I intend to drive my car home this evening I can correctly predict that the trip will wear the tyres slightly, cost me petrol money, and place the setting sun in my eyes during that period. But none of these consequences is part of what I intend to do or bring about. So from the fact that an emergent property is an unintended consequence of an action, it does not follow that the property is an unforeseeable consequence of it. And if the property is not unforeseeable or unpredictable, then its occurrence and nature cannot come as a total surprise. Thus Wisdom is committed to the

297

view that emergent properties are distinctive in having an unpredictable nature. Non-emergent properties have a predictable nature, though their occurrence is sometimes unpredictable.

What, then, does unpredictability of nature amount to, according to Wisdom? He says: 'The supposition of emergence is that neither concepts nor laws are *deducible from* the given constituents (i.e. of course without empirically established discoveries consisting of generalizations bridging the given constituents and the emergent properties).' The position seems to be that emergent (group) properties are to be those whose characterizations are not deducible, without bridging laws, from the given constituents —here, the property characterizations of the individual agents. Because such group-properties are not deducible, in this sense, from the properties of individual agents, knowledge of the latter does not permit us to predict the nature of the former. They are, for this reason, surprising, irreducible, and unforeseeably emergent.

However, if non-deducibility is to be the criterion of emergence, then surprise, and expectation in the sense of regarding something as likely to happen, obviously become inessential. For in any given instance the attempted proof of non-deducibility might be so complicated that we should be surprised to learn that deduction was not possible or to learn that it was. In neither case should we have foreseen or expected the results, though only in the first case would our surprise have been produced by the occurrence of an emergent group-property, and so have the desired connection with it. Emergent group-properties, then, are totally surprising and unforeseeable *only* because they are not reducible to—their characterizations are not deducible from those of—the properties of individual agents. This, for Wisdom, is the strong sense of 'unforeseeable effects'. Yet this information does not, by itself, make unforeseeable effects (in the form of emergent group-properties) any less mysterious in their origin. The non-deducibility criterion does not prevent the actions of individuals from being the causal origin of emergent properties. Within a particular causal field, some actions which are necessary but not sufficient conditions of the occurrence of emergent properties will be describable as causes. And since from the characterization of such necessary conditions no characterization of a property occurrence can be deduced, the distinctive feature of unforeseeable effects—their non-deducibility, as Wisdom now says— remains unexplained. For this reason we must take up the two questions whose answers will account for non-deducibility. The first is 'Are there emergent group-properties?', and the second is 'If so, why are they irreducible to the properties of individuals?'

Both these questions can be clarified if we apply some familiar distinctions to them. We distinguish amongst: (1) Statements about the *psychological properties* of individuals; these are properties which display themselves independently of their owner's membership in any *given* group, properties, that is, which do not arise from membership in a particular group. Examples are: superior memory of interrupted tasks and the effect on shape-perception of a strong contrast between the figure and its ground. It is highly questionable whether a psychological property can arise independently of its owner being a member of *any* group whatsoever. (2) Statements about the *social properties* of individuals; these are properties which display themselves only because their owners are members of certain groups; in the absence of such membership these properties do not appear. Examples are: being British Prime Minister and being best man at a brother's wedding. (3) Statements about the properties of groups themselves, for instance, groups of people. Examples of *group-properties* are numerical size, proportion of English speakers, and being a member of the General Assembly of the United Nations.

Now it is obviously important to know whether Wisdom is suggesting that an *emergent* group-property is irreducible to psychological properties or irreducible to social properties. The first suggestion is indisputably correct but uninteresting; the second is more interesting but also more disputable. The reason for the first suggestion being true is simple. Group properties are, by definition, those properties which characterize a group because it is a collection of members who are related to each other in various ways; it is these relations which make them members of the same group, e.g. the Hell-Fire Club. The existence of some individuals standing in these relationships is a necessary condition for the existence of the group and its properties. Psychological properties, in contrast, are defined as those qualities and relations which characterize an individual independently of his participation in specific, identifiable groups. Hence, given these definitions, it is logically impossible for any statements describing psychological properties to entail, by themselves, any statements describing group-properties, emergent or otherwise. For such a deduction to be possible our premises would have to include statements which told us how to construct groups from their constituents—and such 'bridging' statements are specifically ruled out by Wisdom. Our conclusions can hardly contain expressions (for group-properties) that do not appear in our premises. Thus, no law statements about psychological properties will, in the absence of bridging generalizations, allow us to predict or explain—by means of a deductive schema—the existence of group-properties. This trivial (logical) conclusion is of interest only because it is sometimes confused with another conclusion: that no law statements about *social* properties will, in the absence of bridging generalizations, allow us to predict or explain the existence of group-properties. We have good reasons for thinking *this* conclusion to be false.

Consider Wisdom's own example of Mr Gladstone, as British Prime Minister, signing an order. Here knowledge of a statement about two social properties (being British Prime Minister and being authorized to sign governmental orders) allows us to deduce a statement about a group-property, namely, that the British Government gave legal effect to a decision. Another example is the deduction from 'Each man in this battalion tried to desert during battle' to 'This battalion had low morale during battle'. It is true, of course, that from statements about group-properties, we cannot validly deduce statements about the social properties of particular agents. Because a battalion had low morale it does not follow that every, or indeed any, man in it tried to desert. There are many ways in which low morale can be expressed. But this lack of entailment is not at issue in the question whether there are emergent group-properties. What is of interest in this connection is whether there are some group-properties that are not reducible to the social properties of the members of those groups.

However, since a group-property is emergent if and only if it is not entailed by some set of statements about the properties of the constituents, a group-property can be emergent with respect to one such set and not with respect to another. In brief, as scientific theories about the properties possessed by constituents alter, so do the group-properties which can be predicted from those theories. So in asking whether some emergent group-properties exist we may be asking whether: (*a*) such properties are permanently irreducible to the social properties of their constituents, or if (*b*) such properties are irreducible, at present, to known social properties of their constituents. The latter question cannot be of much interest to Wisdom. For the 'emergent consequences' which, he believes, endow social groups with independent power might become non-emergent when a new theory about the appropriate constituents was established. With this change from being an emergent property to that of being a non-emergent one would have to go a change in the independent power of the social group. This is not a plausible consequence of a change in theory. Wisdom must be arguing for the permanent irreducibility of at least some group-properties. His example of 'psycho-social depression' must be a case of a complex group-property whose characterization will never be entailed by any set of statements about the social properties of its constituents. But can such a strong claim possibly be correct?

To establish that some—but which?—group-properties are permanently irreducible to any social properties we need one of two sorts of argument.

We must show either that the reduction is logically impermissible or that there are empirical reasons which make it forever unlikely. The second alternative would require that we be able to predict rationally the course of future scientific theories about the properties of constituents, and Wisdom surely accepts Popper's argument that prediction of this kind is logically impossible. The first alternative is somewhat better. Either it maintains that there is a special set of group-properties whose irreducibility must be ascribed to certain logical features, or it maintains that each case is distinctive in the logical reasons for its irreducibility. If each case is distinctive, then we must wait upon the appearance of supposed cases of irreducibility. Yet what reasons have we for thinking that each case *is* distinctive? What sorts of logical features could make each case different? The reason why group-properties are not reducible to psychological properties is the same in all cases. Should it not be equally so for social properties?

At this point it is worth remarking that Wisdom's example of social depression does not make clear whether he distinguishes between: (*a*) the logical dependence of statements about group-properties on statements about the properties of constituents, (*b*) the causal dependence of group-behaviour upon the behaviour of the group-members, and (*c*) the methodological thesis that all law-like explanations of group-behaviour can and should be (or should not be) superseded, in the long run, by explanations referring to the law-like behaviour of individual members. Each of these views can properly be held independently of the other two. Wisdom's attempt to identify a supposed syndrome, a group-property that emerges, unheralded and unrecognized, from the properties of group-members, does suggest that he takes psycho-social depression to be, in part, causally independent of the properties of individuals; the latter kind of property will be a necessary but not sufficient condition for the presence of the former. However, the non-deducibility criterion of emergent properties— 'that neither concepts nor laws are *deducible from* the given constituents' without bridging statements—offers us a logical independence whose basis is obscure. The basis cannot be that of (partial) causal independence between group and individual properties. For then the irreducibility of emergent group-properties becomes, once again, an empirical claim based on an ability to predict rationally the future course of certain areas of scientific knowledge. On the other hand, the methodological thesis can only be held if we allow bridging statements. These statements will relate the two classes of predicates, those referring to group-properties and those referring to the properties of constituents, in some law-like way. That is, the generalizations will state causal connections between what are agreed to be two different kinds of properties. But Wisdom excludes bridging

statements, and thus excludes any possible consideration of the methodological thesis. He has to leave unanswered the question his own example
raises: how are *explanations* in terms of group-properties related to those
in terms of individuals' properties? So the basis of irreducibility must
be non-empirical, and we are returned to the problem as exemplified
by such properties as communal depression.

Now Wisdom gives us no logical reason for thinking what may well be
true, that group-depression is not 'a social reflection of widespread
individual depression'. There may be no logical barrier to the truth of this
claim, and so it may be merely an empirical question whether group-
depression can exist in the absence of all individual depression. Perhaps
under certain conditions a group of people who are separately cheerful can
form a group whose group-behaviour is depressive. Yet here we must be
careful. For if there is no logical barrier to group depression *not* being a
'social reflection of widespread individual depression', is there any logical
barrier to group-depression *being* a mere social reflection? Suppose every
member of a newly formed group is known, from his previous history, to be
a depressive. Are there logical considerations that would make it either
self-contradictory or nonsensical for us to claim that the interaction of these
individual depressives produced group-behaviour which was depressive?
Without empirical investigation, have we any way of knowing whether,
and when, group-depression is a mathematical function of the depression
of individuals, and hence a function of one of their *social* properties?

But obviously some properties ascribable to groups (traits of collectives)
cannot be ascribed to individuals, and conversely. Instances of the former are:
crime rate per thousand, degree of bureaucratization, volume of retail trade,
and public ownership of banks. Instances of the latter are: being last person
in line, being Lord Chief Justice, and being the only nurse present. In all these
instances there are logical barriers to interchanging the two sorts of
properties. The sense in which a city can have a high crime rate per thousand people is clearly different from the sense in which an individual might
display a high rate of criminal activity. And being the only nurse present is
not something that can be sensibly said, or verified, of a group of people.

Nevertheless, many properties ascribable only to individuals are *social*
properties; their characterizations embody ineradicable reference to such
membership features as being the sole occupant of a role or the sole agent
in a particular kind of situation. Therefore, some group-properties are
logically irreducible to those same properties when they are ascribed,
improperly, to social individuals: for example, a statement about a city's
crime rate per thousand is not entailed by statements about the crime rate
per thousand of various individuals. What *does* entail the first statement,

of course, is a set of statements completing, for each person in the city, the property ascription 'X number of crimes committed'. So group-properties irreducible to the identical social properties may still be reducible to other social properties. Wisdom seems wrongly to believe that irreducibility of the former kind ensures irreducibility of the latter kind. Even if it were true that communal depression is logically irreducible to individual depression, it would not follow that communal depression is logically irreducible to other social properties.

In point of fact, Wisdom does not supply us with the information necessary for deciding whether communal depression is logically irre-ducible in either of these ways. He does not tell us anything, for example, about the logical relationships between the notions of being a communally depressed group and being a member of such a group. The conditions for being an individual shareholder in a company and being a member of a shareholding group are different. Yet the expression 'shareholder' refers to the same property whether the shareholder is an individual or a group. Is this also true of 'depression'? If it is, then the relation between statements about communal depression and statements about individual depression is one of logical irreducibility: the latter do not entail the former, any more than 'Every member of Group A is a shareholder' entails 'Group A is a shareholder'. But this irreducibility does not prevent entailment of the latter by a statement about other social properties. An example is, 'Every member of Group A has authorized its Financial Secretary to continue purchasing shares on his behalf, and to continue to manage and sell them in common with those of all other members, present and future'. This statement gives something of the necessary and sufficient conditions for an individual to be a member of a shareholding group. It tells us, then, some-thing of the logical relationship between 'shareholding group' and 'member of shareholding group'. We need similar information about group-depression. What, for example, is the difference between a depressed group and a group in which many individuals are depressed? The British disabilities given by Wisdom are all ascribable to individuals. If these properties are also to be ascribed to groups, we need to know how to recognize the differences between group manifestations of the properties and their instantiation by individ-uals. Until we can specify the differences we have no good reason to assert the existence of group-depression as distinct from individual depression.

This conclusion can be applied quite generally. Social scientists often wish to ascribe properties to groups of people. Some of these properties are transferred from those ascribed to individuals: intentions, hopes, fears, and desires are examples referred to by Wisdom. Some properties like social stratification and fertility rate are not. Both kinds, however, raise the question 'How is this group-property related to the social properties of the group's members?' The question is raised because social properties are defined as those which appear if and only if their owners belong to groups, and also because the existence of such individuals standing in certain relationships is a necessary condition for the existence of these groups and their properties. Thus we cannot sensibly answer 'There is no relationship'. There (logically) has to be a relationship. The point of interest, in each case, is whether we have an empirical theory about the interaction of social properties which will entail statements about group-properties. Sometimes we have and sometimes we have not. In the latter case, the logical irreducibility of the group-property arises from an absence of suitable premises about social properties. This absence may be remedied by the recognition of previously unnoticed social properties and their interactions. Quite commonly, this occurs because we notice, or think we recognize, new group-properties and adjust our notions of social properties accordingly. But suppose, as in the case of the proposed property 'group-depression', we are given insufficient information about the social behaviour of individuals belonging to the group. Then we have no way of deciding whether or not the proposed property is a group-trait. The proposal may refer merely to a social property displayed by a prominent part of the group—either by a number of important members or by a significant proportion of members.

The question cannot be settled simply by our learning that the group-trait is observable, or measurable, independently of any observation of the reactions among members. Many group-traits are independently observable or measurable though they are reducible to the social properties of the group-members. Thus group shareholding, speed of committee decision, and the loudness of choir singing are each observable independently. Each, however, is known to be a group outcome and so reducible to the various types of social behaviour displayed by its members. Nor is it a necessary condition for the existence of a group-property that it be independently observable; a group's morale is not observable independently of the behaviour of its members, yet this by itself does not show that morale is not a group-property. Of course, some group-properties, like some properties of constituents, are not thought to be observable at all—given the current development of the subject. Such properties have to be inferred from evidence, as Wisdom suggests that group-depression can be. But a large part of this evidence will usually consist in what is admitted to be observable, namely, the social properties of individuals. (The other part can consist in other social properties, e.g. the crime rate in the case of group-depression.) Thus any lack of information at a given time about these social properties will become a lack of evidence about the existence of the group-property on a particular occasion. And a continuing lack of such information will result in a continuing doubt as to the existence of the supposed group-property.

Until this doubt is removed, it is idle to worry about the testability of statements ascribing the supposed property to particular groups. For we must first know how to test for the presence of the property in groups. If we cannot distinguish group-depression from the depressiveness of individual members there is no point in claiming that we can test the statement 'Great Britain is suffering from group-depression'. We shall not know what it is that we can test. Perhaps we can only test for widespread individual depression. In order for the group-property to be identifiable by us, it must be connected, in a systematic way, with other identifiable properties, and this means that we must be able to discriminate between each of them and the group-property. Again, in order for us to determine whether a supposed property of any kind is genuine or not, we have to discover whether its existence produces any alteration in certain other properties, an alteration which, in the case of a supposedly irreducible property, must be not only characteristically produced by it but also unique to it. If a supposed group-property is genuine, the group to which it belongs must interact differently with other groups, or individuals, from the manner in which a group merely seeming to possess a particular property interacts with them. This difference in result allows us to distinguish, within specific situations, between a genuine and a fictitious property. It also allows us to identify the presence of the property without regard to the errors of any given observer, since a genuine property belongs to a group whether an observer thinks so or not.

The answer to Wisdom's question whether there are emergent group-properties is 'Yes, but emergent in an uninteresting sense'. The related problem which Wisdom's own example brings up seems to me more interesting, and it is certainly more important to the social scientist. That problem is: 'How can I tell, in any given situation, whether an apparent property, or set of properties, is genuine? How can I tell whether communal depression or group-anxiety or status hierarchy really exist? Perhaps I am the victim of non-referring phrases; or perhaps I am merely re-christening properties that belong to individual members of these groups. Assuming there to be group-properties, whether reducible in some sense or not, how can I recognize them as against recognizing the social properties of the members?' Once we seriously confront this problem, we shall find, I suggest, that questions like those raised by Wisdom form a queue of descending priority behind it.

# REPLY

## *by* J. O. WISDOM

### PROFESSOR BROWN'S MAIN CONTENTIONS

Two main criticisms of my paper may be found in Professor Brown's comment, concerning my claim that emergent properties are irreducible to properties of individuals. (1) He mentions rightly that I do not tell how group-properties are related to individual properties; rightly that I give no *logical* reason for holding that a group-depression, for example, is not necessarily a reflection of individual depression, and rightly that I give no way of deciding whether emergent group-properties are logically irreducible to individual properties. Brown makes these correct points because he considers that there is a basic need for us to know how to recognize a group-manifestation, or identify it, as distinct from an instantiation by individuals. For he considers that unless we can do this it is idle to go on to the problem of my paper, concerning whether or not we can test for the existence of a property such as group-depression. In short, before testing for a group-property we have to have some criterion for recognizing it. That is the main point of his comment. (2) Brown finds it obscure which of three possible alternative interpretations I meant; logical, causal, or methodological. He rightly thinks that my thesis concerned the causal interpretation, but certain considerations lead him to think that this leans on the logical interpretation and he proceeds to indicate that this is untenable.

The first criticism rests on a very commonsense metascientific assumption which would seem to fall down when confronted with actual scientific theories. The second criticism concerns a slight misunderstanding of my view, coupled with what would seem to be a metascientific oversight.

The first point concerns something fundamental and of general significance, so I shall discuss it at once and turn to matters of less importance afterwards.

### *Brown's main criticism and basic assumption*

It seems to make good commonsense to suppose that if one is going to have a theory about some property we should have some way of recognizing it, of specifying it, of identifying it, and distinguishing it from other similar things that are not quite the same. This is certainly so on the level of instantiative concepts, i.e. concepts that have observable instances in the

306

world; if we were to speak about some property of an observable kind and not be able to specify this in any way or tell how to recognize it, that would be bizarre. But on the level of non-instantiative concepts, or unobservables, or abstract entities, the practice of science is totally against it. For instance, let us look at the greatest theories in physics. Newton's concept of attractive force is indefinable, unspecifiable, at least in any detail sufficient to tell us how to recognize it or locate it. So is the electromagnetic wave, so is the general wave referred to in quantum mechanics. The Dalton atom is unspecifiable in the sense of being able to tell whether or not you have got one. The concept of strain may fall under the same heading. Again, the radius of curvature of the universe in general relativity is not specifiable like the radius of curvature of a circle or any other geometrical curve.

This may seem so extraordinary as to be unbelievable, but the simple fact remains that theories of physics seem not to contain definitions of entities of this sort. Textbooks give at best an allusive remark or two about them to the effect that they possess some property or other (noticeably shared by a great many other things), and refer to not very close analogies with something familiar. One gets the impression the subject is changed as quickly as possible and the main work proceeded with. Now this consists of deducing tests from hypotheses containing these concepts (at least this is all that concerns us in this context). In other words it is perfectly possible in physics to derive a conclusion which can function as a test, without having specified the entity referred to in the hypothesis. What may be overlooked about this procedure is that *the procedure itself of deducing a test provides a way (an indirect way) of locating the existence of the entity in question.* If the test works, then the test provides evidence for the existence of the entity. It does not, however, provide any way of recognizing the entity in the sense of providing marks of identification in the way that Brown asks for. And the existence ascribed is hypothetical (or more strongly hypothetical than is the case when we can look at something).

Such a situation may be anomalous and give rise to metascientific problems, but it does seem to be the reality; and if so, Brown cannot validly criticize the theory I put forward if it is testable in the manner I claim, even though it does not enable us to specify or recognize the entity in the way he requires.

The main significant difference between Brown and myself has concerned the priority of the questions of locating or identifying an emergent property on the one hand and testing for it on the other. I think I know the source of this difference. I am a staunch admirer of his book, *Explanation in Social Science*, all the way except for the chapter on scientific theories. When I first read this book I thought that Brown had succeeded in giving parallels from the social sciences to the natural sciences, all the way including theories. Later, when I came to check his principal example drawn from Herbert Simon, I found that it was not a high-level abstract theory, but one on a lower level, namely generalization. Hence I trace Brown's view about the need to identify instances of concepts to the fact that he does not regard high-level theoretical explanations as consisting essentially of abstract entities; and this seems to me to be a mistake. If Brown's restricted view of theoretical explanation were correct, then my paper really would be irrelevant.

### Brown's other criticism

Let us turn now to Brown's second criticism concerned with irreducibility. He opens his comment with an accurate statement of my thesis. Nonetheless, he gets into certain difficulties with it, and he spends nearly half his paper drawing certain distinctions to decide exactly what my meaning was. For the most part he ends up with the meaning I intended. There are, however, a couple of misleading features about his discussion.

(*a*) My thesis was in fact intended to be causal. Brown thought I had a foot in the camp of logical irreducibility because I interpreted this to mean that the emergent property was not deducible. What matters is whether one has adequate premises, such as initial conditions, or in this case bridging laws, to enable the deduction to be carried out; and this renders the question of the possibility or otherwise of effecting the deduction empirical. Logic concerns only the validity of the inference. The empirical possibility of obtaining a conclusion depends upon having the appropriate premises.

(*b*) Brown evidently takes it to be out of the question that I could have intended some sort of empirical interpretation of the position about irreducibility. It would seem that he considers such an interpretation not to be fundamental from a metascientific point of view. However, this may appear in a different light if we replace 'empirical' by something else with a family resemblance to it. 'Empirical' is somewhat ambiguous. It is usually taken as meaning *factual*, but in this context I mean *theoretical*, i.e. relating to *empirical theory*. The normal contrast in philosophy for several hundred years has been between analytic and empirical in the sense of factual, and Brown assumes rightly that I cannot be advocating a factual thesis of irreducibility. What, however, I am doing is advocating an irreducibility thesis as a *theory about fact*, and this puts the matter in a wholly different light. When Brown takes it that I am discussing irreducibility in principle, he is contrasting this with the possibility of discussing irreducibility in terms of known properties. Now this latter is exactly what I meant; but *known properties are the terms in which we couch a present theory about what the facts are.* A further development of theory might lead to a situation in which a supposed emergent property became no longer, in relation to this theory, an emergent property. In which case the former irreducibility would be reducibility. We cannot claim to know about irreducibility *in rerum natura*, but only so far as our current theories go. The main point, however, is this. A property, as here conceived, is emergent so long as and only so long as it cannot be deduced from any set of known properties without the aid of a bridging law (and this continues to hold even after the requisite bridging law has been discovered, though the element of surprise vanishes).

## RELATION OF GROUP-DEPRESSION TO INDIVIDUAL DEPRESSION

It would be well to be explicit here about my thesis concerning group-depression as an emergent property. In my opinion, there could be, and probably actually are, some groups manifesting a group-depression that arises because all or most members of the group are depressed (this might be but would not necessarily be an *emergent* property). I had no intention of excluding such a possibility.[1] I was simply not concerned with it, but with the possibility that group-depression might occur even though no members of the group were depressed. I was simply making the hypothesis of an emergent group-property to see whether I could form one that would at the same time be testable.

## MISLEADING IMPRESSIONS

Now for one or two minor matters which, if let go by default, might give a wrong impression. In the course of his analysis Brown deals with certain points in a general way. He does not explicitly ascribe certain positions to me, but since he does not appear to make use of the conclusions he reaches, some readers may get the impression that the point of making these distinctions is because I had not covered them. Two of these are as follows.

Brown rightly says that it is necessary to distinguish between unintended consequences and unforeseeability. I dealt with this not by drawing an analytical distinction between the two but simply by interpreting Popper's conception of 'unintended' in terms of 'unforeseeability' and 'foreseeability'. It is in these terms that I drew the distinction relevant to my paper. (Not that Brown got the answer wrong; but I had virtually made the same point.)

---

[1] It also seems quite possible that a group, *all* of whose members are depressed, might *lack* the group-property of depression.

Brown also remarks that the non-deducibility criterion does not prevent actions from being the causal origin of emergent properties. He does not appear to make an explicit point by means of this remark; it might imply a criticism (for, though it is correct, he may have thought it was incompatible with my view); so it is best if I state the position I had in mind. On my thesis an emergent property is not deducible and therefore not explainable by causal laws to do with individual properties. But this does not mean that emergent properties are not caused by individual actions in a remote way. In fact I asserted that they are. These two propositions are compatible because although a causal chain connects one to the other, we are unable to know what this causal chain is, or make deductions in accordance with it, without a knowledge of certain bridging laws.

Brown took pains to point out the falsity of a general statement that emergent properties are independent of all sets of individual properties. He probably made this point because it is incompatible with a logical interpretation of non-deducibility and therefore constitutes an argument against non-deducibility in this sense. However, I admit, or rather hold, that emergent properties are not necessarily independent of *all* sets of individual properties, but only of *known* sets of them, and not even all of these. It may very well be that, with further empirical discoveries, theories will be developed such that the emergent properties are explainable by them, or more accurately, that further discoveries of bridging laws will be discovered. In which case, emergent properties, though they will remain emergent because of requiring bridging laws to be deducible, will become familiar and lose their character of surprisingness.

(It might be mentioned that Brown's examples, naturally enough, are used to illustrate, or to be counter-examples of, group-properties which are not necessarily emergent properties. I do not, of course, maintain that all group properties are independent of individual properties in the way that emergent properties are.)

One other criticism that Brown makes is that knowledge of deducibility remains unexplained on my view. By this he seems to indicate that I was attempting to explain emergence and that I thought I had succeeded in doing so by accounting for it in terms of irreducibility and therefore in terms of non-deducibility. But I was not trying to *explain* emergence and therefore not trying to explain non-deducibility. This situation seems simply to characterize the world and our knowledge of it or lack of knowledge of it. I was trying rather to bring out the nature of emergent-properties to provide a test for their existence, not to explain why the world should be what it is.

# REPLY BY J. O. WISDOM

## CONCLUSION

From the welter of comments—valid distinctions, correct interpretations, misunderstandings, misleading implications, and criticisms—making up Professor Brown's discussion, I have made two main extracts: (1) A criticism that emergent-properties have to be identifiable before we can properly form theories about them; and (2) a criticism of the thesis that emergent-properties are not deducible from individual properties, on the grounds that this thesis cannot be logically true, and as a matter of empirical fact that the thesis is metascientifically without significance.

The first I have argued is the reverse of the truth, based on a methodological assumption that seems out of accord with the structure of existing scientific theories. The second point is valid in itself but does not hit the target because it misinterprets the thesis, which is concerned not with logical possibility, nor with empirical possibility, but theoretical impossibility.

# THE RELEVANCE OF
# PSYCHOLOGY TO THE EXPLANATION
# OF SOCIAL PHENOMENA

## *by* GEORGE C. HOMANS

Social phenomena will be taken here to mean all phenomena in which the action of at least one man affects directly or indirectly the action of another. Accordingly social phenomena provide by far the largest part of the subject-matter of psychology, history, political science, economics, anthropology, sociology, and probably linguistics.

In discussing the relevance of psychology to the explanation of social phenomena I must first say what I mean by explanation. It is, I think, the view of explanation taken by such philosophers as Braithwaite (1953) and Hempel (1965, pp. 229–489), though I must state it here more briefly, and therefore more crudely, than they do. It has been called, for short, the 'covering law' view of explanation.

First, an explanation consists of a set of propositions, each proposition stating a relationship between properties of nature. To play its part in explanation, it is not enough for a proposition to say that there is *some* relationship between the properties. It must begin to state the nature of the relationship. It is, for instance, not enough to say, in the fashion of many propositions of social science, that a man's status *is related* to his power. One must go at least as far as saying that a man's status *increases* as his power increases, or that the two are *positively associated*. On the other hand, statements about probabilities may be legitimate propositions, and so may statements including theoretical terms, implicitly defined, such as the term *value* in the proposition: the greater the value of a reward, the more likely a man is to perform an action that secures the reward. I shall say no more about the vexed problem of theoretical terms, but refer the reader to the discussion in Braithwaite (1953).

Second, the set of propositions forms a deductive system, such that the proposition to be explained, the *explicandum*, follows as a conclusion in logic from the others in the set. When the *explicandum* is shown so to follow from the others, it is said to be derived or deduced from them, and so to be explained by them. The reason why a proposition must state the nature of a relationship between properties of nature is that little in logic can be deduced from one that does not. But remember that in the social sciences

313

a kind of sketchy logic is as much as we can ask and more than we usually get.

Third, the propositions from which the *explicandum* follows fall into two main classes: general propositions (sometimes called laws) and propositions introducing the given conditions within which the general propositions are to be applied. The former are general in that, unlike the *explicandum*, they cannot be deduced from others in the set, and often also in that they appear in many other explanations besides the one in question. It is the required presence of general propositions that leads the present view of explanation to be called the 'covering law' view.

The given conditions may in turn be explainable, may in turn become the *explicanda* of new deductive systems, or, for lack of factual knowledge, they may not. And so may the general propositions, or, for lack of theoretical knowledge, they may not. That is, the still more general propositions from which they could in turn be deduced have yet to be discovered. This condition is unlikely to last for ever, but it may last for a long time, for example, the two hundred years between Newton and Einstein.

Finally, the propositions must be *contingent* in the sense that data, evidence, fact are relevant to their acceptance as true or false. This condition must be introduced because there are plenty of deductive systems whose general propositions are non-contingent, accepted *a priori*. These are the deductive systems of pure mathematics. Scientific explanations may of course employ mathematics in making their deductions but they must not consist entirely of non-contingent propositions.

An explanation of a relationship is a theory of the relationship. But scholars often use the word *theory* in a broader sense than this—to refer to a cluster of explanations of related phenomena, when the explanations employ some of the same general propositions. Indeed when we speak of the power of a theory, we refer to the fact that a wide variety of *explicanda* (sometimes called empirical propositions) can be derived from a single set of general propositions (the theory) under different given conditions.

There are those who argue that explanation in the social sciences must be essentially different from what it is in the physical sciences. I cannot agree with them. Explanations of social phenomena can certainly be put forward that have all the characteristics of explanation as described above. The process of explanation is the same for all the sciences, though the content of the propositions will naturally differ from science to science. Indeed the efforts to construct a sociological theory by simply translating, for instance, Newton's laws into 'social' terms in the form of 'social physics' have not been notably successful.

Let me now turn to the types of explanation that social scientists have put forward in the endeavour to account for social phenomena. They are well illustrated by the efforts made to explain social institutions. An institution is a rule or set of rules, which some members of a society say should be followed in particular circumstances, and which some actually do follow. Thus trial by jury is an institution of Anglo-Saxon societies and so, in some primitive societies, is a rule that a man ought to marry a woman belonging to some defined category of women. The explanations in question are explanations in the general sense that they are efforts to answer the question why institution $X$ is one of the institutions of society $A$, and that they have obviously given some scholars some degree of intellectual satisfaction. Whether they are also explanations in the more specific senses I began by listing is a question I shall consider later. Four generic types can, I think, be distinguished, though many actual explanations mix the genera (see Homans, 1964a, pp. 951–77).

(1) Structural or 'pattern' explanations. In these the social scientist points out that the institution in question, for example, a particular rule of marriage, occurs in association with other institutions of particular sorts, for instance, matrilineal descent, avunculocal residence, and the vesting of jural authority over a man in his mother's brothers. (To give the meaning of these technical terms is not necessary to my argument.) The social scientist feels that he has explained the institution by showing that it occurs as part of a structure, a pattern, of other institutions; and he may get great intellectual satisfaction out of doing so, especially if the pattern occurs in more than one society.

(2) Functional explanations. In these the social scientist argues that no society can survive unless it meets certain conditions, and that institutions of a particular sort enable a society to meet at least one of these conditions. Then if a particular society is surviving, it must possess an institution of this sort, and if it does in fact possess one, the existence of the institution is explained. Sometimes the term 'equilibrium' replaces the term 'survival' in the argument. None of the types of explanation are confined to the social sciences, and this of course is the type of argument sometimes used in physiology to account, for instance, for the existence of a beating heart in a living human body. And just as the functional explanation of the beating heart presupposes a certain structure of the body: blood, organs to be nourished by blood, arteries for the blood to flow through, etc., so in social science functional explanations are often, though sketchily, associated with structural explanations.

(3) Historical explanations. In these the social scientist explains the existence in a particular society of the institution in question by tracing the

historical process by which the institution, or its progenitor, came into being, developed, and was progressively modified up to the time when its presence among the institutions of the society is to be accounted for. Historical explanation obviously depends on the existence of historical records, and so is more apt to be used to explain the institutions of literate societies than those of non-literate ones, to explain trial by jury rather than cross-cousin marriage. Of course, if one goes back far enough, the historical records of even the literate societies run out, and the adequacy of the records for explanatory purposes runs out even sooner.

(4) Psychological explanations. In these the social scientist explains the institution by showing how its presence in a particular society follows from propositions about the behaviour of men, as men, under specified given conditions. A crude example might be an explanation of the punishment of criminals that pointed out how criminals, persons who violate the more strongly-held rules of a society, present a threat to other members of the society, how men feel anger when threatened, and how they are likely, when angry, to take aggressive action against the source of the threat, provided they perceive they can do so with impunity. Punishment is aggressive action of this sort.

I call this type of explanation psychological because it employs propositions that are presumed to hold good of, to be general with respect to, the behaviour of all men as individual human beings, as distinguished from what I shall call sociological propositions, propositions that are presumed to hold good of societies, groups, or other social aggregates rather than of the individuals who make them up. Functional explanations use sociological general propositions. The relevance of psychology to the explanation of social phenomena is a question of the use of psychological propositions in their explanation.

Now let me apply to each of these types the standards for explanation that I began by setting up. The structural type can usually be cast in a form that looks superficially like a deductive system. Thus institution $X$ is positively associated with institution $Y$; institution $Y$ with institution $Z$, and therefore institution $X$ with institution $Z$. But it is possible to run the deductive system in the other direction and explain an original 'premiss' using the original conclusion as a new 'premiss'. That is, the system contains no proposition more general than the other two. All the explanation does is assert that the institutions in question are positively associated. As such, it may represent a very considerable intellectual achievement, but it is not by the standards used here an explanation at all. It does nothing to answer the question why the institutions are related in the way they are. Though

I have used an institutional example, structural explanation in the social sciences is by no means limited to explanations of the association of institutions.

The functional type of explanation cannot be ruled out for the same reason as the structural, for it characteristically contains a general proposition: If $A$ is a surviving society (or a society in equilibrium), it will possess institutions (or other characteristics) of class $Y$. This purports to be a general proposition, general with respect to societies. That is, it is a sociological general proposition. The difficulties with functional explanation are of another kind. First, it is surprisingly difficult to specify what is meant by the survival of a society. It is even more difficult to define the term equilibrium. This does not mean that something we can recognize as social equilibrium never exists. Rather, no statement of the conditions of social equilibrium has yet been devised that is specific enough to allow definite conclusions in logic to be drawn from it. Perhaps such a proposition will be invented in the future, and philosophy can certainly give no reason why it should not be invented, but it certainly has not been invented so far.

Second, survival is at least a clearer criterion than equilibrium, and the functional propositions purport to state the conditions for survival. It is possible to draw true conclusions from them: if $A$ is a surviving society, it ought to possess institutions of class $Y$, and it often does. But it is also possible to draw false conclusions. For the propositions imply that if $B$ is a non-surviving society, it ought not to have possessed characteristics of class $Y$. There are only a few societies that have clearly not survived, and fewer still whose social characteristics were recorded before their demise. But in all cases for which we have some evidence, the non-surviving societies possessed characteristics of class $Y$ and indeed all the other suggested social requisites for survival, both jointly and severally. The biological requisites, such as resistance to alcohol, measles, or gunfire, are another matter. Functional explanation fails as explanation because its general propositions are not contingent: the social scientists who put them forward do not treat them as if data, fact, evidence were relevant to their acceptance or rejection.

The difficulty with historical explanations is different still. They appear to include, again and again, statements of the following sort. Smith (or the Smiths, persons sharing similar values) wanted a particular result and thought that a particular action was likely to get the result; therefore Smith took the action. Smith might well have been mistaken and might fail to get the result, if only because Jones (or the Joneses) wanted a contrary result and was able to thwart him. Or Robinson wanted still another, if not contrary result, and succeeded in modifying the effects of Smith's action. The final result might well be different from what anyone intended or expected. Intended or not, historical explanation took the form of showing that the final result, the *explicandum*, proceeded from combined actions, over time, of the Smiths, Jones, Robinsons, etc.

Since propositions about particular persons like the Smiths are obviously not very general, it often appears that historical explanations do not meet the standards of explanation because they lack general propositions. But that is because they are characteristically *enthymemic*. That is, the historians leave their general propositions, the major premises of their deductive systems, unstated. But what is missing can readily be supplied if necessary, and it is necessary if we are to understand what historical explanation involves. For the moment I shall consider only one, though the most prominent, of these propositions. It takes this form: not only Smith, but every man, in choosing between alternative actions, is likely to take that one for which, as perceived by him at the time, the value ($v$) of the result, multiplied by the probability ($p$) of getting the result, is the greater; and the larger the excess of $p \times v$ for the one action over the alternative, the more likely he is to take the former action. This has been called the *rationality proposition* or theory, even though actions taken under it may not seem rational in other senses, since a man's perceptions may be mistaken and the results may be 'bad' for him. But, given his perceptions and evaluations, behaviour taken in accordance with this proposition is held to be rational. Historical explanation often takes the form of applying this general proposition to the given conditions represented by the particular values, perceptions, and circumstances of the Smiths, Jones, and Robinsons, and trying to show what the combined and cumulative effects will be of actions, each of which was taken in accordance with this general, though usually unstated, proposition.

One of the reasons why historical explanations leave their general propositions unstated is the following. The separate deductive systems by which the actions of particular persons or groups at particular times are explained are often uninteresting. Only when they are linked together in what have been called genetic chains (on genetic explanations see Hempel, 1965, pp. 447–53) do they begin to get interesting. They become linked when the *explicandum* of one deductive system becomes one of the given conditions of another in the series, as when the action of Smith becomes a given condition for explaining the subsequent action of Jones, and so forth. The separate deductive systems would be apt to contain some of the same general propositions, so that if the genetic chains were spelled out in full, the propositions would be repeated again and again. But that would be boring. Historians subconsciously realize this, and leave the general propositions implied but unstated.

Note now that the rationality proposition is what I have called a psychological general proposition: it is held to be general with respect to the behaviour of men, rather than with respect to societies or social groups as such. Thus the historical type of explanation turns out upon examination to be the same type of explanation as the fourth type, the psychological. The differences between the two are differences of degree, not of kind. If strong and similar forces are at work in many societies and tend to produce similar results, if the societies, that is, are in some respect convergent, then the explanation of the results can afford to neglect the details of the historical chain of events producing the result in a particular society. This might be the case in explaining why in every society criminals are punished. But if these conditions do not obtain, the explanation cannot afford to neglect the details, the particular genetic chain. This would certainly be the case in explaining why trial by jury is an institution of Anglo-Saxon societies. But in either case the explanations employ psychological general propositions.

Depending upon what is to be explained, they may of course contain general propositions of other sorts. For instance, some historical explanations, if spelled out in full, would have to contain the proposition: a man whose head is severed from his body is dead—which is surely a general proposition of human physiology. The explanations may include such propositions; they always include psychological ones.

Let me sum up the argument so far. Of the types of explanation used in the more 'social' of the social sciences, those like anthropology and sociology that are specially interested in institutions, the structural type is not an explanation, and the functional type is non-contingent. There remain the historical and the psychological types. But the historical type turns out to be in fact psychological. By elimination, then, it looks so far as if the only type of explanation that stood a chance in social science were the psychological.

Though I have used it as an illustration, the rationality proposition is certainly not adequate for all psychological explanation. It can stand alone as a general proposition only if men's values and their perceptions can be taken as given, as known or obvious. It is true that they can often, for practical purposes, be so taken. One does not need to explain why a hungry man sets a high value on food or why most of us set some store by money. Nor does one always need to explain why a man perceives a certain action as likely to be successful. If a man is a trained carpenter, it requires physics, so to speak, and not psychology to explain why he takes certain actions in building a house and not others. More generally, what will be successful, or at least the probability of success (the risk), is given by the objective nature of things—and the man in question accurately knows the nature of things. But one does feel the need to explain—whether one can in fact do so or not—why a man finds self-punishment rewarding, or why a man who can have no objective knowledge what his chances of success may be—his situation is one of uncertainty and not of risk—nevertheless goes ahead and takes action. When values and perceptions cannot be taken for granted but themselves cry out for explanation, then the rationality proposition turns out not to be the only general proposition we need for our deductive systems. We also need, as we shall see, propositions about the effects of past history, past experience, on present behaviour. The rationality proposition deals only with the present.

In some ways economics has surely been the most successful of the social sciences. It has developed propositions of its own, which have shown considerable explanatory and predictive power. I speak especially of classical or microeconomics. One reason for the success of economics is that it deals with rather easily measured variables: the quantities of largely material goods and their prices. Another reason has been of a quite different sort. Economic explanation belongs to the psychological type, in the sense that economic propositions, such as the so-called laws of supply and demand, can readily be shown to follow from the rationality proposition (see Homans, 1961, pp. 68–70). Moreover, the latter is the only psychological assumption economics had felt the need to make, precisely because the values of men and their perceptions could largely be taken for granted in explaining the sorts of phenomena economics was interested in. The more obvious of the values of men are the material values, since they are shared by many people; and economics was specially concerned with the exchange of material goods for money. It was also interested in the gross effects, such as inflation, produced through the pursuit by many people of these widely shared values—effects to which the behaviours of individuals pursuing more idiosyncratic goals were irrelevant, or in which they tended to cancel themselves out. Classical economics also studied situations in which the success of action was not problematic: in the market it was assumed that a buyer could always find a willing seller, and the only question was whether the buyer was willing to pay the price. Above all, economics could afford to disregard the permanent or semi-permanent relationships between persons and groups that make up so much of the subject-matter of social sciences such as sociology and anthropology. Indeed, under the conditions assumed to exist in the classic market, no one buyer would have any reason to trade regularly with one seller rather than another. And finally, economics could explain much behaviour provided that certain institutions—the market itself, for instance—were taken as given. Though it was often difficult to explain their details, the institutions were at least the product of the other things economics took for granted: the relatively permanent relationships between people in groups. To put the matter another way: economic explanation was relatively unhistorical, for to explain relatively permanent relationships one needs to know how past behaviour affects present behaviour. For classical economics, a man's present behaviour, his values given, was, or should be, determined by his present options.

The rationality principle, as I have pointed out, is a psychological proposition in the sense that it refers to the behaviour of men and not to the characteristics of social groups or aggregates as such. It is certainly not a psychological proposition in another possible sense: it is not the professional property of persons who call themselves psychologists. The wider set of general propositions, which needs to be added to the rationality proposition for the explanation of human behaviour, is psychological not only in the first sense but also in the second: the propositions are usually stated and tested by professional psychologists. They are sometimes referred to as learning theory, but since they also apply to human behaviour after it has in every usual sense of the word been learned, I prefer to speak of them as the propositions of behavioural psychology.

Obviously I do not have the space here to write a treatise on behavioural psychology. Not all psychologists would agree on which propositions should be included as the really fundamental ones, and even fewer on the terminology in which they should be stated, though I think in substance they would say much the same things. All I want to do is suggest their general nature (see especially Skinner, 1953, and Homans, 1961, pp. 30–82). One of the most important propositions would run as follows: if a man takes an action that is followed by a reward, the probability that he will repeat the action increases. This I call the success proposition; it has also been called the law of effect. It seems obvious enough. Why is it important, and what does it add to the rationality proposition? Let me take an example from history. Suppose we ask why William the Conqueror invaded England. We have independent evidence that he perceived that the conquest would be very rewarding to him, compared with his alternative of remaining as Duke of Normandy—provided the conquest were successful. That is, the value term in the rationality proposition is accounted for: the value of the result was high. But how about the success term? To a contemporary observer his chances of success may well have appeared low. In any event he had no way of assigning a definite probability to his chances: his situation was not one of risk but of uncertainty. Why then did he go ahead with the enterprise? The rationality proposition provides no answer. But at this point it is surely relevant to point out that William had a long record of success in his military enterprises in the past, and accordingly, by the success proposition of behavioural psychology, the probability that he would undertake military action again was apt to be high. In ordinary language, his past experience had given him confidence. To sum up: the rationality proposition would explain his behaviour, provided his perception of success were given. Behavioural psychology explains his perception by relating it to his past experience. Or rather, the perception term, the intermediate term, drops out, and present behaviour is related directly to past experience. It is in this sense that behavioural psychology is an historical theory.

Another important proposition of behavioural psychology is the following: if in the past the occurrence of a particular stimulus-situation has been the occasion on which a person's action was rewarded, the recurrence of the stimuli in the present makes it more probable that the man will repeat the action. The stimulus-situation is, roughly speaking, the set of circumstances surrounding the action. The stimuli may have this effect even if there was no rational connection between the presence of the stimuli and the success of the action in the past. Once again, present behaviour is related to past experience.

The rational theory must take the values of men—what sorts of thing they find rewarding—as simply given. When, as a matter of common knowledge, we can assume the values to be shared by many men, the rational proposition does very well in explanation. When they are somehow queer values, the rational theory is simply at a loss. But behavioural psychology, provided it has enough knowledge of a person's past experience, can sometimes account for the way values are acquired. New values, even queer ones, are acquired by being paired with older and more primordial values, just as money becomes, as we say, valuable in itself, when it is discovered to be a means of getting candy.

A behavioural psychology would also include a proposition about the effect of deprivation on increasing, and of satiation on decreasing, the value of a reward. It should include a proposition about the effect of punishment on behaviour, pointing out that, just as reward includes both positive rewards and the removal of punishments, so punishments include both positive punishments and the removal of rewards. A behavioural psychology would include, finally, propositions about emotional behaviour, such as the frustration-aggression proposition: when a man fails to receive an expected reward, or receives an unexpected punishment, the probability of his taking some form of aggressive action increases. But I have said enough to indicate what I mean by behavioural psychology.

Behavioural psychology leads on the one side towards the explanation of the phenomena of personality—that precariously integrated group of interrelated responses that makes a man an individual. That is to say, modern students of personality, from Freud onwards, assume that the adult personality is a product of a long process of conditioning beginning with earliest childhood. Theorists disagree over the terms in which the gross features of this development and its results should be described: such terms, for instance, as id, ego, and superego. But there is nothing in modern personality theory incompatible with the view that the detailed steps in the conditioning process are those described by the propositions of behavioural psychology. Indeed a good case could be made that Freud was the first great behavioural psychologist. Since this paper deals with the explanation of social phenomena, I shall have nothing further to say about individual personality—which does not mean in the least that the development of personality is not a social process.

Behavioural psychology leads on the other side towards the explanation of the phenomena dealt with by social psychology, especially the phenomena created by the interaction of persons in small groups, phenomena such as co-operation, competition, conformity, deviance, status, power, leadership, and distributive justice (see especially Homans, 1961, and Staats & Staats, 1963). These phenomena also occur, of course, in large organizations and societies, and there is no reason to believe that they are different there—no reason, indeed, to believe that any sharp line can be drawn between a micro- and a macro-sociology. But the phenomena have been most intensively studied in small groups. It is this use of behavioural psychology in explanation that has attracted most criticism, and here I must move especially carefully.

The propositions of behavioural psychology are believed to hold good of all men, and they are stated in terms of the behaviour of a single man: 'If a man takes an action that is followed by a reward . . . etc.' The propositions are particularly concerned with the effects of reward on behaviour. There is nothing whatever in the propositions to suggest that the effect of reward is different when it comes from another man rather than from, for instance, the physical environment. But when two or more men are interacting, when the actions of each reward (or punish) the actions of the other, phenomena of course appear that are different from those that appear when an isolated person is being rewarded by the physical environment. New phenomena appear, which may be called social, but no new general propositions are required to explain them—only the new given condition that two or more men are interacting.

There is absolutely no general philosophical argument that will prove, or disprove, the contention that the propositions of behavioural psychology are adequate general propositions for the explanation of social behaviour. All one can do is take particular phenomena and set up the deductive systems that will explain them. In this sense, behavioural psychology can in fact be used to explain a large number of the grosser social phenomena. But since the actual iteration of the explanations is the only possible argument, the argument can never be exhaustive and conclusive. It is obviously impossible to go through the work here, or for any large number of cases anywhere, and in any event behavioural psychology will never be able to explain everything. Even if one has confidence in one's general propositions, one may still not be able to use them in explanation, because one lacks information about the given conditions in which they are to be applied. For instance, much of the information needed for historical explanation has simply disappeared for ever.

But if the proponents of the position taken here can do no more in its support than provide examples of psychological explanation, they can at least require its opponents to do as much on their side. Again and again since the turn of the century scholars have been asserting that social phenomena can never be explained by the use of psychological propositions. The social whole, they say, is more than the sum of its parts; something new emerges over and above the behaviour of individuals; when many individuals act, they may produce results unintended by any one of them. All the actual facts that 'wholeness', 'emergence', and 'unintended consequences' are supposed to refer to are conceded in advance. The question is how these facts are to be explained. The usual examples of such phenomena are readily explainable by the use of psychological propositions. More important, the persons who cite such phenomena in their favour never produce their own explanations of the phenomena. Let them begin to spell out their explanations. Then we should begin to see whether they in fact possess an alternative type of explanation, and especially whether they use an alternative type of general proposition, presumably what I have called sociological propositions. Until they are ready to explain emergent phenomena themselves, they had better stop throwing them up at us.

The issue considered here has been discussed by philosophers as methodological individualism versus methodological socialism (which has of course nothing to do with political socialism) (see especially Danto,

1965, pp. 257–84). Methodological individualism holds that all social phenomena can be analysed without residue into the actions of individuals, that such actions are what is really fundamental in the social sciences. If they are fundamental, then the general explanatory propositions of the social sciences are propositions about the actions of men: in my terms, they are psychological propositions. That is, methodological individualism entails psychologism. (But Sir Karl Popper (1963, pp. 89–99) says he believes in methodological individualism but not psychologism.) Finally, methodological individualism holds that sociological propositions, pro-positions about the characteristics of social groups or aggregates, can in principle be derived from, reduced to, propositions about the behaviour of individuals. Methodological socialism, on the other hand, would presum-ably hold that social phenomena could not be wholly resolved into indi-vidual actions, and that there are general sociological propositions not reducible to psychological propositions—sociological propositions, indeed, from which propositions about individual behaviour could themselves be derived.

Again, there is no general philosophical argument that will resolve the issue. One can only appeal to current evidence. There are certainly sociological propositions of some generality, but many of them can be derived from psychological ones. On the other hand, there are, I believe, no general sociological propositions at present that meet the two following conditions: they cannot be derived from psychological ones, and from them many features of social behaviour can themselves be derived. In this sense, it is my conviction that there is no current evidence in favour of methodo-logical socialism. But a sociological proposition with the right properties may be discovered tomorrow, and if it is, mere argument will be at an end in face of the fact.

The only very general sociological propositions that have been put forward are those of the functional type. Usually the functionalists do not even begin to spell out their arguments, and so it is impossible to discover whether the arguments explain anything. But on at least one occasion functionalists have sketched out their line of reasoning. It may be instruc-tive to look at a famous functionalist explanation of what is itself a general sociological proposition. It is the explanation offered by Professors Davis & Moore (1945, pp. 242–9) for the fact that all societies are stratified, that in all societies there are differences in status.

Briefly, the explanation goes like this. In order to survive or remain in equilibrium, a society must motivate its members to carry out the activities necessary to its survival. The more important the activities—that is, the more crucial to its survival—the greater the society's need to motivate members to carry them out. But the supply of persons able and willing to carry them out is short. To ensure an adequate supply, the society must make the rewards for filling the more important positions greater than those for filling the less important ones. A stratification system does just this. Accordingly a society that survives or remains in equilibrium will have a stratification system. Q.E.D.

I shall not repeat here the general criticism of functional explanations that I offered earlier, but get down to the specifics of this one. It is not a purely sociological explanation, as it refers from the beginning to the motivation of members, that is, of individuals. Indeed it makes implicit use of a psychological proposition: that individuals will not act without rewards. Beyond that, the explanation raises more questions than it answers. How can one tell which activities are more crucial to the survival of a society than others? It is difficult to show that the peasants were less crucial to the survival of medieval societies than the knights, and yet individually they received lower rewards. Why should the supply of persons able and willing to provide the more crucial activities be short? And why should 'society' have to give them greater rewards to do their jobs?

The functional argument puts matters the wrong way around. If in fact, and for whatever reasons, certain members of a society command capacities to reward others, capacities such as the putative control of magical power or the actual control of land, special skill, or physical force—physical force is a capacity to reward because the power to kill is also the power to spare —and if these rewards are valuable both in themselves and because they are in short supply relative to demand, then these members will be able to get from the other members, and not from 'society' as such, disproportionately high rewards for themselves. The reason they will be able to do so lies ultimately in the psychological proposition: the higher the value of a reward, the more likely a person is to take action to get the reward. It follows from this proposition that, when two men (or groups) are exchanging rewards, the person for whom the reward the other is able to provide is the more valuable will do more to get it. He has, in fact, less power than the other (see Homans, 1964*b*, pp. 113–31). And as, in this way, persons and groups tend to get rewards in proportion to their power, they create and maintain a stratification system (see Lenski, 1966). Under this analysis the references in the functional argument to society, its survival, and the contributions different activities make to its survival simply disappear, and the explanation uses only psychological general propositions. I believe that this sort of fate awaits all sociological efforts at explanation. But no abstract argument will demonstrate that this is the case. Nothing but the analysis of examples will do, and they can be endless in number.

Perhaps I may end by suggesting some of the reasons why some scholars have found the psychological explanation of social phenomena difficult to accept. The first reason is powerful intuitively. Can those great institutional and organizational structures that, especially in the modern world, seem so strong as to dominate mere men and to lead a life of their own—can these great structures really be the product of what seems to be as weak as water: individual human choice? But if we remember that the choices of some men narrow the alternatives of others, and that the choices of men can jointly produce results none of the men, severally, would have envisaged, then it is clear that the structures are so produced and maintained.

The second reason lies in the sociology of knowledge. The scholars who have been most reluctant to accept the psychological explanation of social phenomena have been anthropologists and sociologists. Anthropology and sociology were the most 'social' of the social sciences, peculiarly concerned with the larger institutional structures of society. They were also the latest of the social sciences to be recognized as academic disciplines. Insecure in their status, it was natural for them to insist that they had a unique type of contribution to make, and in view of their subject-matter, to insist that their uniqueness lay in a purely social explanation of social phenomena. Though sociology is now well established academically, some sociologists still fear that their subject will somehow lose its identity if it abandons its special claims.

The last reason is a queer one, but I sometimes think it is the most powerful of all. The propositions of behavioural psychology have been known for a very long time. Though the ordinary citizen may not phrase them as a psychologist would, and though he is often surprised by some of their further implications, in psychopathology for instance, still they do not in themselves strike him as unfamiliar. Indeed he, especially if he is a scholar, is apt to call them obvious or even trivial. And these facts get in the way of their acceptance as the most general propositions of social science. Our view of the nature of science is based on the history of physical science. A science is an enterprise in which the scientists make discoveries, and the later they make a discovery the more fundamental it is apt to be, like the structure of the atom. Now if the fundamental propositions of social science were never discovered by scientists but were for ages part of common knowledge, then (so this unconscious process of reasoning runs) either social science cannot be a science at all or, what is more important, its fundamental propositions must remain to be discovered and must accordingly be other than the propositions of behavioural psychology. Both horns of the dilemma get in the way of our recognizing the relevance of psychology to the explanation of social phenomena.

### REFERENCES

Braithwaite, R. B. 1953. *Scientific Explanation.* Cambridge.

Danto, A. C. 1965. *Anatomical Philosophy of History.* Cambridge.

Davis, K. & Moore, W. E. 1945. Some principles of stratification. *Am. Sociol. Rev.* 10.

Hempel, C. H. 1965. *Aspects of Scientific Explanation.* New York.

Homans, G. C. 1961. *Social Behaviour.* New York.

Homans, G. C. 1964a. Contemporary theory in sociology. *Handbook of Modern Sociology.* Ed. R. E. L. Faris. Chicago.

Homans, G. C. 1964b. A theory of social interaction. *Transactions of the Fifth World Congress of Sociology,* vol. 4.

Lenski, G. 1966. *Power and Privilege.* New York.

Popper, K. R. 1963. *The Open Society and its Enemies,* vol. 2. New York.

Skinner, B. F. 1953. *Science and Human Behaviour.* New York.

Staats, A. W. & Staats, C. K. 1963. *Complex Human Behaviour.* New York.

# COMMENT

## *by* PETER M. BLAU

George Homans has thrown down a gauntlet that I as a sociologist feel obligated to take up. To be sure, there is much I admire in his paper. He presents some excellent criticism of sociological theorizing and outlines a conception of scientific theory that I fully endorse. But he claims that there are no sociological principles that can explain social phenomena, only psychological ones, and challenges opponents of this view to provide specific examples of sociological explanations. I dispute this claim and shall try to meet his challenge.

I largely agree with Homans' criticism of various theoretical explanations in sociology. Structural analysis that merely traces the interdependence between institutions or other characteristics of social structures does not constitute a scientific theory that explains these interrelations. Although functional analysis attempts to explain such interdependence, Homans rightly points out that the criteria of function usually employed—social equilibrium and survival—make it impossible to test functional propositions in research, whereas scientific propositions must be empirically testable, if not directly then indirectly through their implications. The functional theory of stratification has additional shortcomings, notably that no independent criterion of the relative importance of positions is specified, which tends to make the proposition that the more important positions receive higher rewards tautological. I also concede that many historical explanations are implicitly psychological.

In addition, I share Homans' conception of theorizing as having the function of supplying scientific explanations that must meet certain standards, which are taken from the discussions of Braithwaite and other philosophers of science.[1] Our agreement on the basic methodological principles of theorizing is of particular importance, since it should enable us to pinpoint the source of our disagreement concerning sociological explanations. To assure that my understanding of these principles is indeed the same as Homans', I shall summarize them in my own words.

[1] Richard B. Braithwaite, 1953, which I shall use as the primary reference.

Homans stipulates four main criteria scientific explanations must meet. First, a scientific explanation consists of a set of propositions, that is, of statements that specify relationships between two or more variables. In its simplest form, a proposition states, 'if *A*, then *B*', or, 'the more *A*, the more *B*'. A more complex proposition would be, 'only if *C* does *B* increase with increasing *A*'. A variable is a property or characteristic of any unit— be it matter, individuals, organizations, or societies—that is not constant for all units. Thus, 'living' is not a variable of individuals, since it charac- terizes all of them, though it is one of matter, which can be organic or inorganic.

Second, the set of propositions constitutes a deductive system or hierarchy. This means that the less general propositions in the system can be logically derived from the more general ones. Although 'in the social sciences a kind of sketchy logic is all that we can ask', as Homans notes, this does not alter the fundamental character of the system, only its degree of rigour. The fact that the more general propositions imply the less general ones, if not in the strictest logic then at least in a rough approxima- tion of it, is all that is meant by saying that the former explain the latter.

Third, there are three types of propositions in the system: (1) The *explicanda* are the less general propositions to be explained which can be logically deduced from other propositions. (2) The premises are the more general propositions that cannot be deduced from any other propositions in the system and from which the *explicanda* are deducible. (There may also be one or more intermediate sets of propositions that can be deduced from the premises and from which the *explicanda* can be deduced.) (3) Propositions are necessary that indicate the conditions for applying the premises, notably by specifying the connections between the general terms of the premises and the specific variables in the propositions derived from them. Thus, to deduce from the premiss 'all men are mortal' the conclusion 'all Americans living today will die' requires the specifying proposition 'Americans are men'.

Fourth, some of the propositions in the system must be contingent, that is, their acceptance as valid is contingent on empirical evidence that confirms them. The lower-level propositions, in particular, must meet two independent conditions: they must logically follow from the higher-level propositions, and they must also be validated by empirical evidence. It is the latter condition that distinguishes scientific theories from mathematical ones, which do not contain propositions contingent on empirical evidence.

Another point Homans makes deserved comment, because it has

important implications which may reveal differences in our methodological views. He states that propositions may include 'theoretical terms, implicitly defined' (p. 313). I take this to mean theoretical concepts that are not directly observable. Indeed, Braithwaite shows that a theory that is more than a mere description of the empirical findings on which it is based must include such theoretical terms that are not explicitly defined (Braithwaite, 1953, p. 76). This implies that theoretical propositions, inasmuch as they include abstract terms, cannot be directly tested. Hence, Homans' statement that scientific explanations 'must not consist entirely of non-contingent propositions' (p. 314) is not sufficiently precise. The theoretical premisses will include, if the theory goes beyond description, abstract terms not explicitly defined, and these premisses can therefore not be directly tested; they are not, strictly speaking, contingent propositions. But the propositions logically derived from the premisses must not include such abstract terms, precisely because they must be propositions whose acceptance is contingent on empirical evidence, and only propositions whose terms are directly observable variables can be empirically tested. The probable validity of the premisses is inferred from the fact that they logically imply propositions that have been empirically validated, which makes the premisses *to a limited degree* contingent propositions.

Scientific theory is built epistomologically from the lower-level propositions up to the higher-level ones, although its logical structure is from the top down, in contrast to mathematical theory, which proceeds epistomologically as well as logically from top to bottom (Braithwaite, 1953, p. 352). Scientific theory starts with empirical findings concerning relationships between variables. (Even speculative theory which is not based on systematic research findings derives its generalizations from assumptions about empirical relationships, the only difference being that these assumptions are not empirically validated.) From these empirical propositions several theoretical propositions are inferred from which the empirical ones can be deduced, ideally in accordance with strict logic. If the theoretical propositions are in fact more general than the sum of the empirical findings from which they were derived, they will imply new empirical propositions, hypotheses that can be tested. Thus, the probable validity of the theoretical generalizations can be established by the successful empirical predictions they make. I consider it an essential requirement of scientific theory that several new as well as several old empirical propositions logically follow from it. While Homans may disagree, I do not think that disagreement on this point has much bearing on the main issue under discussion.

II

The basic issue is whether it is possible to develop sociological explanations that conform to the criteria reviewed. I shall attempt to meet Homans' challenge to illustrate that this is possible by suggesting two sociological propositions that explain some empirical relationships observed in formal organizations. No claim is made that the theoretical generalizations advanced are original, profound, or universally valid; only that they fulfil the requirements. Imaginative theory is not created on the spur of the moment for purposes of illustration. Of course, I might have selected as my example theoretical principles developed by one of the great sociologists, like Weber or Durkheim, instead of choosing my own. But Weber's theory of the development of capitalism or Durkheim's theory of anomie is so complex that it would be difficult to use it to illustrate all the points under consideration clearly without becoming embroiled in the discussion of various questions that are not directly pertinent, such as whether empirical evidence supports Weber's thesis or which ones of Durkheim's propositions are strictly sociological rather than implicitly psychological. It seemed preferable, therefore, to construct an example deliberately designed to show in fairly simple form how sociological generalizations can meet the criteria of scientific explanation.

The subject-matter is bureaucracy, specifically, the interdependence among several characteristics of American government agencies. The first question is *how* the variables that characterize bureaucracies are related, and the second question is *why* these observed relationships obtain. The empirical data necessary to answer the first question come from a study of all state employment security agencies in the United States, which are responsible for administering unemployment compensation and providing public employment service. The empirical propositions reported below are based on correlations, with a case base of fifty-three (the agencies in all fifty states and three territories).[1] Whether the same associations are observable in formal organizations of all kinds is not known, although similar findings have been obtained for some other organizations. The data are presented in simplified form to highlight the principles of explanation and exclude all results that are not pertinent for this particular purpose.

Three types of proposition in an explanatory system have been distinguished. Seven empirical findings represent the first type, the relationship between variables to be explained. Two theoretical generalizations inferred from these findings represent the second type, the premisses that imply, if

[1] I gratefully acknowledge grant GS 1528 of the National Science Foundation, which supports this research.

only by a sketchy logic, the empirical propositions and that consequently explain them. Two indications of what the empirical manifestations of the theoretical terms employed are represent the third type, propositions specifying the conditions of application. Here is the list of these three types of propositions:

1. Empirical Findings (*Explicanda*)
   (*a*) The larger an agency, the more pronounced is the division of labour in it (as indicated by the number of different occupational positions).
   (*b*) The larger an agency, the greater is the number of functional divisions (whose heads report to top management).
   (*c*) The larger an agency, the greater is the number of managerial levels.
   (*d*) The more pronounced the division of labour, the larger is the relative size of the clerical apparatus (proportion of clerks).
   (*e*) The greater the number of functional divisions, the larger is the relative size of the clerical apparatus.
   (*f*) The greater the number of levels, the greater is the likelihood of the automation of operations.
   (*g*) The greater the number or levels, the larger the relative size of the staff.

2. Theoretical Generalizations (Premisses)
   (*a*) Increasing size of a formal organization gives rise to structural differentiation along various lines.
   (*b*) Structural differentiation in a formal organization increases the need for mechanisms of co-ordination.

3. Empirical Manifestations of Theoretical Terms (Specifying Propositions)
   (*a*) Structural differentiation in a formal organization is manifest in the number of different occupational positions, the number of functional divisions, and the number of managerial levels.
   (*b*) Mechanisms of co-ordination in a formal organization are exemplified by the automation of operations through computers, the extensive use of clerks to maintain channels of communication, and a proportionately large staff.

I claim that the propositions under (2), together with those under (3), logically imply those under (1). To be sure, the two theoretical generalizations are not very abstract, not very far removed from merely summarizing

333

the empirical statements. But they are sufficiently general to be more than descriptive summaries.[1] They imply other propositions besides the seven from which they were inferred, for example, the hypothesis that factories that have assembly lines should have more managerial levels than factories that do not have this mechanical means of co-ordination. Thus, the possible validity of the generalizations can be established by testing the new predictions that can be deduced from them. Since the seven original empirical propositions, as well as new ones, are logically implied by the two theoretical generalizations, given the two specifications, the empirical findings are explained by these two generalizations that large size promotes structural differentiation in formal organizations and that structural differentiation requires more extensive co-ordinating mechanisms.

## III

The objection may be raised that the explanation offered is not complete. For instance, no attempt was made to account for the fact that different co-ordinating mechanisms are associated with various aspects of structural differentiation. Moreover, the two generalizations themselves have not been explained. Why does large size promote differentiation? Why do differentiated structures necessitate more co-ordinating mechanisms than others? However, the criteria stipulated do not require explanations to be complete, nor do they require them to be final, in need of no further explanation. As a matter of fact, all scientific explanations are subject to complementation and revision, to being made more complete by supplementary propositions and to being in turn explained by more general propositions from which they can be deduced.

Granting that the two general propositions in the system of nine propositions outlined meet the criteria of scientific explanations (and I think Homans would grant it), however, this would only resolve the immediate issue of whether *any* sociological explanations are possible and not the fundamental one of whether these in turn cannot always be explained in terms of more general psychological propositions. For in another paper Homans has already agreed that sociological theories that meet his criteria

---

[1] There is a reason why the theoretical propositions are not further removed from the empirical findings. A difficulty with more interesting and abstract theoretical principles in sociology today is that their connection with the empirical propositions they explain tends to be quite tenuous. I do not think, as I have recently emphasized elsewhere, that this should deter us from constructing such theories. But since my task here is to demonstrate that sociological generalizations are possible from which empirical propositions follow in fairly strict logic, I thought it wise to confine myself to sufficiently low levels of abstraction to make the deduction of the *explicanda* from the generalizations rather clearly apparent.

exist, but he added that they are 'open at the top', that is, reducible to more basic psychological theories that explain them (Homans, 1964, p. 968). Let us see whether an attempt to explain why structural differentiation in formal organizations intensifies the need for co-ordinating mechanisms—one of our two general propositions—does not ultimately involve psychological principles.

Structural differentiation makes an organization more complex and the accomplishment of its mission less straightforward and more difficult. Consequently, a variety of problems probably arise in a highly differentiated organization that were not encountered or encountered less frequently before it had become so differentiated. Executives in formal organizations are held responsible for coping with problems that arise and assuring effective operations. These executives may have learned, by trial and error or in business school or from colleagues, that hiring more clerks or installing a computer helps solve the problems of co-ordination in highly differentiated organizations, and they therefore tend to introduce such co-ordinating mechanisms in differentiated organizations in order to improve operations. In short, co-ordinating mechanisms are disproportionately often found in differentiated organizations, because executives in these organizations are most likely to achieve results they value—improved operations—by instituting them. Here indeed we apparently have explained our sociological generalizations by one of Homans' psychological propositions—the rationality principle.

But have we really explained anything? Surely not in the sense that the rationality proposition, complemented by some specifying propositions but *no* additional *sociological* premisses, logically implies the proposition that structural differentiation is associated with co-ordinating mechanisms in organizations. Besides, the allocation of responsibilities and the incentive system in formal organizations are explicitly designed to assure that managers, who are assumed to act rationally, will deal with problems that arise, and the emphasis on impersonal detachment in formal organizations maximizes the likelihood that its personnel will in fact act rationally. These considerations call attention to the admitted sketchiness of the logic in the explanatory system illustrated. Among the many conditions implicitly assumed as given in this system are (1) that managers in formal organizations are held responsible for dealing with problems; (2) that an incentive system rewards them for discharging their responsibilities; (3) that an emphasis on detachment minimizes the likelihood that such factors as emotions impede rational decision making; and (4) that men under these conditions tend to adapt means rationally to the organizational ends for which they are held responsible.

Every scientific explanation must take many conditions as given in order to explain the connections between a few conditions, as no other than Homans has emphasized (Homans, 1961, pp. 205–31), although the most relevant conditions accepted as given should be made explicit in the interest of increasing the rigour of logical deductions. An investigator cannot question everything at once; he needs the firm footing of given conditions to raise questions about some other conditions and try to explain them. Normal behaviour, which the psychologist seeks to explain, is assumed to be given in sociological investigations, just as prevailing social conditions, which the sociologist seeks to explain, are assumed to be given in psychological investigations, and both take as given normal physiological processes. The question arises whether the givens cannot be transformed into premisses that explain the original explanatory proposition. This may be possible, but not for all conditions at once, and which ones are selected for further investigation depends on the conceptual framework of the discipline.

A number of sociological questions might be raised in an endeavour to explain the proposition that differentiated organizations are more likely than others to have extensive co-ordinating mechanisms. What distinguishes those organizations that are highly differentiated but have only rudimentary co-ordinating mechanisms from others? Do certain personnel policies or incentive systems, for example, discourage the introduction of such needed innovations? What other conditions in organizations promote the expansion of co-ordinating mechanisms? What other consequences does structural differentiation in organizations have? What are the consequences of structural differentiation in other collectivities? If a few general propositions could be discovered that logically imply not only the original proposition but also all those that answer these questions, these premisses would constitute a theory that explains why structural differentiation promotes co-ordinating mechanisms and much else as well, and I have little doubt that it would be sociological theory and not a theory about individual behaviour.

IV

The raw material of psychology and sociology as well as the other social sciences is the same. Their empirical data ultimately refer to patterns of human behaviour. But this raw material is differently conceptualized, which means that different variables are abstracted from it by the various disciplines, and the theoretical terms in the generalizations explaining associations between these variables are correspondingly different. The variables that enter into strictly sociological explanations characterize collectivities, *not* individual human beings. An example is the degree of division of labour, which is a variable that cannot refer to individuals but only to collectivities, although the information needed to measure it reflects the behaviour of many individuals—how different their tasks are. (Of course, in terms of this analytical criterion many sociologists are engaged in psychological studies.)

Every sociological proposition, since it pertains to human behaviour, can be made the starting point, properly re-conceptualized, of a psychological explanation. But every psychological proposition, since it pertains to living organisms, can, in similar manner, be made the starting point of a physiological explanation. Indeed, every psychological proposition can also, in somewhat different fashion, become the starting point of a sociological explanation. Thus, Homans' rationality proposition that men pursue results they value cannot explain anything without additional propositions indicating that men value certain results more than alternatives, as he recognizes. American men may value $A$ over $B$, whereas Russians value $B$ over $A$, and this together with the rationality proposition may explain differences in their courses of action. But why do Russians and Americans have different values? Propositions about reinforcement from learning theory can only explain the perpetuation of this cultural difference. The general propositions needed to explain the difference itself would have to be largely sociological ones, since psychological propositions about what all men have in common cannot account for variations between societies (unless these variations are entirely attributed to non-social factors, such as physical resources, which hardly seems plausible).

Perhaps the fundamental issue on which Homans and I disagree is how to analyse the behaviour of units that consist of sub-units. His methodological view seems to be (though he does not consistently adopt it in his own work) that the forces that govern the units must be explained in terms of the properties of their constituent elements, whereas my view is that these forces are explained by analysing the ways the elements are organized in the units, and these organizing principles are properties of the units, not of their constituent elements. The objective of psychology is to explain how individual behaviour is organized, which requires not merely accumulating numerous separate explanations but seeking to integrate them into a systematic body of theory by finding higher-level propositions from which they can be derived. Whereas every single psychological proposition can ultimately be explained by the physiological processes underlying it, a theoretical system of psychological propositions cannot be built by proceeding in this fashion. Physiological processes must be taken as given in constructing a system of propositions that explain the organization of human behaviour, and this is precisely what Homans as a good social psychologist does in his own theory. He does not try to explain in physiological terms why men whose sense of justice is violated become angry, for example, but he combines this proposition with other psychological ones to explain individual behaviour (Homans, 1961, p. 75 and *passim*). His own theorizing violates the principle of reductionism he avows. The objective of sociology is to explain how collectivities of men become socially organized, and the underlying psychological processes must be taken as given in developing a system of theoretical propositions for this purpose. The case is exactly parallel to that of psychology.

A basic assumption I make is that the behaviour of organized aggregates follows its own principles, and the discovery of these explanatory principles does not require detailed knowledge of the principles that govern the behaviour of sub-units. The latter principles may be taken as given in investigating the former; indeed, they must be taken as given, because scientific explanation requires that a few pertinent variables be abstracted from the welter of possible information that could be collected about any phenomenon. The combination of physiological processes entailed in any set of psychological propositions or of psychological processes entailed in any set of sociological propositions is so complex that it seems inconceivable that the latter could be derived even from the most systematic theory of the former. A large number of diverse patterns of human behaviour is involved in any sociological proposition specifying the association between two or more properties of organized collectivities. Although a few general psychological propositions may at some future time suffice to explain all these patterns of behaviour, the sociological proposition could still not be deduced from them without specifying a multitude of particular conditions of application, whereas it could be deduced from more general sociological propositions much more parsimoniously.

The conclusion that empirical relationships between characteristics or organized collectivities must be explained by sociological general propositions rather than by psychological ones requires the very conception of scientific explanation Homans and I have adopted, whereas it would be incompatible with some other conceptions of explanation. Thus, Lazarsfeld means by explaining the association between two variables showing through which intervening variables the two are connected.[1] Given this definition of explanation, all sociological theories would have to include some psychological propositions, in my opinion, because the influence of one

---

[1] The term Lazarsfeld uses is 'interpretation', since he refers by 'explanation' to an analysis that uncovers a spurious correlation. See Patricia L. Kendall & Paul F. Lazarsfeld, 1950, pp. 151–58.

social condition on another is undoubtedly mediated by psychological states and patterns of behaviour of individuals. But the view of explanation here adopted is quite different, requiring a set of general propositions from which the *explicandum* can be logically derived. There is no reason to assume that empirical relationships between variables that characterize collectivities are more likely to be deducible from a limited number of general psychological propositions, which refer to connections between properties of individuals and their behaviour, than from a set of general sociological propositions, which refer to connections between various aspects of the organization of collectivities and their consequences. Indeed, there is much reason to assume the opposite.

### REFERENCES

Braithwaite, R. B. 1953. *Scientific Explanation*. Cambridge.
Homans, G. C. 1961. *Social Behavior*. New York.
Homans, G. C. 1964. Contemporary theory in sociology. *Handbook of Modern Sociology*. Ed. Robert E. L. Faris. Chicago.
Kendall, L. & Lazarsfeld, P. F. 1950. Problems of survey analysis. *Continuities in Social Research*. Eds. Robert K. Merton & Paul F. Lazarsfeld. Glencoe, Illinois.

# REPLY

## *by* GEORGE C. HOMANS

In his comments on my main paper, Blau says of me (p. 329): 'He claims that there are no sociological principles that can explain social phenomena, only psychological ones, and challenges opponents of this view to provide specific examples of sociological explanations.' This is not quite what I claimed, but since I may have been guilty of obscurity, this reply will allow me to make my position clear.

The question is not whether sociological propositions exist. There certainly are sociological propositions of some generality, and accordingly they can be used in the explanation of still less general social phenomena (see Homans, 1967, pp. 80–7). The question is rather *how* general these propositions are. I doubt whether they are fully general in the sense that they hold good of all human groups and societies—fully general in the same sense as that in which the propositions of behavioural psychology are believed to hold good of all men. More important is the question whether the sociological propositions can themselves be derived from, explained by, psychological ones. In my main paper I stated (p. 325) that methodological individualism, which is my position, 'holds that sociological propositions ... can in principle be derived from, reduced to, propositions about the behaviour of individuals'. The crucial sentence was the following (p. 325): 'There are, I believe, no general sociological propositions at present that meet the two following conditions: they cannot be derived from psychological ones, and from them many features of social behaviour can themselves be derived.' I emphasize both conditions.

In any event, I welcome Blau's acceptance of my challenge, and in the light of my restatement of my position, let us examine what he has done. On pp. 333 ff. he tries to show that various empirical findings about governmental agencies follow logically from, and in this sense are explained by, two propositions: '2(a). Increasing size of a formal organization gives rise to structural differentiation along various lines. 2(b). Structural differentiation in a formal organization increases the need for mechanisms of co-ordination.' I do not have much trouble with the first proposition, but I do have with the second. The question is: *whose* is the need in question? Moreover, if we look at proposition 2(a) in relation to the specifying propositions and the *explicanda*, we recognize that Blau assumes the 'need' to be automatically met—untouched by human hands, so to speak

—and we know that this is not the case. If organizations have needs, they certainly are not met unless men in the organization themselves feel the need to meet the organizational needs. What this means is that Blau's explanation, like so many explanations in social science, is enthymemic: a crucial proposition or propositions has been left unstated. None of the empirical findings follow in logic from the second proposition, unless statements are added about the effects of 'need' on human behaviour. Accordingly the findings remain at present unexplained, and I will not grant, as Blau so fondly supposes I will, that his system of propositions 'meets the criteria of scientific explanation'. What is more, the propositions that must be supplied will certainly include general psychological propositions such as, to put it crudely: a man who feels a need is apt to take action to meet the need (provided he perceives the action as apt to be successful in doing so).

But let us suppose, as Blau argues and as I am inclined to agree, that some empirical findings could be explained with the help at least of his first proposition and without the addition of general psychological propositions. Still, Blau concedes that this proposition is not very general, 'not very far removed from merely summarizing the empirical statements' (p. 333), and he then goes on to show, pretty sketchily to be sure, that, if this proposition is itself to be explained, psychological general propositions are required. As he puts it (p. 335), 'Here indeed we apparently have explained our sociological generalizations by one of Homans' psychological propositions—the rationality principle.' (It is not *my* principle; I am not its discoverer, but let that pass.)

In conceding this much, Blau has conceded all I ask. There are certainly sociological propositions of some generality, and they can often be used to explain empirical findings, but these sociological propositions can often be shown to follow in turn from general psychological propositions. (There are other sociological propositions which cannot be shown to follow from psychological ones—such as the propositions of sociological functionalism —but then these do not meet the condition of satisfactorily explaining empirical findings. A general sociological theory must meet both conditions.)

But Blau is still not happy. He makes the point (pp. 334 ff.) that a general proposition by itself will not explain an empirical finding. The former must be applied to specified given conditions (the minor premises). Accordingly a general psychological proposition when used to explain a sociological finding must be applied to specific given conditions, and these givens may themselves be sociological in nature. Of course. I find no difficulties in this argument. Moreover, in any particular deductive (explanatory) system one may simply, for convenience, take the givens for granted and not undertake to explain them in turn. Of course again. But my position is that, if for any reason one *does* wish to explain such sociological givens, it often turns out again that one can do so only with the help of psychological general propositions.

Take the four conditions Blau himself asserts (p. 335) are 'implicitly assumed as given' in his own explanatory system. None of them can in turn be explained without the use of psychological propositions. Take, for instance, no. (2), that 'an incentive system rewards them [managers] for discharging their responsibilities'. Let us not worry about whether this is in fact a *sociological* given condition: it certainly refers to the behaviour of men (managers). That is not important. The important point is rather that one cannot account for the effectiveness or even the existence of an incentive system without the use of a general psychological proposition stating how the value of a reward affects action, such as 'discharging responsibility', that gets a man the reward.

Of course there is an enormous number of possible given conditions that cannot be explained. We cannot altogether explain why Russians and Americans, on the average, differ somewhat in their values (Blau, p. 337)—not because we lack general psychological propositions about the processes by which values are acquired and perpetuated, but because we lack adequate information about the social history of the many generations that preceded modern Americans and Russians—information that would be necessary for the application of these propositions. In any case, if psychology cannot explain these given conditions, no alternative type of explanation can explain them either.

Blau deserves much credit for taking up my challenge and actually beginning to spell out his own explanation of certain social phenomena. Few sociologists have been willing to do as much, but it is necessary. As I said in my main paper, we cannot argue effectively in general terms about explanation in social science. We must examine the particular explanations proposed. An examination of Blau's example revealed just what I expected it would. Had I not expected it, I should not have issued my challenge. One part of his explanation failed because it was enthymemic. A major premiss, required if the conclusions were to follow in logic, was left unstated, but when supplied it turned out to be psychological. A sociological proposition used as a premiss in another part of his explanation could in turn be derived from a psychological general proposition. And even the given conditions of the latter derivation needed psychology for their explanation. Blau provided no evidence, as he was challenged to do on p. 324 of my main paper, that he possessed an alternative *type* of explanation to the psychological or used an alternative type of *general* proposition, presumably what I have called a sociological proposition. In short, an examination of Blau's own example bears out the argument of my main paper. I believe the same fate would befall other such attempts, and accordingly my challenge still stands.

I am not concerned with what Blau says at the end about 'analysis'. There are all sorts of ways of analysing data. I am concerned here only with empirical propositions, regardless of the processes of research by which they are arrived at, and with the question how these propositions are to be explained. But I am sure Blau is wrong about one final matter. He says (p. 337) that 'every single psychological proposition can ultimately be explained by the physiological processes underlying it'. I do not know what he means by 'ultimately', but the explanation cannot be carried out now. The fact that human behaviour would not occur without the physiological processes of the living human body does not mean that the propositions of behavioural psychology can be derived from, reduced to, the propositions of physiology. Though some day, no doubt, in the progress of science the feat will be performed, that day has not dawned yet. Accordingly Blau is surely unfair in taunting me with failing to do what no contemporary has managed to do. But perhaps Blau can perform the reduction himself?

### REFERENCE

Homans, G. C. 1967. *The Nature of Social Science*. New York.

# THE SKINNERIAN ANALYSIS OF
# BEHAVIOUR

*by* R. A. BOAKES AND M. S. HALLIDAY

## INTRODUCTION

The history of psychology has been notable for the rise, and decline, of a large number of movements, and for periods of bitter altercation between the adherents of rival approaches. At present experimental psychology at least appears to be passing through a relatively calm period; psychologists tend to be grouped in terms of the particular subject that they are working upon rather than according to the particular philosophy of psychology that they endorse. The glaring exception to this picture of peaceful consensus is a group of psychologists who have adopted or have been given the label 'Skinnerian'. It is a description which does not indicate a person's field of interest but rather a general attitude towards the whole of psychology. Similarly it is the one term that is sure to arouse either bitter attack or impassioned defence. Like all such labels it tends to be ill-defined and its meaning can be widely adjusted for the present purposes of the user. In its loosest form it is often applied to almost any experimentalist who uses automatic equipment, while its most narrow form implies an unshakeable faith in all that Professor Skinner has suggested, let alone affirmed.

It is possible and, we shall argue, extremely worthwhile, to find a position that lies between these two extremes. In this chapter we shall present such a position and examine its most crucial aspects. The attempt will be justified if its sole effect is to persuade psychologists who have reacted violently against the whole approach, possibly because of the sometimes dogmatic way in which it has been presented, to re-examine its basic content.

In outline the important points are as follows. The approach is based on 'radical behaviourism', which means both that any form of mentalistic explanation is totally rejected and also that behaviour is a subject-matter in its own right, and not, for example, a convenient way of getting at the properties of the nervous system. Since the arguments against mentalism are treated elsewhere in this volume this issue will not be discussed in any detail here. The question of whether any explanation of behaviour necessarily requires reference to some other level of discourse, notably one using physiological concepts, is a matter on which Skinner's views differ from those of many other behaviourists and as such will require discussion. Probably the most distinctive aspect of the Skinnerian approach is what Verplanck has termed its 'nihilistic' attitude towards the work of other psychologists and in particular its distaste for theory and for the manipulation of hypotheses.

Almost all the other characteristics that are associated with the label 'Skinnerian', such as the attitude towards statistics, the limited number of concepts, the insistence on extreme experimental control and the use of automated equipment, are closely related to this central point.

Most of the arguments to be discussed here were initially presented in *The Behaviour of Organisms* (Skinner, 1938). This book has had a somewhat unusual history in that its influence has been greater in recent years than in any other period since its publication three decades ago. Many of the points that were novel at that time are now accepted by psychologists who would in no way consider themselves to be 'Skinnerians'. However, there are many areas of psychological research which would still very much benefit from its influence. And indeed there are signs that the mutual isolation between Skinnerians and non-Skinnerians is breaking down and that Skinnerian approaches are being made to essentially theoretical problems, while non-Skinnerian psychologists are beginning to recognize the advantages of Skinnerian procedures.

### THE NATURE OF PSYCHOLOGICAL ENQUIRY

In one sense this chapter is out of place in a volume like the present one, since Skinner has not attempted to present a theory of psychology, although he has been very insistent on the subject of what forms psychological explanations should *not* take. His position is essentially that there is no point in looking for explanations until it is quite clear what it is that needs to be explained. And in the present state of psychology this is often far from clear. Until it has been established on an experimental basis what the important phenomena are within certain areas, theories can deal only with arbitrarily chosen groups of effects. For example, Hull's theory of learning claimed that rats learn to attach responses to stimuli while Tolman said that they learn what leads to what; a crucial test of the theories appeared to be whether a rat learns which response to make or which place to go to in a maze. Innumerable experiments were carried out in an attempt to resolve the 'place *vs* response' problem and a mass of conflicting and confusing results accumulated (for a review see Restle, 1957). It is now fairly generally agreed that the greater part of this effort was wasted, since the question is wrongly posed. The rat is not 'naturally' either a place or a response learner; it makes use of any cues that may be available in the apparatus and will be a place learner if the extra-maze cues are salient and a response learner if they are weak. Thus theoretical preconceptions lead psychologists to ask largely irrelevant questions. It is for reasons such as these that Skinner believes that a research programme in psychology which proceeds on the basis of hypothesis testing is likely to be both misleading —since its choice of topics will tend to be based on personal preconceptions of a non-scientific kind—and also unrewarding, since its results will probably be meaningful only within the context of a transient theory.

Among the most severe critics of the lack of formalism in psychology and in Skinner's work in particular have been a number of linguists; thus Chomsky (1965) has claimed that 'the social and behavioural sciences provide ample evidence that objectivity can be pursued with little consequent gain in insight and understanding'. They have presented to psychologists the example of linguistics in which, it is claimed, the need is not for more empirical data, but for a comprehensive system to account for the abundant knowledge about language that is available upon reflection. While it is true that an adult may know a considerable amount about his fellow beings' behaviour from personal experience and from ideas rooted in his language and culture, this knowledge is irretrievably embedded in a conceptual framework of dualism. Moreover, this knowledge will be most complete in areas that are of personal importance to the knower, but are not likely to be very helpful to a science of behaviour. This point is particularly relevant to intuitions about animal behaviour. Its relevance to human behaviour is illustrated by the implausibility of extracting a coherent system from the observations on human behaviour contained in proverbial injunctions of the kind 'more haste, less speed', 'strike while the iron is hot', etc. In contrast Chomsky has claimed that traditional views on grammar are essentially correct and are of central importance to a theory of syntax (Chomsky, 1965). Whatever the case may be in linguistics, as far as psychology is concerned intuition has proved to be a very inadequate basis on which to construct explanations. There are intelligent and unintelligent ways of pursuing objectivity; but the search for insight into behaviour based entirely on armchair reflection has produced neither agreement nor explanatory power.

The emphasis here has been negative and the question remains as to what is to guide research if hypothesis testing is rejected, and what criteria are to be used to define 'important phenomena'. Skinner argues that the first step in a science is to determine what basic units and fundamental variables are to be used within a description. The one sense in which he can be understood as a theorist is in his specification of what these should be. His early work is much concerned with the problems of defining 'stimulus' and 'response' and of obtaining a unit of behaviour analogous to the reflex of the physiologists. This topic will be discussed later; what is of special interest here is the notion of 'functional relationship' and the criteria used to determine suitable units. The level of definition chosen for a response, for example, is one that will allow the results of well controlled experiments to yield 'smooth and reproducible' curves. The search for variables that are related by simple and continuous functions is held to be the primary goal for psychology at its present stage. Boyle, Hooke or Brahe, not Newton, should be the model for aspiring psychologists.

This brief exposition requires a number of comments. First, little is said about how one should go about looking for such relationships. This vagueness is not surprising since there are relatively few explicit suggestions as to the procedure a successful scientist should follow. The most detailed has been that advanced by logical positivism, which was enjoying its heyday when Skinner began his work. As well as reacting against its whole description of and prescription for science, he completely rejected its account of how a scientist goes about his work. The picture of the superbly rational thinker carrying out a sequence of experiments whose form is dictated by the logical necessities of his theory was to him a complete travesty of the truth. A much truer, though less flattering, image is that of someone stumbling in the dark, groping for one thing but finding another. The idea that a science progresses by means of hunches and lucky accidents (Skinner has reported a number of cases where faulty equipment or a practical constraint, such as the need to make a limited supply of food last out over a week-end, have led to important discoveries) is itself, of course, only one part of the truth. However, it is one that is neglected in most philosophical accounts.

If regular functional relationships are to be the initial goal of a science, what is to be accepted as 'regular'? 'Smoothness' of a curve alone is obviously an insufficient criterion, in that it may reflect the effects of some gross averaging process applied to the results of a bad experiment on some very irregular phenomenon. However, it is not very clear what modification should be made to this criterion to make it more precise, and we shall return to this problem in the discussion of averaging and statistics, and also in connection with the use of rate as a response measure. In practice the question of whether a function is regular does not often arise, since it is usually quite obvious when the apparent smoothness of some relationship is entirely misleading. Verplanck (1954) has criticized Skinner's stress on the search for simple, continuous relationships on the grounds that many advances in science have resulted from the discovery of discontinuities.

The example he gives is that of the 'triple point' of water. In fact this is an argument in favour of Skinner's views, since the discovery of important discontinuities must necessarily be preceded by establishing continuous relationships within a certain range.

A difficulty also occurs in attempting to obtain a precise definition of 'simple relationship'. For a given empirical function a number of alternative mathematical formulations can always be found and, if they do not differ greatly in terms of formal simplicity, additional factors, such as consistency with related results, need to be taken into account. Satisfactory criteria for judgments of this kind appear extremely difficult to determine. However, disputes over functional relationships of this kind are usually over significance and rarely over the degree of simplicity.

The thirty years that have passed since Skinner's viewpoint was first expounded allow attempts to answer two questions that it prompts. How successful has research in psychology guided by these somewhat vague principles been? And how much closer are we now to the point where it is possible to formulate a comprehensive system that will summarize the important features of animal behaviour?

The answer to the first is clear. A whole range of discoveries has been made on the basis of little more than a general drive to explore and analyse closely how animals behave. The second question is more difficult, since the analysis of behaviour still seems to be an enormously complex task. Each step forward to a slightly more complex situation appears to involve fresh problems. For example, the results of studies of situations in which an animal has more than one response alternative have not at the present stage been successfully integrated with the now thoroughly studied case when only a single response is available.

## BASIC CONCEPTS

The first step for a Skinnerian psychologist is to decide on the basic units between which reliable functional relationships may be expected to hold. Skinner explicitly rejects the idea that we should aim to predict the response given the stimulus in any situation to which the organism may be exposed. Such a 'topographical' listing of reflexes is an impossible task given the complexity of the behaviour of even the simplest organism and the variability of even the simplest environment. It is evidently a mistake to embark on such a project. One alternative is to reduce the number of particular instances that one has to deal with by grouping responses in classes with supposedly similar characteristics. Thus we might talk of defensive reflexes or displacement activity. But this is to get the worst of both worlds, for the scheme of classification is more likely to reflect the preconceptions of the theorist than real behavioural distinctions; while, if the number of classes is to be manageable, one is working at such a high level of generality that exactness of prediction in particular cases is virtually impossible. This is one of the more obvious failings of traditional learning theories. Skinner rejects the 'botanizing' of reflexes and advocates radical simplification in order to allow basic relationships to appear and be studied with minimum confusion. It can be argued that Galileo was working on this principle in studying the laws of motion by rolling balls down an inclined plane; though the complexity of the situation studied by the Skinnerian is far greater, and the degree of control is far less, than was the case in early physics.

A Skinnerian, it should be recognized, is not fundamentally interested in the particular situation or the particular organism that he is studying; indeed behaviour is deliberately made as boringly uniform as possible, and the focus of interest is on the identification of the relationships between the response and other events in the environment. This indifference to the individuality of bits of behaviour can be a weakness since, as we shall show, behaviour can often only be understood by taking account of its special character. In general, however, it is hard to escape the logic of the case against 'botanizing', unless one assumes that there are no general factors underlying wide ranges of behaviour, and this is surely a counsel of despair. The strategy of simplification and intensive study over a limited range offers a way forward to the formulation of general laws of behaviour, and, we shall argue, has not so far seriously disappointed expectations.

Skinner was among the first to recognize the fundamental distinction between classical and instrumental conditioning, and he has always considered that the two cases need to be analysed separately. Conditioned reflexes are primarily concerned with the internal economy of the organism and so have little effect on the animal's environment. When we establish a conditioned reflex the unconditioned stimulus follows the conditioned stimulus regardless of what response the animal makes, and we choose as an unconditioned stimulus some event which reliably produces a response. Skinner describes this as the elicitation of the response by the stimulus, thus emphasizing that the response is the inevitable consequence of the stimulus, that it is stimulus bound. This type of response is called a respondent and is of major importance, particularly in understanding motivation and emotion, but Skinner has been almost entirely concerned with the analysis of operant conditioning. In instrumental learning, some of the animal's responses have effects on the environment and may produce changes detectable by the animal. An operant is a class of such responses defined, not by their topographical characteristics, but by their effects on the environment. Thus all the responses which result in the depression of a bar and the closure of a microswitch may be regarded as examples of the same operant although they may differ widely in topography. It is therefore possible to specify a large number of possible operants in any experimental situation and we are free to study any of them that we like. We can treat general activity as an operant (Graf & Bitterman, 1963), or we can demand a bar press of a specific force (Notterman & Mintz, 1965) and in either case we can apply the same analysis. In practice the operant is usually defined at some intermediate level of generality for the sake of experimental convenience and because it is found that this gives smooth and reliable functional relationships. The problems raised by the definition of an operant and the grain of analysis of behaviour will recur during our discussion.

An operant is most conveniently distinguished from a respondent by the conditions under which it occurs. A respondent can be reliably elicited by a particular stimulus; an operant is spontaneous and there need be no stimulus which regularly precedes its occurrence. Of course this does not mean that there are no causal antecedents of an operant, but simply that these antecedents are to be found largely within the organism and not in the environment. An operant is therefore said to be 'emitted', thus emphasizing that the response is not strictly bound to an external stimulus (though it may come under stimulus control) and underlining the 'spontaneity' of the behaviour. The distinction is obviously closely related to the everyday antithesis between voluntary and involuntary responses. Spontaneity should not be equated with randomness. In one of Thorndike's puzzle boxes the set of possible responses is very large; in another situation, for example a running wheel, it may be relatively small. But in both cases, the behavioural possibilities are restricted by the environment. In the extreme case an environment may be so arranged that the subject is almost bound to emit the required response. A recent procedure for automatic shaping is a good example (Brown & Jenkins, 1968): pigeons may reliably be induced to peck at the response key in a Skinner box by always illuminating the key for a few seconds before presenting the reinforcement. In this case the emission of the response is so bound to the stimulus situation that it seems that the pigeon is sooner or later sure to emit the response; the distinction between the spontaneous emission and the forced elicitation of a response breaks down. Nevertheless it seems essential to make some distinction of this sort, even if on some occasions this leads to awkward ambiguities about the status of particular responses. In practice there is a second criterion since respondents are typically responses of the glands and smooth musculature while operants are produced by the striated muscles. Recent evidence (Trowill, 1967) suggests that this rule, too, has exceptions and that responses of the smooth musculature may, under special conditions, follow the laws of operant conditioning. This is a most important and interesting finding, but it in no way invalidates the distinction between the two types of conditioning. It simply shows that some responses may be conditioned in either way.[1]

Although there are difficulties of this sort, it is undoubtedly true that much behaviour has the 'spontaneous' quality described by Skinner and it will not help our understanding to shut our eyes to this. The first theories of learning tried and failed to tie each response to an identifiable external stimulus; later theorists maintained *S–R* orthodoxy by introducing covert responses with external and unobservable stimulus consequences. The resulting systems were ingenious, but the intervening variables were so numerous and their interrelationships so poorly specified that prediction, as opposed to *post hoc* explanation, was well-nigh impossible. Instead we should acknowledge the impossibility of tying all responses to particular stimulus antecedents and look elsewhere for the fundamental regularities in behaviour. This approach raises problems about the measurement of response strength since, in a free responding situation, where there is no specific stimulus to which the response is related, it is not possible to use probability or latency of response. The solution has been to use rate as the basic measure of the response, and this is usually taken to be a necessary feature of the Skinnerian position. It seems to us that using rate as a measure is by no means as fundamental as has commonly been thought, and we shall discuss this point later.

While we cannot reliably control operant behaviour by manipulating its antecedents, we can control it very effectively through its consequences, and this is the core of the Skinnerian position. This control is achieved by adjusting the relationships between responses and reward or punishment; in the accepted terminology these relationships are called the contingencies of reinforcement. A reinforcer is operationally defined in terms of its effects on preceding responses. Presentation of a positive reinforcer increases the probability of emission of the immediately preceding operant. A negative reinforcer is defined as an event whose termination increases this probability. In principle, therefore, we decide whether a stimulus is a positive reinforcer by considering the effects of one event, presentation of the stimulus, while if we want to know whether it is a negative reinforcer we have to consider a different event, withdrawal of the stimulus. In either

[1] A fuller discussion of this interesting question is to be found in Rescorla Solomon (1967).

case, we use an identical criterion, an increase in the probability of the preceding response, to determine whether the event is reinforcing.[1] In this respect, therefore, the termination of a negative reinforcer is identical by definition to the presentation of a positive reinforcer, but only in this respect. In other ways their effects on behaviour may be totally different; this is a matter not of definition but of empirical investigation. These operational definitions identify without ambiguity two classes of stimuli (positive and negative reinforcers) which are of great importance in the analysis of behaviour. It also appears that withdrawal of a positive rein- forcer reduces rate of response, as does presentation of a negative reinforcer (e.g. Ferster, 1958). Whether or not the symmetry of effects is complete is a matter for research, and an answer is not yet forthcoming. The opera- tional definition of reinforcement has enormous advantages; in particular it allows one to tackle questions about the role of reinforcement without disagreement about what is meant by a reinforcer or whether a reinforce- ment has occurred. For example, much of the controversy over latent learning arose out of the question 'Is reinforcement occurring in this situation?', where reinforcement was defined in terms of unobservable non-behavioural events such as drive reduction; experience has shown that this is not a very fruitful form in which to pose the question. Using the operational definition, it is possible to decide from the changes in the probability of the response whether an event was reinforcing, and then to go on to the much more productive question 'Do all reinforcers have particular properties, drive reduction, for instance, in common?'.

Practical difficulties could arise over this sort of definition since we can only identify a reinforcer by reference to its effect on an operant; so the only way of being sure that a stimulus is a reinforcer in a particular situa- tion is to try it out in that situation. Happily it turns out that almost all stimuli which are reinforcers in one situation are reinforcers in a wide variety of situations; so these difficulties which are logically possible are not a practical problem. Skinnerians, who are typically interested in the effects of schedules of reinforcement rather than in the reinforcement itself, naturally confine themselves to a very small set of highly reliable rein- forcers; outside this set things may be more complicated but even here there appears to be no better definition of reinforcement.

For the sake of simplicity we have been disregarding the stimuli that are present at the time of emission of the operant; but we must now take them into account. In a normal environment reinforcement will follow a response

[1] Azrin & Holz (1966) prefer to define negative reinforcers by the effects of their onset (i.e. in a punishment situation) and this leads to a slightly different formulation of reinforcers.

under some conditions but not under others, and these conditions will often be signalled by changes in the environment. In this case the stimuli correlated with the availability of reinforcement come to control the emission of the operant, which will eventually occur only in the presence of these stimuli. The stimulus thus completes Skinner's three-term contingency of stimulus, response and reinforcement.

Even when a stimulus reliably controls an operant it cannot be said to elicit that operant, it is merely the occasion for the emission of the response. The distinction is a fine one and is sometimes difficult to make, but it is a pronounced feature of Skinner's approach. In rather unusual situations where autonomic responses are treated as operants in a discrimination learning situation (Miller & DiCara, 1967) the distinction can be made only in terms of the history of the response.

### THE GENERALITY OF THE CONCEPTUAL SCHEME

This basic conceptual scheme works extremely well for the prediction and control of behaviour in standard operant situations. For the analysis to apply to anything other than rats bar-pressing and pigeons key-pecking we must assume that these situations are representative ones in which fundamental relations are displayed in a fairly uncontaminated form. The variety of species is so great and the differences in their behaviour so overwhelming that it seems presumptious to suppose that findings derived from two particular species can have any generality. But the generality that we are hoping to find is not to be seen in the infinitely varied topography of responses but rather in basic interrelationships between these responses and other events. A large number of species have now been tested in operant situations (though they represent a minute but fairly representative proportion of all vertebrate species). So far it appears that all vertebrates whose behaviour is at all flexible do show the phenomena of operant conditioning in recognizably the same form as the rat or the pigeon. More than this, the detailed relationships prove to be very similar in widely different species. This is particularly true of the effect of different schedules on rate of responding; in some cases the cross-species similarities are astonishing, as in the well-known example of three cumulative records exhibiting the effects of a multiple fixed-ratio, fixed-interval schedule (Skinner, 1959, p. 374) which are so alike that they might have been produced by a single animal, yet one was produced by a rat, one by a pigeon and one by a monkey. In this example Skinner declines to say which animal produced which curve, and it is in a sense irrelevant; what is important is the similarity of the behaviour. This sort of evidence is impressive; as predicted, operant techniques do reveal a high degree of regularity in the behaviour of widely varying species. This shows that the effects are general and must be of importance in understanding behaviour, but it does not go far towards accounting for the effects. It has been argued, notably by Bitterman (1960), that the best way to understand behaviour is to look for the *differences* between species and to use the correlation between phylogenetic differences as a means to understanding the organization of behaviour. This is an ambitious programme whose results do not yet justify the claims made for it; but a more important criticism is the typically Skinnerian one that we are unlikely to be able to deal with the differences between species in some aspect of behaviour until we understand this behaviour in at least one species. Thus, to compare the effect of partial reinforcement on extinction in a variety of species may be an interesting exercise, but it is hardly likely to increase our understanding of behaviour unless we have a fairly full understanding of the effect in at least one species. Nevertheless, the incentive to investigate certain types of behaviour has sometimes come from noticing differences between species, and we should not shut our eyes to this source of information. But it seems advisable to start by investigating the cases in which a variety of species behave in the same way, since differences between species are likely to be best understood against a background of similarity.

While the Skinnerian policy of using one or two 'representative' species does not raise too many difficulties, interesting problems do arise from the assumption that any stimulus, response or reinforcer can be taken as 'representative'. Not all an organism's responses are necessarily equivalent with respect to reinforcement, and this is an area in which differences between species may well turn out to be instructive. For example, it has been found very difficult to train pigeons to peck a key in order to avoid an electric shock (Hoffman & Fleshler, 1959), but it has recently been shown that pigeons will readily avoid in a shuttle box (MacPhail, 1968). The two responses of shuttling and key-pecking are clearly not equivalent with respect to negative reinforcement; and the reason is not far to seek once attention has been directed to the individuality of the response. Attempts to increase rate of key-pecking in a situation in which electric shock is being delivered are working against certain basic features of the organization of behaviour. Running from one box into another suffers from none of these disadvantages, and so the pigeon's capacity to avoid is easily revealed in such an apparatus. Breland & Breland (1966) report that a pigeon can be induced to pull a loop of string for food reinforcement only after prolonged shaping, though the response is clearly within its repertoire; a hen can easily be trained to peck a key or pull the string for food. The differences

355

are attributed to the different innate feeding patterns in the two species. These authors offer a range of other examples of the misbehaviour of organisms and emphasize the importance of considering how animals 'earn their living' in the natural state if we are fully to understand their behaviour in the laboratory; this must be true to some extent of any operant that we may investigate.

Skinnerians also work on the assumption that all discriminable stimuli can become discriminative stimuli. This also appears dubious. Konorski (1967) reports that dogs will easily learn a go–no-go discrimination when the stimuli are two tones of different frequency but that they have the greatest difficulty in learning a left–right discrimination with the same stimuli. He also reports a number of other more complicated examples in which there appear to be innate connections between particular stimuli and particular responses. Jenkins (1968) has found that if pigeons are taught a discrimination using two stimuli, one of which differs from the other by a presence of an additional feature, learning is much faster when the latter is used as the positive rather than as the negative stimulus. It is well established that avoidance learning is usually faster with auditory stimuli than with visual ones (Smith, McFarland & Taylor, 1961), and that a buzzer is more effective than a pure tone although differences in intensity of the stimulus make relatively little difference (Myers, 1962). The point is perhaps a fairly obvious one that need not be laboured, but it is liable to be forgotten.

Reinforcement also may be effective for one response and not for another. Bolles & Seelbach (1964) used a loud noise as a punishment for rats; they found that it was effective in eliminating the response of poking the head through a window but that it had no effect on the frequency of other responses such as standing on the hind legs or grooming. Similarly it seems that food reinforcement is not a very effective reinforcer for vocalization in a number of species, though vocalization can be reinforced in other ways (see e.g. Warren, 1965).

Examples such as these show that, while it may be possible to develop general principles that apply over a wide range of behaviour, we ignore the individuality of a particular situation at our peril. There is no such thing as a completely typical response, any more than there is a typical stimulus, reinforcer or species. But to say this is not to undermine the whole Skinnerian edifice. In the cases given above results failed to generalize to situations to which they might reasonably have been expected to do so, but no relationships have been found which are different from those found in typical operant situations. Although every situation has its own peculiarities, experience has shown that it is possible to choose paradigm cases where the functional relations between stimulus, response and reinforcement are clearly exhibited and easily analysed. The results of this analysis should be applied to new situations with more caution than has sometimes been shown, but with this proviso the Skinnerian programme emerges unscathed.

We have already stressed Skinner's emphasis on finding appropriate response units. So long as one is considering the responses emitted in a standard Skinnerian situation the definition of the operant in terms of its effects raises few problems. When, however, a Skinnerian analysis is applied to experiments carried out in other types of apparatus serious misunderstanding may arise. In a maze it is not obvious how to define the operant (or operants) that result in an animal arriving in the right-hand goal box; are they to be defined by arrival at the goal box, by turning into the arm, by making a right turn, or by approaching the extra-maze stimuli in the region of the goal box? According to which answer we choose we shall get very different results in a variety of experiments. It was partly because of the difficulties of response definition in conventional experimental situations that Skinner developed forms of apparatus in which there is little ambiguity about the definition of a response.

In an earlier section the 'place vs response' controversy was quoted as an example of unprofitable research generated by theoretical issues. This is also a case in which a Skinnerian analysis clarifies what was essentially a pseudo-problem. Though the basic study of operant behaviour should be undertaken in situations where this sort of uncertainty does not arise, the phenomena that occur in other situations cannot be forgotten simply because they do not occur in a Skinner box; nor can they be satisfactorily dealt with on a *post hoc* basis by a suitable definition of the relevant operant. For instance, Deutsch (1960) has shown that a rat's capacity to learn a simple T-maze is profoundly influenced by whether there are two goal boxes or the arms come together in a single goal box; we are not concerned with the theoretical ideas underlying such an experiment but only with the behavioural facts. This kind of phenomenon, which is not well understood at present, cannot be ignored in a comprehensive analysis of behaviour; it is not clear that the use of a Skinner box would lead to the recognition of such problems nor that the present form of behavioural analysis will be helpful in elucidating them.

## PHYSIOLOGICAL PSYCHOLOGY

Physiological knowledge is treated by psychologists with the greatest respect. Many psychologists share a deep-rooted belief that, however much psychological information they may gather about a particular problem, the question will only be completely answered when a physiological explanation is available. Skinner is the strongest critic of this belief, that the final explanations of behaviour are to be found in physiology, whether it is expressed in everyday language or in a sophisticated form by psychologists such as Hebb.

If the aim of a science of behaviour is to establish functional relations between behavioural events then there is no room in its language for concepts drawn from another universe of discourse. A physiological explanation of a piece of behaviour is not an *improvement* on a full behavioural explanation; it is an explanation in terms of concepts which are, in a sense, more fundamental but which may be totally inappropriate to the explanation of behaviour. To describe the movement of a limb in terms of the full physiological mechanisms involved is obviously a fuller explanation than to say that the animal depressed a bar, but such a description is totally unsuitable as a basis for understanding bar-pressing behaviour. In the same way a biochemical explanation of nervous action is not a better explanation than the equivalent physiological one; which one we use depends on what we are doing and what we are interested in. Skinner is interested in behaviour and so can see no advantage in adopting physiological explanations for events which can be dealt with adequately in purely behavioural terms. Indeed there may be serious disadvantages since attention is likely to be diverted from the search for behavioural explanations.

It is plainly a mistake to attempt to establish a *cordon sanitaire* around psychology against the dangerous infection of physiology. In many areas of experimental psychology the two sciences advance best by close cooperation, and the contact between the two is likely to increase. The application of physiological knowledge of the solution of psychological problems cannot in itself be unsound; but this does not justify particular cases of 'physiologizing'. It is instructive to consider the study of perception, where physiological techniques have produced striking increases in understanding. If Skinner's argument was of general validity the proper approach should be to establish a purely behavioural science of perception, based on the verbal and motor behaviour of observers when faced with particular stimuli; only then should physiological information be collected and only then would it be possible to comprehend it. This analysis is obviously mistaken, but it may be that, though his logic is faulty, Skinner's conclusion is correct, at least where the application of physiology to the study of operant behaviour is concerned. There are at least two important differences between the use of physiology in perception and learning. First, when we are studying the reception and analysis of sensory stimuli we usually have anatomical knowledge which enables us to direct our attention to a restricted part of the nervous system. Thus we can make our lesions and place our electrodes in tissues which are very likely to be primarily associated with the phenomena in which we are interested. It seems likely that one of the reasons that we have relatively little physiological information about olfaction for instance, is that the anatomy of the olfactory pathways is complex and poorly understood. In the case of learning we cannot assume that there is any localization and we may reasonably look for effects in any part of the brain. This has the double effect of making it more difficult to know where to start looking for neurophysiological effects of learning, and also to recognize such effects when we do find them. Secondly, there is a very great difference between the kinds of event that we are looking for in the two cases. In studying perception the physiologist presents a stimulus and looks for changes in the physiological activity of the system that he is studying. These changes may occur in any of a large variety of variables, but essentially they are all likely to represent a fairly transient change in a resting level of activity. If this change can be reliably correlated with some class of presented stimuli and is specific to this class, there is good evidence that the event is associated with the reception and analysis of these stimuli; and the investigation can proceed on that basis. With learning the case is quite different. It is of course easy to record any number of neurophysiological events which occur in a variety of learning situations, but it is very hard to be sure that the events are not simply side-effects of the reception of the stimulus or the production of the response? Also, the event for which we are looking is not a simple change in activity, rather it would seem to be a change in pattern of responding at some indeterminate point in the nervous system; but what sort of change and what pattern? Here we return to the Skinnerian argument, for it seems unlikely that we are going to be able to make sense of these complex physiological events until we have a clear understanding of the behaviour which they are supposed to produce. Time and again detailed behavioural analysis has thrown serious doubt on the interpretation of physiological findings; three recent examples—the effect of RNA on handedness, ECS and retention and the effects of puromycin on learning and memory—have recently been reviewed by Booth (1967).

In this context it is interesting to consider the contribution of physiology to the understanding of hunger and thirst. The anatomical problems are less serious than they are for learning, but those of detailed behavioural analysis remain just as acute. For example, though it was originally assumed that lesions that produced hyperphagia simply made the rat more hungry, it now seems that the effect is behaviourally far more complex than this.

Hyperphagic rats do not seem to be as willing to make an effort to get food as normal rats and will not sustain as high ratio schedules; similarly, if the palatibility of the food is varied the hyperphagic animal appears more sensitive to these changes than the normal rat. Clearly the physiological work on hyperphagia has added to our understanding of hunger, but it is also clear that if there had been fuller behavioural information and an adequate battery of tests, then a lot of misunderstandings could have been avoided. It seems that physiological investigation of this sort of problem is only likely to give information which is of interest to the psychologist when a fairly detailed behavioural analysis has already been completed. However, it must be admitted that psychologists have often been stimulated to carry out these analyses by the behavioural naivety of their physiological colleagues.

These arguments apply with even greater force to those areas of behaviour in which Skinner is mainly interested. It is certainly premature to attempt to reach physiological explanations of the phenomena of operant behaviour when the behavioural analysis itself is so far from complete. Physiological psychologists have made extensive use of lesions in their attempts to study this area of behaviour, and this technique raises a number of special problems. It is worth noting that in the study of perception, where physiological psychology has had its greatest success, lesions have not proved a very useful source of information. Evidently if we are in the position to study the actual working of a system this is preferable to investigating the effects of damaging it. In dealing with more complex behaviour the former course does not seem to be open and so physiologists have fallen back on the latter.

It is now very generally accepted that we cannot argue directly from the fact that a lesion produces a certain effect on behaviour to any direct relationship between the function of that part of the brain and the behaviour in question. To put it at its crudest: if we remove a part of the brain and find that a rat no longer presses a bar for food this may be because it is blind, or paralysed, or no longer hungry, or just confused; it does not show that we have abolished a memory. Few physiological studies fall into this sort of absurdity, but equally the true strength of the argument against lesions is not generally appreciated. The brain is a mechanism of almost unimaginable complexity consisting, it appears, of a large number of closely coupled sub-systems, the activity of any one of which profoundly affects the activity of a number of others. The system exhibits great plasticity and flexibility; it is also probably highly redundant in view of the unreliable components from which it is made up. The behavioural output is serial. since an organism is only doing one, or at most a very small number of things at a time;[1] any piece of behaviour is thus a function of the combined working of all these sub-systems. If we remove a component from a machine of this sort it is not likely that the effect will be to cut out some simple feature of the output. Mechanisms with serial output and this degree of complexity do not in general show any straightforward mapping from physical components to simple elements of the output. If we remove a string from a piano we interfere with a discrete feature of the output, but if we remove a transistor from a computer the changes in output will not reflect in any straightforward way the function of the component that we have removed. The same argument may be expected to apply with far greater force to a system as complicated as the brain. Of course when we remove a portion of the brain, however careful we are, we shall not be removing a simple discrete component; rather we shall be removing some components completely, causing others to function abnormally and cutting the connections between still other components. This can only make it even more difficult to reach useful conclusions about the normal functioning of the brain from studying the effects of lesions.

Thus it appears that while Skinner himself has treated the case against physiologizing as a strictly logical one, it is perhaps equally appropriate to stress the practical objections. Psychology is a science in its own right with its own data language, and capable of producing explanations without benefit of physiological help. There is no reason, in principle, why psychology and physiology should not advance better by co-operating, and in many areas such co-operation is evidently a necessity for the physiologist. On the other hand the application of physiological knowledge to areas of psychology cannot precede an adequate behavioural analysis if it is not to be misleading.

### METHODS, STATISTICS AND RESPONSE MEASURES

One of the more reliable indications of Skinner's influence on a particular piece of research is the number of subjects that the investigator has used. If it has reached double figures, then it is highly likely that the author disagrees with many of the arguments in *The Behaviour of Organisms*. The most direct implication of Skinner's approach to scientific enquiry is that rather than test theories about the average rate one should study the behaviour of individual rats. This means that one should endeavour in every way possible to obtain an exact account of changes in the environ-

[1] This is to omit from consideration the innumerable regulatory functions which must be continually carried out in order that the organism may survive. But these functions are not usually of primary importance for the understanding of behaviour in the context in which it is being discussed here.

ment of each individual subject and also to use measures of responding that are sufficiently informative about the subject's behaviour.

Knowledge about the environment of an organism is not obtained by noting down everything that happens to occur, but by controlling the situation so that the only changes are those initiated by the experimenter. The internal environment is kept as constant as possible by maintaining the animal on a fixed feeding schedule and by holding the experimental session at a fixed time of day. Stray sounds outside the experimental chamber are masked by a steady source of noise. Changes in stimuli, including the presentation of reinforcement, are controlled precisely, remotely and, if possible, automatically.

There is nothing controversial about this and it might well be asked whether there is any other way of doing good experiments. The relevant point is the exceptional importance that Skinner attaches to experimental control. While all psychologists might agree that such rigorous methods are an admirable goal, other factors, including economy and convenience, are often allowed to lower the degree of control and thus to increase the experimenter's uncertainty about the subject's environment and behaviour. To compensate for this ignorance—ignorance about when a door was slammed, about what side of a maze the experimenter was standing on a given trial and so forth—large groups of subjects are used and statistical methods are applied to average out all such random effects.

There are two separate issues here. One is that sophisticated statistical techniques do not make up for a poor experiment. There are, of course, many areas in psychology where the use of statistics is mandatory. In research involving human subjects it is out of the question to attain the degree of control possible when working with animals and some kind of averaging is necessary to reduce the variability the volunteer brings with him into the laboratory. Nevertheless the researcher may be rather too eager to reach for the statistics book. In certain fields, psychophysics and reaction time studies for example, it is found possible to work with single human subjects and obtain precise and reproducible results. In other fields of human psychology it may well be argued that far too little attention is paid to the behaviour of individual subjects.

The second issue is formally similar to the argument against theory. An experiment in which large numbers of subjects are used in order to determine whether some variable has a statistically significant effect is one that is likely to mask important aspects of the behaviour. The more subjects that are used, the more remote the experimenter becomes from the behaviour of an individual subject. It appears to be all too easy to identify a variable's statistical significance, which reflects the quality of the experiment and the size of the sample used, with the general notion of the significance of that variable in a certain analysis. And Grant (1962) has pointed out the curious paradox that, when significance tests are employed in the conventional way to test theoretical predictions, acceptance of a theory depends on the quality of the related research: 'the larger and more sensitive the experiment, the more likely it will lead to results opposed to the theory'. Thus when theorist and experimenter are the same person, careful research may have mixed advantages.

The need for caution also applies to the use of averaging. An average result does not always represent a typical result. For example, a learning curve may show a smooth and steady increase in the number of correct trials for the average rat and thus suggest that learning is a steady, incremental process; in contrast the results from a single rat usually display a number of sequential patterns which indicate that learning is a much more complex process than one would infer from the averaged data. A number of other examples in very different areas could equally well be cited, since this is a common problem whose importance has been recognized by many psychologists besides Skinner. It is crucial here because in order to discover whether averaged results are representative one needs to know in some detail about the behaviour of individuals and any sloppiness in the experimental conditions will greatly impoverish this knowledge.

The other way in which this knowledge may be impoverished is to use an uninformative response measure. In his paper 'Are Theories of Learning Necessary?' Skinner has discussed the relative merits of various measures. The three that he compares are by far the most commonly used ones: choice—in the sense of which of two or more alternatives occurred at a certain time—latency and rate. The basic objection to choice measures is that they provide too bare a summary of what the subject was doing at a given time. Under certain conditions where the behaviour studied is relatively uniform for a long period of time this may not be important and a choice measure may well be the most appropriate datum. However, for the study of relatively transient processes, and almost any learning situation comes into this category, choice is a very inadequate response measure. A fundamental weakness of most stochastic models of learning is that they rely almost entirely on choice measures. The primary variables in such models is the probability that a subject will choose a certain response on a certain trial. Since this probability cannot be directly observed and since its value is assumed to be changing so that it cannot be estimated from repeated observations on the same individual, it has to be inferred from averages obtained from large groups of subjects. Thus all the difficulties of parameter estimation and of averaging arise.

The objection to latency is on different grounds: it is a measure both of the behaviour of the subject and that of the experimenter. Latency is the time between some stimulus change and a specified response on the part of the subject. The stimulus change is controlled by the experimenter and thus, for example, if a signal light is switched on when the subject's back is turned an abnormally long latency will be recorded. This source of variability in a trial situation can be removed by having the subject himself present the stimulus. An example is provided by an experiment on auditory discrimination in which, following a press on one lever, one of two tones is sounded; reinforcement follows a response on a second lever when one tone but not when the other is present. What is measured is no longer a latency, but the interval between two responses. If instead of timing the interval between two responses one counts the number of responses occurring in a given time to obtain the rate of responding, one is in effect obtaining the arithmetic mean of the inter-response intervals.

Rate is the third of the response measures considered by Skinner and is, of course, very closely associated with Skinnerian studies of behaviour. The reasons for his preference are those given above: that choice measures are too uninformative and latency measures are too variable. However, rates of responding are open to the same kind of criticism that Skinner has applied elsewhere because they, too, involve averaging. It has been increasingly recognized over the last few years that to express the effect of some variable in terms of changes in response rate is inadequate, since a given change may reflect a variety of different patterns of underlying behaviour on the part of the subject. Stimulus generalization studies provide an excellent example of how our understanding of a transient process has derived enormous benefit from the use of response rate as a measure. Yet to obtain a complete analysis of what is happening when, say, a subject, who emits an average of 100 responses each minute in the presence of the positive stimulus, emits only 80 responses each minute to a slightly different stimulus, one needs to know whether this decrement is due to a tendency to pause more frequently, whether it reflects a general deceleration in the response rate, and so on. In other words one requires a full analysis of the intervals between successive responses.

Recording and analysing inter-response times is a great deal more involved, in terms of technical problems and computation, than using average response rate, and this is probably a major factor contributing to the latter's prolonged use. At present two distinct ways of meeting this problem appear to be developing. One is to reduce the magnitude of the task by considering only a small number of such intervals in each trial. An example of this is the situation described above in which a stimulus change is controlled by the first response and reinforcement is in some way contingent on a second response. Thus only one inter-response time is recorded for each trial. The second approach is to allow a large number of responses to occur within a trial, as when rate is used as a datum, and employ a digital computer to record and analyse the inter-response times.

So far this discussion has been couched in general terms and a detailed description of a paradigm experiment in the Skinnerian tradition may help to clarify some of the main points. The problem is to determine the effect of aversive stimulation, electric shock in this example, upon on-going behaviour; the specific questions are whether the shock is contingent upon a response, i.e. acts as a punisher, or not, is a critical factor, and also what relationship holds between the size of the effects and the intensity and duration of the shock.

Two pigeons are first trained to respond at a constant rate on a schedule which makes positive reinforcement available at irregular intervals. Once these rates have become stable the conditions are changed so that the first subject receives a slight electric shock following each response. Since the experiment is controlled automatically it is a simple matter to arrange that each time that the first subject is shocked a shock is also delivered to the second subject. When the rates of responding have again stabilized their values are recorded and the occurrence of shocks is prevented. The rates of responding when the subjects have settled down for a third time can be compared to the initial values to determine whether there has been any change in the base-line response rate and then the whole procedure is repeated with the same two subjects and new set of shock parameters. The design makes it possible to obtain from a single individual the relationship between suppression, in terms of changes in rates of responding, and punishment measured in terms of various parameters of electric shock. The use of a yoked control enables one to compare this with a second function relating suppression to shock which occurs randomly, but with exactly the same frequency distribution as for the first subject.

This kind of design has enabled Skinnerians to study a wide range of phenomena and, though the results have usually been based on a very limited number of subjects, they have been shown to have general validity. A close parallel exists here with Pavlov's work on classical conditioning, where a similar emphasis on experimental control produced results which, though mainly obtained from no more than two or three dogs, have stood the test of time to an amazing degree. However, success usually entails a certain cost.

A brief comparison between the topics studied in research influenced by traditional learning theory and the topics studied in the majority of

Skinnerian experiments reveals what the cost has been in this case. Experimental techniques which rely on individual subjects acting as their own controls can only be applied to reversible processes. If one wants to study the behaviour of an organism under different doses of a drug whose effects are temporary one can use the same subject repeatedly; if, however, one is interested in the speed of action of different concentrations of a lethal poison one needs a large number of subjects. Learning, like death, is irreversible and on the whole is a process about which Skinnerians have said little.

To a large extent the preponderance of studies on reversible effects is a result of the emphasis on intra-subject designs. Another reason appears to be the view that it is just as important to study the steady-state properties of a system as it is to study transient processes. For example, the effect of delay of reinforcement on the terminal behaviour on some schedule may be just as interesting as its effect on acquisition. Outside of the Skinnerian literature one can gain the impression that the experimental study of animal behaviour is largely confined to the study of learning and of motivation; it is clearly an achievement to have redressed this balance. Moreover, in some circumstances it is far more appropriate to use a steady-state procedure; determinations of the distances between stimuli based on the speed of acquisition of a discrimination involving these stimuli is not a good psychophysical method. (This topic is discussed more fully in the next section.)

However, many irreversible or transient phenomena pose extremely important questions which Skinnerians have not attempted to answer. Instead, there has been some tendency to regard a steady-state design as a sufficient justification for an experiment, so that research becomes a matter of adding to an endless list of the effects of $x$ upon terminal behaviour $y$ under schedule $z$ without any attempt at systematization. This tendency is not a necessary consequence of Skinner's views and appears to have become very much weaker in recent years. In addition we should stress that many important irreversible phenomena have been discovered and investigated in the context of Skinnerian research; the peak-shift (Hanson, 1959) and the work of Honig (1968) on attentional factors provide excellent examples.

Certain ideas about the nature of learning and of extinction were proposed in his early work, but Skinner's later writing on the subject has tended to emphasize only the general concept of reinforcement. *The Behaviour of Organisms* contain some brief comments on discrimination learning, whose significance has been recognized in the last few years. One suggestion is that, since the process of discrimination learning is slowed down by the use of intermittent schedules of reinforcement, it can be studied in great detail under such conditions. If a pigeon is first reinforced intermittently in the presence of a square and subsequently a triangle is occasionally introduced, which is not associated with reinforcement, it takes some time before the rate of responding to the triangle subsides to a low and steady value. This form of discrimination learning is the subject of a great deal of current research. Present results suggest the need to admit an additional concept among the select few, that of 'inhibition', or at least something very similar to the concept of inhibition used in the analysis of classical conditioning (Bloomfield, 1968). At the time of *The Behaviour of Organisms* Skinner decided that there were insufficient grounds for using such a concept in the analysis of operant behaviour.

The study of irreversible processes does not allow intra-subject experimental designs, but this does not mean that the importance of experimental control and the study of individual subjects can be neglected. An experiment by Siegel (1968) demonstrates the benefits to be obtained by looking carefully at the behaviour of individual subjects, even when studying discrimination learning in groups of animals in a very non-Skinnerian situation.

## APPLICATIONS

One common justification proposed for Skinner's analysis of behaviour is the pragmatic one: that it works. The basic conceptual framework and the attendant techniques have been applied to a very large range of problems in a way that is unique for research in animal behaviour. The initial success of these applications has been much greater in some spheres than others; but it seems fair to say that, even where the results of an operant analysis have not yet matched the original optimism, the exercise has always been very worthwhile; if only because it has made necessary the detailed examination of a problem and demanded that preconceived ideas be made explicit.

At the beginning of *Science and Human Behaviour* Skinner points out that people are in general very reluctant to accept the consequences of the widely held belief that a major problem in Western society is the gap between our understanding of the physical world and our understanding of human nature. For Skinner this gap can only be narrowed if the methods of science, so successful in physics, are applied to the study of human behaviour. He has indeed practised what he preached in this book and the bulk of his recent work has been concerned with various critical aspects of society. His interests have included education, mental illness and the design of societies. The range of his enquiry and his general impact have been greater than any experimental psychologist since Watson.

We do not propose to examine these wider aspects of the Skinnerian approach in this chapter. In connection with education the relative merits of different views on how to program teaching machines is a matter of controversy and it is probably wise at this stage only to claim for an operant analysis the merit of getting the whole movement going. Also, we do not feel qualified to judge the practical results of the encroachment of an analysis of animal behaviour into the formerly exclusive preserves of clinical psychologists and psychiatrists. Instead, we shall examine a more limited and academic area where the successful application of operant conditioning has been widely proclaimed.

One such application has been to the study of sensory processes in animals; the result has been that there is now a thriving area of study, termed 'animal psychophysics', where before there was scarcely any research whatsoever. A well-known example of such studies is that of Blough (1955) who used operant conditioning methods to obtain from pigeons a continuous record of changes in visual thresholds with dark adaptation. An equally striking study is that by Scott & Powell (1963) on visual after-effects in monkeys. The subjects were trained to press a right-hand lever to an expanding circle and a left-hand lever to a contracting circle; during this training stage a stationary spiral was projected around the fixation point. During the testing stage the spiral was rotated prior to the presentation of the circle. It was found that the percentage of left responses increased or decreased depending on the direction of rotation of the spiral. This is just one of a number of studies that have both demonstrated and carefully measured a perceptual effect in animals.

There are essentially two kinds of psychophysical research, which for present purposes will be termed 'objective' and 'subjective'. In the objective kind the experimenter knows whether a given response is 'correct'. If we are interested in determining difference thresholds for the detection of a spot of light against an illuminated background there is no problem about defining a correct response. An animal can be trained in a Y-maze, for example, to enter the arm whose far end has a spot projected upon it and to avoid the arm whose end is uniformly illuminated. The intensity of the spot may then be changed and a difference threshold obtained by one of the standard psychophysical procedures. Similarly the ability of rats to discriminate between visual patterns can be studied by training the subject to approach one set of shapes but not to approach a second set. In this type of study there is no problem about scheduling reinforcement; if 'approaching the spot' has been defined as the 'correct' response we can continue to reinforce this response even when the discrimination has become so difficult that the stimulus differences no longer control the animal's behaviour.

In a study of the subjective kind there is no way of defining a correct response and so it is difficult to decide when to reinforce the subject. Consider the experiment on visual movement after-effects described above. During the test procedure there is no way of deciding whether a left or right response is correct, since the experimenter does not know what the animal is seeing. Of course it would have been possible to continue to reinforce left responses only when the circle was *physically* contracting and disregard the prior rotation of the spiral. Yet this would defeat the purpose of the experiment and change it to a procedure for training monkeys to disregard perceptual after-effects. There is a wide range of psychophysical topics of this type, which includes any study of scaling or subjective equality.

It is perfectly feasible to obtain precise results on an objective problem by using equipment and techniques that differ little from those used by Yerkes (e.g. Muntz & Sokol, 1967). Furthermore, for certain problems a Skinner box is by no means a useful piece of equipment; the second example of an objective study given above, that of pattern discrimination in rats, is a case in which a simple, old-fashioned item (in this instance, the Lashley jumping stand) has no automated rival (Sutherland, 1961).

The main contribution of Skinnerian methods and analysis has been to the second form of psychophysics. There appear to be three reasons why the Skinner box has been very successful in the study of subjective problems. The first is the degree of control over the subject's behaviour that can be achieved. Blough's experiment on dark adaptation is a dramatic demonstration of the effective use of stimulus control and reinforcement contingencies in shaping complex sequences of behaviour. The second lies in the use of intermittent schedules of reinforcement to establish sufficient resistance to extinction; then the problem of how to reinforce 'correct' responses no longer occurs, since reinforcement can be withheld without disrupting performance. The last reason is that it permits the use of automatic control and recording; thus the precision and amount of data collection required by most psychophysical studies can be very easily handled.

The first two reasons clearly support Skinner's claim that his analysis has conferred tremendous practical benefits. The third, the use of automatic methods, raises some intriguing questions. Within psychology Skinner was a very early prophet of the age of automation and one of the first to realize the enormous potential of electro-mechanical relays and switching circuits in behavioural research. How much of the success of his analysis and the belated impact of *The Behaviour of Organisms* has depended on this technical insight? If Skinner had pursued his original literary ambitions and the development of automated methods had been in other hands, would one have now the same analysis but under another name? Such questions are more amusing to ask than to answer. On the whole we are inclined to the view that technical skills have been of great benefit to Skinnerians, but that the development of automated methods in a variety of other hands would probably have led to decades of muddle.

## THE EXPLANATION OF HUMAN BEHAVIOUR

We have left until last the most provocative part of Skinner's work, namely the manner in which he makes confident assertions about human behaviour on the basis of direct extrapolations from the laboratory rat or pigeon. The uneasiness that these extensions evoke appears to be based on more than a lingering resistance to the idea of continuity between animals and man; after all, analogies between the behaviour of animals in the wild and various aspects of human affairs appear to be enjoying considerable popularity at present. A partial explanation may well be that, while we do not mind very much when our sexual habits and aggressions are compared to those of baboons or even tropical fish, direct comparisons between what we feel are purposeful acts and the key-pecking of pigeons are too close to the bone.

Skinner's explanation of purpose in human behaviour is very similar to Darwin's explanation of purpose in evolution (Herrnstein, 1964). Responses appear to be uncannily directed towards some goal, because those responses have been most successful in reaching that goal in the past, while less successful responses have become extinct. This kind of account is not a truism which follows immediately from the definition of reinforcement. The classes of events that are reinforcing for human beings can be discovered empirically just as they can for animals. There is no reason to suppose that various intermittent schedules immediately lose their properties at the human level; thus the fact that a particular instance of what we term a 'purposive' act may occur frequently, without being followed by an obvious reinforcement, may be very adequately analysed in terms of a history of very intermittent reinforcement. There are now a number of studies showing that the performance of human subjects on certain schedules differs very little from that of animal subjects (e.g. Baron & Kaufmann, 1966). Furthermore, the analysis of reinforcement depends only on temporal contiguity and not on any causal link between the response and the reinforcing event that follows it. Thus responses may be sustained by 'accidental' reinforcement; there is much to be said for labelling work with this kind of situation 'studies of superstitious behaviour' (Herrnstein, 1966).

Much of the criticism of this account is like that of a contemporary of Galileo pointing out in derisive fashion that in the real world lead weights do fall more quickly than feathers. Everyday behaviour is determined by an immense number of interacting factors, which make the rate of descent of a feather a trivial problem in comparison. The fact that behaviour outside the confines of a laboratory appears to be of a different order of complexity to that within does not mean that a different set of principles holds for each situation.

However, there are very basic objections to the kind of extrapolation that Skinner makes. The four most important ones are discussed briefly below and at greater length—though accompanied by a blind eye towards the strong points of Skinner's position—in Chomsky's (1959) review of *Verbal Behaviour*.

The first point concerns the definition of units. Whereas Skinner's initial work on animal behaviour concentrated on this problem and stressed the need for empirical solutions, his explanations of human behaviour are framed in terms of units defined on intuitive grounds. This is particularly crucial in the analysis of language behaviour and Chomsky's major argument is that the solution of this problem must precede anything else. This criticism is really part of a general one against the abandonment of experimental enquiry for speculation. For this reason the present section of this chapter takes the form of the assertion of opinions rather than the assessment of evidence.

Secondly, the extrapolations are made from rather limited aspects of the behaviour of rats and pigeons. As we have seen above, preference for situations that allow refined methods of experimental control has influenced the kind of problem that Skinnerians have tackled. Thus, for example, almost every experiment is performed upon adult animals and the extrapolations to learning in human childhood are not based on comparative developmental evidence. If one is interested in the way that language is acquired and wishes to compare this with the behaviour of birds, then the acquisition of bird-song seems as good a place to start as the control of pecking in pigeons.

The final two points concern the analysis of purpose described above and are illustrated by the following quotation from *Science and Human Behaviour*: 'Instead of saying that a man behaves because of the consequences which *are* to follow his behaviour, we simply say that he behaves because of the consequences which *have* followed similar behaviour in the past. This is, of course, the Law of Effect or operant conditioning.'

One thing that is wrong with this account is that men sometimes behave in a way that is appropriate with respect to its consequences, but is not similar to any previous behaviour. It is possible to bribe a man to do something that he has never done in his life before. Chomsky has pointed out that this 'creative' aspect is a fundamental property of human language —we are continually speaking and understanding novel sentences—and that studies of animal behaviour have nothing to say on the subject.

Chomsky has a second objection to the kind of quotation that has been given above and in this case the criticism is to a large extent misdirected. For him the whole process of re-wording colloquial descriptions of behaviour in terms of Skinnerian concepts is completely unproductive. This is clearly mistaken, since if there were very strong grounds for making identifications of the kind 'consequences equal reinforcement' experimental studies of the properties of reinforcement would be a very important source of insight into human behaviour. It seems to us that such identifications can be made very profitably in many cases and that the explanatory power that this confers is not just confined to oddities like nervous tics. A great deal of care is required in this process. One difficulty in identifying 'consequences' with 'reinforcement' is the creativity problem. A second stems from the fact that reinforcing events have their most powerful effect on the responses that *immediately* precede them. The behaviour of animals is very insensitive to reinforcement when even a relatively short delay is interposed between a response and the reinforcing event that is contingent upon it. A very important problem in human behaviour arises when consequences are not immediate and there is no apparent chain of intermediate responses and secondary reinforcement to bridge the gap.

CONCLUSION

The point of view that we have put forward in this chapter does not agree in every way with what Skinner himself has said. A doctrinaire Skinnerian might refuse to recognize it as an exposition of his creed. In places we have been critical of Skinner's position and in others we have suggested substantial modifications or expressed strong reservations. We have concentrated upon the points of dispute because Skinner himself has presented his case in such a readable and lucid manner that there is little that we can say to add to it. But although we have been somewhat eclectic, we believe that the line we have taken is essentially Skinnerian and that it includes all the really crucial features of Skinner's position. These essential characteristics are a distrust of theorizing or physiologizing in advance of the behavioural facts, an emphasis on experimental control, and confidence in a science based purely on behavioural data. The fundamental functional units on which this analysis is based are operationally defined; the spontaneity of operant behaviour is emphasized as is the crucial role of contingencies of reinforcement. This type of analysis has been remarkably successful and offers, we believe, by far the most promising approach to the understanding of behaviour.

## REFERENCES

Azrin, N. H. & Holz, W. C. 1966. Punishment. *Operant Behaviour: Areas of Research and Application.* Ed. W. K. Honig. New York.

Baron, A. & Kaufman, A. 1966. Human free-operant avoidance of 'time out' from monetary reinforcement. *J. exp. Analysis Behav.* **9**, 557–65.

Bitterman, M. E. 1960. Toward a comparative psychology of learning. *Am. Psychol.* **15**, 704–12.

Bloomfield, T. M. 1968. Discrimination learning in animals: an analysis through side-effects. *Nature* **217**, 929.

Blough, D. S. 1955. Method for tracing dark adaptation in the pigeon. *Science* **121**, 703–4.

Bolles, R. C. & Seelbach, S. E. 1964. Punishing and reinforcing effects of noise onset and termination for different responses. *J. comp. physiol. Psychol.* **58**, 127–31.

Booth, D. A. 1967. Vertebrate brain nucleic acids and memory retention. *Psychol. Bull.* **68**, 149–77.

Breland, K. & Breland, M. 1966. *Animal Behaviour.* New York.

Brown, P. L. & Jenkins, H. M. 1968. Auto-shaping of the pigeon's key-peck. *J. exp. Analysis Behav.* **11**, 1–8.

Chomsky, N. 1959. Review of B. F. Skinner's *Verbal Behaviour. Language* **35**, 26–58.

Chomsky, N. 1965. *Aspects of the Theory of Syntax.* Cambridge.

Deutsch, J. A. 1960. *The Structural Basis of Behaviour.* Cambridge.

Ferster, C. B. 1958. Control of behaviour in chimpanzees and pigeons by time out from positive reinforcement. *Psychol. Monogr.* **72** (whole no. 461).

Graf, V. & Bitterman, M. E. 1963. General activity as instrumental; applications to avoidance training. *J. exp. Analysis Behav.* **6**, 301–5.

Grant, D. A. 1962. Testing the null hypothesis and the strategy and tactics of investigating theoretical models. *Psychol. Rev.* **69**, 54–61.

Hanson, H. M. 1959. Effects of discrimination training on stimulus generalization. *J. exp. Psychol.* **58**, 321–34.

Herrnstein, R. J. 1964. Will. *Proc. Am. phil. Soc.* **108**, 455–8.

Herrnstein, R. J. 1966. Superstition: a corollary of the principles of operant conditioning. *Operant Behaviour: Areas of Research and Application.* Ed. W. K. Honig. New York.

Hoffman, H. S. & Fleshler, M. 1959. Aversive control with the pigeon. *J. exp. Analysis Behav.* **2**, 213–18.

Honig, W. K. 1969. Attentional factors governing the slope of the generalization gradient. *Animal Discrimination Learning.* Eds. R. M. Gilbert and N. S. Sutherland. London.

13-2

Jenkins, H. M. 1968. Discriminative learning with the distinctive feature on positive or negative trials. Technical Report No. 19, McMaster University.

Konorski, J. 1967. *Integrative Activity of the Brain*. Chicago.

Macphail, E. 1968. Avoidance responding in pigeons. *J. exp. Analysis Behav.* **11**, 629–32.

Miller, N. E. & DiCara, L. 1967. Instrumental learning of heart rate changes in curarized rats; shaping, and specfiicity to discriminative stimulus. *J. comp. physiol. Psychol.* **63**, 12–19.

Muntz, W. R. A. & Sokol, S. 1967. Psychophysical thresholds to different wavelengths in light adapted turtles. *Vision Res.* **7**, 729–41.

Myers, A. K. 1962. Effects of CS intensity and quality in avoidance conditioning. *J. comp. physiol. Psychol.* **55**, 57–61.

Notterman, J. M. & Mintz, D. E. 1965. *Dynamics of Response*. New York. Wiley.

Rescorla, R. A. & Solomon, R. L. 1967. Two-process learning theory: relationships between Pavlovian conditioning and instrumental learning. *Psychol. Rev.* **74**, 151–83.

Restle, F. 1957. Discrimination of cues in mazes: a resolution of the 'place *vs* response' question. *Psychol. Rev.* **64**, 217–28.

Scott, T. R. & Powell, D. A. 1963. Measurement of a visual motion after-effect in the rhesus monkey. *Science* **140**, 57–9.

Siegel, S. 1969. Discrimination, overtraining and shift behaviour. *Animal Discrimination Learning*. Eds. R. M. Gilbert and N. S. Sutherland. London.

Skinner, B. F. 1938. *The Behaviour of Organisms*. New York.

Skinner, B. F. 1950. Are theories of learning necessary? *Psychol. Rev.* **57**, 193–216.

Skinner, B. F. 1953. *Science and Human Behaviour*. New York.

Skinner, B. F. 1957. *Verbal Behaviour*. New York.

Skinner, B. F. 1959. A case history in scientific method. *Psychology: A Study of a Science*, vol. 2. Ed. S. Koch. New York.

Smith, O. A., McFarland, W. L. & Taylor, E. 1961. Performance in a shock-avoidance conditioning situation interpreted as pseudoconditioning. *J. comp. physiol. Psychol.* **54**, 154–7.

Sutherland, N. S. 1961. The methods and findings of experiments on the visual discrimination of shape by animals. *Exp. Psychol. Soc. Monogr.* no. 1.

Trowill, J. A. 1967. Instrumental conditioning of the heart rate in the curarized rat. *J. comp. physiol. Psychol.* **63**, 7–11.

Verplanck, W. S. 1954. Burrhus F. Skinner. *Modern Learning Theory*. Eds. W. K. Estes *et al*. New York.

Warren, J. M. 1965. The comparative psychology of learning. *A. Rev. Psychol.* **16**, 95–118.

# COMMENT

## *by* KARL H. PRIBRAM

This chapter on 'Skinnerian Analysis of Behaviour' by R. A. Boakes and M. S. Halliday highlights the occasionally clear and useful as well as the more ubiquitous detrimental prejudices of today's operant behaviourists. The chapter thus provides me with a vehicle for undertaking a delightful journey in criticism on which I have long wanted to embark. My trip is made in two stages. One surveys the larger issues of psychological enquiry and the stance of operant behaviourism within those issues. This topic is handled by the authors in the first and last thirds of their manuscript. In between they address themselves to the specific relationship of operant behaviourism to physiology and to measurement. Let me begin the voyage with these specifics, in part to get them out of the way of the more interesting vista.

In view of the fact that my own investigative endeavour centres on brain–behaviour studies, what I have to say about the Boakes–Halliday statements on the relationship between psychology and physiology may come as a surprise. In general I *agree* with their main propositions: behavioural science can function (though at a limited level) independently of physiology; *physiologizing is* often deleterious to clear thinking; the neurological and the behavioural are two languages which describe universes of events which display considerable overlap. My position is that in order to do justice to this overlap, both neurological (physiological) and behavioural data are demanded—however, one need not necessarily be concerned with the overlap. My prejudice is that one misses a good deal of the fun and richness of the field by ignoring the brain–behaviour interface. For an operant behaviourist perceptual constancy poses little in the way of a problem. For one concerned with the variations of the retinal image and the reconstructions that have to be made from them in order to perceive at all, there is a world of investigating to do and operant techniques can help in the doing as shown by Tom Bower's classic studies on infants (Bower, 1966). But *chacun à son goût.*

I do, however, object to some of the details in the Boakes–Halliday exposition which expose their typical operant behaviourists' ignorance in this field. For instance I do *not* consider Hebb and Deutsch as examples of the 'sophisticated use of neurological language'. Quite the contrary: both have repeatedly and specifically *disavowed* any such claim. Hebb has stated

375

that he is talking about a CNS—a conceptual nervous system—that his language is analogical and derived from behavioural data and *not* from neural. Deutsch has said that his interest is in devising a class of machines that will do the job necessary to explain the behavioural facts—that he is only mildly interested in the actual neural machinery within that class which does indeed do the job.

I also object to the usual mouthings against brain lesion work which has taught us at least as much as any other technique about brain function in behaviour—the neurological language with which to approach the brain–behaviour overlap. True, there are grave limitations to the technique—but then there are to every technique used in isolation: even, I dare say, to that of operant conditioning.

This parochialism of the operant behaviourist is what is so dismaying to encounter. Why, for instance, are references given, with few exceptions, exclusively from the operant literature? When neurological, cognitive psychological or the results obtained by other disciplines are discussed they are almost never referenced. Is the work on the fixed interval schedule really so overridingly precise, holy and important that when hypothalamic hyperphagia is at issue, the details of procedure, analysis and interpretation need not be evidenced?

Again, what makes these behaviourists so sure that physiology's greatest success has been in the area of perception and that none has been achieved in the area of learning? Unit analysis by electrophysiologists, in my opinion, has so far contributed little to our understanding of *perception*. The at-the-moment generally accepted view that a hierarchy of detectors is involved in pattern recognition I have criticized elsewhere. There are alternatives and they are being explored (Pribram, 1969a). But only naivety allows one to consider investigations of the physiology of perception to have been successfully concluded.

Also, only naivety would allow the statement that nothing has *or can* be learned from a physiological approach to learning. Perhaps *the* most important area of neurological enquiry which has been opened during the past decade is that of memory storage. Though controversial as to detail, there can be no question that for the first time changes, anatomical and chemical, are being found in the brain as a consequence of experience (Pribram, 1966). Nor would I discount my own labours which have shown that memory is distributed in the input systems and how remembering is effected by way of a process dependent on the functions of the so-called association systems (Pribram, 1969a).

But so much for physiology. The operant behaviourists' views of the role of measurement are equally confused between an excellent grasp of what measurement means in their discipline and what other disciplines are about when they measure. Certainly response measures of choice, latency and rate are useful and I was delighted to read that at least Boakes and Halliday have loosened the usual operant shackles sufficiently to allow the use of all three and not only the canonized rate. But nonsense does creep in. Why do choice measures any more than others require statistics? Just because mathematical modellers have used statistical sampling models as being the most readily manipulable mathematically? I have had an automated discrimination apparatus for discrete trial analysis (DADTA) working for almost ten years (Pribram *et al.* 1963; Pribram 1969*b*). It is a computer-controlled device which measures response choices, latencies and, if it is so wished, rate. Each response is recorded at the time it is performed and a summary is collated at the end of each run by the computer. If one wishes to perform statistical manipulations on the data one can; if there are obvious conclusions to be drawn without the need for statistics (such as the development or lack thereof of position habits) one draws them and goes on. Further, latency is not to be discarded as completely useless provided one has the wit to use it. Lindsley, for instance, has made this measure pay off in his studies of attention (Lindsley, 1961); my colleagues and I have found latency an effective gauge of distraction (Douglas & Pribram, 1969). I grant the operant behaviourists their fad for inter-response times (I use such histograms in studying neural unit activity) but ask in return that they occasionally look around the world of behavioural science and see what effective use is made of other response measures, yea, even occasionally of statistics and of models. E.g., response operator characteristic curves (ROC) have been extremely useful adjuncts in psychophysical studies and more recently in those involving verbal learning.

ROC analysis might prove equally fruitful when applied in physiological work (as investigators in my laboratory are doing) and in operant learning. Finally, whether one uses one or several subjects depends on how many variables one is juggling, not only on how good is one's control over these variables. If, as in my case, I am interested in the relationship between brain and behaviour variables I do usually need more subjects than one just to be reasonably sure of the generality of my results. Mechanization and statistics are two types of technique by which control can be enhanced —they are neither inimical, exclusive of one another nor infallible.

I want now to venture to the larger issue of psychological enquiry. Operant behaviourism plays two roles in psychological research and these two roles so often become confused in the minds of the practitioners that the appellation of cultism can be fairly levelled against them. Boakes and Halliday unfortunately do not escape this confusion. Operant conditioning encompasses a set of techniques and as such operant conditioners serve as behavioural engineers. In this capacity they have served well, as is admirably detailed by Boakes and Halliday in their discussion of comparative studies. There is here still a tinge of restrictiveness but this is hardly noticeable compared to an incident which occurred not so long ago, when during a meeting which was to lay the foundations for the *Journal of the Experimental Analysis of Behaviour*, someone suggested that perhaps a joining up between operant behaviourism and ethology might prove fruitful: this suggestion elicited a stony silence, some polite chit-chat and the break-up of the meeting for the time being.

Despite their current excellence as behavioural engineers I foresee some dangers for operant behaviourists even here. Already the equipment they use so fluently is completely out-of-date. Unless they catch up with computer technology with its enhanced flexibility through facile hierarchical programming, better and more varied input–output equipment and the like, operant behaviourists are likely to become obsolete, good technicians of an outmoded technology. And by catching up I mean more than using a Linc-8 to analyse inter-response times. With the precipitous drops in the price of general purpose computers which continue from year to year there will remain little excuse for failing to seize the opportunity to use fully these powerful and flexible instruments.

But operant behaviourism purports to be more than a technology. It is also addressed to a set of problems. Yet it is typical of operant behaviourism that, in the Boakes–Halliday treatment of the topic, one must run through two-thirds of their chapter before encountering just what that set of problems is. And when one does, the confusion is woefully compounded. One must continuously work one's way through a blinding array of often brilliant applications of a technology. Animal psychophysics is the stellar example. The application to human behaviour—good even in studying *subjective* [!!!] problems, the authors admit—has proved valuable. But what is the core of interest, the conduct with which operant behaviourism is concerned?

The answer is, of course, the problem of reinforcement. Boakes and Halliday take the usual operant behaviourists' pride in the cleanliness of the operational definition of reinforcement in terms of its effect on preceding responses. But I must say that the high point of my journey came when Boakes and Halliday openly and without shame displayed their scrubbed and sterilized conception with the comment 'Learning is a process about which Skinnerians have said little'!

I might point out to them that they here do a grave disservice to Skinner (as so many Freudians have to Freud, etc.). Only a month ago I heard a superb talk by Skinner (1968) on the occasion of an IBRO meeting on Brain and Human Behaviour sponsored by UNESCO. Skinner discussed his *theory* (his term). He stated that it is *not* an S–R theory. It is, in fact, an R theory and therefore much more defensible since stimulus and response mutually imply each other unless one is talking exclusively about correlations among 'distal', i.e. environmental, events operated upon by the organism. (See Estes (1959) for a detailed discussion of this issue.) He also portrayed his *model* (his term) of reinforcement: he described this as an attempt to so *arrange the contingencies of environmental events that reinforcement can and will occur*. Note that by this approach reinforcement becomes a process internal to the organism, a process which can legitimately be studied by neurological methods.

I was, of course, delighted to hear this since my own interest lies in the neurological process produced by behavioural reinforcement. I have elsewhere tried to make the proposal that behavioural reinforcement is an organizing process which takes place when sequences, i.e. temporal patterns, of behavioural outcomes fit into the neurological context (memory) created by prior such sequences. This view of reinforcement as consequence is dismissed by Boakes and Halliday with a phrase and without discussion so I don't know just why it failed to make the grade with them, unless they mean by a consequence only the stimulus aspect of the event *per se*, or the reference to a central organizing process is so distasteful to them that further mention would make them ill. But now they must deal with Skinner himself on this issue and so I wonder whether he or they would object to my treatment of the problem on another occasion when I suggested that neurologically, reinforcement may well proceed by a brain mechanism not unlike that which produces embryological differentiation, the mechanism of induction (Pribram, in press).

The point is that neuropsychology (by contrast to what Boakes and Halliday say about operant behaviourism) has a great deal to say about learning—both in data and in fascinating possibilities that need exploring. Yet, neuro—or in general—physiological psychology would hardly be able to make its contributions if it paid no heed to those contributions which operant behaviourism has to offer, both technical and intellectual. Perhaps operant behaviourism would have more to say about psychological problems—which are, of course, of a piece and do not care for our arbitrary distinctions—were its practitioners to attend wholeheartedly to the explorations of their non-operant colleagues.

The experimental analysis of behaviour within the framework of operant behaviourism, as is detailed by Boakes and Halliday, has a great deal to offer to students immersed in psychological enquiry. As of the moment, with few exceptions, however (e.g. the work of Premack, 1965), contributions are coming from those who only use operant behaviourism and stand solidly outside it. This is to my mind largely due to the abysmal provincialism and cultivated bigotry of so many who remain within the operant confines. I do *not* advocate any abandonment of operant behaviourism as a scientific enterprise in its own right. Rather I want to see it strengthen its core by admitting the contributions which other disciplines within psychology are making and by usefully incorporating and improving them whenever they are relevant. The problem to which operant behaviourism is addressed is reinforcement. Reinforcement is a central problem in psychology (by whatever name it is called: outcome, consequence, law of effect, feedback, stamping in, etc.). Thus the demand for a sophisticated operant behaviourism pervades psychology. But the converse also holds or should hold—psychology must pervade operant behaviourism if either is to remain viable. Boakes and Halliday appear to be not unaware of this need in their treatment of the subject but as yet their probings are timid and constrained by the behaviourists' stereotypes. But an auspicious beginning can be felt in their 'substantial modifications' and 'strong reservations' of the doctrinaire Skinnerian view, especially since Skinner himself is making such dramatic revisions.

## REFERENCES

Bower, T. G. R. 1966. The visual world of infants. *Scient. Am.* **215**, 80–92.

Douglas, R. J. & Pribram, K. H. 1969. Distraction and habituation in monkeys with limbic lesions. *J. comp. physiol. Psychol.*, vol. **3**, 473–80.

Estes, W. K. 1959. The statistical approach to learning theory. *Psychology: A Study of A Science*, vol. 2, 380–491. Ed. S. Koch. New York.

Lindsley, D. B. 1961. The reticular activating system and perceptual integration. *Electrical Stimulation of the Brain*, pp. 331–49. Ed. D. E. Sheer. Austin.

Premack, D. 1965. Reinforcement theory. *Nebraska Symposium on Motivation*, pp. 123–88. Ed. D. Levine. Lincoln, Nebr.

Pribram, K. H. 1966. Some dimensions of remembering: steps toward a neuropsychological model of memory. *Macromolecules and Behavior*, pp. 165–97. Ed. J. Gaito. New York.

Pribram, K. H. 1969 a. The Four R's of remembering. *Biological and Biochemical Bases of Learning*, pp. 127–57. Ed. K. H. Pribram. New York.

Pribram, K. H. 1969 b. DADTA III: An on-line computerized system for the experimental analysis of behaviour. *Perceptual and Motor Skills*, vol. **29**, 599–608.

Pribram, K. H., Gardner, K. W., Pressman, G. L. & Bagshaw, M. 1963. Automated analysis of multiple choice behavior. *J. exp. Analysis Behav.* **6**, 123–4.

Skinner, B. F. Presentation at *Symposium on Brain Research and Human Behavior*. UNESCO House, Paris, 11–15 March 1968.

# REPLY

## *by* R. A. BOAKES AND M. S. HALLIDAY

In writing our chapter we had hoped to show that it was possible to be Skinnerian without being doctrinaire, and that the Skinnerian position was more than a set of dogmas and clever techniques. It is, rather, an approach to the whole field of psychology with the distinctive features that we have already described. Unfortunately, reasoned examinations of the Skinnerian point of view have often been buried under a mountain of polemics. It appears from Pribram's comments that he considers that we too are guilty of naivety, dogmatism and narrow-mindedness that sometimes characterize such polemics. We shall answer some of his more detailed points about methodology and the status of physiology at the end of this reply, but we can make no answer to the main tenor of Pribram's discussion, except to emphasize that we attempted to avoid provincialism and bigotry.

We were particularly surprised that Pribram made no reference to the final section, where we quoted, with overall approval, doubts raised by one of Skinner's arch-critics. This part was expected to evoke a reaction of glee or of regret that we had not gone far enough, but not of silence. We had hoped that this discussion would clarify the arguments about Skinner's views on human behaviour; the brevity with which the topic was discussed in our paper appears to have led to misunderstanding, since reactions to this section have ranged from the impression that it left a gaping hole in the Skinnerian edifice to that of Pribram, who apparently regards it to be as consistently 'Skinnerian' as the remainder. Consequently despite the lack of comment on this section we wish to begin this reply by adding a little to what we have already written.

The crucial point in the section on human behaviour is what we have termed 'creativity'. The views of Skinner and Chomsky differ on two aspects of this problem. The first is its importance. For Chomsky an explanation of creativity is *the* fundamental problem in the study of language. Skinner recognizes that it is a common aspect of human behaviour and has discussed topics such as imitation, obeying verbal instructions and following 'rules' in *Science and Human Behaviour* and *Verbal Behaviour*. In some ways our example of a bribe (p. 372) is misplaced: it is relevant to the quotation concerning 'consequences', but is not meant to suggest that Skinner has never considered this kind of behaviour. Indeed he has described many similar instances of human behaviour as examples of

381

'repertoires', which have 'the advantage of freeing men from the necessity of acquiring all their behaviour through operant conditioning' (Skinner, personal communication). This applies equally to language, where he accepts that it is impossible for a person to acquire all his verbal behaviour through operant conditioning. Nevertheless he does not appear to find that these repertoires raise any fundamental problems that involve new principles.

The second issue is the origin of such behaviour and on this Chomsky and Skinner face each other from the opposite ends of the nativist–environmentalist continuum. According to Chomsky all the important properties of human language are determined by human genes and only a relatively brief and unstructured exposure to an actual language at a critical age is required in order for a child to speak English instead of Urdu or Swahili. He appears to predict that a child who watched television solidly for the first seven years of his life and had no other exposure to language, would master the essentials of producing and understanding human speech despite the complete lack of any reinforcement contingencies. On the other hand Skinner accepts only that there may be some small variation from species to species in genetic predispositions towards, for example, learning by imitation or learning a language, and maintains that in general the repertoires described above are themselves acquired through operant conditioning.

The reason for presenting Chomsky's criticisms in our paper was that we agree with him that the problem of creativity is of central importance in understanding most aspects of human behaviour. The main reason we have for finding Skinner's solution to this problem in terms of repertoires, established by operant conditioning, to be unsatisfactory is that it is not all clear in what way such repertoires have the status of responses.

This inadequacy in the Skinnerian position does not arise because the analysis of animal behaviour is wrong or is misdirected. 'Creativity' is not an important feature of the behaviour of any species below the level of primates. Given the interest in 'sub-human thinking' and the many ingenious experiments on such topics as problem-solving that this interest has produced, it is remarkable how little good evidence there is for 'creative' behaviour even in monkeys.

Speculation about the general question of the origin of this kind of behaviour does not seem very interesting to us, since the answers to specific questions are likely to consist of varying mixtures of nativism and empiricism. Differences of opinion on language acquisition seem to be based largely on whether one is impressed by the fact that children have mastered the basic principles in the short span of two years or by the fact that children, with little else to do and with the full-time encouragement of parents, take as long as two years to acquire the rudiments of language. The recent increase of interest in this topic (see Smith & Miller, 1966) promises to provide answers to many of these questions in the near future. However, even if the importance of learning is found to confirm Skinner's views rather than those of Chomsky, it seems unlikely that such learning can be analysed in terms of the limited set of concepts derived from animal studies (see the discussion between Braine, 1963, and Bever et al. 1965). Before leaving this point we would like to note in passing that Bower's work, to which Pribram refers, provides an example of an application of operant techniques which came up with a nativist answer.

A general criticism made by Pribram is that we confuse technology with science. In planning the paper we decided that a discussion of the various ways in which operant conditioning had been applied to practical problems was not very relevant to its main purpose and that we would ignore Skinnerian technology. The one exception made to this decision was the section on animal psychophysics (p. 368). This was examined in some detail because the claim that a certain successful application of Skinnerian principles supports the truth of these principles is often made without any examination of why the application succeeds. Our general rule to exclude technology has led to a slightly one-sided view of Skinnerians.

The reason that Pribram appears to detect so much technology, despite our efforts, is because he has an entirely different view of what the term 'Skinnerian' means. According to him the label describes anyone who studies the role of reinforcement in operant conditioning. If this were so, there would be no more reason to include this chapter in the present volume than to include chapters on short-term memory or habituation. The point of the chapter was to show in what ways 'Skinnerian' characterizes more than a tendency to study a certain kind of conditioning. It is true that Skinnerians have addressed themselves to a certain range of problems within the general context of learned behaviour, but this range includes very much more than the problems of reinforcement, even if this problem is given the extraordinarily wide interpretation supplied by Pribram.

The most serious charge levelled by Pribram is that we have misrepresented Skinner and have given a distorted account of his present views, since, while we have confined ourselves to timid probings, Skinner himself has recently made drastic revisions. Skinner's own reaction to our paper (personal communication) contained several criticisms, notably on the section on human behaviour, but no hint that we had done a grave disservice to him. Since the proceedings of the symposium that Pribram attended have not been published, we are unable to comment on the content of Skinner's talk. However, following our enquiries Skinner (personal communication) stated that this had not contained startling changes in his views and that it was an abbreviated version of material presented in the first part of his recent book *Contingencies of Reinforcement*. This book does not seem to represent a change in Skinner's views and its contents are fully consistent with our presentation. As to the discovery that Skinner is not an *S–R* theorist, we have already attempted to establish this point in the discussion on the spontaneity of behaviour (p. 352).

*Methodology*

There appears to be little disagreement between Pribram and ourselves over methodological questions. The criticisms he makes in this connection arise from misunderstanding our position. The following brief notes on these criticisms essentially repeat what has been said before, but hopefully in a more precise fashion.

The comparison of response measures was intended to explain why response rate is used so frequently in operant conditioning; it was not intended as an argument for abolishing either choice or latency. The selection of a response measure must take into account the purpose for which it is to be used and the particular situation. We provided an example of a situation in which choice probability was probably the most suitable dependent variable and Pribram has now pointed out situations in which latency is a very useful measure.

In our discussion of the use of statistics we were not claiming that one can trade experimental control against statistical expertise. In fact we were arguing against this point of view and were attempting to stress the pitfalls of accepting a large amount of noise in an experiment in the hope that this will be compensated by using a sufficiently large group of subjects. This section was presenting a very different point of view to that of Sidman (1960) who has argued against making any form of comparison between individual subjects or between different groups. It is not clear to us why ROC analysis in some way provides an example to refute our points, especially since this is almost always applied to the performance of individual subjects and becomes very confused when applied to averaged results. Incidentally we thoroughly agree about the potential of this kind of analysis; one of us is at present using it to study steady-state discrimination behaviour (Boakes, in preparation).

We would share Pribram's concern at the prospect of operant conditioning grimly resisting enticement by new-fangled technologies away from its relays and snap-leads, if it seemed at all likely. In fact transistorized logic and computer control are rapidly replacing electro-mechanics. For over a year a 'general purpose computer with enhanced flexibility' has been used at the University of Sussex for the control of operant conditioning experiments (see Halliday (in preparation) as an example of a study which would be impossible without computer control) and it seems probable that this will be the case in most laboratories in the next few years.

## *Physiology*

It is encouraging to find that Pribram agrees with our main point about physiology: that the behavioural and physiological realms of discourse are separate, though they describe many of the same events, and that physiologizing is unlikely to help the behavioural scientist to think clearly. He also appears to agree that it is possible to carry out a fruitful study of behaviour without dealing with the area in which physiological and behavioural data overlap, but feels that to do this is to impoverish psychology. However, we explicitly rejected the extreme anti-physiological position sometimes adopted by Skinnerians; we too believe that the 'brain–behaviour interface' is well worth studying, but we are not at all happy about the assumptions made by the many of those who study it. These seem to be that it is possible to do the physiology without bothering overmuch with the behavioural analysis, since the successful solution of the physiological problems would automatically explain the behaviour. We have already explained why we think that this is a mistaken approach, and we do not think that Pribram has put forward any arguments that tell against this point of view. For example, there is no reason why a Skinnerian should deny himself the knowledge that the retinal image of a tilted object does not correspond with its perceived shape, when he is studying constancy.

To answer a more specific point, we did not say that Hebb or Deutsch make 'sophisticated use of neurological language'; what we do claim is that they seem to believe that the ultimate explanation of behaviour is to be made in neuro-physiological language, and that their use of this language is more sophisticated than that of the ordinary man, who often appears to share their belief. To demonstrate the type of argument to which we object it is enough to quote a few sentences from Hebb's well-known paper on the Conceptual Nervous System:

Physiologically, we may assume that cortical synaptic function is facilitated by the diffuse bombardment of the arousal system. When this bombardment is at a low level an increase will tend to strengthen or maintain the concurrent cortical activity; when arousal or drive is at a low level, that is, a response that produces increased stimulation and greater arousal will tend to be repeated (Hebb, 1955, p. 250).

Whether or not this is a good way of explaining behaviour, it can hardly be denied that it is a paradigm example of the type of physiologizing to which we have raised objections.

Pribram is, of course, logically correct in saying that we cannot be sure that the present advances in sensory physiology will not turn out to be a blind alley; however, we venture to think that most physiologists would agree that physiology has so far contributed far more to our understanding of perception than to our understanding of learning. However, we never suggested, as Pribram seems to believe, that the physiological study of perception had in any sense been successfully concluded; this would obviously be absurd. Here, as elsewhere, it seems that Pribram read what he expected to find rather than what we wrote![1] Similarly we specifically rejected the argument that there are logical difficulties which mean that physiology *cannot* tell us about learning.

Before concluding we should like to stress that the above notes on method and physiology form a relatively large part of this reply because it is mainly on these points that misunderstanding has arisen between Pribram and ourselves. We do not intend to convey the impression that these are the most central aspects of Skinner's views. The importance of Skinnerian methods and, in particular, of the attitude towards physiology has possibly been exaggerated in the past and we hope that the space allotted to them here is not taken as a measure of the emphasis we wish to attach to them.

REFERENCES

Bever, T. G., Fodor, J. A. & Weksel, W. 1965. Acquisition of syntax: a critique of contextual generalization. *Psychol. Rev.* **72**, 467–82.

Boakes, R. A. Analysis of errors in a successive discrimination. In preparation.

Brain, M. D. S. 1963. On learning the grammatical order of words. *Psychol. Rev.* **70**, 467–82.

Halliday, M. S. Temporal patterns in the acquisition of Sidman avoidance. In preparation.

Hebb, D. O. 1955. Drives and the CNS (conceptual nervous system). *Psychol. Rev.* **62**, 243–54.

Sidman, M. 1960. *Tactics of Scientific Research*. New York.

Skinner, B. F. 1969. *Contingencies of Reinforcement*. New York.

Smith, F. & Miller, G. A. 1966. *The Genesis of Language*. Cambridge.

[1] This impression is reinforced by Pribram's comments on our references; out of the thirty-six references which are not to Skinner's own work, only twelve can reasonably be considered to be to Skinnerian work—surely not an unreasonable balance in a chapter on 'The Skinnerian Analysis of Behaviour'.

# EXPLANATION AND THE CONCEPT
# OF PERSONALITY

## *by* H. J. EYSENCK

The verb 'to explain' has two dictionary meanings; much confusion may arise through the common practice of not distinguishing between the two. One meaning is defined by the Oxford Dictionary as: 'To make known in detail', the other as 'To make intelligible'. Description and greater understanding are the two goals which science in its various forms has always striven for, and 'explanation' might thus be regarded simply as synonymous with 'scientific', were it not for the fact that there are scientific and unscientific explanations. Newton's explanation of the movements of the planets in terms of the law of universal gravitation was clearly an instance of the former; his explanation of the deviations of these movements from the prescribed paths in terms of active interference by God was equally clearly an instance of the latter. Explanation by appeal to gravitation is descriptive, but appeal to God is an aid to understanding for those who believe in God; could it be that explanation through description is essentially scientific, explanation as an aid to understanding non-scientific?[1] Some credence might be given to this notion by the division of science into

---

[1] A more purely psychological differentiation between description and understanding which cuts across the differentiation made above relates to a quality of theories which may best be conceptualized as *giving rise to descriptions which can be visualized*. We seem to feel that we *understand* descriptions which can be visualized, in a way that we do not understand descriptions which cannot be so visualized, although the only function of both types of description may be descriptive. A good example is the theory of heat, where we have side by side the thermodynamic and the kinetic theory. Thermodynamics deals with unimaginable concepts of a purely quantitative kind: *temperature*, measured on a thermometer, *pressure*, measured as the force exerted per unit area, and *volume*, measured by the size of the container. Nothing is said in the laws of thermodynamics about the nature of heat. Bernoulli, in his famous treatise on hydraulics, postulated that all 'elastic fluids', such as air, consist of small particles which are in constant irregular motion, and which constantly collide with each other and with the walls of the container. This was the foundation stone of the kinetic theory of heat, which results in a picture of events which is eminently visualizable, and which gives to many people a feeling of greater 'understanding', of better and more thorough 'explanation', than do the laws of thermodynamics. Consider for example the 'insight' which we seem to gain in looking at Cailletet's famous experiment, which originated cryogenic research, by considering his cooling device as part of a single stroke of an expansion engine! Nevertheless, many phenomena are quite intractable to kinetic interpretations even today, which yield easily to a thermodynamic solution. It seems that visualizability is a kind of bonus which may make a theory more easily acceptable, perhaps particularly to people who are visualizers, but which is of psychological interest only, not of general scientific importance.

two parts, so prominent on the Continent, particularly in Germany: *Geisteswissenschaft* and *Naturwissenschaft*. The former aims at understanding; it is ideographic and does not follow the methods of the exact sciences. The latter is nomothetic; its aim is descriptive, and it does follow the methods of the exact sciences. Windelband, Dilthey, Spranger and many others have argued the philosophical case for such a distinction, with the implied or explicit deduction that psychology in general, and personality study in particular, should be *geisteswissenschaftlich*, rather than *naturwissenschaftlich*. Allport is perhaps the best-known exponent of this view of the Anglo-American world.

It will be clear that even if we could come to an agreement as to the nature of explanation in respect to the natural sciences, there would still be complications and difficulties in respect to psychology, and in particular to personality study; it is the purpose of this chapter to discuss these in some detail. But first of all some clarification of the nature of 'explanation' beyond dictionary definition is required. Nagel (1961) suggests that 'explanations are answers to the question "Why?"' and goes on to distinguish four distinct types of explanation. The first is the *deductive model*.

A type of explanation commonly encountered in the natural sciences, though not exclusively in those disciplines, has the formal structure of a deductive argument, in which the explicandum is a logically necessary consequence of the explanatory premises. Accordingly, in explanations of this type the premises state a sufficient . . . condition for the truth of the explicandum. This type . . . has been widely regarded as the paradigm for any 'genuine' explanation, and has often been adopted as the ideal form to which all efforts at explanation should strive.

The second model deals with probabilistic explanations.

Many explanations in practically every scientific discipline are *prima facie* not of deductive form, since their explanatory premises do not formally imply their explicanda. Nevertheless, though the premises are logically insufficient to secure the truth of the explicandum, they are said to make the latter 'probable'. Probabilistic explanations are usually encountered when the explanatory premises contain a statistical assumption about some class of elements, while the explicandum is a singular statement about a given individual member of that class.

Most generalizations in the fields of personality research and abnormal psychology clearly fall under this heading. Others are more likely to fall into yet another category, that of functional or teleological explanations.

In many contexts of enquiry . . . explanations take the form of indicating one or more functions (or even dysfunctions) that a unit performs in maintaining or realizing certain traits of a system to which the unit belongs, or of stating the instrumental role an action plays in bringing about some goal ... In many functional explanations there is an explicit reference to some still future state or event, in terms of which the existence of a thing or the occurrence of an act is made intelligible.

Homeostatic types of explanations are of this type, as are those postulating instincts and drives.

A fourth and last type of explanation relies on genetic considerations.

Historical enquiries frequently undertake to explain why it is that a given subject of study has certain characteristics, by describing how the subject has evolved out of some earlier one ... The task of genetic explanations is to set out the sequence of major events through which some earlier system has been transformed into a later one.

Nagel's discussion of genetic explanations deals largely with social events, but the concept covers equally well the laws of inheritance, and for psychologists this meaning will be more relevant, as we shall see later. These four types of explanation are distinguished for the sake of convenience; they do not form absolute, qualitatively different groupings, and more than one may be involved in any actual application.

If these are the major ways in which science 'explains' natural phenomena, can it in truth be said that science furnishes us with genuine explanations? Many philosophers believe that the answer must be no. Thus Hobson (1923) argues that:

the very common idea that it is the function of Natural Science to explain physical phenomena cannot be accepted as true unless the word 'explain' is used in a very limited sense. The notions of efficient causation, and of logical necessity, not being applicable to the world of physical phenomena, the function of Natural Science is to describe conceptually the sequences of events which are to be observed in Nature; but Natural Science cannot account for the existence of such sequences, and therefore, cannot explain the phenomena in the physical world, in the strictest sense in which the term explanation can be used. Thus Natural Science describes, so far as it can, *how*, or in accordance with what rules, phenomena happen, but it is wholly incompetent to answer the question *why* they happen.

This objection is true, but not perhaps of very great concern to us; it is based on linguistic usage, and does not restrict the meaning of the term unduly. It might be argued that 'why' questions, as defined by Hobson, are meaningless and cannot, by definition, have any answer; 'how' questions do have answers, discoverable by science. The notion of a 'necessary order' in things, against which Hume already argued, is not one which has any meaning for modern science, and to say that science cannot explain facts by derivation from such an order is not to agree with Hobson that this restricts the term 'explanation' to a relatively unimportant function.

This discussion will have shown that of the two meanings of the term 'explain', the one relating to description is the one which has proved useful in science, while the one relating to understanding has gone by the board. Consider such questions as: 'Why do the planets move as they do?', or 'Why do the tides occur?' or 'Why do rotating masses assume circular shapes?' The answer, of course, is to be found in terms of Newton's theory, but this is essentially descriptive; we gain an illusion of understanding because the descriptive power of the theory links together phenomena usually considered in separation. Newton was under no such illusion; he saw that he had gained descriptive consistency by adopting *ad hoc* notions which were quite repugnant to him, and to other scientists of his time, such as 'action at a distance' (Hesse, 1961). If 'to understand' is merely taken as a psychological state of acquiescence in the familiar order of things, then we may say that Newton's theory helps us to understand, because even the schoolboy of today feels secure in ascribing many divergent phenomena to gravitation; but if we mean anything more cognitive than such a primitive emotional state by the term then we can hardly say that Newton's theory helps us to 'understand' the facts referred to in our three questions. Scientific theories, even the most far-reaching, are essentially circular; our laws and theories are derived from observations which become ever more precise and consistent, and in turn summarize these observations in terms of formulae, laws and theories covering larger and larger numbers of at first sight divergent phenomena. This is an important point to which we shall return later on in our discussion of factor analysis and type- and trait-theories of personality.

Of the four models of explanation it is the deductive one which is usually regarded as the most useful, or even the only properly 'scientific' one. It is also often argued that this model has no place outside the 'hard' sciences, and that psychology in particular cannot progress to wide-ranging theories by means of the hypothetico-deductive method. The reasons given for this belief vary, but they tend to imply that the complexity of phenomena is too great in psychology, as compared with physics or chemistry, and that regularities, such as those found in the physical sciences, are absent in the biological ones. These arguments assume altogether too much; those who argue in this way are hardly in a position to compare the complexity of phenomena in physics and psychology when we have only begun to unravel the former, and have hardly even looked at the latter. Does any real meaning attach to a statement that theories about the goings-on at sub-atomic levels are in some fundamental sense simpler than those in the brain of a rat at a choice point? As we don't know the appropriate level of complexity of either set of phenomena, any attempt at an answer would be little more than a projective test revealing the subject's prejudices. Nor is it reasonable to deny the presence of regularities simply because not many of these have been found in the present early stage of development. It will be argued that the hypothetico-deductive method has an important part to play in psychological research, although it requires to be supplemented by the other methods of explanatory research mentioned. It will also be argued that many of the objections which have been voiced are based on an idealistic (and unrealistic) version of how this method really works in physics. Most psychologists are familiar with physical examples of the hypothetico-deductive method only from hopelessly simplified and inaccurate quotations in popular books by philosophers, logicians and popularizers. These idealize the method, and falsify the historical account, to such an extent that the achievements of the method, which are very real, appear in a superhuman light, so that psychologists easily feel that their subject will never be able to furnish anything similar.

Consider the case of the discovery of the planet Neptune, as it is usually presented. Herschel had recently discovered the planet Uranus by accident, and this was found to show certain aberrations from Newtonian predictions in its orbit. John Couch Adams and Urbain Jean LeVerrier set themselves the task of finding by theoretical calculation the position of the hypothetical outer planet which was thought to have been responsible for these orbital perturbations. Adams finished first, but the British astronomers to whom he gave his calculations were dilatory in their search; LeVerrier sent his results to J. G. Galle in Berlin, who almost immediately picked up the new planet just where prediction said it would be. Could anything illustrate better the incredible accuracy of Newtonian calculations, or the predictive value of Newton's theory? Consider now the story with some of the warts left in.

The first wart is that in fact Adams and LeVerrier used in their computations as a major building stone, indeed an indispensable one, an empirical generalization known as Bode's Law.

The rule is expressed by the following simple formula. For each planet first write a four, then add a number that varies from planet to planet; for Mercury, the innermost planet, the number is zero; for Venus, next nearest to the Sun, it is three. After Venus the number is simply doubled each time. For the Earth it is six, for Mars twelve, and so on. The numbers obtained in this way run in the series 4, 7, 10, 16, 28, 52, 100, 196 and 388. If the actual mean radii of the planetary orbits are measured by a scale on which ten units represent the radius of the Earth's orbit, then the planetary orbits run in the sequence 3·9, 7·2, 10, 15·2 and so on. These figures lie strikingly close to the series suggested by Bode's Law. (Hoyle, 1962.)

Table I shows the figures comparing law and observation; it will be seen that for Neptune there is a huge discrepancy between the two, amounting to something like 25%! Few psychologists would tolerate such huge errors in their generalizations. Adams and LeVerrier happened to be lucky in that Neptune at that time passed through that part of its orbit which permitted of a solution of the problem; otherwise, as Hoyle points out, 'the efforts of both Adams and LeVerrier would have been doomed to failure'. Thus this exhibition piece illustrating the power of the hypothetico-deductive method used as a major prop a 'law' which was purely empirical, had not

TABLE I. *Bode's Law of Planetary Orbits*

|  | Mercury | Venus | Earth | Mars | Ceres |
|---|---|---|---|---|---|
|  | 4 | 4 | 4 | 4 | 4 |
|  | 0 | 3 | 6 | 12 | 24 |
| Law | 4 | 7 | 10 | 16 | 28 |
| Observation | 3·9 | 7·2 | 10·0 | 15·2 | 27·7 |

|  | Jupiter | Saturn | Uranus | Neptune |
|---|---|---|---|---|
|  | 4 | 4 | 4 | 4 |
|  | 48 | 96 | 192 | 384 |
| Law | 52 | 100 | 196 | 388 |
| Observation | 52·0 | 95·4 | 191·9 | 300·7 |

then, and does not have now, any rational basis, and is in any case erroneous. By a piece of luck this error was compensated at one particular moment of time, but of course their calculations of Neptune's orbit and future positions turned out to be entirely wrong. Indeed, had Adams and LeVerrier assumed that the orbit of Neptune was circular, their calculations would have been far simpler and far more accurate, as R. A. Lyttleton was to point out recently. And as an afterpiece, we may just mention that LeVerrier later on tried to account for perturbations of Mercury's orbit by postulating another undiscovered inferior planet; the perturbations are quite genuine, but there is in fact no such inferior planet. This, then, is the true story of the discovery of Neptune, a mixture of genuine deduction from established scientific law, chance, error, luck and farce. Psychologists may be relieved to note that this sort of thing resembles much more the sort of deduction-testing that goes on in their field than does the traditional textbook story.[1]

But can it not at least be said that when there is a strikingly good fit between theory and observation, then the verification of the prediction strongly supports the theory? Consider the case of Kepler and the harmony of the planets.

He suggested that the planets emit some sort of harmony analogous to musical notes, the pitch of the note being proportional to the speed of the planet. By using the known size of the orbits of the planets, their eccentricities and their periods, he obtained a system of notes . . . Of course it was not the case that the calculated notes agreed precisely in frequency with the musical notes (obtained by Kepler) . . . Suppose we make the notes come out exactly as they should be on a properly tempered scale, and suppose we then infer from this the maximum and minimum distances of the planets from the sun. (Hoyle, 1962.)

A comparison of values derived from this nonsensical theory, and those given by the actual observations that Tycho Brahe, Kepler's chief, had made, give the results shown in Table 2. Values are given for the aphelion (maximum distance) and perihelion (minimum distance), and it will be seen that, as Hoyle puts it, 'the agreement is frighteningly good—

TABLE 2. *Kepler's Law of Planetary Harmony, compared with empirical observations by Tycho Brahe*

(The mean distance of Earth from Sun is taken as 1·000)

| | Harmony | | Tycho Brahe | |
|---|---|---|---|---|
| Planet | Aphelion distance | Perihelion distance | Aphelion distance | Perihelion distance |
| Mercury | 0·476 | 0·308 | 0·470 | 0·307 |
| Venus | 0·726 | 0·716 | 0·729 | 0·719 |
| Earth | 1·017 | 0·983 | 1·018 | 0·982 |
| Mars | 1·661 | 1·384 | 1·665 | 1·382 |
| Jupiter | 5·464 | 4·948 | 5·451 | 4·949 |
| Saturn | 10·118 | 8·994 | 10·052 | 8·968 |

[1] It should perhaps be mentioned at this point that ephemeral as are theories in psychology, they are at least as ephemeral in the natural sciences; Newton's theory is the exception, not the rule. Take as an example research in cryogenics; Mendelssohn writes: 'As was inevitable, ever since superconductivity was first discovered many different theories for its explanation have been proposed; roughly at the rate of two or three per annum, and for the better part of half a century . . . Eventually Felix Bloch who had done so much for our understanding of electrons in metals enunciated an axiom of his own which ran "every theory of superconductivity can be proved wrong". And for a long time this axiom turned out to be the only correct one.' (From K. Mendelssohn, *The Quest for Absolute Zero*, Weidenfeld and Nicolson, 1966.)

frightening because the idea has no physical relevance whatever. One wonders how many modern scientists faced with a similar situation would fail to be impressed by such remarkable numerical coincidence.'

Given, then, that the usual arguments against the use of hypothetico-deductive method in psychology are invalid, and that a more historical and less starry-eyed view of its actual working suggests a less inhuman perfection than might have been expected in its application to physical problems, thus making it more relevant to the muddled state of research in psychology—equally muddled, it may perhaps be said, as ongoing research in any science at any time, no more and no less; given all this, we may perhaps hope that psychologists just as much as other scientists will be able to use this type of explanation, as well as the others listed, in their efforts to bring order into their particular domain.

We must now turn to the application of these general considerations to the explanation of individual differences in conduct and behaviour, which we may provisionally identify as the domain of personality research. The first question to arise, of course, relates to the necessity of postulating personality variables at all as essential parts of a scientific explanation of behaviour. Some neo-behaviourists believe that all psychological problems are of a functional kind, following the $S$–$R$ sequence, and can be solved without paying any attention to conceptions of organismic potentialities or 'personality'. Experimental psychologists traditionally throw variance properly belonging to personality variables into the 'error' term of their analysis, thus throwing away what is often the most important, and may be the largest, part of the complex of variables which go to determine the value of the dependent variable in addition to the selected independent variable. Indeed, personality variables thus neglected often make a more important contribution than the selected independent variable itself; the absurdly inflated error terms which are so characteristic of many experimental studies are witness to the failure in experimental control which is implied in the neglect of experimentalists to measure and use personality variables in their research. To take but one example: if it is true that introverts condition well, and extraverts poorly, on the eye-blink conditioning task (Eysenck, 1965), then any experiment which investigates the effect of some independent variable on eye-blink conditioning, and which does not control this personality variable, will add the considerable variance due to introvert–extravert differences to the error variance, thus making the discovery of significant effects of the independent variable that much more difficult. Many examples of such effects of personality variables on experimental trials have been given elsewhere (Eysenck, 1966).

Even more impressive are examples where personality acts as an interaction variable, in such a way that there is no overall effect of the main independent variable at all; the effects of this variable on different personality types cancel each other out. Consider the following problems, all of which might be considered to come under the usual $a = (f)b$ formula of the experimental psychologist. (1) Is performance on a simple crossing-out task affected by time of day (a.m. $vs$ p.m.)? (2) Does Meprobamate affect the performance of subjects on a variety of experimental tasks? (3) Do subjects perform better in groups or individually on a simple crossing-out task? (4) Does the introduction of conflict into a simple muscular movement task increase or decrease the amount of movement produced? (5) Do various drugs affect the mood of subjects positively or negatively? (6) Does the inhalation of alcoholic fumes increase or decrease the activity of mice? The answer in each case was that if personality or strain variables were disregarded, the independent variable (time of day, drug, individual $vs$ group performance, conflict, alcoholic fumes, etc.) had no significant effect. The reason for this lack of effect, however, lay hidden in an interaction term with certain personality variables (strain variables in the case of mice). Thus introverts work better in the morning, extraverts in the afternoon; the two effects cancel out if personality is not considered (Colquhoun & Corcoran, 1964). Meprobamate affects the performance of subjects high and low on neuroticism differently, improving that of the latter, and worsening that of the former (Munkelt, 1965). Extraverts perform better in groups, introverts under individual conditions (Colquhoun & Corcoran, 1964). The various drugs used by Janke (1964) depressed the mood of subjects low on neuroticism and improved that of subjects high on neuroticism. The introduction of conflict changed the performance of introverts and extraverts in different directions (Eysenck, 1967). And the inhalation of alcoholic fumes increased activity in two strains of mice, decreased it in two others, and made no change on two further strains (McClearn, 1962). Even rewards and punishments act differently on persons of different personality; Weisen (1965) has shown that introverts will perform an operant task more assiduously in order to turn off light and noise, while extraverts will perform more assiduously in order to turn on light and noise! These and other examples have been discussed in some detail elsewhere (Eysenck, 1966, 1967), and the conclusion has been drawn that it is likely that many failures to obtain significant results in studying the effects of an independent variable on a dependent one have been due to a failure to take into account relevant personality variables. Similarly, many failures to replicate other experimenters' work may have been due to similar causes. It has even been suggested that major theoretical controversies, such as that between the Hull–Spence school and the Tolman group may have been due simply to the fact that one school used emotionally non-reactive animals in its experimentation while the other used emotionally reactive ones (Jones & Fennell, 1965)!

The upshot of these considerations would appear to be that explanations of psychological events, whether experimental or in real life, require the introduction of personality variables in order to prevent the overloading of the error term to such an extent that results obtained become meaningless, non-replicable, and misleading. Even such apparently emotionally neutral tasks as nonsense syllable learning and paired associate learning have been found to be closely related to personality functions (Eysenck, 1967), as have sensory thresholds, perceptual phenomena, vigilance, and many other traditional experimental subjects. Personality is all-pervasive, and different individuals even react differently to various psychophysical experimental methods. The failure of experimentalists to pay attention to such an all-pervasive variable is difficult to understand, except in historical terms; as long as personality variables were poorly defined and lived in a mythical country outside the boundaries of traditional science, experimentalists might with propriety decline to have any dealings with such strange bed-fellows. However, recent advances in a more scientific understanding of such variables, and a more rational explanation of their mode of working, makes this excuse inapplicable. After all, physicists and chemists, after whose methods of working we may with advantage model our own, do not deal with some unanalysed 'stuff' in their experiments; they know that gold has a different specific gravity to lead, that some metals show super-conductivity when cooled to within a few degrees of absolute zero, while others do not, and that different elements, or even isotopes of the same element, differ in important ways in their sub-atomic structure, ways which profoundly affect their reactions to experimental manipulation. Why should psychologists assume that all their subjects react uniformly to their experiments, as if they were all uniovular twins? The evidence is so conclusive in showing the error of such a view, held perhaps not at a conscious and rational but at an unconscious and emotional level, that one may hope for the imminence of a change in view, and a determined effort to include personality variables in experimental designs of all kinds.

The crucial problem in this connection, of course, is the proper choice of personality variables, i.e. the correct choice of a system of explanation which will render intelligible past findings and mediate the discovery of future ones. It will be argued that there are three main levels of explanation in the field of personality; that these correspond to rather different types of theoretical and experimental work; and that progress in personality research consists (a) in refining concepts and experimental findings at each level, and (*b*) in forging theoretical connections between these levels. Illustrations of these three levels will be given largely from the writer's own work, for two reasons. In the first place, the details are more familiar to the writer; in the second place, no other system seems capable of illustrating all three levels simultaneously. There is no intention of slighting other systems in thus restricting our survey in this manner; this chapter does not pretend to serve as a textbook of personality research, and the examples are merely offered as illustrations of certain points of view. It is the latter which constitute the main contribution of this chapter.

The first level of explanation is the lowest; it corresponds to the probabilistic model of Nagel's system. This model itself, being hierarchical, has several levels; its main characteristic is that it describes uniformities in conduct in terms of such concepts as types and traits. The model is not dissimilar to usages in common parlance; the postulation of a trait of sociability, aggressiveness, or impulsiveness, or of an emotional type, is not far removed from common sense. Models of this kind go far back in history; Galen, a Greek physician writing in the second century A.D., put forward the theory of the four temperaments (sanguine, choleric, melancholic, phlegmatic) which was to become almost universally accepted in Europe, and which gained a second burst of popularity when Kant adopted it in his *Anthropologie*, giving personality descriptions of the four types which, as we shall see, still have relevance to modern theories of personality. This Galen–Kant model was *categorical*; in other words, each type constituted a category between which there was no overlap. Kant specifically excluded any such possibility; everyone could be assigned to one type or the another, and none could have characteristics which arose from a blending of the types. This notion of type is still sometimes criticized by American writers who prefer the *dimensional* kind of model, i.e. one in which personality characteristics are found to lie on a continuum, and are distributed normally. Actually such a model was first introduced by Wundt, who noted that the choleric and the melancholic type share in common their great emotionality, which set them off from the less emotional phlegmatics and sanguines. Similarly, he thought, the cholerics and the sanguines were characterized by being changeable, while the phlegmatics and the melancholics were not. This scheme resulted in a two-dimensional model which is illustrated in Fig. 1; the traits around the circumference are taken from Kant's description of the four types, which now occupy the four quadrants produced by the two dimensions of emotionality–stability and changeableness–unchangeableness. Typologists since Wundt, such as Jung, Kretschmer, Burt and others have followed him in postulating some form of dimensional system; the often-heard criticism of their systems as being categorical has no basis in fact, as the writer has pointed out in some detail elsewhere (Eysenck, 1960 *a*).

Many writers do not distinguish at all between types and traits, using these concepts as synonymous. The writer has suggested that 'type' be used as a supraordinate concept in a hierarchical description of personality,

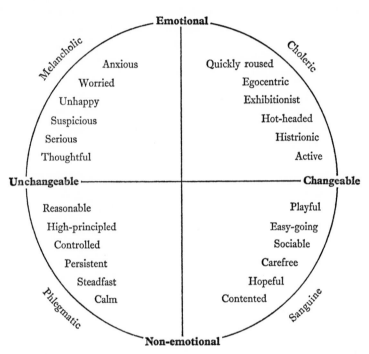

Fig. 1. Galen–Wundt model of personality, showing both *categorical* and *dimensionless* hypotheses. The emotionality axis is nowadays often referred to under the name 'neuroticism' or 'anxiety' by Eysenck and Cattell respectively; the 'changeable or unchangeable' axis as 'extraversion–introversion' or 'exvia–invia' by the same authors. (From H. J. Eysenck, 1967.)

'trait' as a subordinate concept; this notion is illustrated in Fig. 2, taken with some modifications from Eysenck (1947). He has also suggested that the terms 'emotional' and 'changeable' be dropped, and that instead the dimensions be named 'neuroticism' (N) and 'extraversion–introversion' (E). There is a very large body of evidence, based on questionnaire data, ratings, life histories, experimental tests, projective tests, physiological and constitutional measures, all of which support in greater or lesser detail the general theory depicted in Fig. 1 (Eysenck, 1960*a*, 1967), which will consequently provide the provisional basis for our further discussion.

Note that this model is purely descriptive; it asserts, as shown in Fig. 2, that certain traits (sociability, impulsiveness, activeness, etc.) tend to go together in the same person, i.e. that over random samples positive correlations will be found between these traits. Such correlations would then normally be subjected to factor analysis or some similar form of latent structure analysis, but such mathematical manipulations are apparently not essential; Galen, Kant and Wundt arrived at a scheme very much like that widely accepted nowadays by factor analysts (Eysenck & Eysenck,

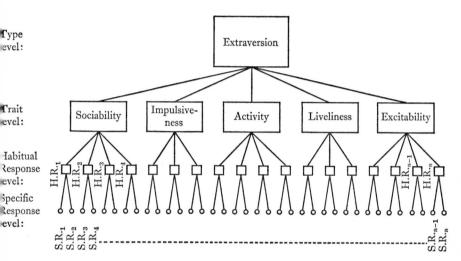

Fig. 2. Hierarchical model of personality dimension 'extraversion'. The concept is built up on the empirically observed correlations between traits such as sociability, impulsiveness, etc. (From H. J. Eysenck, 1947.)

1967) on the basis of simple observation and what one might perhaps call 'implicit factor analysis', i.e. a simple structuring of observations in the mind without the intervention of an explicit mathematical model, and without the use of numerical data. The traits involved in the scheme, too, are based on explicit (or in the case of the older writers, implicit) correlational analysis; to posit a trait of 'sociability' implies that over a random sample of people different occasions involving the possibility of sociable conduct will call forth consistently sociable or non-sociable behaviour from different members of the group.

From the point of view of explanation, we may say that this descriptive model uses Nagel's second model, namely that of *probabilistic* explanation. By assigning a person to one or other of the two classes: 'sociable' or 'non-sociable', we attribute to his future behaviour in social situations certain probabilities which can be calculated with some precision. (We are of course not restricted to two classes; we can have as many classes as we like, thus dealing in effect with a continuum of scores, as in a questionnaire.) Or we may say that an extravert is more likely to act in an impulsive fashion than an introvert in a certain situation; that a person high on emotionality will be more likely than one low on that variable to have a neurotic breakdown; and that a person high on both E and N will have a higher likelihood of becoming a criminal than a person low on both E and N (Burt, 1965). Although such probabilities do not approach certainty very closely, neither are they too close to zero to make this approach practically useless.

The whole trait and type approach is often criticized along the following lines (Lundin, 1961). We start out by trying to explain a person's sociable behaviour, so it is said, and we do this by positing a trait of sociability; when we are asked to furnish evidence for this trait of sociability, all we do is to point to the sociable behaviour! Here, surely, is a prime example of a circular explanation, which is quite useless from both the theoretical and the practical points of view. Trait and type 'explanations' resemble concepts like McDougall's instincts, to which the same objection has been made.

For McDougall, people were social not so much because of the social influences resident in the culture but because they were all born the same way, sharing instincts to do certain things. If a particular behaviour needed explaining, he simply posited some instinct for it. The bully fights because of a strong instinct for pugnacity. The very social or party-loving person has a strong gregarious instinct, and the miser's instinct for acquisitiveness accounts for his hoarding. (Lundin, 1961.)

Criticisms of this kind are partly right, partly wrong. There is obviously a certain circularity involved in any probabilistic explanation, because the thing to be explained must in the past have been assessed in relation to the predicting category. But is there less circularity in the notion of universal gravitation? Why does this stone fall—because of universal gravitation. How do you know about universal gravitation—because this stone falls. There is some circularity, but there is also a good deal more; the falling stone is not identical with, or alone responsible for, such generalizations as the inverse square law or the formula: $s = \frac{1}{2}gt^2$. In the case of trait and type theories, also, the observed behaviour is not the only thing that is taken into account when positing the existence of a given trait; there is much more, and this additional empirical material is left out of the equation by the critic.

Consider the hypothesis: all variations in social behaviour are due to individual differences in a general trait of sociability. The first point to note about this hypothesis is that it may be falsified empirically; it is not logically inconsistent to maintain that there are more than one kind of sociability, independent of each other. In fact, the writer has shown that items from a typical sociability questionnaire, carefully selected to measure Guildford's trait 'S' (social shyness), break up into two quite separate groups when administered to large samples of subjects; some of the items correlate with introversion and define an *introverted* kind of social shyness, while other items correlate with emotionality and define a *neurotic* kind of social shyness. To put it slightly differently, some persons are unsociable because they don't *care* much for other people, while other persons are unsociable because they are *afraid* of other people. If this vital differentiation is neglected, prediction of social behaviour would be very haphazard and inaccurate indeed (Eysenck & Eysenck, 1967).

Now we have surely gone beyond the simple stage of positing a trait of sociability in order to account for sociable behaviour; we have instead analysed sociable behaviour and discovered that this behaviour is determined by two (and possibly more) quite distinct causes. This is still descriptive, and may still be circular, but the circle is widening; the argument now involves more than a mere tautology. Similarly, consider the hypothesis involved in the diagram shown in Fig. 2; this hypothesis could be disconfirmed by empirical research, and is therefore more than a mere tautology. It is a (descriptive) hypothesis about the structure of human personality, and as such deserving of critical attention; there are many different (descriptive) hypotheses of the structure of human personality, and not all of these be right. Factor analysis offers us a means of deciding between them on an empirical basis, and it does not seem reasonable to object to work of this kind on the grounds that it involves circular reasoning. Reality breaks into this circle on all sides, and can (and often does) prove the theory wrong. It is a fact that traits correlate in certain patterns, and generate higher order concepts like N and E; these offer a rather low order of explanation of individual differences in conduct and behaviour, but there is a perceptible gain over common sense; at least we have numerical estimates of the degree of generality and structure of behaviour, and methods for deciding between alternative descriptive schemes. In the same way might one say that the Copernican and Ptolemaic theories of planetary orbits were purely descriptive; yet the discovery of a method of deciding between them, and the gain in finally having a heliocentric theory which corresponded descriptively to reality, will hardly be gainsaid. Such descriptive phases must precede causal analysis; we cannot seek higher-order explanations while we are still unsure about lower-order ones. Accurate description, typology, nosology, or taxonomy must precede historically causal explanation, aetiology, genetics or evolution.

Starting with this descriptive model, the writer has attempted to construct a causal model which would be of the kind called *deductive* by Nagel. In fact, we may distinguish two sub-divisions of this category; deductive models may be reductionist or non-reductionist. By reductionist is meant here simply that the explanatory concepts are sought in a different science to that in which the quest for an answer started. Suppose that the answer to the psychological question: 'Why is John more intelligent than Jim?', were to be found in the greater excretion of cholinesterase in the cortex of one of the boys as compared with the other; this would be a reductionist theory of the deductive kind. If the answer were to be found in some psychological property of the environment, then we might have a deductive theory which was non-reductive. More will be said about the whole problem of reductionism later on; here let us merely note that the model to be considered next is a non-reductionist one, while the third model will be a reductionist one.

In this second model, the N factor is related to a conception of differential emotional activation, high N scorers having low thresholds, low N scorers high thresholds. This connection, of course, is already implicit in the Wundtian terminology, where the N factor is called 'emotionality', a nomenclature later followed by Burt in his pioneering factor-analytic studies of personality. Of more interest perhaps is the hypothesis that introversion and extraversion are related to excitation and inhibition, respectively (Eysenck, 1957). Put briefly, the writer's hypothesis states that introverts are characterized by high cortical excitation and low cortical inhibition, while extraverts are characterized by low cortical excitation and high cortical inhibition. The terms 'excitation' and 'inhibition' in this context are used very much in the Pavlovian sense, although Hull's theoretical contribution (particularly his notion of reactive inhibition) was not neglected. This theory gave rise to many deductions which were experimentally tested, deductions linking the E factor to laboratory studies of conditioning, vigilance, satiation, figural after-effects, time error, reminiscence, critical fusion frequency, apparent movement, dark vision, motor reactions, time judgments, pain tolerance, spiral after-effect, flutter fusion, and many others; whether correct or not, the theory could certainly claim to be deductive in nature, and in fact the majority of findings seemed to be in line with prediction (Eysenck, 1957; 1960*b*).

One point should be noted, however, which is not covered in Nagel's discussion of deductive-type explanations. A theory such as the present one provides a major premiss for a logical deduction, but such a deduction also requires a minor premiss. Take as an example the following argument:

Major premiss: Extraverts build up inhibition quickly and strongly.

Minor premiss: The bowing of the serial learning curve is due to the build-up of inhibitory potential (Hull–Lepley hypothesis).

Deduction: Extraverts will show a more marked bowing of the serial learning curve.

When the experiment was done, no differences were in fact observed between extraverts and introverts with respect to the bowing of the serial learning curve. This may be interpreted in several different ways. (1) The major premiss is wrong, and the results disconfirm Eysenck's personality theory. (2) The minor premiss is wrong, and the results disconfirm the Hull–Lepley theory. (3) Experimental parameters were wrongly chosen, so that the results are irrelevant to the deduction. Special experiments were carried out to test the second of these possibilities, and it does in fact seem that the Hull–Lepley theory is false (Eysenck, 1959; Jensen, 1962). In other cases where the prediction failed, the third possibility was found to be the correct interpretation of the finding; thus for example in the case of eye-blink conditioning Eysenck has shown that certain parameter conditions are favourable to introverts, others to extraverts, and that these conditions can be deduced from the general theory (Eysenck, 1965b; 1967). Introverts benefit from short CS–UCS intervals, weak UCS, and partial reinforcement, while extraverts benefit from long CS–UCS intervals, strong UCS and complete reinforcement. Many apparently negative findings in this field have been due to neglect of this point.

The type of explanation given in our second model is representative of what the writer has called 'weak' theories, as opposed to 'strong' theories, such as the Newtonian (Eysenck, 1960b). This distinction cannot be discussed in detail here, but it is related to the emphasis placed in the last paragraph on the importance of the minor premiss. In a strong theory failure of prediction is very serious, and may lead to the abandonment of the theory; in a weak theory, failure of prediction is less serious because a weak theory has to make assumptions regarding the minor premiss, and cannot specify in advance the most suitable parameter values for successful results. Thus in psychology, and indeed in most sciences where research is carried on at the edge of the unknown, weak rather than strong theories are the rule, and successful prediction is of greater importance than unsuccessful prediction; the former suggests important follow-up investigations, and leads to the conclusions that the theory, though not necessarily correct, may be leading in a promising direction, while the latter is open to many divergent explanations which do not necessarily imply a failure of the theory represented by the major premiss. In fact, even a strong theory, such as Newton's, is not usually overthrown by failures of prediction; such failures could be observed in Newton's system right from the very beginning, and as we have noted he was forced to imply the active interference of the Almighty in the movements of the planets in order to save his theory. It is interesting to speculate how the editor of the *Journal of Experimental Psychology* would react to such a plea on the part of a contributor whose predictions had been found to be falsified in some important detail!! Newton's lack of rigour, even in his mathematical analysis, was criticized severely, especially by French mathematicians, and it was not until Cauchy published his *Cours d'Analyse* 150 years after the appearance of the *Principia Mathematica*, that Newton's methods were given a properly rigorous substructure. Psychologists sometimes seem in danger of throwing out promising ideas of wide applicability through premature insistence on a degree of deductive and experimental rigour which is quite out of place in a young science just beginning the hard task of building up its foundations.

The second model of explanation just discussed was obviously destined to be short-lived because from the very beginning evidence was available to suggest that Nagel's fourth explanatory model, the genetic, could with advantage be applied to this field. The evidence regarding the importance of heredity in the genesis of extraversion–introversion, and of neuroticism–stability, is so extensive that we may refer to a detailed summary of it which has been given elswhere (Eysenck, 1967); here we can only note the major conclusion, namely that individual differences in E and N can be ascribed to constitutional and hereditary factors, as opposed to environmental factors, in the proportion of about 3 to 1. This figure is very similar to that found to obtain in the cognitive field too, where individual differences in intelligence are determined by hereditary factors to the extent of about 75%. It should be noted that this finding does not conflict in any way with our second model; deductive and genetic models of explanation can obviously coexist quite peacefully. All we have to posit is that individual differences in the excitation–inhibition balance are largely determined by genetic factors; the models are not alternative to each other, but may be used to supplement the one the other. The difficulty which arises from these genetic findings is of quite a different kind.

It is obvious that what is inherited must be *structure*, not *function*; we cannot inherit behaviour, but only some physiological–neurological–anatomical substratum which underlies behaviour and interacts with environmental stimulation to generate those reactions which we observe and classify as 'introverted' or 'emotional'. Psychological concepts such as 'excitation' and 'inhibition' clearly will not do; we cannot inherit psychological notions which arose primarily as explanatory concepts in experimental psychology, to act as intervening variables. What is required is clearly an anchoring of these notions outside the psychological field altogether, i.e. a reductionist approach of one kind or another. To some psychologists the view identified by Boring as that of the 'empty organism' has much appeal, but it violates the primary law of science, namely that we must not go counter to fact in our theorizing. Whatever else may be true or false, there can be no doubt that organisms are not 'empty'; they are filled with neurons, cells, CNS and autonomic systems, ganglia, reticular formations, and all the paraphernalia that forms the subject-matter of the physiologist and the neurologist. Our genetic findings point clearly and unmistakably to the nervous system as the *locus* for our E and N factors, and for our explanatory concepts of excitation and inhibition.

The factor of neuroticism, fortunately, does not present any unsuperable difficulties in this connection, in view of the close and well-established relationship between emotion and the autonomic system. The sympathetic and the parasympathetic systems, and in particular the controlling and co-ordinating centres in the limbic system, particularly the 'visceral brain', are the obvious locus for the source of individual differences in emotionality (N), and there is much evidence in the literature to suggest that this connection between descriptive factor, theoretical concept and physiological entity is probably along the right lines (Eysenck, 1967). As before, it is the extraversion–introversion factor which poses the more difficult and serious problem.

The writer has suggested that the concept of 'arousal', which has been attracting more and more attention in physiological circles, should be formally identified with the psychological notion of 'cortical excitation', and that the seat of individual differences in arousal should be sought in the Ascending Reticular Activating System (ARAS). The same system also provides a mechanism for cortical inhibition, and the theory may be formally stated in the following form: introverts are characterized by a reticular formation which has a low threshold to incoming stimulation, and consequently produces strong arousal in the cortex of the individuals concerned. Extraverts are characterized by a reticular formation which has a high threshold to incoming stimulation, and consequently produces weak arousal in the cortex of the individuals concerned. Conversely, the synchronizing portion of the reticular formation has high thresholds for introverts and low thresholds for extraverts, thus producing little inhibition in introverts and much inhibition in extraverts. The details of these systems have been worked out elsewhere (Eysenck, 1967), and references will be found in the same place to direct electrophysiological studies (particularly with the EEG) furnishing some direct evidence for these suggestions. Thus it appears that introverts typically have low alpha amplitude and high alpha frequency, while extraverts typically have high alpha amplitude and low alpha frequency, EEG characteristics respectively of high and low arousal.

The deductions on to experimental fact from these hypotheses are for the most part similar to those made on the basis of our second model, but in some cases the fit is clearly better. Attempts to account for vigilance experiments, for instance, succeeded in linking the decrement of perfor-mance with concepts like reactive inhibition, but failed to account for the occasional finding that extraverts were inferior in performance right from the beginning; reference to greater arousal of introverts would satisfactorily account for this finding. The lower sensory thresholds of introverts would have been rather more difficult to account for in terms of lack of reactive inhibition; again greater arousal gives a natural and fitting explanation.

A reductionist account, like the present one, essentially throws the burden of explanation onto another discipline. Genetics, physiology, neurology, anatomy and biochemistry are all implicated in our attempt to account for individual differences in N and E. It will be clear that this is not the end of the chain, however, and much further research will be required in these various fields. Precisely how many genes are involved in the transmission of the biological bases of extraverted or emotional behaviour? Are they recessive, or what is their degree of dominance? Of penetrance? These and many other questions will immediately occur to the geneticist. The physiologist will enquire after the precise mode of action of the reticular formation in its interaction with the cortex, and the visceral brain, and so forth. The theory suggests fruitful ways of conducting future research; being a weak theory it does not give us a complete blueprint, nor does it furnish us with the quantitative detail which the scientist requires before considering a problem closed. This referral of the problem to physiologists and others will undoubtedly offend some psychologists, particularly those with a firm belief in the 'empty organism', but to most people it will simply serve as a reminder of the obvious fact that the distinctions between psychology, physiology, genetics and other biological specialities are man-made, serve administrative and other practical purposes, but have no counterpart in nature. It may be convenient to cut the cloth in this way, but to make a suit it has to be sewn together again. Most psychological problems transcend psychology, and co-operative research is the only sensible answer to the problem posed by the administrative dissection of the unified science of biology. From this point of view it would appear wrong to characterize our third model as 'reductionist'; all that one might be tempted to say would be that it includes in its scope physiological and other elements that should never have been left out of account, and that belong indissolubly with psychological reality. As T. H. Huxley put it: 'No psychosis without a neurosis', meaning 'no psychic event without a corresponding physical or neurological event'.

Thus far we have forged a chain from anatomical detail of the structure of the reticular formation and the visceral brain, through their functional properties in keeping the cortex in a state of arousal and activating the autonomic system, to the results of laboratory experiments of various kinds. How is the chain of deduction and explanation lengthened to encompass such life-space phenomena as extraversion and introversion, or neuroticism, and how can the theory help us to account for the facts of criminality and neurosis? Many sub-hypotheses are, of course, required but by and large the chain of argument may be considered like this.

High cortical arousal leads to greater facility in forming conditioned responses, and these have been implicated theoretically in the formation of neurotic symptoms, particularly of a dysthymic character (Eysenck & Rachman, 1965), and also in the formation of an individual's 'conscience', which in turn prevents him from succumbing to temptation and indulging in crime, or in any case in anti-social and asocial activity (Eysenck, 1965 a). Other factors of course help in producing these various effects; the stimulus hunger of the extravert, due to his less efficient reticular formation, leads him to search for bright lights, loud music, and other less innocuous sources of sensory stimulation. At the same time, his greater pain tolerance, due to the same cause, makes him less susceptible to discouragement along traditional lines. These and many other chains of cause and effect, mediated through the various laboratory phenomena discussed, lead the individual equipped with an over-functioning reticular formation to the behaviour patterns which are characteristic of introversion, the individual equipped with an under-functioning reticular formation to the behaviour patterns which are characteristic of extraversion. Similarly with the highly emotional and the unemotional; low and high thresholds of autonomic arousal predispose the individual to react in certain organized and recognizable ways to stimulation from his environment. The cumulative effects of environmental stimulation, mediated through learning and conditioning, must of course not be under-estimated or left out of account; but equally should we be careful not to neglect the innate personality characteristics predisposing a person to react in this manner or that.

Some chains of argument and deduction from laboratory phenomena to real-life situations may seem at first sight to be little more than analogues. Is it just fanciful to see similarities in the laboratory behaviour of the extravert, who succumbs easily to monotony, who creates many 'blocks' or involuntary rest pauses for himself, and who craves for change, and his real-life behaviour, in which he does poorly at monotonous types of work, moves easily from job to job and from one place of residence to another, and is more prone than most to divorce and other modes of changing his sexual partners? There clearly are considerable difficulties in proving a direct connection, but on the other hand both sets of facts, the experimental and the real-life ones, can be deduced with some rigour from the underlying hypothesis, and thus 'explained' in terms of some general hypothetico-deductive system. In so far as falsification is clearly possible along empirical lines, we may perhaps say that even Popper would agree that in principle this is a scientific argument, unlike some other attempted 'explanations' in this field.

Little has in fact been said about alternative explanations, primarily because these either clearly fail to 'explain' in any real sense (e.g. the idiographic type of 'understanding' theory), or else are not in principle falsifiable, and hence outside the realm of science altogether (e.g. the psychoanalytic or 'dynamic' type of theory). The arguments regarding these different types of pseudo-explanations have been presented *in extenso* elsewhere (Eysenck, 1953) and will not be repeated here; they are more widely accepted nowadays than when they were first presented, and might seem quite obvious and self-evident nowadays. These systems constitute, in fact, purely verbal solutions of no predictive or explanatory relevance. (Note that the terms 'predictive' and 'explanatory' may be regarded as being almost synonymous in a scientific sense; 'explanation' is merely 'post-diction', i.e. the deduction of facts already known from certain laws and principles, whereas prediction refers to the same process of deduction as applied to facts not yet ascertained. It is the process of deduction which is the hall-mark of science in any but the most elementary sense, and this process can be applied with equal force to facts known and as yet unknown—although of course the latter application is more impressive!)

In summary, we may perhaps restate our general position. It is felt that the term 'explanation' should be used in psychology with exactly the same connotations as it is used in the exact sciences. This means essentially that *description*, quantitative if possible, is the appropriate synonym, rather than *understanding*; understanding, in some non-scientific, humanistic, idiographic sense is a function of many things, including present levels of arbitrary beliefs and superstitions, and cannot be checked against reality. To revert to an example already given, if you feel that the orbital aberrations of Mercury can be 'explained' by reference to God's intervention, then no more need be said; such feelings are outside science. The two meanings of explanation are clearly mirrored in the two senses in which the term 'to know' is used in these two sentences: 'I know that $V = H \times R$, where $V$ is the velocity at which a galaxy is receding from the solar system, $R$ is its distance and $H$ is Hubble's constant', and 'I know that my Redeemer liveth'. The first meaning is embodied in a sentence which may be wrong, but is scientific because it is amenable to falsification; the second meaning is embodied in a sentence which may be right, but is unscientific because it is not amenable to falsification. Explanation, in order to have any value in psychology, must emulate the former and not the latter; it is unfortunate that historically explanation in the personality field has so often chosen the wrong way and opted for the primrose path of dalliance with 'understanding'. Understandable though this may be in historical and human terms, it has been a misfortune for the scientific explanation of human conduct, which has been delayed considerably because of this erroneous choice. Fortunately there are signs that better counsels are beginning to prevail.

### REFERENCES

Burt, C. 1965. Factorial studies of personality and their bearing on the work of the teacher. *Br. J. educ. Psychol.* **35**, 368–78.

Colquhoun, W. P. & Corcoran, W. J. 1964. The effects of time of day and social isolation on the relationship between temperament and performance. *Br. J. soc. clin. Psychol.* **3**, 226–31.

Eysenck, H. J. 1947. *The Dimensions of Personality*. London.

Eysenck, H. J. 1953. *The Scientific Study of Personality*. London.

Eysenck, H. J. 1957. *The Dynamics of Anxiety and Hysteria*. London.

Eysenck, H. J. 1959. Serial position effects in nonsense syllable learning as a function of interlist rest pauses. *Br. J. Psychol.* **50**, 360–2.

Eysenck, H. J. 1960a. *The Structure of Human Personality*. London.

Eysenck, H. J. 1960b. *Experiments in Personality*. London.

Eysenck, H. J. 1965a. *Crime and Personality*. Boston.

Eysenck, H. J. 1965b. Extraversion and the acquisition of eyeblink and GSR conditioned responses. *Psychol. Bull.* **63**, 258–70.

Eysenck, H. J. 1966. Personality and experimental psychology. *Bull. Br. psychol. Soc.* **62**, 1–28.

Eysenck, H. J. 1967. *The Biological Basis of Personality*. Springfield.

Eysenck, H. J. & Eysenck, S. B. G. 1967. *The Description and Measurement of Personality*. London.

Eysenck, H. J. & Rachman, S. 1965. *The Causes and Cures of Neuroses*. London.

Hesse, M. B. 1961. *Forces and Fields*. London.

Hobson, E. W. 1923. *The Domain of Natural Science*. London.

Hoyle, F. 1962. *Astronomy*. London.

Janke, W. 1964. *Experimentelle Untersuchungen zur Abhängigkeit der Wirkung psychotroper Substanzen von Persönlichkeitsmerkmalen.* Frankfurt.

Jensen, A. R. 1962. An empirical theory of the serial-position effect. *J. Psychol.* **53**, 127–42.

Jones, M. B. & Fennell, R. S. 1965. Runway performance by two strains of rats. *Q. Jl Fla Acad. Sci.* **28**, 289–96.

Lundin, R. W. 1961. *Personality.* New York.

McClearn, G. E. 1962. Genetic differences in the effect of alcohol upon behaviour of mice. *Proc. Third Internat. Conf. Alcohol and Road Traffic.* London.

Munkelt, P. 1965. Personlichkeitsmerkmale als Bedingungsfaktoren der Psychotropen Arzneimittelwirkung. *Psychol. Beiträge* **8**, 98–183.

Nagel, G. 1961. *The Structure of Science.* New York.

Weisen, A. 1965. Differential reinforcing effects of onset and offset of stimulation on the operant behaviour of normals, neurotics and psychopaths. Unpublished Ph.D. thesis, Univ. of Florida.

# COMMENT

## *by* D. BANNISTER

Three aspects of the essay will not be dealt with in any detail for the reasons given below.

First, the concept of explanation as defined by 'understanding' and thus including 'arbitrary beliefs and superstitions' and so forth is clearly a straw man put up to be knocked down. Similarly the closing references to 'dynamic' theories seems to be a conjuring of demons so that they may be ritually exorcized. Neither are relevant to the primary argument.

Secondly, the series of anecdotes about the use of the hypothetico-deductive method in the natural sciences will not be discussed since the whole argument hinges on a clear *non sequitur*. Even if it were a fact that there has been a confused use of the hypothetico-deductive method in the natural sciences this would not make the method 'more relevant to the muddled state of research in psychology'. The fact that a method is used in a muddled way in some areas does not make it either more or less relevant to other 'muddled' areas.

Thirdly, the argument that 'neo-behaviourists' should pay attention to personality concepts does not seem a profitable one to pursue. The aim of science is not prediction at any price. A scientist uses predictive tests to explore the explanatory capacity of a group of linked concepts. To bring into play bundles of different types of concept might increase accuracy of prediction within a specific experiment, but it would hinder the systematic elaboration of the concepts selected as a primary focus. To a psychologist intent on experimentally elaborating $S-R$ concepts, personality concepts are irrelevant and the effect of personality variables is legitimately part of the error term—just as the effect of sociological, physiological and other types of variable is irrelevant. It is a frequent practice in psychology to call a concept a 'variable', thereby implying that it is a 'thing' of which others *must* take note. Additionally this issue is confused by argument by analogy. The physicist knowing that 'gold has a different specific gravity to lead' is arbitrarily equated with a psychologist knowing that individual differences exist, when it could equally arbitrarily be equated with a knowledge of functional $S-R$ relationships. Nor does the logic become clearer when the section concludes with an *argumentum ad hominem*—the assertion that these neo-behaviourists hold their views at an 'unconscious' level. Incidentally, what type of 'explanation' is this?

# EXPLANATION AND THE CONCEPT OF PERSONALITY

One final problem which the essay presents to any would-be commentator lies in its lack of relation to its title. It is entitled 'Explanation and the Concept of Personality' but it deals with 'Explanation and the *Trait* Concept of Personality'. The nature of the explanations yielded by other types of personality theory are not dealt with. It is beyond the scope of a commentary to present such theories but by way of passing illustration the standpoint taken in this commentary is that of personal construct theory (Kelly, 1955).

A specifically relevant aspect of this theory is that it provides a concept of 'explanation' integral to the theory itself. Construct theory is based on the assumption that all men may be thought of as 'scientists' in the sense that each is concerned with the prediction and control of his environment. Each individual develops his own personal network of constructs by means of which he structures (interprets, conceptualizes, attaches meaning to) his world and tries to anticipate events. These constructs may be thought of as the elements of a system by means of which the individual codifies his experience. Thus to 'explain' an event is to place a construction on it which (by virtue of the links between the construct newly applied and other constructs within the network) generates a series of expectations and proffers an interpretative account of the origins of the event. The 'explanations' of scientists may be more systematically articulated and formally tested than those of ordinary mortals but they are not, in construct theory terms, of an essentially different kind. The procedures of science developed into a formal mode out of 'lay' construing processes.

## TRAIT PSYCHOLOGY

In the essay, the argument that trait/type psychology is tautological is rebutted by equating it with the simple assertion that trait psychology consists only of the circular statement that a person behaves in, say, an extraverted way because he is a person of extraverted type. It is then fairly claimed that various testable elaborations can be formulated for a trait theory. This does not, however, deal with the argument that trait psychologies are tautological in two senses—they inhibit the development of concepts of process and change and they produce unelaboratable concepts of original cause.

In a non-trait personality theory it is possible to have concepts of change occupying a definitive position. Thus construct theory argues that man can be viewed as a form of motion, not as a static object that is occasionally kicked into movement. Thus intra-person change can be subsumed under the thesis that an individual's superordinate constructs change slowly relative to rapid subordinate change; intra-person differences in change rate are described in terms of differing degrees of permeability in construct systems and this again may vary intra-person across construct sub-systems. In a trait theory we are committed to identifying hypothesized samenesses in an individual and change is not so much a focus for scientific exploration as a source of experimental 'error'. Curiously enough trait psychology has yielded massive numbers of cross-sectional studies and very few of the longitudinal studies which would seem an appropriate empirical foundation for it. Perhaps any lengthening of the experimental time line renders trait theory so rapidly unviable that the habit died early. Linked to the static categories of trait psychology is the intriguing assertion (usually presented as a sort of categorical imperative) that 'Accurate description, typology, nosology or taxonomy must precede historically causal explanation, aetiology, genetics or evolution'. Presumably Harvey should have left alone the idea of a circulatory process system for the blood until the question of blood groups had been properly settled. Concepts of process can be grouped to define condition, concepts of condition serve as markers against which to define process. To say that taxonomy must precede causal explanation means as little as saying the reverse. Categories of kind and transition are necessary contrasts, integral to all thinking, not an intellectual horse and cart with a necessary sequence.

To construe a person as having some fixed psychological nature seems inevitably to make the 'cause' of this nature something equally fixed and ultimately unanalysable. Thus traits tend to be seen as 'constitutional' and thus 'genetic'. The question of the logical viability of the reductionist argument will be discussed later but for the moment it is stressed that the 'genetic' argument is ultimately untestable. Quite apart from any dispute as to the present state of evidence all that could possibly be shown would be an observable relationship between operational definitions in psychology (e.g. questionnaires) and operational definitions in biology (e.g. kinship definitions). The superordinate concepts to which each of these operational definitions are attached are cast in different language systems and cannot be included as parts of a single, integral argument. Hence no argument exists which could be tested. The mismatching of the two language systems is explicit within the essay. Nagel is quoted as saying that 'The task of genetic explanations is to set out the sequence of major events through which some earlier system has been transformed into a later one.' It is pointed out that Nagel saw genetic explanations as dealing largely with social events and then it is claimed that 'the concept covers equally well the laws of inheritance'. The argument that people of type $X$ inherit the structure which causes type $X$ behaviour does not 'set out the sequence of major events through which' the earlier system has been transformed into a later one. What are the differences between the earlier and the later systems? In what language do we specify what major events? The source difficulty may be Eysenck's confusion between 'genetic' meaning causal and 'genetic' meaning inherited.

## REDUCTIONISM

The essay specifically argues that explanation in psychology ultimately requires that concepts be anchored into other sciences and specifically into neurophysiology. A fuller account of the problems posed by reductionism is available in Jessor (1958); here the difficulties are briefly indicated.

Kelly (1955) suggested that the core of the reductionist fallacy lay in speaking of psychological *events* and physiological *events*. It is certainly likely that if you scratch a reductionist you will find a naive realist. Psychology and physiology are *modes of construing* and events are no more intrinsically psychological or physiological than they are sociological, political, chemical or geographical. On reality we impose structure and any event can be alternatively construed in many ways. A girl crossing a bridge can be construed as 'a series of moments of force about a point', 'a likely dish', 'a mass of whirling electrons', 'a poor credit risk' and so forth. She is not *really* any one of these things, yet each construction can have predictive value. Each construction is derived from a network of constructs which defines it and generates predictions. These sub-systems are not necessarily cross-referrable, they may not translate or 'reduce'. Thus the question of reductionism is basically one of construct relationships and while psychologists have massed together psychological and physiological concepts there is, as yet, no sign that they can be integrated. In a technology, as contrasted with a science, assorted types of construing can be used together. For example, sailing ship navigators successfully used concepts from astronomy, the physics of leverage and oceanography but they did not thereby integrate these disciplines. A brief inspection of the construct sub-systems of physiology and psychology suggests that cross-translation is unlikely to be achieved. Psychology traffics in concepts (e.g. stimulus and response) whose definitions inevitably involve reference to an external environment while physiology traffics in 'sub-dermal' concepts; physiology as a mode of construing can envisage a deterministic system while psychology cannot (see Scriven's (1965) argument that given equal conceptualizations of the situation by subject and experimenter unlimited prediction is not even theoretically possible). Physiology envisages systems which are independent within each individual while psychology envisages interactive individuals. Physiological arguments need not be reflexive while psychology may need to be reflexive (see later discussion); and so forth. The problem presents itself in fairly sharp form if we stop and ponder the oft and here quoted *dictum* that there is 'no psychic event without a corresponding physical or neurological event'. If we abandon the notion that we can identify an event and accept the idea that we can only interpret it then this statement becomes not only unfalsifiable in current practice, it seems doubtful if it can carry any useful meaning. Out of all the myriad of constructions we can place on a piece of behaviour (to say nothing of the alternative neurological constructions) which are going to be the 'corresponding' ones, on what basis would we select them? Into what 'bits' should we section these events?

If 'Paul shoots a gun' do we tally our 'corresponding physiological event' with this behaviour or with 'Paul kills a man' or with 'Paul thought he was shooting a moose' or with 'Paul pointed a straight object'? Are these temporally the wrong 'units' of behaviour?

An equally semantically puzzling *dictum* is the statement 'most psychological problems transcend psychology'. If we erect a problem in purely empirical terms (e.g. how can I get to Chapel-en-le-Frith) then the answer may involve data from several disciplines. If we are working as scientists then our purpose can be argued to be the elaboration of the construct sub-system (couched in an integral language) which is our science and the extension of its range of convenience by checking the predictive capacity of various alternatives within it. From this point of view a *psychological* problem cannot transcend psychology unless we originally stated it in sloppy (and, for our purposes, non-psychological) terms—analogies about men sewing suits together notwithstanding.

Perhaps the key to misconceptions in this type of reductionism is the notion that problems (or phenomena or areas of study) exist somehow independently of the sciences which define them. A chemical 'problem' is one which is stated *in chemical terms* and a psychological problem is so because it represents alternative lines of implication for a group of *psychological* concepts (this is why it is a problem). As such it cannot be solved in non-psychological terms. What may happen is that some other problem involving similar operational definitions is set up in other (e.g. neurophysiological) terms. This can be solved in such neurophysiological terms but the psychologically defined problem has not thereby been 'transcended' or 'reduced', it remains to be solved in its own terms.

415

# EXPLANATION AND THE CONCEPT OF PERSONALITY

Throughout the essay it is fairly assumed that psychology is, by intention, a science. This seems to be taken to mean that we must accept the natural sciences as an exact model—except possibly for the 'weak' theory argument.

This weak theory proposal in itself seems no major issue since it can be argued that all theories are in a most important sense 'weak'—it is hypotheses that are sharply falsifiable while theories are necessarily to some degree elastic since they are designed to cover events not specifically envisaged at the time of their construction. Rather than posit a formal weak–strong dichotomy we might note that the clarity with which hypotheses can be derived and tested is a function of the degree to which the network of implications constituting the theory is articulated. In turn the process of hypothesis-derivation and testing is itself designed to articulate the network. (Incidentally, the syllogism presented to illustrate the weak theory argument is not merely weak—it is downright sickly, in that it contains no indication as to whether its premises and conclusions are universals or particulars.) It is not so much that particular theories or sciences are weak or strong as that any theory, as it is progressively elaborated, may be alternatively presented weakly or strongly in terms of the observed relationship between modification and evidence. Thus in the course of experimental elaboration a theory may be presented in alternatively more committed and cautious terms as the evidential going gets rougher or smoother. Compare the specificity of some of the contentions in the essay with, say, Eysenck's (1962) declaration that evidence '*seems* sufficient *to indicate* that the *suggestion* of a significant relation between introversion and conditioning *may not be altogether mistaken*' (present author's italics).

Of the requirements which make a concretistic imitation of the natural sciences unprofitable in psychology, probably the most serious is that of reflexivity. Formulating theories, conducting experiments and modifying explanatory concepts in terms of outcome is a part of human behaviour and any discipline which purports to be 'the science of human behaviour' should account for its own construction. The arguments for reflexivity have been well canvassed (e.g. Oliver & Landfield, 1963) and here only one or two salient points will be dealt with.

The essay under consideration does not raise the issue of reflexivity, much less deal with it, and in this it follows the tradition of advocating a simple imitation of the natural sciences. The natural sciences face no such issue as reflexivity—there is no onus on the chemist presenting an acid–alkali theory to account, in terms of an acid–alkali distinction, for his behaviour in writing a paper on it (Bannister, 1966).

The implications of making psychological explanations reflexive are manifold and three major consequences are briefly referred to.

First, from a reflexivist standpoint the psychological experiment is not a logical copy of the natural science experiment. It is inevitably a social situation in which one professional, formally qualified theorizer and predictor tries to predict the behaviour of non-formally qualified theorizers and predictors. The nearly invariable practice of trying to keep the subject in ignorance of the purpose of the experiment is a fore-doomed attempt to deal with this problem (cf. the experimenter bias effect—Rosenthal, 1964 *et seq.*).

Secondly, if the reflexive argument is accepted the psychologist cannot present a picture of man which patently contradicts his behaviour in presenting that picture.

Thirdly, from a reflexive viewpoint no theory in psychology can be final, even in a speculative sense, since once a theory has been formulated and has become general property, behaviour influenced by cultural awareness of the theory must additionally be accounted for.

The acceptance of the need for a reflexive stance in psychology does not mean that psychologists cannot be 'scientific', it means that they cannot routinely copy other sciences but must face the problem of what 'being scientific' in psychology involves.

Two arguments which seek to justify a kind of wilful simple mindedness in psychology might be mentioned in this context.

First, the reductionist view that 'lower order', 'simpler' activities must be studied first and more 'complex' ones later. What is lower order and simpler in this context, of course, are not the activities but the concepts used to subsume them. If, as seems to be the case, these simpler concepts cannot be hierarchically linked to higher levels of abstraction then they represent an intellectual cul-de-sac.

Secondly, the argument used in the essay that psychology has special problems because it is a 'young' science is a fallacious platitude. Psychology, in any practical sense, is not a young science. The vast majority of scientists who have ever lived are alive today, and the addition to the labour force of the natural sciences of the few isolated figures who represented them in earlier centuries gives them little edge over psychology. Enough psychologists have been at work to make us, in man-years expended, quite an old science. Additionally, psychology developed on the basis of a sophisticated mathematics and philosophy of science not available in the earlier life of the natural sciences. If psychology is unimpressive it is not because it is 'young' but because though we have amassed an enormous literature, we have signally failed to solve certain problems fundamental to the discipline.

## CONCLUSION

Within the limits of a brief commentary it is argued that the essay is important mainly for its modal quality. For some time this mixture of trait-thinking and physiologizing, linked to a concretistic and self-congratulatory view of 'science' has been a staple of personality psychology and of concepts of explanation current in the field. Its supersession may be as significant an event in the development of psychology as the original break with lay psychology which this type of personality theory represents.

## REFERENCES

Bannister, D. 1966. Psychology as an exercise in paradox. *Bull. Br. psychol. Soc.* **19**, 21.

Eysenck, H. J. 1962. Conditioning and personality. *Br. J. Psychol.* **53**, 299.

Jessor, R. 1958. The problem of reductionism in psychology. *Psychol. Rev.* **65**, 170.

Kelly, G. A. 1955. *The Psychology of Personal Constructs*, vols. 1 and 2. New York.

Oliver, W. D. & Landfield, A. W. 1963. Reflexivity: an unfaced issue of psychology. *J. indiv. Psychol.* **20**, 187.

Rosenthal, R. 1964. The effect of the experimenter on the result of psychological research. *Progress in Experimental Personality Research*, vol. 1. Ed. B. A. Maher. New York.

Scriven, M. 1965. An essential unpredictability in human behaviour. *Scientific Psychology*. Ed. B. J. Wolman and E. Nagel. London.

# REPLY

## *by* H. J. EYSENCK

Bannister's commentary on my paper shows his usual sparkling, coruscating style, but is rather spoilt by a certain crash-bang-wallop attitude which relies on assertion, rather than on argument. As an example, consider his assertion that my concept of explanation by 'understanding' is 'clearly a straw man put up to be knocked down'; furthermore, it is not 'relevant to the primary argument'. Now terms such as 'clearly' are seldom used when arguments or facts are at hand to rebut an opponent's point of view; it is when arguments and facts are lacking that their absence is disguised by such weasel words. Even if it were clear to Bannister that this was an irrelevant man of straw, it is not clear to the writer, and may not be clear to the reader; some form of argument would seem needed. As it is, no answer is possible to Bannister as he gives no reasons for his statement; *obiter dicta* cannot easily be argued about. Thus much of what he says has to remain without a reply as it consists of nothing but *ex cathedra* statements; the reader must ascertain their value by reference to my original article.

At other times Bannister resorts to innuendo to cover a failure to look at the literature, or give a true impression of its contents. Consider his argument in favour of longitudinal studies; he says: 'Perhaps any lengthening of the experimental time line renders trait theory so rapidly unviable that the habit died early.' Here we have the (factually incorrect) suggestion that a few early studies failed to disclose consistency over time, and the (personally offensive) suggestion that trait theorists for that reason discontinued this line of work. It is only necessary to look at the work of Burt, mentioned in my original paper, or the Californian studies of Nancy Bailey, or the studies mentioned in my *Biological Basis of Personality*, to see that Bannister is quite wrong in suggesting that longitudinal studies rendered trait theory 'unviable'; quite the opposite is true. Trait theory finds some of its strongest support in this field. Readers familiar with the field will know how to evaluate the accuracy of Bannister's criticisms in the light of this monumental error, which is not untypical of his writing. His innuendo, based upon this error, that trait theorists abandoned this line of work because it gave evidence against their view is an *argumentum ad hominem* which hardly needs rebuttal; responsible scientists just don't behave like this. Even if Bannister's assessment of the situation had been a true one, there are other reasons for not undertaking longitudinal studies than the wish to save a useless theory. Such studies are extremely difficult to mount, impossibly expensive to run, and take some thirty years to mature; few scientists are in a position to undertake this sort of work. What is surprising is rather that in spite of the difficulties so many studies do in fact exist.

When Bannister does consent to argue, his argument often seems arbitrary and difficult to follow. It is an integral part of my view that experimental psychology, in the narrow sense of that term, and personality theory must learn to live together and pay attention to each other's findings and concepts, as otherwise experimentalists will find themselves left with unwieldy and large error terms, while personality theorists will not have available proper concepts in terms of which to 'explain' their findings. Take a recent example. There is a large literature on verbal reminiscence, summed up by one reviewer in the words: 'Now you see it, now you don't'—in other words, the phenomenon is rather difficult to pin down. I have suggested that under certain circumstances introverts might show reminiscence, extraverts forgetting; overall there would then be just no apparent effect at all. The prediction was arrived at by way of a lengthy chain of theorizing; briefly it is suggested that introverts are habitually working in a state of higher cortical arousal, that arousal facilitates consolidation of the memory trace, and hence that introverts would in the long run remember better. In the short run, however, consolidation interferes with reproduction, so that in the short run introverts would show poor recovery of the memory trace. Over time, therefore, introverts would improve their performance, extraverts would show the opposite trend. Figure 3 shows the results of an experiment carried out by Howarth and myself to test this hypothesis, using paired CVC nonsense syllables, learned to one perfect repetition, as the material; various introverted and extraverted groups were retested at various intervals after learning. It will be seen that the predicted cross-over does in fact take place; introverts show a reminiscence effect, extraverts a forgetting effect. Is it really sensible to say, as Bannister does, that 'to a psychologist intent on experimentally elaborating S–R concepts, personality concepts are irrelevant and the effect of personality variables is legitimately part of the error term'? By not paying attention to personality variables we might be faced with a situation in which one writer (using a somewhat introverted group of subjects—unknowingly of course, because Bannister forbids him to look at personality!) finds reminiscence in such an experiment, while another (using a somewhat extraverted group of subjects—equally unknowingly) finds forgetting instead. This is precisely what has apparently happened time after time in the literature; one person's results were difficult or even impossible to replicate by another. The theory I have put forward may of course be wrong, but at least it does give some promise that we can now bring these phenomena under proper experimental control —by measuring personality as well as other experimental variables. It just does not make sense to omit a relevant variable knowingly, simply because of some odd preconception about the purity of $S$–$R$ connectionism—which in any case has long since been replaced, even for behaviourists, by the

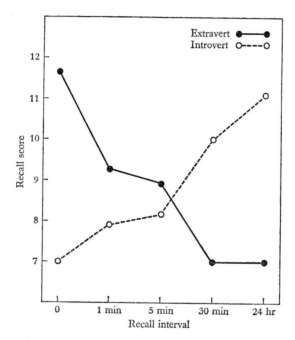

Fig. 3. Recall scores of introverted and extraverted subjects after different recall intervals. Ten groups of eleven subjects were used in all.

$S$–$O$–$R$ formula. Properties of the organism do unquestionably determine the precise mode of operation of the stimulus–response contingencies, and they must be taken into account; I see nothing in Bannister's argument that would cause me to change my mind on this point.

Another point to which Bannister takes exception is the fact that I have dealt with personality in a rather restricted sense; 'the nature of the explanations yielded by other types of personality theory are not dealt with'. Trait theory appears to rouse Bannister's ire, although why he should feel that way about a theory he seems to consider defunct is not clear. Perhaps he objects to the fact that it refuses to lie down, and seems to go on to bigger and better triumphs. As I have pointed out in *The Biological Basis of Personality*, it is now giving us the promise of a unification of personality theory, experimental psychology, and physiological theory; one could hardly ask for more. Bannister apparently prefers Kelly's 'personal construct theory', but this can hardly be regarded as a theory in the scientific sense; it does not explain anything in the sense in which I have used that term, and consequently I could see no reason for discussing it (and many others like it) in any detail. The reader is invited to consider the degree to which Kelly's theory can be integrated with our current knowledge of other areas of psychology, or physiology, or genetics; or the predictions which it makes possible; or the degree to which it relates to modern methods of changing personality, such as behaviour therapy. Such a review will make him somewhat doubtful of the rather excessive advocacy of this theory which is offered by Bannister.

One of Bannister's central complaints is that the theory I am advocating is static; it does not allow for change. This is a curious charge; I entitled my 1957 book *The Dynamics of Anxiety and Hysteria* precisely because in going beyond simple factor analysis and description I thought I had gone beyond a static picture and had arrived at a formulation which gave equal value to statics and dynamics. What could be more dynamic, more prone to change and the mediation of change, than conditioning and learning? These are two of the concepts I have tried to integrate with personality theory; the effort may not have been as successful as might have been desired, but Bannister does not argue on experimental or factual grounds— he simply dismisses the whole approach, without paying attention to the actual details involved. All this is therefore quite beside the point, even where it is intelligible; some of his comments are not even that. What is one to say to the statement that 'the genetic argument is ultimately untestable'? I have reviewed the evidence in some detail in *The Biological Basis of Personality*, and will not attempt to do so here; in the same book I have also gone in detail into the possible criticisms of the methods used, and the answers to these criticisms. Bannister's statement baffles me, and the state of being baffled is not conducive to producing counter-arguments to what is essentially another assertion, rather than an argument.

Bannister's treatment of reductionism and reflexivity seems to me philosophical rather than useful. If introverts and extraverts, to take just one example, are characterized by physiological differences (such as EEG patterns), then I do not see why the psychologist should disregard this evidence, just because Bannister feels that 'cross-translation is unlikely to be achieved' between psychology and physiology. In particular, when the observed differences are in a direction predicted by theory, and extremely relevant to that theory, then it seems to me an unreasonable act of self-denial to keep these facts in splendid isolation from 'pure' psychology. There is in psychology far too much of this tendency to restrict oneself to a narrow and unrealistically circumscribed set of data; important findings are much more likely to occur on the borderlines between such arbitrary divisions.

As far as the difficulties presented by reflexivity go, these seem largely imaginary, or at best to depend on what one is trying to do. If I wish, as Burt did in his paper referred to a number of times, to predict which of a group of children is likely to end up as a neurotic, or as a habitual criminal, then I do not see how reflexivity comes into the picture. If I wish to construct a psychological or a physiological theory to account for Burt's findings, again reflexivity is irrelevant. If I wish to test a theory of behaviour modification for criminals or neurotics, based on the concepts involved in account for their development, then I fail to see why reflexivity should prevent me. In other words, as far as the explanation of behaviour, or its experimental change are concerned, it is difficult to understand Bannister's concern with reflexivity; we might just as well worry about solepsism! These are philosophical notions which are essentially irrelevant to experimental psychology, and its search for explanatory concepts; even modern philosophers may come to the conclusion that they are verbal traps, rather than genuine problems. Until their relevance can be experimentally demonstrated, I cannot take them too seriously.

Perhaps the major difficulty in finding a common meeting ground with Bannister is his tendency to argue in abstract terms, rather than talk in concrete instances. Consider his statement that a psychological problem cannot be solved by, or reduced to, physiological terms, 'it remains to be solved in its own terms'. Now this sounds fine, but has it any meaning? Take a specific example. Extraverted children who are also high on emotionality tend to grow up to become habitual criminals. Why? This is a problem. My explanation is that these children have inherited a reticular formation which has high thresholds to incoming stimulation; hence it provides the cortex with too little arousal, and causes it to be rather slow and weak in mediating conditioned responses which I hold to be fundamental in elaborating a 'conscience'. Other aspects of extraverted behaviour have been traced to the same source, i.e. lack of cortical arousal. It is assumed that the thresholds of the reticular formation are determined largely by genetic causes, i.e. an individual inherits a predisposing neural structure which, in interaction with the environment, determines his criminal conduct. This theory may, like all scientific theories, be erroneous, and it is almost certainly oversimplified, and does not account for all the facts, and may not account entirely for any of the facts. But what does it mean to say that 'the problem remains to be solved in its own terms?' The words are there to be read, but the sentence as such carries no meaning at all. One keeps wishing that Bannister had taken the original argument sufficiently seriously to use even one of the many examples given as a vehicle to make his generalizations less abstract, and more relevant to the theory which he is purportedly criticizing. Abstract arguments may sound splendid, but their meaning often eludes the grasp of the scientist— whether because of his inability to understand, or their insubstantial nature must remain a point at issue.

# PROBLEMS OF EXPLANATION IN LINGUISTICS†

## *by* NOAM CHOMSKY

One difficulty in the psychological sciences lies in the familiarity of the phenomena with which they deal. A certain intellectual effort is required to see how such phenomena can pose serious problems or call for intricate explanatory theories. Rather, one is inclined to take them for granted as necessary or somehow 'natural'.

Effects of this familiarity of phenomena have often been discussed. Wolfgang Köhler, for example, suggests that psychologists do not open up 'entirely new territories' in the manner of the natural sciences 'simply because man was acquainted with practically all territories of mental life a long time before the founding of scientific psychology . . . because at the very beginning of their work there were no entirely unknown mental facts left which they could have discovered' (Köhler, 1940). The most elementary discoveries of classical physics have a certain shock value—man has no intuition about elliptical orbits or the gravitational constant. But 'mental facts' of even a much deeper sort cannot be 'discovered' by the psychologist, because they are a matter of intuitive acquaintance, and, once pointed out, obvious.

There is also a more subtle effect. The familiarity of phenomena can be so great that we really do not see them at all, a matter that has been much discussed by literary theorists and philosophers. For example, Viktor Shklovskij, in the early 1920s, developed the idea that the function of poetic art is that of 'making strange' the object depicted. 'People living at the seashore grow so accustomed to the murmur of the waves that they never hear it. By the same token, we scarcely ever hear the words which we utter . . . We look at each other, but we do not see each other any more. Our perception of the world has withered away; what has remained is mere recognition.' Thus the goal of the artist is to transfer what is depicted to the 'sphere of new perception'; as an example, he cites a story by Tolstoy in which social customs and institutions are 'made strange' by the device of presenting them from the viewpoint of a narrator who happens to be a horse.[1]

† This work was supported in part by the U.S. Air Force (ESD Contract AF19(628)-2487) and the National Institutes of Health (Grant MH-13390-01). I am indebted to John Ross for helpful comments.
[1] Cf. Ehrlich, 1955, pp. 150 f.

The observation that 'we look at each other, but we do not see each other any more' has perhaps itself achieved the status of 'words which we utter but scarcely ever hear'. But familiarity, in this case as well, should not obscure the importance of the insight.

Wittgenstein makes a similar observation, pointing out that 'the aspects of things that are most important for us are hidden because of their simplicity and familiarity (one is unable to notice something—because it is always before one's eyes)' (*Philosophical Investigations*, 129). He sets himself to 'supplying . . . remarks on the natural history of human beings: we are not contributing curiosities however, but observations which no one has doubted, but which have escaped remark only because they are always before our eyes' (*Philosophical Investigations*, 415).

Less noticed is the fact that we also lose sight of the need for explanation when phenomena are too familiar and 'obvious'. We tend too easily to assume that explanations must be transparent and close to the surface. The most serious defect of classical philosophy of mind, both rationalist and empiricist, seems to me to be its unquestioned assumption that the properties and content of the mind are accessible to introspection; it is surprising to see how rarely this assumption has been challenged, in so far as the organization and function of the intellectual faculties are concerned, even with the Freudian revolution. Correspondingly, the far-reaching studies of language that were carried out under the influence of Cartesian rationalism suffered from a failure to appreciate the abstractness of those structures that were 'present to the mind' when an utterance is produced or understood, and the length and complexity of the chain of operations that relate the mental structures expressing the semantic content of the utterance to the physical realization.[1]

A similar defect mars the study of language and mind in the modern period. The structuralist and behaviourist approaches to these topics are grounded in a faith in the shallowness of explanations, a belief that the mind must be simpler in its structure than any known physical organ and that the most primitive of assumptions must be adequate to explain whatever phenomena can be observed. Thus it is taken for granted without argument or evidence (or is presented as true by definition) that a language is a 'habit structure' or a network of associative connections, or that knowledge of language is merely a matter of 'knowing how', a skill expressible as a system of dispositions to respond. Knowledge of language, accordingly, must develop slowly through repetition and training, its apparent complexity resulting from the proliferation of very simple elements rather than

---

[1] See my *Cartesian Linguistics* (Chomsky, 1966), for some discussion of rationalist linguistic theory and the theories of mind that provided a background for it.

from deeper principles of mental organization which may be as inaccessible to introspection as the mechanisms of digestion or co-ordinated movement. Although there is nothing inherently unreasonable in an attempt to characterize knowledge of language in these terms, this attempt also has no particular plausibility or *a priori* justification. There is no reason why one should react with uneasiness or disbelief if study of the knowledge of language and the exercise of this knowledge should lead in an entirely different direction.

I think that in order to achieve progress in the study of language and human cognitive faculties in general it is necessary first to establish 'psychic distance' from the 'mental facts' to which Köhler made reference, and then to explore the possibilities for developing explanatory theories, whatever they may suggest with regard to the complexity and abstractness of the underlying mechanisms. We must recognize that even the most familiar of phenomena require explanation and that we have no privileged access to the underlying mechanisms, no more so than in physiology or physics. Only the most preliminary and tentative hypotheses can be offered concerning the nature of language, its use and acquisition. As native speakers, we have a vast amount of data available to us. For just this reason it is easy to fall into the trap of believing that there is nothing to be explained, that whatever organizing principles and underlying mechanisms may exist must be 'given' as the data is given. Nothing could be farther from the truth, and an attempt to give a precise characterization of the system of rules that we have mastered, the system that enables us to understand new sentences and produce a new sentence on an appropriate occasion, will quickly dispel any dogmatism on this matter. The search for explanatory theories must begin with an attempt to determine these systems of rules and to reveal the principles that govern them.

The person who has acquired knowledge of a language has internalized a system of rules that relate sound and meaning in a particular way. The linguist, constructing a grammar of a language, is in effect proposing a hypothesis concerning this internalized system. The linguist's hypothesis, if presented with sufficient explicitness and precision, will have certain empirical consequences with regard to the form of utterances and the ways in which they are interpreted by the native speaker. Evidently, knowledge of language—the internalized system of grammatical rules—is only one of the many factors that determine how an utterance will be used or under-stood or perceived in a particular situation. The linguist who is trying to determine what constitutes knowledge of a language—to construct a correct grammar—is studying one fundamental factor that is involved in performance, but not the only one. This idealization must be kept in mind when considering the problem of confirmation of grammars on the basis of empirical evidence. There is no reason why one should not also study the interaction of several factors that are involved in complex mental acts and that underlie actual performance, but such a study is not likely to proceed very far unless the separate factors are themselves reasonably well understood.

In a good sense, the grammar proposed by the linguist is an explanatory theory; it suggests an explanation for the fact that (under the idealization mentioned) a speaker of the language in question will perceive, interpret, form or use an utterance in certain ways and not in other ways. One can also search for explanatory theories of a deeper sort. The native speaker has acquired a grammar on the basis of very restricted and degenerate evidence; the grammar has empirical consequences that extend far beyond the evidence. At one level, the phenomena with which the grammar deals are explained by the rules of the grammar itself and the interaction of these rules. At a deeper level, these same phenomena are explained by the principles that determine the selection of the grammar on the basis of the restricted and degenerate evidence available to the person who has acquired knowledge of the language, who has constructed for himself this particular grammar. The principles that determine the form of grammar and that select a grammar of the appropriate form on the basis of certain data constitute a subject that might, following a traditional usage, be termed 'universal grammar'. The study of universal grammar, so understood, is a study of the nature of human intellectual capacities. It tries to discover the innate organization that determines what constitutes linguistic experience and what knowledge of language arises on the basis of this experience. Universal grammar, then, constitutes an explanatory theory of a much deeper sort than particular grammar, although the particular grammar of a language also can be regarded as an explanatory theory. In practice, the linguist is always involved in the study of both universal and particular grammar. When he constructs a descriptive, particular grammar in one way rather than another on the basis of what evidence he has available, he is guided, whether consciously or not, by certain assumptions as to the form of grammar, and these assumptions belong to the theory of universal grammar. Conversely, his formulation of principles of universal grammar must be justified by the study of their consequences when applied in particular grammars. Thus at several levels, the linguist is involved in the construction of explanatory theories, and at each level, there is a clear psychological interpretation for his theoretical and descriptive work. At the level of particular grammar, he is attempting to characterize knowledge of a language, a certain cognitive system that has been developed, of course unconsciously, by the normal speaker–hearer. At the level of universal grammar, he is trying to establish certain general properties of human intelligence. Linguistics, so characterized, is simply the sub-field of psychology that deals with these aspects of mind.

I will try to give some indication of the kind of work now in progress that aims on the one hand to determine the systems of rules that constitute knowledge of a language and, on the other, to reveal the principles that govern these systems. Obviously, any conclusions that can be reached today regarding particular or universal grammar must be quite tentative, and restricted in their coverage. And in a brief sketch such as this, only the roughest outlines can be indicated.

The general framework that seems most appropriate for the study of problems of language and mind was developed as part of the rationalist psychology of the seventeenth and eighteenth centuries, and then largely forgotten as attention shifted to different matters. According to this traditional conception, a system of propositions expressing the meaning of a sentence is produced in the mind as the sentence is realized as a physical signal, the two being related by certain formal operations which, in current terminology, we may call *grammatical transformations*. Continuing with current terminology, we can thus distinguish the *surface structure* of the sentence, its organization into categories and phrases as a physical signal, from the underlying deep structure, also a system of categories and phrases, but with a more abstract character. Thus the surface structure of the sentence 'a wise man is honest' might analyse it into the subject 'a wise man' and the predicate 'is honest'. The deep structure, however, will be rather different. It will, in particular, extract from the complex idea that constitutes the subject of the surface structure an underlying proposition with the subject 'man' and the predicate 'be wise'. In fact, the deep structure, in the traditional view, is a system of two propositions, neither of which is asserted, but which interrelate in such a way as to express the meaning of the sentence 'a wise man is honest'. We might represent the deep structure, in this simple case, by the formula (1), and the surface structure by the formula (2), where paired brackets labelled with the symbol $A$ bound a phrase of the category $A$.

(1) $[_S [_{NP}$ *a man* $[_S [_{NP}$ *man* $]_{NP} [_{VP}$ *is wise* $]_{VP} ]_S ]_{NP} [_{VP}$ *is honest* $]_{VP} ]_S$.

(2) $[_S [_{NP}$ *a wise man* $]_{NP} [_{VP}$ *is honest* $]_{VP} ]_S$.

If we understand the relation 'subject-of' to hold between a phrase of the category noun phrase (NP) and the sentence (S) that directly dominates it, and the relation 'predicate-of' to hold between a phrase of the category verb phrase (VP) and the sentence that directly dominates it, then the structures (1) and (2) specify the grammatical functions of subject and predicate in the intended way. The grammatical functions of the deep structure (1) play a role in determining the meaning of the sentence. The phrase structure indicated in (2), on the other hand, is closely related to its phonetic shape. Specifically, it determines the intonation contour of the utterance represented.

Knowledge of a language involves the ability to assign deep and surface structures to an infinite range of sentences, to relate these structures appropriately, to assign a phonetic interpretation to the surface structure and to construct a semantic interpretation on the basis of the grammatical relations of the deep structure. This formulation slightly restates a traditional analysis, without, I think, doing violence to it.[1] Furthermore, this formulation seems to be quite accurate, as a first approximation to the characterization of 'knowledge of a language'.

How are the deep and surface structure related? Clearly, in the simple example just given, we can form the surface structure from the deep structure by performing such operations as the following:

(3)  (i)  Assign the marker *wh-* to the most deeply embedded NP, 'man';
    (ii)  replace the NP so marked by 'who';
    (iii) delete 'who is';
    (iv)  invert 'man' and 'wise'.

Applying just the operations (i) and (ii), we derive the structure underlying the sentence 'a man who is wise is honest', which is one possible realization of the underlying structure (1). If, furthermore, we apply the operation (iii) (deriving 'a man wise is honest'), we must, in English, also apply the subsidiary operation (iv), deriving the surface structure (2) which can then be phonetically interpreted.

If this approach is correct, then a person who knows a specific language has control of a grammar that *generates* (i.e. characterizes) the infinite set of potential deep structures and maps them on to associated surface structures, and determines the semantic and phonetic interpretations of these abstract objects. As far as information is now available, it seems accurate to propose that the surface structure determines the phonetic interpretation completely, and that the deep structure expresses those grammatical functions that play a role in determining the semantic interpretation, although certain aspects of the surface structure may also participate in determining the meaning of the sentence, in ways that I will not discuss here. A grammar

---

[1] For a detailed development of this point of view, see Katz & Postal, 1964, and Chomsky, 1965. These contain references to an earlier work that they extend and modify. Some criticisms are discussed in Chomsky, 1966.

of this sort will therefore define a certain infinite correlation of sound and meaning. It constitutes a first step towards explaining how a person can understand an arbitrary sentence of his language.

Even this artificially simple example serves to illustrate some properties of grammars that appear to be quite general. An infinite class of deep structures much like (1) can be generated by very simple rules that express a few rudimentary grammatical functions, if we assign to these rules a recursive property, in particular, allowing them to embed structures of the form $[_S...]_S$ within other structures. Grammatical transformations will then iterate to form, ultimately, a surface structure which may be quite remote from the underlying deep structure. The deep structure may be highly abstract; it may have no close point-by-point correlation to the phonetic realization. Knowledge of a language—what is now often called 'linguistic competence', in a technical use of this term—involves a mastery of these grammatical processes.

With just this much of a framework, we can begin to formulate some of the problems that call for analysis and for the development of explanatory theories. One major problem is posed by the fact that the surface structure generally gives very little indication in itself of the meaning of the sentence. There are, for example, numerous sentences that are ambiguous in some way that is not indicated by the surface structure. Consider the sentence (4):

(4)  I disapprove of John's drinking.

This sentence can refer either to the fact of John's drinking or to its character. The ambiguity is resolved, in different ways, in the sentences (5) and (6):

(5)  I disapprove of John's drinking the beer.
(6)  I disapprove of John's excessive drinking.

It is clear that grammatical processes are involved. Notice that we cannot simultaneously extend (4) in both of the ways illustrated in (5) and (6), giving (7):

(7)  *I disapprove of John's excessive drinking the beer.[1]

Our internalized grammar assigns two different abstract structures to (4), one of which permits the extension to (5), the other of which permits the extension to (6). But it is only at the level of deep structure that the distinction is represented; it is obliterated by the transformations that map the deep structures into the surface form associated with (4).

The processes that are involved in examples (4)–(6) are quite general in

---

[1] I use the asterisk, in the conventional way, to indicate a sentence that deviates in some respect from grammatical rule.

English. Thus the sentence 'I disapprove of John's cooking' may be understood as indicating either that I think his wife should cook or that I think he uses too much garlic, and so on. Again, the ambiguity is resolved if we extend the sentence in the manner indicated in (5) and (6).

The fact that (7) is deviant requires explanation. The explanation would be provided, at the level of particular grammar, by formulation of the grammatical rules that assign alternative deep structures and that in each case permit one but not the other of the extensions to (5) or (6). We would then explain the deviance of (7) and the ambiguity of (4) by attributing this system of rules to the person who knows the language, as one aspect of this knowledge. We might, of course, try to move to a deeper level of explanation, asking how it is that the person has internalized these rules instead of others that would determine a different sound-meaning correlation and a different class of generated surface structures (including, perhaps, (7)). This is a problem of universal grammar, in the sense described earlier.

Notice that the internalized rules of English grammar have still further consequences, in such a case as the one just discussed. There are transformations of great generality that permit or require the deletion of repeated elements, in whole or in part, under well-defined conditions. Applied to the structure (8), these rules derive (9).[1]

(8)   I don't like John's cooking any more than Bill's cooking.
(9)   I don't like John's cooking any more than Bill's.

The sentence (9) is ambiguous. It can mean either that I don't like the fact that John cooks any more than the fact that Bill cooks, or that I don't like the quality of John's cooking any more than the quality of Bill's cooking.[2] It cannot, however, mean that I don't like the quality of John's cooking any more than I do the fact that Bill cooks, or conversely, with 'fact' and 'quality' interchanged. That is, in the structure underlying (8) we must understand the ambiguous phrases 'John's cooking' and 'Bill's cooking' in the same way, if the deletion operation that produces (9) is to apply. It seems reasonable to assume that what is involved is some general condition on the applicability of deletion operations such as the one that

---

[1] Henceforth I shall generally delete brackets in giving a deep, surface, or intermediate structure, where this will not lead to confusion. One should think of (8) and of (9) as having a full labelled bracketing associated with them. Notice that (8) is not, of course, a deep structure, but rather the result of applying transformations to a much more primitive abstract object. I will also speak of deriving sentences from other sentences, though more properly one should speak of these (and their surface structures) as derived from the structures underlying these other sentences.

[2] There may also be other interpretations, based on other ambiguities in the structure 'John's cooking', specifically, the cannibalistic interpretation and the interpretation of 'cooking' as 'that which is cooked'.

gives (9) from (8), a rather abstract condition that takes account not only of the structure to which the operation applies, but also of the history of derivation of this structure.

Other examples can be found where a similar principle is at work. Thus consider the sentence (10), which is presumably derived from either (11) or (12), and is therefore ambiguous:

(10)   I know a taller man than Bill.
(11)   I know a taller man than Bill does.
(12)   I know a man who is taller than Bill is.

Evidently, the ambiguity of (10) is not represented in the surface structure; deletion of 'does' in (11) leaves exactly the same structure as deletion of 'is' (with application of (3iii) and (3iv)) in (12). But now consider the sentence (13):

(13)   I know a taller man than Bill, and so does John.

This sentence, like (9), is two ways ambiguous rather than four ways ambiguous. It can have either the meaning of (14) or of (15), but not (16) or (17):[1]

(14)   I know a taller man than Bill does and John knows a taller man than Bill does.

(15)   I know a man who is taller than Bill is and John knows a man who is taller than Bill is.

(16)   I know a man who is taller than Bill is and John knows a taller man than Bill does.

(17)   I know a taller man than Bill does and John knows a man who is taller than Bill is.

But now a problem arises. The sentence (13) is formed, by deletion of the repeated element, from something like (18):

(18)   I know a taller man than Bill and John knows a taller man than Bill.

The operation is the same as the one that forms 'I know Bill and so does John' from 'I know Bill and John knows Bill'. However, the co-ordinated elements in (18) do not indicate whether the underlying forms are (14), (15), (16) or (17), the distinction having been eliminated already in the operations that give (10). But the operation giving (13) can apply to (18)

---

[1] Furthermore, it cannot have the meaning 'I know a taller man than Bill and John likes ice cream'. Hence if deep structure determines meaning (in so far as grammatical relations are involved), it must be that something like (14) or (15) is the immediately underlying structure for (13). It is a general property of deletion operations that some sort of recoverability is involved, a non-trivial matter, with interesting empirical consequences. For some discussion, see Chomsky, 1964, §2.2, and Chomsky, 1965, §4.2.2. The problem posed by such examples as (9) and (13) was pointed out to me by John Ross.

only if the underlying structure was (14) or (15), as we see from the meaning of (13); and (18) does not, in its associated structure, contain this information.[1] It appears, once again, that some general condition on applicability of deletion transformations must be involved, a principle that somehow brings into consideration the history of derivation of strings, perhaps certain properties of the deep structure from which they ultimately derive.

In fact, the correct principle is unknown in such cases as these, although some of the conditions it must meet are clear. The problem posed by these examples is a typical one. Attention to linguistic fact reveals certain properties of sentences, relating to their sound, their meaning, their deviance, and so on. Evidently, no explanation for these facts will be forthcoming so long as we restrict ourselves to vague talk about 'habits' and 'skills' and 'dispositions to respond'. We do not have the 'habit' of understanding sentences (4), (9) and (13) in a certain way; it is unlikely that the reader has ever encountered sentences closely resembling these, but he understands them in a highly specific way nevertheless. To explain such phenomena we must discover the rules that relate sound and meaning in the language in question—the grammar that has been internalized by the person who knows the language—and the general principles that determine the organization and the functioning of these rules.

The misleading and inadequate character of surface structure becomes evident as soon as even the most simple patterns are studied. Consider, for example, the sentence (19):

(19)  John was persuaded to leave.

The deep structure underlying this sentence must indicate that the subject–predicate relation holds in an underlying proposition of the form (20) (assuming grammatical functions to be represented in the manner suggested earlier), and that the verb–object relation holds in an underlying proposition of the form (21):

(20)  $[_S [_{NP} \text{John}]_{NP} [_{VP} \text{leave}]_{VP}]_S$.
(21)  $[_S [_{NP} \ldots]_{NP} [_{VP} \text{persuade} [_{NP} \text{John}]_{NP}]_{VP}]_S$.

Although the deep structure must be constituted of such propositions, if the approach loosely outlined earlier is correct, there is no trace of them in the surface structure of the utterance. The various transformations that produce (19) have thoroughly obliterated the system of grammatical relations and functions that determine the meaning of the sentence.

The point becomes still more obvious if we take note of the variety of sentences that seem superficially to resemble (19), but that differ widely in

---

[1] If (18) itself is only two ways ambiguous, the problem in fact arises even at an earlier point. The unnaturalness of (18) makes it difficult to determine this.

the way they are understood and the formal operations that apply to them. Suppose that 'persuaded' in (19) is replaced by one of the following words:[1]

(22)   expected, hired, tired, pleased, happy, lucky, eager, certain, easy.

With 'expected' replacing 'persuaded', the sentence means roughly that the fact of John's leaving was expected; but it is impossible to speak of the fact of John's leaving being persuaded. With 'hired', the sentence has an entirely different meaning, roughly, that the purpose of hiring John was so that he would leave—an interpretation which becomes more natural if we replace 'leave' by 'fix the roof' etc. When 'tired' is substituted, we derive a non-sentence; it becomes a sentence if 'too tired' replaces 'persuaded', the sentence now implying that John didn't leave. The word 'pleased' is still different. In this case we can have 'too pleased', implying that John didn't leave, but we can also extend the sentence to 'John was too pleased to leave to suit me', which is impossible in the earlier cases. 'Happy' is rather like 'pleased', though one might argue that the verb–object relation holds between 'please' and 'John' in the latter case. The sentence 'John was lucky to leave' is interpreted in still a different way. It means, roughly, that John was lucky in that he left, an interpretation that is impossible in the earlier cases; furthermore, we can construct such sentences as 'John was a lucky fellow to leave (so early)', but none of the earlier examples can replace 'lucky' in such sentences. Furthermore, extension to 'John was too lucky to leave to suit me' is now impossible. 'John was eager to leave' differs from the earlier cases in that it is formally associated with such expressions as 'John was eager for Bill to leave' and 'John's eagerness (for Bill) to leave'. 'John was certain to leave' can be paraphrased as 'it was certain that John would leave'; of the other examples, only 'expected' is subject to this interpretation, but 'expected' obviously differs from 'certain' in numerous other respects, e.g. it appears in such sentences as 'they expected John to leave', and so on. The word 'easy' is of course entirely different; in this and only this case the verb–object relation holds between 'leave' and 'John'.

It is clear, in short, that the surface structure is often misleading and uninformative, and that our knowledge of language involves properties of a much more abstract nature, not indicated directly in the surface structure. Furthermore, even such artificially simple examples as these show how hopeless it would be to try to account for linguistic competence in terms of 'habits', 'dispositions', 'knowing how', and other concepts associated with the study of behaviour, as this study has been circumscribed, quite without warrant, in recent years.

---

[1] See Lees, 1960, pp. 207–21, for a discussion of such structures.

Even at the level of sound structure there is evidence that abstract representations are formed and manipulated in the mental operations that are involved in language use. Space does not permit a detailed discussion, but one simple illustrative example, which is quite typical, may give some idea of the nature of this evidence and the conclusions to which it points.

Consider such words as 'sign'–'signify', 'paradigm'–'paradigmatic', and so on. These alternations suggest that English contains a phonological rule which has the consequences of (23):

(23)　The phonological segment /i/ followed by a velar is realized phonetically as [ay] in certain contexts.[1]

Thus the underlying element /sign/ is realized as phonetic [sign] before the affix *-ify*, but as [sayn] in isolation.

Next, consider such alternations as 'ignite'–'ignition', 'contrite'–'contrition', 'expedite'–'expeditious', and so on. Such examples indicate that the phonology of English contains rules with the effect of (24) and (25):

(24)　The segment realized phonetically as [ay] is realized phonetically as [i] before such affixes as *-ion* and *-ious*.

(25)　The segment /t/ is realized as [š] before high front vowels.

Thus the segments realized as [ayt] in 'expedite' are realized as [iš] in 'expeditious'.

But now consider the alternation 'right'–'righteous', phonetically, [rayt]–[raycəs]. The latter form appears to deviate from the regular pattern in two respects; namely, in vowel quality (we would expect [i], not [ay], by rule (24)) and in the final consonant of the stem (we would expect [š], not [č], by rule (25)). If 'right' were subject to the same processes as 'expedite', we would have [rišəs] rather than [raycəs] as the phonetic realization. What is the explanation for this double deviation?

Notice first that rule (25) is not quite exact; there are, in fact, other cases in which /t/ is realized as [č] rather than [š], for example 'question' [kwesčən], as contrasted with 'direction' [dərekšən]. A more accurate formulation of (25) would be (26):

(26)　When it is followed by a high front vowel, the segment /t/ is realized as [č] after a continuant and as [š] elsewhere.

---

[1] This and the other rules given below are quite inexact. For details, see Chomsky & Halle (1968), in this case, Chap. 5, §4.

　　Several very different processes are collapsed in the rough formulation (23). A more accurate description would be that a velar becomes a continuant (i.e. /g/ becomes /γ/) before a word-final nasal, and a vowel is tensed (i.e. /i/ becomes [ī]) before a velar continuant. Velar continuants drop pre-consonantly, and all instances of /ī/ are realized phonetically as [ay]. Thus /sign/ becomes [siγn], then [sīγn], then [sīn], then [sayn]. For the most part, these separate processes are independently motivated. For example, the same vowel alternation is involved in what we state here informally as (24).

Returning now to the form 'right', we see that the final consonant would be correctly determined as [č] rather than [š] if in the underlying representation there were a continuant preceding it—that is, if the underlying representation were /riɸt/, where ɸ is some continuant. But the continuant ɸ must, furthermore, be distinct from any of the continuants that actually appear phonetically in this position, namely, the dental, labial or palatal continuants in the italicized position of 'wris*t*', 'ri*f*t', or 'wi*sh*ed'. We may assume, then, that ɸ is the velar continuant /x/, which does not, of course, appear phonetically in English. The underlying form, then, would be /rixt/. But we already have a rule, rule (26), which gives [ay] as the phonetic realization for /ix/ (that is, for /i/ followed by a velar) under certain conditions; and a precise analysis of the relevant conditions (see note 1, p. 436) shows that they do, in their simplest formulation, cover the cases we have so far discussed. Thus by rule (26) and rule (23), which are independently motivated, the underlying representation /rixt/ will be realized phonetically as [rayt] in isolation and as [rayč] in 'righteous', exactly as required.

These facts strongly suggest that the underlying phonological representation must be /rixt/ (in accord with the orthography, and, of course, the history). A sequence of rules which must be in the grammar for other reasons gives the alternation 'right'–'righteous'. Therefore this alternation is not at all exceptional, but rather perfectly regular. The underlying representation is quite abstract; it is connected with the superficial phonetic shape of the signal only by a sequence of interpretive rules.

Putting the matter differently, suppose that a person knows English but does not happen to have the vocabulary item 'righteous'. Hearing this form for the first time, he must assimilate it to the system he has learned. If he were presented with the derived form [rišəs], he would, of course, take the underlying representation to be exactly like that of 'expedite', 'contrite', etc. But hearing [rayčəs], he knows that this representation is impossible; although the consonantal distinction [š]–[č] might easily be missed under ordinary conditions of language use, the vocalic distinction [i]–[ay] would surely be obvious. Knowing the rules of English and hearing the vocalic element [ay] instead of [i], he knows either that the form is a unique exception, or that it contains a sequence /i/ followed by velar, and is subject to rule (26). The velar must be a continuant,[1] that is, /x/. But given that the velar is a continuant, it follows, if the form is regular (the

---

[1] If it were a noncontinuant, it would have to be unvoiced, i.e. /k/, since there are no voiced–voiceless consonant clusters in final position, by general rule. But it cannot be /k/, since /k/ remains in this position (e.g. 'direct', 'evict', etc.).

null hypothesis, always), that the consonant must be [č], not [š], by rule (26). Thus the hearer should perceive [rayčəs] rather than [rayšəs], even if the information is lacking in the received signal. Furthermore, the pressure to preserve regularity of alternations should act to block the superficial analogy to 'expedite'–'expeditious', 'ignite'–'ignition', etc., and preserve [č] as the phonetic realization of underlying /t/ as long as [ay] appears in place of expected [i], exactly as we observe to have occurred.

I do not mean this as a literal step-by-step account of how the form is learned, but rather as a possible explanation of why the form resists a superficial (and in fact incorrect) analogy, and preserves its status. We can explain the perception and preservation in the grammar of the [č]–[š] contrast in 'righteous'–'expeditious' on the basis of the perceived distinction between [ay] and [i] and the knowledge of a certain system of rules. Furthermore, the explanation rests on the assumption that underlying representations are quite abstract; and the evidence cited suggests that this assumption is, in fact, correct.

A single example can hardly carry much conviction. However, a careful investigation of sound structure shows that there are quite a number of examples of this sort, and that, quite generally, highly abstract underlying structures are related to phonetic representations by a long sequence of rules, just as on the syntactic level, abstract deep structures are in general related to surface structures by a long sequence of grammatical transformations. Assuming the existence of abstract mental representations and interpretive operations of this sort, we can find a surprising degree of organization underlying what appears superficially to be a chaotic arrangement of data; and, in some cases, we can also explain why linguistic expressions are heard, used, and understood in certain ways. One cannot hope to determine either the underlying abstract forms or the processes that relate them to signals by introspection. There is, furthermore, no reason why one should find this consequence in any way surprising.

The sketch of an explanation just given is at the level of particular rather than universal grammar, as this distinction was formulated earlier. That is, we have accounted for a certain phenomenon on the basis of the assumption that certain rules appear in the internalized grammar, noting that these rules are, for the most part, independently motivated. Of course, considerations of universal grammar enter into this explanation in so far as they affect the choice of grammar on the basis of data. This interpenetration is unavoidable, as noted earlier. There are cases, however, where explicit principles of universal grammar enter more directly and clearly into a pattern of explanation. Thus investigation of sound systems reveals certain very general principles of organization, some quite remarkable, governing phonological rules.[1] For example, it has been observed that certain phonological rules operate in a cycle, in a manner determined by the surface structure. Recall that the surface structure can be represented as a labelled bracketing of the utterance, such as (2). In English, the very intricate phonological rules that determine stress contours and vowel reduction apply to phrases bounded by paired brackets in the surface structure, applying first to a minimal phrase of this sort, then to the next larger phrase, etc., until the maximum domain of phonological processes is reached (in simple cases, the sentence itself). Thus in the case of (2), the rules apply to the individual words (which, in a full description, would be assigned to categories and therefore bracketed), then to the phrases 'a wise man' and 'is honest', and finally to the whole sentence. A few simple rules will give very varied results, as the surface structures that determine their cyclic application vary.

Some simple effects of the principle of cyclic application are illustrated by such forms as those of (27):

(27)　(i) Rel*a*xation, em*e*ndation, el*a*sticity, conn*e*ctivity;
　　　(ii) ill*u*stration, dem*o*nstration, dev*a*station, an*e*cdotal.

The italicized vowels are reduced to [ə] in (27ii), but they retain their original quality in (27i). In some cases, we can determine the original quality of the reduced vowels of (27ii) from other derived forms (e.g. 'illustrative', 'demonstrative'). The examples of (27i) differ from those of (27ii) morphologically in that the former are derived from underlying forms that contain primary stress on the italicized vowel when these underlying forms appear in isolation; those of (27ii) do not have this property. It is not difficult to show that vowel reduction in English, the replacement of a vowel by [ə], is contingent upon lack of stress. We can therefore account for the distinction between (27i) and (27ii) by assuming the cyclic principle just formulated. In the case of (27i), on the first, innermost cycle, stress will be assigned by general rules to the italicized vowels. On the next cycle, stress is shifted,[2] but the ideal stress assigned on the first cycle is sufficient to protect the vowel from reduction. In the examples of (27ii), earlier cycles never assign an ideal stress to the italicized vowel, which therefore reduces. Observe that it is an *ideal* stress that protects the vowel from reduction. The actual, phonetic stress on the italicized non-reduced vowels is very

---

[1] For discussion of this matter, see Chomsky, 1967. More detail appears in Chomsky & Halle (1968), where references to earlier work are given.

[2] In 'connectivity', it is on the third cycle that the stress is shifted. The second cycle merely reassigns stress to the same syllable that is stressed on the first cycle.

weak; it would be stress 4, in the usual convention. In general, vowels with this weak a stress reduce, but in this case the ideal stress assigned in the earlier cycle gives a value to the syllable in the abstract underlying form that is lost in the phonetic realization.

In this case, we can provide an explanation for a certain aspect of perception and articulation in terms of a very general abstract principle, namely, the principle of cyclic application. It is difficult to imagine how the language learner might derive this principle by 'induction' from the data presented to him. In fact, many of the effects of this principle relate to perception and have little or no analogue in the physical signal itself, under normal conditions of language use, so that the phenomena on which the induction would have to be based cannot be part of the experience of one who is not already making use of the principle. In fact, there is no principle of induction or association that offers any hope of leading from such data as is available to a principle of this sort (unless, begging the question, we introduce the principle of cyclic application into the inductive procedure, in some manner). Therefore, the conclusion seems warranted that the principle of cyclic application of phonological rules is an innate organizing principle of universal grammar that is used in determining the character of linguistic experience and in constructing a grammar that constitutes the acquired knowledge of language. At the same time, this principle of universal grammar offers an explanation for such phenomena as were noted in (27).

There is some evidence that a similar principle of cyclic application applies also on the syntactic level. John Ross has presented an ingenious analysis of some aspects of English pronominalization illustrating this (cf. Ross, 'On the cyclic nature of English pronominalization', forthcoming). Let us assume that pronominalization involves a process of 'deletion' analogous to those discussed earlier in connection with examples (8)–(18). This process, to first approximation, replaces one of two identical noun phrases by an appropriate pronoun. Thus the underlying structure (28) will be converted to (29), by pronominalization.

(28)  John learned that John had won.

(29)  John learned that he had won.

Abstracting properties of (28) that are essential to this discussion, we can present it in the form (30), where $x$ and $y$ are the identical noun phrases and $y$ is the one pronominalized, and where the brackets bound sentential expressions.

(30)  $[...x...[...y...]]$.

Notice that we cannot form (31) from (28) by pronominalization:[1]

(31)   He learned that John had won.

That is to say, we cannot have pronominalization in the case which would be represented as (32), using the conventions of (30):

(32)   $[...y...[...x...]]$.

Consider next the sentences of (33):

(33)   (i)   That John won the race surprised him
              $[[...x...]...y...]$;
       (ii)  John's winning the race surprised him
              $[[...x...]...y...]$;
       (iii) that he won the race surprised John
              $[[...y...]...x...]$;
       (iv)  his winning the race surprised John
              $[[...y...]...x...]$.

Continuing with the same conventions, the forms are represented below in each case. Summarizing, we see that of the possible types (30), (32), (33 i, ii) and (33 i, iv), all permit pronominalization except for (32). These remarks presumably belong to the particular grammar of English.

Notice that alongside of (33 iv) we also have the sentence (34): .

(34)   Winning the race surprised John.

Given the framework we have been assuming throughout, (34) must be derived from the structure 'John's winning the race surprised John'. Hence in this case, pronominalization can be a full deletion.

Consider now the sentences (35) and (36):

(35)   Our learning that John had won the race surprised him.
(36)   Learning that John had won the race surprised him.

The sentence (35) can be understood with 'him' referring to John, but (36) cannot. Thus (35) can be derived by pronominalization from (37), but (36) is not derived from (38):

(37)   [[Our learning [that John had won the race]] surprised John].
(38)   [[John's learning [that John had won the race]] surprised John.

What might be the explanation for this phenomenon? As Ross observes, it can be explained in terms of the particular grammar of English if we

---

[1] Of course (31) is a sentence, but 'he' in this sentence does not refer to John as it does in (29). Thus (31) is not formed by pronominalization from (28). Notice also that (28) will not undergo pronominalization if the two occurrences of 'John' are intended to be different in reference. We exclude this case from discussion here. For some remarks bearing on this problem, see Chomsky, 1965, pp. 144 f.

assume, in addition, that certain transformations apply in a cycle, first to innermost phrases, then to larger phrases, and so on—that is, if we assume that these transformations apply to the deep structure by a process analogous to the process by which phonological rules apply to the surface structure.[1] Making this assumption, let us consider the underlying structure (38). On the innermost cycle, pronominalization does not apply at all, since there is no second noun phrase identical to 'John' in the most deeply embedded proposition. On the second cycle, we consider the phrase '[John's learning [that John had won the race]]'. This can be regarded as a structure of the form (30), giving (39); it cannot be regarded as of the form (32), giving (40), because the particular grammar of English excludes (32) from pronominalization, as we have noted:

(39)   John's learning [that he had won the race].
(40)   His learning [that John had won the race].

But (40) would have to be the form underlying (36). Hence (36) cannot be derived by pronominalization from (38), although (35) can be derived from (37).

In this case, then, a principle of universal grammar interacts with an independently established rule of particular English grammar to yield a certain rather surprising empirical consequence, namely, that (35) and (36) must differ in the referential interpretation of the pronoun 'him'. Once again, as in the formally somewhat analogous case of vowel reduction discussed earlier in connection with examples (27i) and (27ii), it is quite impossible to provide an explanation in terms of 'habits' and 'dispositions'. Rather, it seems that certain abstract and in part universal principles governing human mental faculties must be postulated to explain the phenomena in question. If the principle of cyclic application is indeed a regulating principle determining the form of knowledge of language for humans, a person who has learned the particular rules governing pro-

---

[1] That transformational rules may be supposed to function in this way, itself a non-trivial fact if true, is suggested in Chomsky, 1965, Chap. 3. Ross' observation suggests that this principle of application is not only possible but also necessary. Other interesting arguments to this effect are presented in Jacobs & Rosenbaum, 1967, Chap. 28. The matter is far from settled. In general, understanding of syntactic structure is much more limited than of phonological structure; descriptions are much more rudimentary, and, correspondingly, principles of universal syntax are much less firmly established than principles of universal phonology, though the latter too, needless to say, must be regarded as tentative. In part, this may be due to the inherent complexity of the subject matter. In part, it results from the fact that universal phonetics, which provides a kind of 'empirical control' for phonological theory, is much more firmly grounded than universal semantics, which should, in principle, provide a partially analogous control for syntactic theory. In modern linguistics, phonetics (and in part, phonology) has been studied in considerable depth and with much success, but the same cannot be said as yet for semantics.

nominalization in English would know, intuitively and without instruction or additional evidence, that (35) and (36) differ in the respect just noted.

The most challenging theoretical problem in linguistics is that of discovering the principles of universal grammar that interact with the rules of particular grammars to provide explanations for phenomena that appear arbitrary and chaotic. Probably the most persuasive examples at this time (and also the most important ones, in that the principles involved are highly abstract and their working quite intricate) are in the domain of phonology, but these are too complex to present within the scope of this paper.[1] Another syntactic example that illustrates the general problem in a fairly simple way is provided by the rules for formation of *wh*-questions in English.[2]

Consider such sentences as the following:

(41)   (i)  Who expected Bill to meet Tom?;
      (ii)  who(m) did John expect to meet Tom?;
      (iii)  who(m) did John expect Bill to meet?;
      (iv)  what (books) did you order John to ask Bill to persuade his friends to stop reading?

As examples (i)–(iii) show, a noun phrase in any of the three italicized positions in a sentence such as '*John* expected *Bill* to meet *Tom*' can be questioned. The process is essentially this:

(42)   (i)  *Wh-placement*: assign the marker *wh* to a noun phrase;
      (ii)  *wh-inversion*: place the marked noun phrase at the beginning of the sentence;
      (iii)  *auxiliary attraction*: move a part of the verbal auxiliary to the second position in the sentence;
      (iv)  *phonological interpretation*: replace the marked noun phrase by an appropriate interrogative form.[3]

---

[1] See the references of note 1, p. 439. The issue is discussed in a general way in Chomsky, 1962; Chomsky, 1964, §2; Chomsky, 1965, Chap. 1; and in other publications referred to in these references.

[2] This matter is discussed in Chomsky, 1964. There are several versions of this monograph. The first, presented at the International Congress of Linguists, 1962, appears in the *Proceedings of the Congress* with the title of the session at which it was presented, 'Logical basis of linguistic theory', in H. Lunt (ed.), Mouton, 1964; a second appears in Fodor and Katz, 1964; the third, as a separate monograph. These versions differ in the treatment of the examples discussed here; none of the treatments is quite satisfactory, and in fact the general problem remains open. New and interesting ideas on this matter are presented in Ross, Doctoral Thesis (1967). I follow here the general lines of the earliest of the three versions of Chomsky (1964), which, in retrospect, seems to me to offer the most promising approach.

[3] Actually, it seems that only indefinite singular noun phrases can be questioned (i.e. 'someone', 'something', etc.), a fact that relates to the matter of recoverability of deletion mentioned in note 1, p. 433. See Chomsky, 1964 for some discussion.

All four of these processes apply non-vacuously in the case of (ii) and (iii). Sentence (41 ii), for example, is formed by applying *wh-placement* to the noun phrase 'someone' in 'John expected someone to meet Tom'. Application of *wh-inversion* gives 'wh-someone John expected to meet Tom'. *Auxiliary attraction* gives 'wh-someone did John expect to meet Tom'. Finally, *phonological interpretation* gives (41 ii). Sentence (41 iv) illustrates the fact that these processes can extract a noun phrase that is deeply embedded in a sentence, without limit, in fact.

Of the processes listed in (42), all but *auxiliary attraction* apply as well in the formation of relative clauses, giving such phrases as 'the man who John expected to meet Tom', and so on.

Notice, however, that there are certain restrictions on the formation of questions and relatives in this manner. Consider, for example, the sentences of (43):

(43)  (i) For him to understand *this lecture* is difficult;
  (ii) it is difficult for him to understand *this lecture*;
  (iii) he read the book that interested *the boy*;
  (iv) he believed the claim that John tricked *the boy*;
  (v) he believed the claim that John made about *the boy*;
  (vi) they intercepted John's message to *the boy*;

Suppose that we try applying the processes of interrogative and relative formation to noun phrases in the italicized positions of (43). We should derive the following interrogatives and relatives from (43 i)–(43 vi), respectively:

(44)  (i I)  *What is for him to understand difficult?;
  (i R)  *a lecture that for him to understand is difficult;
  (ii I)  what is it difficult for him to understand?;
  (ii R)  a lecture that it is difficult for him to understand;
  (iii I)  *who did he read the book that interested?;
  (iii R)  *the boy who he read the book that interested;
  (iv I)  *who did he believe the claim that John tricked?;
  (iv R)  *the boy who he believed the claim that John tricked;
  (v I)  *who did he believe the claim that John made about?;
  (v R)  *the boy who he believed the claim that John made about;
  (vi I)  *who did they intercept John's message to?;
  (vi R)  *the boy who they intercepted John's message to.

Of these, only (ii I) and (ii R) are grammatical. It is not at all obvious how the speaker of English knows this to be so. Thus sentences (43 i) and (43 ii) are synonymous, yet only (43 ii) is subject to the processes in question. And although these processes do not apply to (43 iv) and (43 vi), they can be applied, with fairly acceptable results, to the very similar sentences (45 i) and (45 ii):

(45) (i) He believed that John tricked *the boy* (who did he believe that John tricked?—the boy who he believed that John tricked);

(ii) They intercepted a message to *the boy* (who did they intercept a message to?—the boy who they intercepted a message to).

The principles of (42) are devised in some unknown way on the basis of examples by the speaker of English; still more mysterious, however, is the fact that he knows under what formal conditions they are applicable. It can hardly be seriously maintained that every normal speaker of English has had his behaviour 'shaped' in the indicated ways by appropriate rein-forcement. The sentences of (43)–(45) are as 'unfamiliar' as the vast majority of those that we encounter in daily life, yet we know intuitively, without instruction or awareness, how they are to be treated by the system of grammatical rules that we have mastered.

It seems, once again, that there is a general principle that accounts for many such facts. Notice that in (43 i), the italicized noun phrase is contained within another noun phrase, namely, 'for him to understand *this lecture*', which is the subject of the sentence. In (43 ii), however, a rule of *extraposition* has placed the phrase 'for him to understand this lecture' outside of the subject noun phrase, and in the resulting structure this phrase is not a noun phrase at all, so that the italicized phrase in (43 ii) is no longer contained within a noun phrase. Suppose we were to impose on grammatical transformations the condition that no noun phrase can be extracted from within another noun phrase—more generally, that if a transformation applies to a structure of the form $[_S...[_A...]_A...]_S$, for any category A, then it must be so interpreted as to apply to the *maximal* phrase of the type A.[1] Then the processes (42) would be blocked, as required, in the cases (43 i), (43 iii)–(43 vi), but not in (43 ii).

There are other examples that support a principle of this sort, which we will refer to as the *A-over-A* principle. Consider the sentences (46):

(46) (i) John kept the car *in the garage*;

(ii) Mary saw the man walking towards *the railroad station*.

Each of these is ambiguous. Thus (46 i) can mean that the car in the garage was kept by John, or that the car was kept in the garage by John. In the

---

[1] We might extend this principle to the effect that this transformation must also apply to the *minimal* phrase of the type S (sentence). Thus the sentence $[_S$ John was convinced that $[_S$ Bill would leave before dark $]_S]_S$ can be transformed to 'John was convinced that before dark Bill would leave' but not to 'before dark John was convinced that Bill would leave', which must have a different source. This extension is, like the original principle, not without its problems, but it has a certain amount of support nevertheless.

first case, the italicized phrase is part of a noun phrase, 'the car in the garage'; in the latter case it is not. Similarly, (46ii) can mean that the man walking towards the railroad station was seen by Mary, or that the man was seen walking towards the railroad station by Mary. Again, in the first case the italicized phrase is part of a noun phrase, 'the man walking to the railroad station'; in the latter case, it is not. But now consider the two interrogatives (47):

(47)  (i)  What (garage) did John keep the car in?
      (ii) what did Mary see the man walking towards?

Each of these is unambiguous, and can have only the interpretation of the underlying sentence in which the italicized phrase is not part of another noun phrase. The same is true of the relatives formed from (46). These facts too would be explained by the *A-over-A* principle. There are many similar examples.

A slightly more subtle case that might, perhaps, be explained along the same lines is provided by such sentences as (48), (49):

(48)  John has the best proof of that theorem.

(49)  What theorem does John have the best proof of?

In its most natural interpretation, under normal intonation, sentence (48) describes a situation in which a number of people have proofs of that theorem, and John's is best. The sense thus suggests that 'best' modifies the nominal phrase 'proof of that theorem', which contains another nominal phrase 'that theorem'.[1] The *A-over-A* principle would therefore imply that the phrase 'that theorem' not be subject to the processes of (42). And in fact, sentence (49) has a rather different interpretation. It is appropriate to a situation in which John has proofs of a number of theorems, and the questioner is asking which of these proofs is the best. The underlying structure, whatever it may be, would associate 'best' with 'proof', not with 'proof of that theorem', so that 'that theorem' is not embedded within a phrase of the same type and is therefore subject to questioning (similarly, relativization).

The general principle just proposed has a certain explanatory force, as such examples illustrate. If postulated as a principle of universal grammar, it can explain why the particular rules of English operate to generate certain sentences while rejecting others, and to assign sound-meaning

---

[1] Space does not permit a discussion of the distinction implied here in the loose terminology, 'noun phrase'–'nominal phrase', but this is not crucial to the point at issue. See Chomsky, 'Remarks on nominalization', in Jacobs & Rosenbaum, *Readings in English Transformational Grammar*, forthcoming. There are other interpretations of (49) (e.g. with contrastive stress on 'John'), and many open problems relating to such structures as these.

relations in ways that appear, superficially, to violate regular analogies. Putting the matter in different terms, if we assume that the *A-over-A* principle is a part of the innate schematism that determines the form of knowledge of language, we can account for certain aspects of the knowledge of English possessed by speakers who obviously have not been instructed or trained and who have not even been presented with data bearing on the phenomena in question in any relevant way, so far as can be ascertained.

Further analysis of data of English reveals, not unexpectedly, that this account is oversimplified. Consider, for example, the sentences (50), (51):

(50) John thought (that) Bill had read *the book*.
(51) John wondered why Bill had read *the book*.

In the case of (50), the italicized phrase is subject to interrogation and relativization, but not in the case of (51). It is unclear whether the phrases 'that Bill had read the book' and 'why Bill had read the book' are noun phrases. Suppose that they are not. Then sentence (50) is handled in accordance with the *A-over-A* principle, but not (51). To explain the blocking of the processes (42) in the case of (51), we would have to assign the phrase 'why Bill had read the book' to the same category as 'the book'. In fact, there is a natural suggestion, along these lines. Sentence (51) is typical in that the phrase from which the noun phrase is to be extracted is itself a *wh*-phrase, rather than a *that*-phrase. Suppose that the process of *wh*-placement (cf. (42i)) assigns the element *wh*- not only to 'the book' in (51) but also to the proposition immediately containing it. Thus both '*wh*-the book' and 'why Bill had read the book' belong to the category *wh*, which would now be regarded as a syntactic feature of a sort discussed in *Aspects*, Chapter 2 (see note 1, p. 430). Under these assumptions, the *A-over-A* principle will serve to explain the difference between (50) and (51).

Suppose that the phrases in question are noun phrases. Now it is (50), not (51), which poses the problem. Assuming that our analysis is correct so far, there must be some rule that assigns to the proposition 'that Bill had read the book' a property of 'transparency' that permits noun phrases to be extracted from it even though it is a noun phrase. There are, in fact, other examples that suggest the necessity for such a rule, presumably, a rule of the particular grammar of English. Thus consider the sentences (52)–(54):

(52) Who would you approve of my seeing?
(53) What would you approve of John's drinking?
(54) *What would you approve of John's excessive drinking of?

Sentences (52) and (53) are formed by applying the processes of interrogation to a noun phrase contained in the larger noun phrases 'my seeing——', 'John's drinking——'. Hence these larger noun phrases are transparent to the extraction operation. However, as (54) indicates, the italicized noun phrase in (55) is not transparent to this operation:

(55)  You would approve of *John's excessive drinking of the beer.*

These examples are typical of many which suggest what the rule might be that assigns transparency. Earlier we discussed sentence (56) (=(4)), pointing out that it is ambiguous:

(56)  I disapprove of John's drinking.

Under one interpretation, the phrase 'John's drinking' has the internal structure of a noun phrase. Thus the rule that inserts adjectives (cf. (3 iv)) between a determiner and a noun applies, giving 'John's excessive drinking'; and, in fact, other determiners may replace 'John's', e.g. 'the', 'that', 'much of that', etc. Under this interpretation, the phrase 'John's drinking' behaves exactly like 'John's refusal to leave', 'John's rejection of the offer', and so on. Under the other interpretation, 'John's drinking (the beer)' does not have the internal structure of a noun phrase, and is handled analogously to 'John's having read the book', 'John's refusing to leave', 'John's rejecting the offer', and so on, none of which permit adjective insertion or replacement of 'John's' by other determiners. Suppose that we postulate a rule of English grammar that assigns transparency, in the sense just defined, to noun phrases that are also propositions lacking the internal structure of noun phrases. Thus the phrases 'that Bill had read the book' in (50), 'my seeing——' in the structure underlying (52), and 'John's drinking——' in the structure underlying (53), would be assigned transparency; more precisely, the dominant noun phrase in these examples would not serve to block extraction by the *A-over-A* principle. In sentence (51), extraction would still be blocked by the category *wh-*, along the lines indicated earlier. And the sentence (54) is ruled out because the relevant noun phrase of the underlying structure, 'John's excessive drinking of——', does have the internal structure of a noun phrase, as just noted, and there-fore is not subject to the special rule of English grammar that assigns transparency to the category NP when this category dominates a proposi-tion that lacks the internal structure of an NP.

There are a few other cases that suggest the need for rules of particular grammar assigning transparency, in this sense. Thus consider the sentences (57), (58):

(57)  (i)  They intercepted John's message to *the boy* (=(43 vi));
      (ii) he saw John's picture of *Bill*;
      (iii) he saw the picture of *Bill*.

(58)   (i)   They intercepted a message to *the boy* ($=(45\text{ii})$;
       (ii)  he saw a picture of *Bill*;
       (iii) he has a belief in *justice*;
       (iv)  he has faith in *Bill's integrity*.

The noun phrases in the italicized positions in (57) are not subject to the processes of interrogation and relativization, in accordance with the *A-over-A* principle, as we have already noted. In the case of (58), interrogation and relativization seem much more natural in these positions, at least in informal spoken English. Thus the noun phrases containing the italicized phrases must be assigned transparency, in these dialects. It seems that what is involved is indefiniteness of the dominating noun phrase; if so, then for certain dialects, there is a rule assigning transparency to a noun phrase of the form $[_{\text{NP}}\text{ indefinite...NP}]_{\text{NP}}$.

Along these lines, we might develop on the one hand a system of general principles of universal grammar,[1] and on the other, particular grammars that are formed and interpreted in accordance with these principles. The interplay of universal principles and particular rules leads to empirical consequences such as those we have illustrated; at various levels of depth, these rules and principles provide explanations for facts about linguistic competence, the knowledge of language possessed by each normal speaker, and about some of the ways in which this knowledge is put to use in the performance of the speaker or hearer.

The principles of universal grammar provide a highly restrictive schema to which any human language must conform, and specific conditions determining how the grammar of any such language can be used. It is easy to imagine alternatives to the conditions that have been formulated (or those that are often tacitly assumed). These conditions have in the past generally escaped notice, and we know very little about them today. If we manage to establish the appropriate 'psychic distance' from the relevant phenomena, and succeed in 'making them strange' to ourselves, we see at once that they pose very serious problems that cannot be talked or defined

---

[1] Notice that we are interpreting 'universal grammar' as a system of conditions on grammars. It may involve a skeletal substructure of rules that any human language must contain, but it also incorporates conditions that must be met by such grammars and principles that determine how they are interpreted. This is something of a departure from a traditional view which took universal grammar to be a substructure of each particular grammar, a system of rules at the very core of each grammar. This traditional view has also received expression in recent work. It seems to me to have little merit. As far as information is available, there are heavy constraints on the form and interpretation of grammar at all levels, from the deep structures of syntax, through the transformational component, and in the rules that interpret syntactic structures semantically and phonetically. In fact, the deepest results so far obtained, in my opinion, involve those principles of universal grammar that govern the phonological component, as remarked earlier.

out of existence. A serious consideration of such problems as those illustrated very sketchily here indicates that to account for knowledge of language we must postulate that the speaker–hearer has developed an intricate system of rules that involve mental operations of a very abstract nature, applying to representations that are quite remote from the physical signal. Furthermore, if we seriously face the question of how this knowledge arises, we are led to postulate a system of innate structures and principles that determine the form of the acquired knowledge and that explain how degenerate and restricted data serve as the basis for highly specific knowledge.

If a scientist were faced with the problem of determining the nature of a device of unknown properties that operates on data of the sort available to a child and gives as 'output' (i.e. as a 'final state of the device', in this case) a particular grammar of the sort that it seems necessary to attribute to the person who knows the language, he would naturally search for inherent principles of organization that determine the form of the output on the basis of the limited data available. There is no reason to adopt a more prejudiced or dogmatic view when the device of unknown properties is the human mind; in particular, there is no reason to suppose that the general empiricist assumptions that have dominated speculation about these matters have any particular relevance to the problem. No one has succeeded in showing why the highly specific empiricist assumptions about how knowledge is acquired should be taken seriously. They appear to offer no way to describe or account for the most characteristic and normal constructions of human intelligence, for example, linguistic competence. On the other hand, certain highly specific assumptions about particular and universal grammar give some hope of accounting for the phenomena that we face when we consider knowledge and use of language. Speculating about the future, it seems not unlikely that continued research along the lines indicated here will bring to light a highly restrictive schematism that determines both the content of experience and the nature of the knowledge that arises from it, thus vindicating and elaborating some very traditional thinking about problems of language and mind.

### REFERENCES

Chomsky, N. 1962. Explanatory models in linguistics. *Logic, Methodology and Philosophy of Science*. Stanford.
Chomsky, N. 1964. *Current Issues in Linguistic Theory*.
Chomsky, N. 1965. *Aspects of the Theory of Syntax*.
Chomsky, N. 1966. *Cartesian Linguistics*.
Chomsky, N. 1966. *Topics in the Theory of Generative Grammar*.

Chomsky, N. 1967. Some general properties of phonological rules. *Language* **43**.

Chomsky, N. & Halle, M. 1968. *The Sound Pattern of English.*

Ehrlich, V. 1955. *Russian Formalism.*

Jacobs, R. & Rosenbaum, P. 1967. *English Transformational Grammar.*

Katz, J. & Fodor, J. 1964. *Structure of Language: Readings in the Philosophy of Language.*

Katz, J. & Postal, P. 1964. *An Integrated Theory of Linguistic Descriptions.*

Köhler, W. 1940. *Dynamics in Psychology.* New York.

Lees, R. B. 1960. A multiply ambiguous adjectival construction in English. *Language* **36**.

Ross, J. 1967. *Constraints on Variables in Syntax.*

Ross, J. Forthcoming. On the cyclic nature of English pronominalization. *Essays in Honor of Roman Jakobson.* Ed. Peter de Ridder. The Hague.

# COMMENT

## *by* MAX BLACK

Noam Chomsky's stimulating and provocative views about linguistic theory deserve more elaborate and systematic discussion than I can here provide. I offer the following remarks, in no spirit of carping criticism, with the hope of highlighting some of the main questions that are likely to need attention in a sympathetic evaluation of the position of transformational linguistics. Many of these questions have, of course, already been considered by Chomsky and his associates.

1. *What is the task of a formal grammar of a language?* One way of conceiving it is as follows. We start with an infinite class of sentences or, more generally, an infinite class of linear 'strings', constructed from a finite stock of atomic constituents (words, or other elements, treated as indivisible for the sake of the grammatical analysis). We now look for a simple and necessarily finite way of characterizing the members of that class of strings. That is to say, we search for some set of necessary and sufficient conditions for membership in the given class. More specifically, we hope to end with some finite set of rules for generating just those strings that belong to the class. For simplicity, I will here ignore the equally important and fundamental task of generating the 'phrase structure' of the strings, i.e. of showing the internal structures of the sub-elements of a given sentence and the morphological categories to which those sub-elements belong.

A simple analogy may help. Suppose the initial class of strings to consist of linear sequences of the symbols *a* and *b*, with repetitions allowed, limited only by the condition that 'triple sequences' of the form $XXX$, where $X$ stands each time for the same sub-sequence of symbols, shall not occur. Thus *a*, *bb*, *aaba*, *ababa* are members, while *bbb*, *baaa*, *abababa* and *bbabbabba* are not. We ask for a set of rules to generate just those strings of *a*'s and *b*'s that belong to the class described. (The solution, even in this very simplified illustration, is by no means obvious.) Here the occurrence of a 'triple sequence' corresponds to the occurrence in a natural language of *ungrammaticality*. Criteria for ungrammaticality are, of course, immeasurably more complex than the presence of a triple sequence.

2. *What kind of task is set by a specific formal grammar?* In my simple analogy, the task is plainly one for pure mathematics. The problem of

finding an adequate set of generative rules can be solved without concern for any possible use for the strings of $a$'s and $b$'s, or any attention to physical representations of the abstract structures investigated. In writing a generative grammar proper, however, matters are less simple. Whether a string of morphemes shall be regarded as grammatical (admissible) is determined not by a simple formal criterion, as in the illustration, but rather by the verdict of a suitably competent user of the language in question—often the investigator himself, acting as his own informant. Here, there is a link with some kind of extraneous verification.

The resulting grammar can therefore, quite plausibly, be regarded as a mathematical theory of a class of well-defined empirical phenomena. Some grammarians like to compare a descriptive grammar with a simple physical theory. Grammar is to speech behaviour as, say, Kepler's laws are to planetary motions. Both are theories about the actual world; if both traffic in idealizations and abstractions, they are equally subject to strict, though indirect, control by empirical facts.

3. *How much idealization is involved?* Chomsky wishes to characterize the behaviour of an ideal speaker, having perfect command of syntax and morphology, and free from the memory lapses, slips of tongue, and other irregularities and inconsistencies that distinguish actual 'performance' from ideal 'competence'. This much idealization seems no more suspect in principle than physicists' talk about ideal gases or perfect fluids. Yet it may be important to recall how far from grammatically people speak, even in such relatively formal performances as speeches and lectures. To listen to a typical conversation is to be vividly reminded of the great gap between ideal competence and actual performance.

A more troublesome kind of idealization enters through the simplifying assumption that allows sentences to be arbitrarily long and to have any degree of internal complexity. The extent of what might be called 'theoretical distance' thus introduced may be important. The very concept of a sentence is already something of a grammarian's artifact: it is by no means easy to chop up speech into sentences without relying upon some antecedently available grammar. And actual utterances are bounded, however, indeterminately, even when produced by speakers as long-winded as Faulkner. A speaker can no more produce an indefinitely long utterance than a man can walk an indefinitely long distance. Pragmatic constraints limit the length and complexity of speech episodes: if you make your utterances too long you will violate the tacit understandings needed to maintain a viable speaker–hearer relationship. Were I to say, in any natural context, 'Caesar and Nelson and Atilla and . . .', with the extraordinary intention of using a hundred proper names in the subject-phrase, my hearer would soon fail to understand what I was doing. It may be a useful theoretical fiction to say that the grammar 'describes' the speech-behaviour of idealized speakers, but we need to be clear about how far we are thereby departing from direct empirical confirmation. The assumption of un-bounded length of sentences might be compared with an assumption of the infinite indivisibility of matter (or the assumption, in certain forms of probability theory, of infinitely long runs of trials with pennies or other random devices). Both are useful as simplifying the initial taks of theory; both raise similar problems of verification.

4. *How is a descriptive grammar verified?* In the light of what has been said, it would be wrong to think that the generative rules of a grammar straight-forwardly predict speech-behaviour. Kepler's laws can properly be taken to predict certain planetary motions, because the orbits are there to be observed. But statements about indefinitely long sentences are like statements in probability theory about the 'long run'. In the long run, as Keynes reminded us, we are all dead; in the long run, also, we are all silent.

The position seems to be that the empirical controls for a grammatical theory are not so much what people say in the object-language in which they talk about ships and sealing wax as what they say in a grammatical meta-language, speaking as amateur grammarians. The question about my centi-pedal sentence's grammaticality has to be settled by appeal to an informant acting as a grammatical judge: the test is not how he *would* respond to the long sentence itself, but how he *does* respond to its short description. The appeal is to a referee, not to a player—even if the same man plays both parts.

It seems to me, therefore, that in such cases we are checking the gram-marian's proposed set of generative rules against another untidier set of embryonic and half-explicit rules. The verification route is from rules to rules, not from rules to hypothetical and idealized performance. Now of course we should like to do more than this, by providing a theory of the language itself, not a theory of users' ideas about the language. But how is this to be done? Is there perhaps the risk of merely reviving and extending a set of rules previously taught in the classroom? The kind of verification involved seems to me interestingly different from what arises in the case of physics. We cannot ask a planet how it thinks it ought to behave in situations that don't arise: yet this seems to be what we are doing when we consult a native 'informant'.

5. *Does a particular grammar explain anything?* Whatever their value, Kepler's laws do not explain the planetary motions, in any useful sense of

'explain': they replace a crude and unsystematic description ('those orbits out there', or 'the orbits conforming to these readings') by another description concisely presenting some mathematical properties of the orbits. The same applies, *mutatis mutandis*, to the rules constituting a specific generative grammar. Our initial crude 'intuitions' as to what should count as grammatical or the reverse are replaced by a set of precise and explicit rules that (approximately and with idealization) generate a corresponding classification. This provides valuable insight into structural connections: it may be said to provide intelligible reasons for what we previously seemed to be doing by a kind of instinct. But 'explanation' hardly seems the right tag. If we use that word here, what are we to reserve for our attempts, at deeper theoretical levels, to account for the superficial grammatical rules themselves? Are we to talk about explaining an explanation?

A generative grammar can, from another standpoint, be regarded as equivalent in theoretical efficacy to an axiom-system. I would not myself regard an axiom-system for plane euclidean geometry or, for that matter, for the voting behaviour of committees, as *explaining* anything. But if we do call this explaining, we shall do well to recognize that it is not of the disparity-reducing kind that arises from the attempt to solve a puzzle, nor the causal sort common in the natural sciences.

The process, I would suggest, is really one of *codification*. The descriptive grammarian is like somebody undertaking to provide the rules of a game, whose players have no explicit code to which to appeal: he is the Hoyle of speech behaviour. Or he is like the legal theorist who undertakes to provide a legal code for some land where claimants and prosecutors have long behaved *as if* they were controlled by explicit rules.

I do not know how pregnantly Chomsky would wish to use the terms 'rule' and 'rule-governed behaviour'. A radical distinction between a rule and a regularity, which seems to me desirable on other grounds, would reinforce the point I have been making about the special types of verification and explanation in question.

6. *What are the constraints on the choice of a grammar?* It should be immediately obvious that the illustrative puzzle of the *a–b* strings can be solved in indefinitely many ways. For any set of rules can always be transformed into some other set, having the same logical power. Some solutions will appear simpler than others, but nobody has yet given a plausible analysis of what simplicity means in such connections.

Let us call two different sets of rules that generate the same sets of admissible sentences *productively equivalent*. We can see that the codification of the grammar of a given language might produce indefinitely many different but productively equivalent systems of grammatical rules. It is well known that an axiom system for plane euclidean geometry can be formulated by using as a primitive only the notion of a circle, or alternatively, only the notion of a line (with corresponding changes in the axioms); analogously, since the reasons are quite general, there will be many alternative but different ways of presenting a grammar. In particular, the very notion of a rule is dispensable: in order to segregate the grammatical from the ungrammatical and to reveal the grammatical structure of admissible sentences, we could use sets, or functions, or a number of other organizing ideas that a professional mathematician could produce with ease. Mathematical organization is never uniquely determined by extra-mathematical data.

So long as we confine ourselves to the theoretical interests of grammarians, the choice between these alternatives will be dictated by convenience, ease of use, and heuristic fertility. The linguist will choose a grammar that he finds easy to work with, useful in field work, suggestive of experiments, and so on. How we describe celestial motions is, of course, a matter of indifference to the planets themselves: if we find it less cumbersome to use conic sections rather than circles and epicycles, or integral equations rather than differential ones, that is our affair. But in speech, where rules are, in some sense hard to render precise, acknowledged by the speakers themselves, the case is different. Some rules, that might provide elegant codifications of grammatical relations, would be unusable by normal speakers. Chomsky has never supposed that the normal speaker constructs his sentences by following the canonical representation of grammatical structure. The ordinary speaker does not, except in very special cases, start with the category of sentence, proceeding by bifurcation into subject and predicate, and onwards to noun-phrases, verb-phrases, and so on, to end only after many intermediate steps with desired words in a proper order.

We seem to be faced then with the ticklish problem of reconstructing grammars that are actually used by speakers and hearers. The nature of such a task is by no means sufficiently clear. Yet the attempt will need to be made if we want the theorist's grammar to be something more than a device for 'saving the appearances'. We shall need to make the transition from the algebra to some psychological 'reality', whatever we mean by that dubious but indispensable word.

7. *How can we pass from mathematical grammar to psychology?* A transformational grammar, I have argued, can be regarded as a piece of mathematics, suggested by selected grammatical verdicts of qualified informants.

So, Chomsky's main premisses, which I accept on trust as well-founded, look to me like mathematical ones, having the form: such-and-such will just suffice to generate such-and-such a class of sentences (abstract structures, suggested by utterances that competent speakers would certify as 'correct'). Yet Chomsky ends with *psychological* and *epistemological* conclusions. If his reasoning is sound, some psychological and epistemological premisses must have been introduced in order to warrant the transition: I am unclear as to what these additional premisses are. No doubt they could be found in Chomsky's numerous discussions of his position, but I cannot see them as sharply as I should like.

8. *Must we be rationalists?* The most challenging conclusion that Chomsky draws from his conception of language and the appropriate methodology for linguistic science is the need to return to something like the position of classical rationalism. But we never swim in the same tradition twice: the rationalism that Chomsky recommends differs in important respects from the position of such paradigmatic rationalists as Descartes, Leibniz, or Kant.

As I understand them, these ancient advocates of a doctrine of 'innate ideas' typically regarded the intellect as a source of *necessary* principles, supposedly required for the organization and cognitive mastery of the external world. Just because such principles were held to be 'intuited' with certainty, they were judged to be logically independent of experience. If experience yields only the tentative and corrigible, and if we find ourselves in the possession of apodictic truths, there must be some other non-experiential and infallible source of knowledge. The rationalists' conviction of irrefutable knowledge of the necessary truth of such principles as universal causation, the three-dimensionality of space, etc., was their main reason for denying the sovereignty of experience. This is far removed from Chomsky's climate of thought.

Chomsky is a rationalist with a difference: he cannot think that the principles of 'universal grammar' (still waiting, so far as I can see, to be laid bare) are known, in anything like the classical sense. They are supposed to operate at a level so far below awareness that we have a hard time formulating them, let alone recognizing them as authoritative. Nor does Chomsky regard the basic principles as necessary in the logical sense, since he insists upon the existence of many conceivable, but unused, alternatives.

Given such far-reaching differences in doctrine, it seems to me somewhat wilful, from a historical point of view, to appropriate the tattered label of 'rationalism' for a philosophical standpoint so transformed in content and intention. I suspect that this unnecessary bit of pseudo-historical affiliation is mainly introduced *pour épater les empiristes*.

457

The more genuine dispute between Chomsky and his critics is as to the proper allocation of the responsibility for language acquisition between learning and antecedent biological endowment. (Roughly speaking, how much can we expect from the genes?) Chomsky is a 'nativist' rather than a rationalist.

Now, expressed in the most general terms, the old dispute between nativists and empiricists is quite without interest. Everybody can agree that the notion of a perfect *tabula rasa* is an absurdity and the notion of a *machina immaculata* as much so. Even a blank sheet of paper will not permit anything whatever to be inscribed on it! The only question of interest is *how much* a child can learn from experience and *how much more* must be imputed as biological endowment in order for such learning to be possible.

I must say that it looks to me as if Chomsky pits an old-fashioned and unresourceful empiricism (roughly speaking, the sensationalism of Hume, or its modern survival in stimulus–response psychology) against an up-to-date and sophisticated 'rationalism' that Leibniz would disown. Of course, the poorer one's conception of experience, the more tempting it will be to succumb to nativism or rationalism. But learning from experience need not be construed on the simplistic model of association, by contiguity and resemblance, of logically independent sense-data; nor need the generalizing power of the organism necessarily be restricted to Baconian induction. Perception of spatial relations, and the emergence of concepts, inherently productive, because applicable to novel instances, are at least as much to be found in 'experience' as immediate recognition of patches of colour. I should hesitate to locate any aspects of experience, thus generously construed, 'in the organism' without substantial evidence.

Such evidence will certainly be hard to find, since on Chomsky's views the latent dispositions, already at least one remove from direct observation, themselves need to be activated by suitable experience. Since we can at best observe certain complex patterns of behaviour (including a productive generalizing capacity) demanding previous training, it is hard to see what work the assumption of innate structures really does. In natural science, reference to dispositions and powers is usually a promissory note, to be cashed by some identifiable internal structure: if we seriously attribute elasticity to a piece of metal, it is because we expect to find inside it some relatively stable configuration that will reveal the dispositional term as something more than convenient shorthand for if–then connections between impressed forces and reactions. But there is no serious question at present of finding such internal configurations in human organisms: we hardly know what we should be looking for, or where to look. So far as stimulation of research goes, 'nativism' looks to me like a dead end, while 'empiricism' (the provisional attribution of learning to something, however complex, detectable in antecedent training and learning) suggests programmes of investigation that may be expected to uncover interesting data. Chomsky may be sinning against that special version of Occam's razor that runs 'Innate endowments are not to be multiplied beyond necessity'. I am yet to be convinced of the necessity.

9. *How good is the evidence for linguistic nativism?* The evidence offered by neo-nativists is classifiable as follows: (*a*) the presence in all languages of certain 'universals'; (*b*) the ease with which young children learn language (including a command of grammar) and the comparative difficulty with which adults learn it; (*c*) the difficulty of accounting for these facts upon any 'inductive' conception of the learning process.

Now as to the existence of language universals, I am somewhat sceptical about the strength of the available evidence. For one thing, there is risk of tautology in the argument from examination of known languages. The 'universals' we fancy we find may simply be part of our criteria of what is to *count* as language. Consider the strikingly particulate character of all known language, at the levels of phonology and morphology. We can choose to exclude other modes of communication, e.g. by continuously varying gestures, as non-linguistic: but is that more than arbitrary definition? It would certainly be implausible to argue from some component of a definition, or from the partial analysis of a concept, to a general empirical truth about human beings. And the fact that human beings do without exception, it seems, use particulate linguistic systems may perhaps be explained more plausibly on the ground of the obvious efficiency of such systems than by invoking a corresponding biological predisposition. The use of levers is almost universal among mankind, but it would be a hardy thinker who postulated a corresponding specific endowment.

The examples of linguistic universals that Chomsky, at least tentatively, regards as innate, seem to me rather unplausible in themselves. One might expect any such principles that might emerge to be special cases of more general ways of responding to and structuring experience. Are there perhaps universals of symbolism?

The evidence for childish facility in acquiring language is no doubt stronger. But neglecting such inessentials as correctness of accent, is the evidence as good as it is usually supposed to be? Very little is known at present about how much grammar children really acquire in the formative years. But suppose, for the sake of argument, that there is a period in early infancy when children can learn to talk more easily than in later years; what would this show? Young children can probably learn to swim faster than

459

adults; yet in this case the nativist inference looks unpalatable. An anti-nativist might try to argue that variation in external facilitating conditions, such as the obvious advantages of learning the first language, the influence of adults, etc., suffice to account for the imputed difference in learning rates. That an anti-nativist could reply in this way, or in some elaborated version of such a defence, suggests that the dispute is not straightforwardly empirical after all. It may be a conflict between two methodologies rather than a dispute between two theories.

Finally, it might be held that the difficulty of accounting for language acquisition on the basis of inductive analysis of data (given the highly fragmentary and 'degenerate' examples of linguistic performance that the child is exposed to) makes any explanation but the possession of innate endowment impossible. This argument has all the weakness of any conclusion *ab ignorantia rationis*. 'What else *could* explain this?' is too often a sign of the questioner's lack of imagination to inspire much confidence. Suppose we found a stone so intricately streaked that we were at a loss to conceive how environmental influences could have produced it: it would be a bold leap to hold something in the stone itself responsible. Now the argument from incapacity to understand how induction could serve the learner's needs may be no stronger than this: Is the fault perhaps with the crudity of our conception of induction or, more generally, with our conception of experience?

10. *A possible incoherence?* I will end with a qualm, connected with a point I have already made about the logical contingency of the imputed principles of universal grammar. Chomsky is able to tell us some of the alternative principles (or modes of organization) that are *not* used in the languages we know. Let us call such principles *unnatural*. Are we asked to believe that human beings cannot *in fact* use unnatural principles, at however much inconvenience, and still make themselves understood? If so, I doubt whether we can know it, without a trial. But suppose it turned out, as it might, that we could speak and make ourselves understood by employing unnatural principles. Then, on Chomsky's approach we would have to postulate some higher-order disposition for modifying the first-order dispositions that produce the 'natural' linguistic capacities. Which looks suspiciously like a *reductio ad absurdum*. Chomsky seems to me to be in the position of other theorists (such as Karl Mannheim and Whorf) who have claimed to discover general limitations upon human powers: in the course of describing the limitation, they show by their own discussion how to escape from it.

It may be possible, therefore, to find an empirical disproof of Chomsky's position, by teaching children or adults to violate the principles of universal grammar. Since human beings seem to be rather good at breaking grammatical rules, I think it not impossible that such an experiment would succeed. At least, it would be interesting to try.

11. *No simple resolution?* My list of questions has probably made the issues between transformational grammarians and their critics look simpler than they really are. Chomskyites will, I suspect, retort that I am expecting too much of a theoretical reorientation, in asking for piecemeal empirical verification. The test must be global theoretical adequacy, as against available alternatives. I am sympathetic with this reply. The current vogue of transformational grammar is symptomatic of a revolt among linguists, at least in America, against the somewhat doctrinaire neglect of theory by an older generation strongly under the influence of such giants as Boas, Sapir and Bloomfield. For all the enormous merits of the empirical research inspired by these pioneers in scientific linguistics, it cannot be denied that it too often emphasized the descriptive at the expense of the explanatory. Whatever may be its weakness in details, the point of view outlined in Chomsky's paper has had the great merit of reviving interest in such long-neglected but fundamental issues.

# REPLY

## *by* NOAM CHOMSKY

Professor Black raises a number of searching questions that go to the heart of what seems to me the basic issue, namely, how the study of language can illuminate problems of mental structure and function. I think a rather stronger case can be made for a connection between these studies than is suggested by his remarks. I will try to indicate briefly why I think this is so.

The account that Professor Black gives of the task of a formal grammar is consciously oversimplified, but the oversimplification is perhaps misleading in what it suggests as the natural scope of descriptive grammar. A grammar is, in principle, a finite system of rules for generating the strings constituting a language and also the structural descriptions of these strings (not just their phrase structure, which is only one aspect—the surface structure—of this structural description). A complete structural description of a string should specify its phonetic and semantic representation (each in some universal system) as well as the array of syntactic structures that serve to interrelate them. In practice, we can only approach complete structural descriptions in a limited way, but it is important, nonetheless, to keep the general problem in mind when questions of empirical relevance arise. In addition, a grammar should indicate, for each phonetically admissible string (a meaningful notion, given a universal phonetic theory), just how it deviates from well-formedness if it is not *directly generated* by the grammar, in the sense just described.

To take a standard example, the grammar of English must deal with the fact that 'mârking pápers can be a nuisance' has a certain phonetic, syntactic, and semantic interpretation, different in important respects from 'struggling artists can be a nuisance'; with the fact that 'visiting relatives can be a nuisance' has two structural descriptions, each parallel to one but not the other of the two preceding examples in its syntactic and semantic aspects; and that 'papers can be a' does not have the same status, in fundamental respects, as the preceding examples. These observations can be elaborated and sharpened in many ways and extended to a mass of other cases. They are, however, factual observations about the knowledge of English that has been acquired by the normal speaker–hearer. We might, if we wished, consider experimental techniques to provide factual information of this sort. Or, we may concentrate on the theoretical problem of how to account for such facts. In the present state of the subject, the latter

462

seems to me far the more important and interesting pursuit, and I will therefore concentrate solely on this aspect of the problem.

To use Professor Black's example, Kepler's laws describe planetary orbits (under some idealization); analogously, the grammar of English describes a variety of facts of a sort just illustrated. Suppose that the facts regarding planetary motion were obtained by a man with a telescope. We might then intelligibly, but misleadingly, say that Kepler's laws are about the judgments of a man with a telescope. Similarly, suppose that the facts regarding linguistic expressions were obtained by an appeal to informant judgment. We might then intelligibly, but misleadingly, say that a grammar is about the informant's judgments. But I do not accept Professor Black's view that in the case of grammar, the appeal is to a referee rather than to the facts themselves, unless the same conclusion is accepted in the case of the planetary orbits. In both the astronomical and linguistic case there are certain facts and certain judgments (accurate or not) about these facts. A theoretical description tentatively assumes the accuracy of the judgments, and describes the facts. The logic of the situation is not altered by the fact that the judgments, in the case of astronomy, are more easily reducible to repeatable experimental procedures. In either case, the judgments can be in error, in which case the theory may well have to be revised, the facts being other than what they were thought to be. But I see no problem of principle that distinguishes verification of a grammar from verification of Kepler's laws, although, naturally, the analogy should not be pressed too far in the case of theories of such different character.

I think everyone would accept without question that a theory of the speaker–hearer's knowledge of his language (a generative grammar, in other words) must contain recursive devices. This being so, I do not see why Professor Black regards the assumption of unbounded length of sentences to be 'a more troublesome kind of idealization'. We could eliminate this 'idealization' by assigning an arbitrary upper bound to sentence length, say, a million words. Speech behaviour would provide no counter-evidence to this hypothesis, but it would still be rejected by ordinary canons of scientific method, just as would an addendum to Kepler's laws stating that they only apply when a system of planets does not exceed $10^{10^{10}}$ in number. In each case, we have a simpler theory that excludes the addendum, which is supported by no evidence. In the language case, it seems to me that ordinary canons of scientific method lead us to reject the assumption that in addition to a recursive system of rules, there is a principle that blocks assignment of structural descriptions past a certain point (which, in practice, is never reached).

The more important problem, I think, is the one that Professor Black discusses when he refers to the 'great gap between ideal competence and actual performance'. This gap poses numerous problems for the psychologist, who must explain how competence (knowledge of a language represented by a generative grammar) is used in production and understanding and how it is acquired in the first place, on the basis of the degenerate data of performance. This is a serious and difficult empirical problem, but I see no conceptual or logical issue that stands in the way of research.

I would be prepared to accept Professor Black's argument that a grammar, like Kepler's laws, provides a codification of certain facts. Like Kepler's laws, this codification can play a role in explanation of phenomena. We can use Kepler's laws to explain why a planet occupies a certain position (given an earlier state of the system), and we can use the rules of English grammar to explain why a person understands a certain sentence to be ambiguous. We may say that the explanation is not very 'revealing', that Newton's laws, for example, give a deeper explanation of the phenomena codified in Kepler's laws. In a partially similar manner, we can say that the principles of universal grammar (in the sense outlined in my paper) give a deeper explanation for certain facts about knowledge of language, namely, that as a hypothesis about innate mental structure it explains why, given certain empirical 'boundary conditions' (the primary linguistic data), a grammar has been constructed that (in a less deep sense) explains the phenomena in question. I have suggested elsewhere that the term 'level of descriptive adequacy' be used for the study of the relation between grammars and data and the term 'level of explanatory adequacy' for the relation between a theory of universal grammar and these data. I think that this terminology is in accord with Professor Black's suggestion, if I understand it correctly.

Point 6 of Professor Black's comments deals with the problem of distinguishing grammars that express 'psychological reality' from others consistent with whatever empirical data can be obtained. It is surely true that (exactly as in the case of celestial mechanics, quantum physics, or whatever) there will be non-equivalent theories (grammars) compatible with any mass of evidence that we have about a given language. The problem can be approached in two ways: by discovering a wider range of crucial evidence; by refining the criteria that determine the choice of theories (considerations of simplicity, and so on). I agree with Professor Black that there has been no plausible or useful analysis of these criteria; in fact, in any discipline, it is only the first approach that is productive, so far as our understanding of these matters now extends. In the study of language, one can enrich the evidence by extending the domain of facts for a particular language, by considering evidence on language use or acquisition, by considering evidence from other languages, which becomes relevant to the extent that one has a precise general theory of language that can be used to choose among alternative 'codifications'. In principle, neurological evidence might also be relevant to this study, at some future time. It is important to be clear, however, about the actual scale of the problem. I think that our problem in study of language is not that we have too many competing 'codifications' all compatible with the available evidence, but rather that we find it difficult to construct even one grammar (or, *a fortiori*, one universal grammar) that even begins to encompass the available evidence. To be sure, we know that in principle there will always be non-equivalent alternatives compatible with finite evidence, but this is not a real problem in linguistics or, to my knowledge, any other field.

With respect to point seven, I do not see, again, why the status of language study is different from that of any other discipline. A grammar must face a mass of empirical evidence of a sort illustrated above, in my paper, and in many other publications. As in all other fields, there are elements of idealization that must be kept in mind when a theory is confronted with evidence. In any domain, when a theory becomes sufficiently precise it is possible and sometimes useful to study it as a mathematical object, apart from any consideration of confirmation. The temptation to do so is perhaps greater in linguistics than in other fields, because the fragmentary theories that have so far been developed do have certain interesting properties, when studied abstractly. But I see no logical problem here.

Professor Black argues that the reconstructed rationalism that I have proposed differs from the original in that the principles of 'universal grammar' are not known and are not necessary in the classical sense. As noted in my paper, I do depart from classical rationalism (as well as from empiricism) in denying that the 'contents of the mind' are necessarily available to introspection. Nevertheless, it should be emphasized that Leibniz, at least, emphasized that 'we have an infinite amount of knowledge of which we are not always conscious' and that we may not at all be aware of the innate general principles that 'enter into our thoughts, of which they form the soul and the connection'; and so on, repeatedly. There is, of course, no objective answer to the question whether the 'essence' of rationalism is lost when the doctrine is developed in the manner that we are discussing. As to the matter of necessity, it is not easy to interpret the classical doctrine in modern terms; Descartes, for example, assumed that much of what we would regard as empirical is 'necessary', in his special sense. Still, I would certainly agree that there has been a departure in this respect. But it seems to me historically plausible (and, furthermore, heuristically useful) to regard the reconstructed rationalism that we are discussing as an outgrowth of classical rationalism which rejects some of it (for example, Descartes' fallacious proof of the existence of God, which plays a central role in his theory), modifies, extends and clarifies other parts, but maintains some fundamental insights and framework, in particular, the assumption that innate properties of mind provide a schematism that determines what counts as experience, and that restricts the knowledge (in our case, the grammar) that is based on experience. There is an important difference between this approach, as developed in classical thinking and again in modern work, and the empiricist alternative. I have tried to argue this elsewhere, and won't go into it here.

Nevertheless, a few points of clarification are in order. I do not think it quite fair to say that I am pitting 'an old-fashioned and unresourceful empiricism' against a rationalism that shares only the name with the classical doctrine. For one thing, I do not know of a more up-to-date and resourceful empiricism than the ones that I and others have discussed in this connection. Secondly, I think that there are some interesting and essential doctrines expressed in classical empiricism that have continued to dominate modern thinking about the nature of mind and that can, when made precise in what seems to me a natural way, be refuted by linguistic evidence. Thus I believe, and have tried to argue, that one can save 'empiricism' only by denuding it of its interesting and important content, thus saving the word and abandoning the doctrine, and with it most of the speculation and thinking that has dominated the study of mind. Furthermore, many of the classical rationalist doctrines seem to me to be carried over in the modern version under discussion. For example, it was Descartes and later Leibniz who insisted that innate ideas and principles must be 'activated by suitable experience'. Descartes' interesting ideas about the limits of 'mechanical explanation' and about how the mind interprets degenerate data in terms of innate structure (e.g. in seeing a physical shape as a triangle, 'which can be more easily conceived by the mind than the more complex figure of the triangle drawn on paper') reappear, in a more elaborate and refined form, in recent proposals regarding universal grammar and its function in language acquisition, these proposals themselves being an extension and elaboration of the linguistic theories that were developed by rationalist grammarians and their successors. Although this is not relevant here, I think one can argue that much recent work in perception also receives a natural formulation within the rationalist framework, for example, the work of Hubel and others on innate organization of the lower cortical centres (which degenerates, unless 'activated by suitable experience'), the experiments of Bower that suggest an innate basis for the perceptual constancies, the work of Held and his associates on the role of voluntary action in activating the mechanisms of perception, and much else. It seems to me both enlightening and historically reasonable to interpret this work in classical terms, sharpened and revised in certain ways. I might mention that Konrad Lorenz proposed many years ago, with much justification, it seems to me, that the findings of comparative ethology can receive a natural interpretation within a Kantian framework (cf. Lorenz, 1941, 94–125); I think that Lorenz errs, however, in assuming that there is a theory of natural selection with empirical content that changes the terms of the discussion in a fundamental way.

Quite apart from the historical question, I disagree with Professor Black on the matter of 'emptiness' of nativistic assumptions. There is clear empirical content to the view that principles of universal grammar or of perceptual organization restrict and determine the nature of the knowledge and beliefs that we acquire. Contrary to what Professor Black suggests, such assumptions, when made precise, do a great deal of work and can easily be refuted if wrong. Nativism and empiricism are, furthermore, quite on a par when it comes to the matter of neurological interpretation; the neurological basis for learning is as obscure as the neurological basis for universal grammar, and I think it can be argued that the little that is known about neurological structure (as in the work of Hubel mentioned above) lends support to the view that there are fixed, highly restrictive interpretive mechanisms even in mammals, as there are known to be in lower organisms. Conceivably, humans are unique in lacking such mechanisms, but I see no empirical reason for supposing that humans have some unique sort of 'indefinite plasticity', or that they acquire knowledge by the empiricist principles that have been stipulated in classical and modern speculation. Hence I would deny strongly the claim that nativism is a dead end, whereas empiricism suggests programmes of investigation.

I do not see how Occam's razor figures in this discussion. In investigating acquisition of language, we are concerned with the problem of determining the characteristics of a system of unknown properties which gives as 'output' (i.e. final state, in this case) a generative grammar of a language, given sufficient data as input. For the moment, we cannot formulate a hypothesis regarding this system which attributes to it a structure sufficiently rich so as to bridge this gap between data and knowledge. At the same time, the postulated structure must not be so restrictive as to be refuted by the linguistic facts from some natural language. When we reach the happy state of having before us several alternative hypotheses that meet the empirical conditions just sketched, we can apply Occam's razor and other general considerations that in some poorly understood way are used to select among competing theories when crucial evidence is unavailable (although, as noted earlier, I see no reason to anticipate that this will be a real problem, in the foreseeable future). It is an empirical question whether the structure attributed to the mind as an innate property should involve inductive or heuristic procedures, as the empiricist might speculate, or a restrictive schematism that determines the form of the knowledge to be acquired, perhaps along the lines of the reconstructed rationalism under discussion (or some combination, or some entirely different idea). There is no *a priori* sense in which one of these approaches is 'simpler' or 'more natural' or more 'parsimonious'. It seems to me that the problem is analogous to that of determining the factors that lead to the organization of any biological system, on the basis of native endowment and organism-environment interaction.

I do not see how the particular assumptions of universal grammar that seem to have some empirical support can be explained on the basis of 'efficiency' or why they are 'universals of symbolism'. Even the discrete character of language to which Professor Black alludes is not necessary for efficient communication; gestural systems or the so-called 'bee language' operate in terms of continuous dimensions (in the only sense in which a physical system can be said to operate in this way). And it is certainly difficult to imagine that some universals of symbolism or considerations of efficiency require that phonological rules should apply in a linear ordering or a cycle, or that grammatical transformations should be structure-dependent. Furthermore, even if such considerations could be formulated, we could rely on them to explain the construction of a grammar of a particular sort only if we were willing to attribute to the mind, as an innate property, the ability to apply these considerations to the choice of a grammar on the basis of what data is available to the child, a conclusion that is no more compatible with empiricism than is the reconstructed rationalism discussed earlier.

The particular assumptions of universal grammar have empirical consequences in terms of which they must be evaluated. I agree that these assumptions are 'unplausible in themselves', but regard this as their scientific merit. It is just this fact that gives importance to the evidence that is offered in support of these principles. There is, for example, no *a priori* consideration that lends plausibility to a theory of syntactic operations that precludes the formation of interrogatives by left–right inversion of the corresponding declarative, or innumerable other perfectly conceivable, and quite efficient operations. It is just this fact that constitutes the scientific interest of such a theory.

468

# REPLY BY NOAM CHOMSKY

I do not see any great significance, in this particular connection, in the fact (if it is a fact) that there is a critical period for language learning (for evidence in support of this view, see Lenneberg, 1967). If this were not true, we would face just the same empirical problem of explaining how a particular grammar is devised by the learner on the basis of the scattered and degenerate evidence available to him. Nor do I agree that 'the nativist inference looks unpalatable' in the case of the ability to learn to swim. Why so? Can anyone doubt that there is a specific biological basis for this ability, with, perhaps, a critical period for acquisition of the skill. Nor do I see why it is the nativist, rather than the empiricist, who is using the argument from ignorance. In particular, one who tries to account for language acquisition by providing a specific 'black box' theory that incorporates a schema of universal grammar has no responsibility to demonstrate that all other conceivable alternatives are incorrect. He should try to demonstrate that other clearly formulated alternatives are refuted by the evidence, but this, I think, has been fairly well shown. And I also think that a plausible, if not conclusive argument has been outlined suggesting that no approach that restricts itself to the particular assumptions of classical or modern empiricism is likely to be of any particular relevance to this problem. I would not argue against the empiricist that he has not demonstrated the impossibility of any rationalist approach; and, correspondingly, I do not agree that the burden of proof rests on the rationalist (or nativist) to demonstrate that his particular assumptions are the only ones thinkable. It is enough for him to give evidence in support of them, and against clearly formulated alternatives.

The final point that Professor Black raises is an interesting one, but I do not think that it suggests any incoherence. Consider the following line of argument (I do not, for the moment, argue that it is correct, but only that it is coherent). A theory of mind will postulate a number of faculties, one, a language faculty with the properties sketched in my paper. (Apparently, this faculty will 'degenerate' if not utilized at a particular stage of maturation, like other known biological structures, but this is not here relevant.) Given appropriate stimulation, this faculty will organize itself into a grammar that expresses a certain linguistic competence, in the technical sense, which, of course, goes far beyond the given evidence and even rejects much of it as irrelevant. But the knowledge of language provided by this faculty is not the totality of human knowledge, and its mechanisms do not exhaust the devices by which knowledge (better, belief) can be acquired. For example, these mechanisms preclude an operation of interrogation by left–right inversion, but permit an operation of interrogation of the highly complex sort that we have in English. Making use of other faculties of the mind, a person could in some sense 'learn' a system of communication that forms interrogatives by left–right inversion; thus the experiment that Professor Black suggests would surely succeed. But it would not disprove the assumption that the language faculty precludes such an operation. However, it should follow that the acquisition and use of this new system of communication should be qualitatively different, in a number of respects, from the acquisition and use of the system containing the English rule. It should be much more difficult (or perhaps even impossible) to learn under conditions that suffice for ordinary language-learning. It should be impossible (or very difficult) to use with anything like the facility of normal English or in a similar range of situations. And so on. If we were interested in distinguishing in a precise and detailed way among the various faculties of mind, we could go on to refine and elaborate such criteria as these. Of course, I have no doubt that the other faculties of mind also have their intrinsic limitations and structure. But nothing is known of any importance about this matter, and we cannot, therefore, say anything significant about the limits of human intelligence and the bounds on human knowledge. There is, so far as I can see, no incoherence in supposing such bounds to exist or in expecting that some day we will be able to say something interesting about them. But in any event, there is surely no incoherence in supposing that one of the components of mind—the language faculty— has particular properties, abilities, and limitations. I do not, therefore, see that there is any *reductio ad absurdum* lurking dangerously in the background. What we face, rather, is the empirical problem of determining the precise character of the faculty of language, and the combined conceptual-empirical problem of placing this faculty properly in a general theory of mind.

REFERENCES

Lenneberg, E. H. 1967. *Biological Foundations of Language.*
Lorenz, Konrad. 1941. Kants Lehre vom apriorischen im Lichte gegenwärtiger Biologie. *Blätter für Deutsche Philosophie* **15**.

# FREUD AND THE IDEA OF
# A PSEUDO-SCIENCE

## *by* FRANK CIOFFI

### I

A successful pseudo-science is a great intellectual achievement. Its study is as instructive and worth undertaking as that of a genuine one. In this paper I shall maintain that psychoanalysis is such a pseudo-science; that the character of this claim has often been misunderstood; and that when it is understood its intractability is less surprising. Psychoanalysis may be described as an attempt to determine the historicity and/or pathogenicity of episodes in a person's infantile past and the character of his unconscious affective life and its influence over his behaviour, by the manner in which he responds to assertions or speculations concerning these—not however, just *any* attempt, but a particular, historically identifiable one which issued in a body of aetiological and dynamic theses, the abiding core of which is the claim that 'only sexual wishful impulses from infancy are able to furnish the motive force for the formation of psycho-neurotic symptoms' (Freud, 1949*b*, pp. 605–6).

In attempting to assess the genuinely empirical character of such an enterprise the statements which we must subject to scrutiny are not merely those in which the claims that are the ostensible object of investigation are advanced, but also those which describe, or enable us to infer, what the procedures of investigation are. A pseudo-science is not constituted merely by formally defective theses but by methodologically defective procedures. We could express this mnemonically by saying: the notion of a pseudo-science is a pragmatic and not a syntactic one.

The contrary view is partly the result of preoccupation with a narrow and untypical range of examples (e.g. the dormitive powers of opium, etc.) and partly a function of the calculus-dominated assumption that the logical character of a thesis can always be determined by inspection.

In his paper 'Can psychoanalysis be refuted?' (Farrell, 1961) B. A. Farrell argues that when we ask 'whether a generalization about some *specific* unconscious process or state is true or not—for example, the generalization about unconscious Oedipal wishes . . . we presuppose that these generalizations about specific unconscious processes can be refuted in principle. We have discovered no reason so far for saying that this presupposition is false in any way that matters.'

But the real difficulty is that it can't be *true* in any way that matters. Refutability in principle is not an adequate criterion of the genuinely empirical character of an enterprise. (Nor is it the criterion advanced by Popper in the paper which Farrell alludes to—Popper, 1957.) If it were, none of the specimens we would naturally proffer as examples of pseudo-diagnostic, pseudo-therapeutic or pseudo-explanatory claims would qualify. The efficacy of the high-potency microdose, the propensity to melancholy of those born under Saturn, the immunity to appendicitis of people ignorant of the existence of the vermiform appendix; these claims are eminently refutable and, as far as most of us are concerned, have been refuted. But they are nonetheless pseudo-scientific. For an activity to be scientific it is not enough that there should be states of affairs which would constitute disconfirmation of the theses it purports to investigate; it must also be the case that its procedure should be such that it is calculated to discover whether such states of affairs exist. I use the word 'calculated' advisedly. For to establish that an enterprise is pseudo-scientific it is not sufficient to show that the procedures it employs would *in fact* prevent or obstruct the discovery of disconfirmatory states of affairs but that it is their function to obstruct such discovery. To claim that an enterprise is pseudo-scientific is to claim that it involves the habitual and wilful employment of methodologically defective procedures (in a sense of wilful which encompasses refined self-deception). The necessity for this provision becomes clear if we consider cases like the following.

Lind's failure to employ a placebo control group when he tested the efficacy of lemons in the prevention of scurvy would, had a placebo effect been operative and the lemons prophylactically or therapeutically inert, have prevented him from discovering this. But this would not justify our characterizing his procedure as pseudo-scientific, since he was not aware of the placebo effect and therefore could not be said to have been attempting to avoid the discovery that his theory was mistaken. If today, however, someone failed to employ placebo control groups in making therapeutic tests this might well justify the characterization 'pseudo-science'. Our hesitation in so characterizing it would be due to doubts as to whether there might not be moral or practical reasons for failure to undertake the necessary tests.

Consider, as an illustration, the following piece of apologetic offered by Brian Inglis on behalf of homoeopathic medicine:

The difficulty has been to *prove* that the homoeopathic micro-dose works . . . It is no use the homoeopaths replying that the proof lies in the results they have achieved; these are dismissed by the profession as self-deception or, at best, as placebo-effect. Why not—doctors ask—allow the homoeopathic contention to be proved or disproved by experiments, double blind, with controls? This is a challenge the homoeopaths have been unwilling to take up, and they would have been foolish if they had, for it would have been a negation of their founder's third principle: that no two patients are alike for purposes of treatment . . . Two patients with the same symptoms, or the same patient with the same symptoms but in a different mood—may require very different prescriptions. Homoeopathic treatment cannot be given 'blind' (Inglis, 1964, pp. 83–4).

If we find this an unconvincing defence against the charge of 'pseudo-science' this is not because of any formal deficiencies in the assertion that the homoeopathic micro-dose works, but because of non-syntactic considerations like the failure (which Inglis' excuses, even if valid, don't explain) to rule out the effect of suggestibility by seeing whether the micro-dose is effective even when administered surreptitiously; to say nothing of the non-existence of homoeopathic veterinarians. [Though Inglis argues that 'homeopathic treatment cannot be given blind' it is piquant to note that it is over a century since the most eventful occasion on which it was. In *Science and Health* Mary Baker Eddy relates the following incident:

A case of dropsy, given up by the faculty, fell into my hands . . . as she lay in her bed, the patient looked like a barrel. I prescribed the fourth attenuation of *Argentum nitratum* with occasional doses of a high attenuation of *Sulphuris*. She improved perceptibly . . . learning that her former physician had prescribed these remedies, I began to fear an aggravation of symptoms from their prolonged use, and told the patient so; but she was unwilling to give up the medicine while she was recovering. It then occurred to me to give her unmedicated pellets and watch the result. I did so, and she continued to gain. Finally she said that she would give up her medicine for one day, and risk the effects. After trying this, she informed me that she could get along two days without globules; but on the third day she again suffered, and was relieved by taking them. She went on in this way, taking the unmedicated pellets—and receiving occasional visits from me—but employing no other means, and she was cured (Eddy, 1875, p. 156).

If Mrs Eddy had always reasoned this well we should never have heard of her.]

In what follows I shall attempt to show that there are a host of peculiarities of psychoanalytic theory and practice which are apparently gratuitous and unrelated, but which can be understood when once they are seen as manifestations of the same impulse: the need to avoid refutation. I shall proceed by arguing that the apparent diversity of the ways in which the correctness of psychoanalytic claims may be assessed—observation of the behaviour of children, enquiry into the distinctive features of the current sexual lives or infantile sexual history of neurotics, awaiting the outcome of prophylactic measures based on Freud's aetiological claims—all resolve themselves into one which itself ultimately proves illusory: *interpretation,* or, as Freud variously formulates it, 'Translating unconscious processes into conscious ones' (Freud, 1950, p. 382); 'Filling in the gaps in conscious perception' (Freud, 1950, p. 382); '... constructing a series of conscious events complementary to the unconscious mental ones' (Freud, 1949*b*, p. 24); '...[inferring] the unconscious phantasies from the symptoms and then [enabling] the patient to become conscious of them' (Freud, 1924*b*, p. 54).

## II

It is characteristic of a pseudo-science that the hypotheses which comprise it stand in an asymmetrical relation to the expectations they generate, being permitted to guide them and be vindicated by their fulfilment but not to be discredited by their disappointment. One way in which it achieves this is by contriving to have these hypotheses understood in a narrow and determinate sense before the event but a broader and hazier one after it on those occasions on which they are not borne out. Such hypotheses thus lead a double life—a subdued and restrained one in the vicinity of counter-observations and another less inhibited and more exuberant one when remote from them. This feature won't reveal itself to simple inspection. If we want to determine whether the role played by these assertions is a genuinely empirical one it is necessary to discover what their proponents are prepared to call disconfirmatory evidence, not what *we* do.

An example of this is provided by what Freud calls his 'Libido Theory' of the neuroses. Freud makes many remarks whose bearing seems to be that the sexual nature of the neuroses is an inference from the character of the states which predispose towards them, or of the vicissitudes which induce them, e.g. that their causes are to be found 'in the intimacies of the patient's psycho-sexual life'. The importance of this claim for our present purpose is that it would relieve our doubts about the validity of the psychoanalytic method if the inferences to which it leads as to 'which the repressed impulse is, what substitutive symptoms it has found and where the motive for repression lies' were corroborated by independent investigation of the patient's sexual life, as, for example, Freud's aetiology of the actual neuroses was (presumably) corroborated. But the claims which constitute Freud's libido theory are only apparently assessable by an investigation of the relation between the patient's sexual life and his accesses of illness.

Here are some assertions by means of which the impression that they are so assessable is produced.

Whenever a commonplace emotion must be included among the causative factors of the illness, analysis will regularly show that the pathogenic effect has been exercised by the ever present sexual element in the traumatic occurrence (Freud,

1924*a*, p. 281). In the ensuing remarks, which are based on impressions obtained empirically, it is proposed to describe those changes of conditions which operate to bring about the onset of neurotic illness in a person predisposed to it. The following view . . . connects the changes to be described entirely with the libido of the person concerned (Freud, 1924*b*, p. 113). . . . human beings fall ill when . . . the satisfaction of their erotic needs in reality is frustrated (Freud, 1962, p. 80). . . . patients fall ill owing to frustration in love—(owing to the claims of the libido being unsatisfied . . .) (Freud, 1950, p. 87). People fall ill of a neurosis when the possibility of satisfaction through the libido is removed from them—they fall ill in consequence of 'frustration' . . . in all cases of neurosis investigated the factor of frustration was demonstrable (Freud, 1956, pp. 310, 353).

These statements certainly look like hypotheses. But our hopes that Freud might be placing a limit on the kinds of events or states which are conducive to the onset of neurosis and might then go on to tell us what these are, are dashed when we read: 'We see people fall ill who have hitherto been healthy, to whom no new experience has presented itself, whose relation to the outer world has presented no change . . .' Though they rise again when Freud goes on to say: 'Closer scrutiny of such cases shows us nevertheless that a change has taken place . . . the quantity of libido in their mental economy has increased to an extent which by itself sufficed to upset the balance of health and establish the conditions for a neurosis . . . this warns us never to leave the quantitative factor out of consideration when we are dealing with the outbreak of the illness.' But what are these changes in the mental economy which 'closer scrutiny' reveals? Once again Freud keeps the word of promise to our ears and breaks it to our hopes: 'We cannot measure the amount of libido essential to produce pathological effects. We can only postulate it after the effects of the illness have manifested themselves' (Freud, 1924*b*, p. 119).

For instance, this is how the apparent counter-examples constituted by the war neuroses are assimilated to the libido theory. Freud says that those who had observed 'traumatic neuroses, which so often follow upon a narrow escape from death, triumphantly announced that proof was now forthcoming that a threat to the instinct of self-preservation could by itself produce a neurosis without any admixture of sexual factors' but that 'any such contradiction has long since been disposed of by the introduction of the concept of narcissism, which brings the libidinal cathexis of the ego into line with the cathexis of objects and emphasises the libidinal character of the instinct of self-preservation . . .' (Freud, 1961, p. 43). In any case 'Mechanical concussions must be recognised as one of the sources of sexual excitation' (Freud, 1922, p. 39).

Consider in this connection Freud's account of the relation of neurosis to the perversions: 'Neuroses are related to perversions as negative to positive. The same instinctual components as in the perversions can be observed in the neuroses as vehicles of complexes and constructors of symptoms' (Freud, 1962, p. 76). And elsewhere: 'The path of perversion branches off sharply from that of neuroses. If these regressions do not call forth a prohibition on the part of the ego no neurosis results; the libido succeeds in obtaining a real though not a normal satisfaction' (Freud, 1956, p. 368).

We might take this to imply that Freud is ruling out the occurrence of a condition in which perverted sexual impulses are being gratified and the pervert is nevertheless suffering from neurotic symptoms. But no. Freud tells us that we are not to be surprised at the existence of such states of affairs: 'Psycho-neuroses are also very often associated with manifest inversion' (Freud, 1938, p. 575). The symptoms may then express the patient's repressed conviction of the unacceptability of his perverted practices (Freud, 1925a, p. 335).

And this is how Freud reconciles his view that delusional attacks of jealousy are due to surplus libido, with the fact that he came across a case in which the attacks 'curiously enough appeared on the day following an act of intercourse'—: '. . . after every satiation of the heterosexual libido the homosexual component likewise stimulated by the act forced for itself an outlet in the attack of jealousy' (Freud, 1924b, p. 235).

Freud also maintains that 'homosexual tendencies . . . help to constitute the social instincts': 'It is precisely manifest homosexuals and among them again precisely those that struggle against an indulgence in sensual acts who distinguish themselves by taking a particularly active share in the general interests of humanity—' (Freud, 1925a, p. 447).

Does it follow that homosexuals indulge their tastes at the expense of their philanthropic impulses? That Casement would have been more solicitous on behalf of the exploited natives of Putomayo and the Congo had he been more chaste? Or does it not follow? The following formula could cope with either contingency: 'In the light of psychoanalysis we are accustomed to regard social feeling as a sublimation of homosexual attitudes towards objects. In the homosexual person with marked social interests, the detachment of social feeling from object choice has not been fully carried out' (Freud, 1924b, p. 243).

But on occasions on which counter-observations are too vividly present to him, Freud's claim that the neuroses have sexual causes takes this form:

It did not escape me . . . that sexuality was not always indicated as the cause of neurosis; one person would certainly fall ill because of some injurious sexual condition, but another because he had lost his fortune or recently sustained a severe organic illness . . . every weakening of the ego from whatever cause must have the same effect as an increase in the demands of the libido; viz., making neurosis possible . . . the fund of energy supporting the symptoms of a neurosis, in every case and regardless of the circumstances inducing their outbreak is provided by the libido which is thus put to an abnormal use (Freud, 1956, p. 394).

And on a later occasion, in connection with the case of a businessman in whom 'The catastrophe which he knows to be threatening his business induces the neurosis as a by-product', Freud says that nevertheless 'the dynamics of the neurosis are identical' with those in which 'the interests at stake in the conflict giving rise to neurosis are . . . purely libidinal . . . Libido, dammed up and unable to secure real gratification, finds discharge through the repressed unconscious by the help of regression to old fixations' (Freud, 1950, p. 471).

It is fair to conclude that though the introduction of the term 'libido' permits Freud to give the impression that claims are being advanced as to the nature of the vicissitudes which precipitate, or the states which predispose to, the development of neurotic disorders, in fact a convention has been adopted as to how these vicissitudes and states are to be described. 'Sexual trauma' has been extended in the direction of pleonasm; 'nonsexual conflict' no longer has a use. Whenever necessary these terms are employed in a 'typically metaphysical way, without an antithesis'.

### III

Consider some typical psychoanalytic scenery as it occurs in Freud's reconstructions of infantile sexual life. One sequence shows the child complacently fondling his penis while entertaining obscure projects of doing something in relation to his mother with it, subsequently discovering that this activity is frowned upon, but meeting with incredulity the threat that persistence in it will result in his sexual organ being taken from him until, on observing with horror the wound where his sister's penis should be, he abandons his defensive scepticism and initiates a struggle against his masturbatory impulses and the incestuous phantasies which accompany them. In another sequence the child is speculating as to the nature of the sexual transactions between his parents and on its relation to the puzzle of birth, deciding that the orifice of parturition and intercourse is the anus, yearning to usurp his mother's sexual place with his father and to bear him a child, but once again rejecting these wishes when the facts of female anatomy convince him that a precondition of their fulfilment would be his own submission to castration.

These accounts of infantile sexual life are startling, but, once we have got over being thrilled by them, ultimately trivial contributions to natural history, with as much human relevance as the fact that we once had gills. Their claim to our attention rests on two things: their explanatory value and the evidence they afford of the validity of psychoanalytic method.

Our confidence in Freud's reconstructions of his neurotic patients' infantile sexual life, and therefore in his claim that adult neuroses are continuations or recrudescences of infantile ones, might be justified by the endorsement of the validity of psychoanalytic method afforded by the accuracy of those portions of the reconstructions which are held to characterize childhood in general and can thus be confirmed by the contemporary observation of children. That Freud recognizes that at least some of their significance resides in this latter fact is indicated by this remark in the *Three Essays on Sexuality*: 'I can point with satisfaction to the fact that direct observation has fully confirmed the conclusions drawn from psychoanalysis and thus furnished good evidence for the reliability of the latter method of investigation' (Freud, 1938, p. 594). And on many occasions Freud does say that his clinically derived theses regarding the infant's sexual life could be tested by systematically observing the behaviour of children. In the case history of Hans he refers to the observation of children as a 'more direct and less roundabout proof of these fundamental theories' and speaks of 'observing upon the child at first hand, in all the freshness of life, the sexual impulses and conative tendencies which we dig out so laboriously in the adult from among their own debris'. He even implies that the facts to which he has called attention are so blatant that one must take pains to avoid noticing them. For example in his paper 'The sexual theories of children' he says 'one can easily observe' that little girls regard their clitoris as an inferior penis. In his paper 'The resistances to psychoanalysis' he writes of the Oedipal phase: 'At that period of life these impulses still continue uninhibited as straightforward sexual desires. This can be confirmed so easily that only the greatest efforts could make it possible to overlook it' (Freud, 1925 *b*, p. 172). And again

In the beginning my formulations regarding infantile sexuality were founded almost exclusively upon the results of analysis in adults . . . It was, therefore, a very great triumph, when it became possible years later to confirm almost all my inferences by direct observation and analysis of children, a triumph that lost some of its magnitude as one gradually realised that the nature of the discovery was such that one should really be ashamed of having to make it. The further one carried these observations on children, the more self-evident the facts became and the more astonishing was it too that so much trouble was taken to overlook them.

But on occasions when Freud is under the necessity of forestalling dis-confirmatory reports he forgets the so-easily-confirmable character of his reconstructions of infantile life and insists on their esoteric only-observable-by-initiates status. In the preface to the fourth edition of *Three Essays on Sexuality* we are told that 'none, however, but physicians who practise psychoanalysis can have any access whatever to this sphere of knowledge or any possibility of forming a judgment that is uninfluenced by their own dislikes and prejudices. If mankind had been able to learn from direct observation of children these three essays could have remained unwritten' (Freud, 1953, p. 133). This retreat to the esoterically observable in the face of disconfirmatory evidence is a general feature of psychoanalytic apologetic. A reviewer of the NYU symposium on psychoanalysis and scientific method meets the challenge to demonstrate the empirical status of these claims by offering to provide a list of child-analysts who will vouch for them (Waelder, 1962).

B. A. Farrell has made the following defence of this mode of dealing with disconfirmatory reports: 'We have to remember that almost any negative finding—such as that over female penis envy—can be countered by analysts claiming that it is necessary to postulate such girlish envy in order to account for the material thrown up in analysis. The strength of this reply can only be resolved, presumably, by settling the validity of psycho-analytic method' (Farrell, 1961). Something seems to have gone wrong with this argument. The question of the bearing of empirical observation of children on reconstructions based on the use of psychoanalytic method cannot be postponed until the validity of that method is resolved, since the only way of resolving the validity of that method is by determining whether, and to what extent, it accords with empirical observation of children. Farrell's argument illustrates a tendency to implicitly dehistoricize psycho-analytic reconstructions, since it is obvious on reflection that what women felt or underwent as infants cannot depend upon what 'it is necessary to postulate in order to account for the material thrown up in analysis'. But aside from the intrinsic objectionableness of such a procedure for deter-mining the character of infantile mental life, it renders the observation of children futile for the purpose of validating psychoanalytic method. Freud's peripheral awareness of this would account for a lack of candour in his expositions. The expression 'direct observation' alternates with 'direct analytic observation' as if they were synonymous, so that it only becomes clear after several rereadings that when Freud speaks of 'the direct observation of children' he is referring to the psychoanalytic interpretation of infantile behaviour. That is, Freud in attempting to dispel our doubts as to the validity of psychoanalytic method by appeals to 'direct observation', proffers us a copy of the same newspaper, this time with his thumb over the banner.

But there is apparently another way of testing the validity of psycho-analytic method. It might seem that there can be no question of the genuinely empirical–historical character of those clinical reconstructions which incorporate references to the external circumstances of the patient's infantile life, such as that he had been threatened with castration or been seduced, or seen his parents engaged in intercourse. These at least are straightforwardly testable and their accuracy would therefore afford evidence of the validity of psychoanalytic method; for if the investigation into the infantile history of the patient revealed that he had had no opportunity of witnessing intercourse between his parents (the primal scene), or that he not been sexually abused, or not threatened with castration, this would cast doubt on the validity of the interpretative principles employed and on the dependability of the anamnesis which endorsed them.

But Freud occasionally manifests a peculiar attitude towards indepen-dent investigation of his reconstructions of the patient's infantile years. In 'From the history of an infantile neurosis' he writes: 'It may be tempting to take the easy course of filling up the gaps in a patient's memory by making enquiries from the older members of the family: but I cannot advise too strongly against such a technique . . . One invariably regrets having made oneself dependent on such information. At the same time confidence in the analysis is shaken and a court of appeal is set up over it. Whatever can be remembered at all will anyhow come to light in the course of further analysis' (Freud, 1925a, footnote on p. 481). In the same paper he even expresses misgivings about the value of child analysis: '. . . the deepest strata may turn out to be inaccessible to consciousness. An analysis of childhood disorder through the medium of recollection in an intellectually mature adult is free from these limitations' (Freud, 1925a, p. 475).

This preference is expressed as early as *The Interpretation of Dreams* (Freud, 1949b, p. 258), where Freud remarks of the death wish of children against the same-sexed parent: 'though observations of this kind on small children fit in perfectly with the interpretation I have proposed, they do not carry such complete conviction as is forced upon the physician by the psychoanalysis of adult neurotics.' (This is as if Holmes, having concluded from the indentation marks on his visitor's walking stick that he was the owner of a dog smaller than a mastiff and larger than a terrier, instead of glancing with interest in the direction from which the animal was approach-ing were to turn once again to a more minute inspection of the stick.)

Finally Freud makes assurance double sure by dispensing with the patient's anamnesis altogether. In the case history of the Wolf Man he says: '... it seems to me absolutely equivalent to a recollection if the memories are replaced ... by dreams, the analysis of which invariably leads back to the same scene, and which reproduce every portion of its content in an indefatigable variety of new shapes ... dreaming is another kind of remembering ...' (Freud, 1925a, p. 524).

These remarks suggest that the foundering of the seduction theory on the failure of independent investigation of the alleged seductions to authenticate their occurrence had an influence on the development of psychoanalytic practice similar to that exerted on Percival Lowell, the astronomer of Martian Canals fame, by his failure to detect the canals under especially favourable conditions with a more powerful telescope than he customarily used.

Finding out that the seeing of canals occurred oftener and less ambiguously when the smaller telescopes were used, Lowell used these instruments far oftener in his work than the more powerful ones, claiming that magnification of atmospheric perturbation with the larger telescopes vitiated their use for the observation of the canals ... Lowell never used Mexico as an observation post again and he never again relied upon the splendid 24-inch telescope for observation of Mars ... (Hofling, 1964, p. 33).

Freud's reliance on dream interpretation to determine the historicity or pathogenicity of an individual's infantile sexual past ('dreaming is another kind of remembering') plays a role in his practice similar to that of the small telescope in Lowell's.

However if, by some chance, circumstances from the patient's infantile past which were at variance with Freud's reconstructions did come to light, the validity of his interpretative principles would not thereby be imperilled. This is how Freud deals with the fact that according to his clinical experience 'it is regularly the father from whom castration is dreaded, although it is mostly the mother who utters the threat': 'We find that a child, where his own experience fails him, fills in the gap in individual truth with prehistoric truth; he replaces occurrences in his own life with occurrences in the life of his ancestors. Wherever experiences fail to fit in with the hereditary schema they become remodelled in the imagination. We are often able to see the schema triumphing over the experience of the individual ...' (Freud, 1925a, pp. 557 and 603).

Nor does Freud restrict this device to paternal castration threats. He extends it to 'memories' of seduction and of the primal scene as well.

'These primal phantasies (i.e. of seduction, castration and witnessing parental intercourse) are a phylogenetic possession ... If they can be found in real events, well and good; but if reality has not supplied them they will be evolved out of hints and elaborated by phantasy ... the individual, where his own experience has become insufficient, stretches out beyond it to the experience of past ages' (Freud, 1956, p. 379).

If 'memories' of infantile events which prove never to have occurred are to be taken as due to an 'analogy with the far-reaching instinctive knowledge of animals', how could Freud ever discover that the discrepancy between his reconstruction of the patients' infantile life and the independently ascertained facts of his infantile history was not an instance of 'the phylogenetically inherited schema triumphing over experience', but was due to the invalidity of his reconstructive procedures?

Freud sometimes offers a therapeutic rationale for his conviction as to the authenticity of his reconstructions, asserting that it rests on the fact that anamnesis of the reconstructed scenes, fantasies, impulses or what not, dissipates the symptoms which are held to be the distorted manifestations of their repression. In 1909 he wrote: 'Starting out from the mechanism of cure, it now became possible to construct quite definite ideas of the origin of the illness.' And 'it is only experiences in childhood that explain susceptibility to later traumas' since '... it is only by uncovering these almost invariably forgotten memory traces and making them conscious that we acquire the power to get rid of the symptoms' (Freud, 1962, pp. 48 and 71).

But we are also told that 'marked progress in analytic understanding can be unaccompanied by even the slightest change in the patient's compulsions and inhibitions' (Freud, 1924 b, pp. 221–2). And that he often 'succeeded in ... establishing a complete intellectual acceptance of what is repressed—but the repression itself is still unremoved' (Freud, 1950, p. 182), and that 'Quite often we do not succeed in bringing the patient to recollect what has been repressed. Instead ... we produce in him an assured conviction of the truth of the construction, which achieves the same therapeutic result as a recaptured memory' (Freud, 1950, p. 368).

So that as well as patients who do not recall their infantile sexual impulses and retain their symptoms and patients who do recall their infantile sexual impulses and relinquish their symptoms, we have patients who do not recall their infantile sexual impulses but nevertheless relinquish their symptoms, and patients who do recall their infantile sexual impulses and nevertheless retain them. And since this is just what we might expect to find if there were no relation between the anamnesis of infantile sexuality and the remission of the neurotic symptoms, it is fair to conclude that there is no support from this source for the authenticity of Freud's reconstructions.

I conclude that the reconstructions of infantile sexual history which Freud proffers are pseudo-narratives: the space in which he locates the incidents which comprise them is *a priori* unoccupied.

IV

In his stimulating introduction to the Pelican edition of Freud's *Leonardo*, B. A. Farrell writes: 'What distinguishes Freud's essay about Leonardo from the usual run of narratives is that it uses the technical generalisations of psychoanalytic theory' (Farrell, 1963, p. 13). I shall argue that what distinguishes Freud's narrative, as it distinguishes all his narratives, is its abstention from determinate inference. For Freud does not use his 'technical generalisations' to infer the character of Leonardo's adult sexual life from his knowledge of Leonardo's infantile sexual history, nor does he use them to infer Leonardo's infantile sexual history from his knowledge of Leonardo's adult sexual life. What he does is to link, through the use of a variety of idioms and a plethora of mechanisms and interpretative principles, what is already 'known' of Leonardo's childhood with what is already 'known' of his adult character, selecting from the conflicting traditions about Leonardo what best serves this purpose.

Consider how Freud proceeds to demonstrate 'the existence of a causal connection between Leonardo's relation with his mother in childhood and his later manifest, if ideal (sublimated), homosexuality'. He works backwards from Leonardo's adult attitude towards sex to see what in his infantile years might yield a determinant consonant with the psychoanalytic thesis of the overwhelming predominance of infancy and settles on Leonardo's early years of uninterrupted possession of his mother as a determinant of his homosexuality. Because, Freud says, 'we . . . know from the psychoanalytic study of homosexual patients that such a connection does exist and is in fact an intimate and necessary one' (Freud, 1963, pp. 137–8).

But in the first two editions of *Three Essays on Sexuality* (1905 and 1910) Freud had this to say of the infantile determinants of male inversion:

in the case of men a childhood recollection of the affection shown them by their mother and others of the female sex who looked after them when they were children contributes powerfully to directing their choice towards women . . . The frequency of inversion among present-day aristocracy is made somewhat more intelligible by . . . the fact that their mothers give less personal care to their children (Freud, 1957, p. 229).

483

The Leonardo essay was published in 1910. In the next edition of 'Three essays on sexuality', published in 1915, this passage was supplemented as follows: '. . . on the other hand, their early experience of being deterred by their father from sexual activity and their competitive relation with him deflect them from their own sex . . .' (Freud, 1957, p. 229). So it would seem that it was only after Freud had enrolled the (supposed) absence of Leonardo's father among the determinants of his homosexuality that this became a homosexuality-inducing mechanism. And a counter-intuitive one at that. For our natural expectation on being told that a child has escaped the castration-anxiety inspiring presence of a father during his crucial years is that this would minimize the possibility of his erotic feelings towards his mother undergoing a pathogenically intense repression. That this is an expectation which Freud intermittently encourages, and further evidence of the *ad hoc* character of his post-Leonardo emendation of the circumstances favouring the development of homosexuality, is indicated by his remarks on 'retiring in favour of the father': '. . . another powerful motive urging towards the homosexual object-choice [is] regard for the father or fear of him; for the renunciation of women means that all rivalry with him (or with all men who may take his place) is avoided' (Freud, 1924b, p. 241).

What then is there which, had he found or failed to find it in the accounts of Leonardo's infancy, would have shaken Freud's view that the 'accidental conditions of his childhood had a profound and disturbing effect on him'? Whenever advances in Leonardo studies deprive some environmental circumstance to which Freud had assigned an explanatory role of its historicity, we are told that this does not necessarily impugn Freud's account of Leonardo's infantile reveries and preoccupations. This is true, but it then becomes difficult to see what revelation short of his having died at birth could do so.

As another instance of the ease with which the variety of mechanisms at Freud's disposal enables him to press into the service of the thesis of infantile pathogenicity, whatever parental circumstances the childhood history of his subject happens to provide, consider his account of how inevitable it was, given the character of Dostoevsky's father, that he should have come to possess an over-strict super-ego:

If the father was hard, violent and cruel, the super-ego takes over these attributes from him, and in the relations between the ego and it, the passivity which was supposed to have been repressed is re-established. The super-ego has become sadistic, and the ego becomes masochistic, that is to say, at bottom passive in a feminine way. A great need for punishment develops in the ego, which in part offers itself as a victim to fate, and in part finds satisfaction in ill-treatment by the super-ego (that is, the sense of guilt) (Freud, 1950, p. 231).

This is not at all implausible. But neither is this:

The unduly lenient and indulgent father fosters the development of an over-strict super-ego because, in the face of the love which is showered on it, the child has no other way of disposing of its aggressiveness than to turn it inwards. In neglected children who grow up without any love the tension between ego and super-ego is lacking, their aggressions can be directed externally . . . a strict conscience arises from the co-operation of two factors in the environment: the deprivation of instinctual gratification which evokes the child's aggressiveness, and the love it receives which turns this aggressiveness inwards, where it is taken over by the super-ego (Freud, 1930, footnote on p. 117).

That is, if a child develops a sadistic super-ego, either he had a harsh and punitive father or he had not. But this is just what we might expect to find if there were no relation between his father's character and the harshness of his super-ego.

As a final illustration of the use which Freud makes of the indefiniteness and multiplicity of his pathogenic influences and of the manner in which they enable him to render any outcome whatever an intelligible and apparently natural result of whatever circumstances preceded it, consider the contrast between his accounts of two children—Hans, who was almost five and Herbert, who was four. Herbert figures as a specimen of enlightened child-rearing in a paper of 1907, 'The sexual enlightenment of children', where he is described as 'a splendid boy . . . whose intelligent parents abstain from forcibly suppressing one side of the child's development'.

Although Herbert is not a sensual child he shows 'the liveliest interest in that part of his body which he calls his weewee-maker' because 'since he has never been frightened or oppressed with a sense of guilt he gives expression quite ingenuously to what he thinks'. On the other hand the unfortunate Hans was a 'paragon of all the vices'—his mother had threatened him with castration before he was yet four, the birth of a younger sister had confronted him 'with the great riddle of where babies come from' and 'his father had told him the lie about the stork which made it impossible for him to ask for enlightenment upon such things'. Thus, due in part to 'the perplexity in which his infantile sexual theories left him' he succumbed to an animal phobia shortly before his fifth year.

We learn from Jones' biography that Hans and Herbert are the same child, the account of Hans written *after* and that of Herbert *before* he had succumbed to his animal phobia (but not before the events to which Freud later assigned pathogenic status). Freud even decides as an afterthought that Hans'/Herbert's 'enlightened' upbringing would naturally have contributed to the development of his phobia: 'Since he was brought up without being intimidated and with as much consideration and as little coercion as possible his anxiety dared to show itself more boldly. With him there was no place for such motives as a bad conscience or fear of punishment which with other children must no doubt contribute to making the anxiety less' (Freud, 1925 a, p. 284). This belongs with Falstaff's account of why he ran away at Gadshill.

Freud's general accounts of his infantile theory show the same elusiveness and adjustability in relation to counter-example. He makes claims which appear to commit him to a distinctive infantile sexual history for neurotics, and thus seem vulnerable to refutation, while at the same time insisting on the universality, or at least the undetectability of the pathogenic features invoked.

'At the root of the formation of every symptom are to be found traumatic experiences from early sexual life' (Freud, 1950, p. 117). But since 'Investigation into the mental life of normal persons then yielded the unexpected discovery that their infantile history in regard to sexual matters was not necessarily different in essentials from that of the neurotic' (Freud, 1924a, p. 279), the occurrence of 'traumas' in the childhoods of neurotics can hardly afford us grounds for belief in their causal relevance. This must lie in some pathological repercussion of the 'traumas' which distinguishes neurotic childhoods from normal. What are those repercussions? 'The important thing ... was how he had reacted to these experiences, whether he had responded to them with repression or not' (Freud, 1924a, p. 279). Then is it in repression that the differentia between neurotic and non-neurotic childhoods lies? It seems not. For not only is 'no human being spared such traumatic experiences' but 'none escapes the repression to which they give rise' (Freud, 1949a, p. 52). 'Every individual has gone through this phase but has energetically repressed it and succeeded in forgetting it' (Freud, 1950, p. 172). So once again the differentia must lie elsewhere.

Let us suppose it to have been the case of those early patients of Freud's whose illness he mistakenly attributed to their having been seduced in infancy, that their illness was not a function of their infantile sexual history at all, and not merely not a function of their having been seduced. How could he have discovered the one as he discovered the other? What would have stood to his view concerning the infantile fantasy theory of the neuroses as the discovery that not all his patients had had 'a passive sexual experience before puberty' (Freud, 1924a, p. 142) stood to the seduction theory of the neuroses? But it must be conceded that this consideration is insufficient to establish that Freud's infantile sexual aetiology is without content. For the universality of a putative causal factor does not necessarily impugn its status as a pathogenic agency.

If Freud had discovered features of infantile life whose pathogenic power depended on the presence of an inherited predisposition to respond pathologically to normal contingencies this would have been an important discovery even though those features were not abnormal interventions. Though it would not be a theoretically satisfying answer to the question why certain individuals are neurotic it would be of great practical and humanitarian interest. For we might, by adopting extraordinary measures, prevent the predisposition from having pathological outcomes, as we do, for example, in the case of phenyl-pyruvic oligophrenia, where infants with a metabolic disposition to make poisons of ordinarily innocuous nutrients are provided with a special diet which enables them to escape the usual pathological outcome of such a predisposition; or as we do in the case of children with a tendency to protruding teeth, where braces are fixed to the teeth to ensure normal development. But with Freud's post-seduction theory of the neuroses it is impossible to say what he believes stands to the neuroses as the failure to put braces on the teeth of a child with a tendency to protrusion stands to his adult mouth deformation.

At first it looks as if precautions against castration fears might. But because of the ubiquity of castration surrogates, because, that is, any attempt to control the child's sexuality (of which infantile incontinence is, according to Freud, a manifestation) can induce castration anxiety and thus initiate pathological developments and because the failure to intervene in the child's sexual life is in itself a pathogenic agency—('Complete freedom . . . would do serious damage to the children themselves' (Freud, 1933, p. 191)) and because furthermore, 'the power of refractory instinctual constitutions can never be got rid of by education' (Freud, 1933, p. 192)—there is no ostensibly psychoanalytic child-rearing regimen whose failure to protect against neuroses could be said to have disconfirmed Freud's claim of infantile pathogenesis.

Freud is well aware that when a putatively pathogenic factor is discovered to be widely distributed this casts doubt on the genuineness of its relation to the illness. Before his abandonment of the seduction theory he said: 'Of course if infantile sexual activity were an almost universal occurrence it would prove nothing to find it in every case' (Freud, 1924a, p. 205). His awareness of this may account for his increased tendency to treat the patient's infantile sexual vicissitudes as *grounds for* rather than *causes of* his neurotic illness, since the explanatory status of a motive is not called into doubt because of the absence of a general connection between it and the behaviour it is invoked to explain.

It finally becomes clear that Freud's assertion that 'the factors which go to form neuroses are to be found in the patient's infantile sexual life'

(Freud, 1925a, p. 303) is not a causal claim in the sense in which the seduction theory of the neuroses was a causal claim, since the contingencies about which Freud is explicit are neither manipulable nor differentiating. We would convey more of its real character if we characterized it as crypto-verstehen (although, as I hope to show, it is really pseudo-verstehen).

The explanation of these equivocations, evasions and inconsistencies is that Freud is simultaneously under the sway of two necessities: to seem to say and yet to refrain from saying which infantile events occasion the predisposition to neuroses. To seem to say, because his discovery of the pathogenic role of sexuality in the infantile life of neurotics is the ostensible ground for his conviction that the neuroses are manifestations of the revival of infantile sexual struggles and thus for the validity of the method by which this aetiology was inferred; to refrain from saying, because if his aetiological claims were made too explicit and therefore ran the risk of refutation this might discredit not only his explanations of the neuroses but, more disastrously, the method by which they were arrived at. Only by making these prophylactic and pathogenic claims can his preoccupations and procedures be justified, but only by withdrawing them can they be safeguarded. So the 'quantitative factor' to which Freud invariably alludes in his exposition of the aetiology of the neuroses, and his insistence on the importance of the inherited constitution, are not examples of scientific scrupulousness. They are devices for retaining a preoccupation long after any reasonable hope of enhancing one's powers of prediction and control by means of it have been exhausted.

It could be argued that the counterfeit character of Freud's infantile aetiology could be conceded without substantially reducing our estimate of his contribution to the understanding of neuroses. For there is an alternative construction to be put on Freud's claim that the neuroses are sexual disorders: that it be understood as a claim not about the causes but about the character of neurotic disorders. In other words, that Freud's theory of the neuroses be treated as an answer not to a 'what happened to him?' question but to a 'what's wrong with him?' question. On this view, infantile sexuality stands to the neuroses somewhat as the upright posture of the human species does to the fallen arches of some of its members, and neurotics are persons who are so constituted that for unknown reasons, out of memories of their infantile sexual life, they make illnesses. And that this is so would be demonstrated by the fact that the neurotic symptomatology serves, expresses, or alludes to sexual interests, interests which ultimately date from infancy. Or to use Freud's own formulations: 'The aetiological importance of sexual factors' would rest on 'the delicate but firm inter-relationship of the structural elements of the neurosis' (Freud, 1924a,

p. 150); 'the logical structure of the neurotic manifestations' (p. 159); 'the inevitable completion of the associative and logical structure' (p. 201); 'the structural connections between symptoms, memories and associations' (p. 179); 'the copious and intertwined associative links' (Freud, 1949*b*, p. 191).

## V

Let us see how these rather figuratively expressed criteria work in practice. Freud says of Dora's attack of acute abdominal pain accompanied by a high fever and succeeded by a dragging right foot that it is 'a typical example of the way in which symptoms arise from exciting causes which seem to be entirely unconnected with sexuality'. This is how the appearance of unconnectedness is dissipated: 'What then was the meaning of this condition, of this attempted simulation of perityphlitis?... I asked Dora whether this attack had been before or after the scene by the lake. Every difficulty was resolved at a single blow by her prompt reply. "Nine months later...". Her supposed attack had thus enabled the patient with the modest means at her disposal (the pains and the menstrual flow) to realise the phantasy of childbirth.' But 'What was all this about her dragging her leg ...? That is how people walk when they have a twisted foot. So she had made a "false step": which was true indeed if she could give birth to a child nine months after the scene by the lake.' If this is what Freud considers 'a typical example of the way in which symptoms arise from exciting causes which seem to be entirely unconnected with sexuality', we must ask how he could ever discover that the symptoms had arisen from causes which were in fact, and not just apparently, unconnected with sexuality.

Since in this instance Dora did not confirm Freud's account of the nature of her limp (assuming for the moment that it was the kind of thing which she could confirm) and since the symptoms were no longer there to confirm it by 'entering into the discussion', on what does Freud rest his conviction that in Dora's 'simulated attack of perityphlitis' and its sequelae there was an allusion to her feelings of regret at the outcome of the scene by the lake? On the fact that both menstrual discharges and babies come out of the vagina, on the coincidence between the period required for gestation and that which elapsed between Herr K's overtures and her abdominal attack and on the possibility of placing a sexual construction on an idiom ('false step') which might be used in connection with her limp. How is the probative, evidential value of such considerations to be assessed? How do we assess them elsewhere?

What we do in assessing such claims is to make intuitive judgments of the likelihood of the events, between which the relation of allusion is said to hold, occurring independently of each other. Let us examine some examples of blatantly spurious allusiveness and see on what our conviction that they are spurious rests.

In *La Vita Nuova*, Dante argues that the date of Beatrice's death, 9 June 1290, was determined by her relation to the Trinity and other significant numerical values:

according to the computation used in Italy, her most noble spirit departed hence in the first hour of the ninth day of the month; and, according to the computation used in Syria, she died in the ninth month of the year, for there the first month is Tismin, which is our October. And, according to our computation, she died in that year of our calendar (that year of our Lord, to wit) in which the perfect number was nine times completed, within that century wherein she was born into the world, she being a Christian of the thirteenth century. Why this number was so propitious to her may be possibly explained thus. According to Ptolemy, and according to Christian truth, the heavens that move are nine, and, according to the commonly received belief among astrologers, these heavens exert a con-current influence on mundane things, each according to its peculiar position; so this number was propitious to her, indicating, as it did, that at her birth all the nine moving heavens were in the most perfect conjunction. This is one reason; but when the matter is scanned more closely, and in conformity with infallible truth, this number was her very self. I speak by way of similitude meaning thus: —The number three is the root of nine, because without any other number, multiplied by itself, it makes nine, it being obvious that three times three makes nine. If, then, three is by itself the author of nine, and the author of miracles is in Himself three, Father, Son and Holy Ghost, which are Three and One, this lady was accompanied by the number nine, in order to show that she was a Nine, in other words a miracle, whose only root is the adorable trinity (Dante, 1871, pp. 59–60).

In the sixth chapter of Newton's *Observations on the Prophecies of Daniel*, Newton argues that the identity of the fourth of the beasts mentioned by Daniel with the Roman Empire and therefore his miraculous anticipation of its downfall is attested by the coincidence between the number of horns of the fourth beast and the number of kingdoms into which the Empire was fragmented, since if to the seven kingdoms listed by Sigonious you '. . . add the Franks, Britons and Lombards . . . you have the ten; for these arose about the same time with the seven' (Newton, 1841, p. 48).

As a last illustration of spurious allusiveness consider the proof produced by a cleric encountered in India by Thomas Babington Macaulay that Napoleon was the Beast mentioned in the thirteenth chapter of 'Revelations': if 'Bonaparte' is written in Arabic omitting only two letters it yields the number 666.

A felicitous characterization of what is going on in these specimens and of the manner in which they differ from bona fide enquiries is provided by Pareto. I have adapted it slightly to give it greater generality.

Case 1. We have a (datum D) which is assumed to (allude to) certain facts $A$ ... Our purpose is to determine $A$. If our effort is successful, we shall be following the line $DA$ ... But, if our venture ... chances to fail, we get not to $A$, but to $B$, and imagine, though mistakenly, that $B$ is the source of $D$.

Case 2. From a (datum D), the idea is to draw certain conclusions, $C$, which are generally known in advance ... the quest is not for $C$ ($C$ being already known), but for a way of getting to $C$. Sometimes that is done deliberately. A person knows perfectly well that $C$ does not follow from $D$, but he thinks it desirable to make it seem to ... But more often, the search for a road that will lead from $D$ to $C$ is not consciously premeditated ... Quite without conscious design he brings the two sentiments together over the path of $DC$ ... The person who is trying to persuade others has first of all persuaded himself. There is no trickery.

In the first case ... the search is for $A$. In the second case ... the search, deliberately or unconsciously, is for the route $DC$ ... the search for the path $DC$ is represented in all sincerity as a quest strictly for $A$ ... the person ... is using an *interpretation DC* suitable for getting him to the desired goal ... Those two termini are fixed. The problem is simply to find a way to bring them together (Pareto, 1963, pp. 385-7).

This gives us an account of the character of spurious allusions. But how is it done? How is the impression of allusiveness created? Martin Gardner has some enlightening remarks on this problem in his examination of the claims of Pyramidologists to have found allusions to scientific truths, such as the axis of the earth, its polar radius, its mean density, the mean temperature of its surface and the period of precession of its axis, etc., in the dimensions of the Great Pyramid.

It is not difficult to understand how (Pyramidologists) achieved these astonishing scientific correspondences. If you set about measuring a complicated structure like the Pyramid, you will quickly have on hand a great abundance of lengths to play with. If you have sufficient patience to juggle them about in various ways, you are certain to come out with many figures which coincide with important figures in the sciences. Since you are bound by no rules, it would be odd indeed if this search for Pyramid 'truths' failed to meet with considerable success.

Take the Pyramid's height, for example. Smyth multiplies it by ten to the ninth power to obtain the distance to the sun. The nine here is purely arbitrary. And if no simple multiple had yielded the distance to the sun, he could try other multiples to see if it gave the distance to the Moon, or the nearest star, or any other scientific figure (Gardner, 1957, pp. 176-7).

How do these specimens differ from the genuine allusiveness which psychoanalytic apologetic is compelled to maintain is exemplified by

Freud's accounts of symptom-formation, dreams, errors, etc.? 'How would it be possible', asks Merlau-Ponty, 'to credit chance with the complex correspondences which the psychoanalyst discovers? How can we deny that the psychoanalyst has taught us to notice echoes, allusions, repetitions from one moment of life to another?' These remarks express a typical reaction to Freud's interpretations and it must be conceded that on a great number of occasions Freud does produce an overwhelming impression of cogency in his demonstrations that some contemporary item of behaviour contains an allusion, to a remote event of infancy, say. But could it be that this cogency is illusory and that it arises, as for the credulous in the cases cited, because we overlook the enormous discretion he enjoys in the selection and characterization of the data he explains and because of our assumption that he undoes the work of distortion and arrives at his interpretation by the application of antecedently formulated rules?

Consider in the light of these suspicions one of the devices which Freud employs to get an allusion to fellatio into Leonardo's vulture reminiscence: 'Its most striking feature was after all that it changed sucking at the mother's breast into being suckled, that is into passivity and thus into a situation whose nature is undoubtedly homosexual.' Thus Leonardo's account of the bird inserting its tail between his lips and beating about with it could not be merely a distorted reminiscence of nursing at his mother's breast but must be a fantasy of sucking at a penis, since sucking a nipple is active but sucking a penis passive. I will forbear asking where Freud derives his assurance that sucking a penis is passive rather than active since it might be felt that I was polemically exploiting the fact that it is not a question on which anyone is likely to pronounce authoritatively. But I will point out that it doesn't sort well with the view, also expressed by Freud, that fellatio is a revival of the infantile demand for the breast. And this suggests that the decision as to the passivity of the breast in relation to the nursing child's mouth is due to Freud's prior knowledge of the tradition according to which Leonardo was homosexual. For elsewhere he writes: 'The mother is in every sense active in her relations to her child; it is just as true to say that it is she who gives suck to the child as that she lets it suck her breast' (Freud, 1933, p. 148). And 'The first sexual or sexually tinged experiences of a child in its relation to the mother are naturally passive in character. It is she who suckles, feeds . . .' etc. (Freud, 1950, pp. 264–5).

Consider, too, one of the mechanisms which Freud employs to derive Leonardo's homosexuality from his infantile relation to his mother. 'By repressing his love for his mother he preserves it in his unconscious and from now on remains faithful to her. While he seems to pursue boys and to be their lover he is in reality running away from the other women who might cause him to be unfaithful'. So the unconscious which is driven by incestuous urges is nevertheless, like a good bourgeois, respectably monogamous. But does not this sexual attachment of the homosexual to his mother contradict the revulsion which Freud says her penisless state inspires in him? And if the reply to this is that it is the phallic mother to whom his unconscious is attached and to whom he is under the compulsion of remaining faithful, how is this end served by 'seeking his sexual object in men who through other physical and mental qualities remind him of women' (Freud, 1924*b*, p. 66)?

But Freud has an answer to these philistine objections which brings us to another feature of psychoanalytic theory which has been found puzzling but is easily accounted for on the assumption that its function is to avoid refutation: Freud's insistence that the unconscious contains no contradictions. Its effect is to allow Freud to reserve the right to determine *a posteriori* what the logical implications of any interpretative claims advanced are. An extended exposition of this property of the unconscious occurs on the following occasion: Freud is anticipating an objection that his account of a patient's symptoms as due to an unconscious wish to be penetrated by his father's penis and bear him a child plus an unconscious conviction that castration was a precondition of this wish being fulfilled conflicted with another interpretation which presupposed that the patient was unconsciously convinced that the anus was the orifice of intercourse and birth; 'That it should have been possible . . . for a fear of castration to exist side by side with an identification with women by means of the bowel admittedly involved a contradiction. But it was only a logical contradiction —which is not saying much. On the contrary, the whole process is characteristic of the way in which the unconscious works' (Freud, 1925*a*, p. 557). But the first of the two accounts which Freud attempts to reconcile by this device itself presupposes that the unconscious is aware of contradiction, for it is this which brings the desire to play a passive sexual role *vis-à-vis* the father in conflict with the desire to retain the penis. And one cannot help noticing that the unconscious, which 'has no logic' and knows nothing of contradictions, is nevertheless constantly involved in the drawing of inferences. For example, in the case history of Hans, Freud describes the following as a 'typical unconscious train of thought': 'Could it be that living beings really did exist which did not possess widdlers? If so, it would no longer be so incredible that they could take his own widdler away and, as it were, make him into a woman' (Freud, 1925*a*, p. 179). But if there is no 'No' in the unconscious ('In the unconscious No does not exist and there is no distinction between contraries' (Freud, 1925*a*, footnote on p. 559)), then there can be no 'If', 'So', or 'Thus' either.

## VI

With the principle that 'it is not necessary that the various meanings of a symptom be compatible with one another' (Freud, 1925a, p. 65) ready to be invoked in case of need the risk involved in the proffering of an interpretation would seem to be minimal.

But all these arguments may be beside the point, for Freud's interpretations may belong to a class of claims for which arguments from independent probability and related considerations have no relevance; and which are only confirmed or otherwise by the event. On this view the interpretative principles are rules of thumb whose justification is that in conjunction with imponderables, which can only be discriminated through talent and experience, they lead to correct inferences about unconscious processes. The congruity of the patient's behaviour and associations with the proffered interpretation would then be symptoms and not criteria of its correctness, so that production of spurious allusions, which show just as intricate an interrelatedness as Freud's putatively genuine ones, would have no weight.

When Alastair Fowler, a literary critic, investigating numerological allusions in Renaissance poetry, is considering the claim that 'The Amorous Zodiac' embodies an allusion to the lunar month, he concedes that 'the probability against its having the same number of stanzas as there are days in the lunar month . . . is not very great', but goes on to argue:

> while a routine caution about probabilities should certainly be observed, we should also remember that, unless it is observed in moderation, many intentional patterns will be neglected. For a Renaissance poet was not in any way obliged to construct statistically improbable configurations. The pattern he intended might well be one that had a good chance of occurring at random . . . imponderables ought to influence our statement about numerological patterns . . .

It is on this model that we are sometimes invited to construe Freud's interpretations. The associative patterns produced or elicited by the analyst would then not themselves constitute the content of his interpretative inferences but would merely be symptomatic of their correctness. But their correctness as to what? As to—'something that was then going on in him'. And this construal of Freud's interpretations has, for him, an additional advantage, in that many of them not only fail to stand up to tests of independent probability but are difficult to unpack in dispositional terms at all.

Dora, Paul, etc., could no more *act* as if they had entertained the unconscious thoughts which figure in Freud's reconstructions and to which he attributes their symptoms, than Freud could have *acted* as if he had dreamt the dream of the Botanical Monograph. Such phenomena are intrinsically linguistic. Having is here tied to telling. But aren't the surroundings which would allow Freud's reconstructions to bear this sense missing too?

John Wisdom has remarked that the insistence on a behavioural elucidation of Freud's claims concerning a patient's unconscious is mistaken and has made the following suggestion as to how they might be construed: 'If someone modifies the use of "there is in X's mind hatred of Y" so that X's introspection at the time is not given its normal weight we may still give it a sense in which X's subsequent introspection, what appears later in X's mind, is still what finally settles the truth of the statement. A statement about a hidden hatred is still a statement such that (the patient) has a way of verifying it which no other could have' (Wisdom, 1961). We might well allow a belated avowal or even a belated disavowal-repudiating avowal to confirm an imputation of the kind Wisdom mentions. But the bulk of Freud's interpretations is not susceptible of this sort of elucidation. Wisdom's are conceptually bowdlerized examples of the sort of questions which confront us in attempting to understand them; questions such as whether a belated avowal could transform a happening, like falling, say, into an action. Or a physiological event like menstruation into a performance. Or confirm that a pain or a fear or a delusion was effected and not suffered. Could Paul (the man with the rats) belatedly confirm that, when in a mood of mingled resentment and longing, due to his lady's being away looking after her sick grandmother, he was overcome by a compulsion to cut his throat succeeded by the thought 'No, you must go and kill the old woman', the underlying *unconscious* train of thought was really 'I should like to kill that old woman for robbing me of my love. Kill yourself, as a punishment for these savage and murderous passions', only on reverse order, à la White Queen, the punitive command coming first and the outburst which provoked it afterwards? (Freud, 1925a, p. 326).

The impression that Freud's assertions concerning a patient's unconscious mental processes are such that the patient's 'subsequent introspection' might confirm them is an impression which Freud needs to create. For if we can be persuaded to locate the differentia between phenomena which are mental and those which are not inside the agent we won't be troubled by our inability to give an account of just what it is that is implied by the ascription to someone of an unconscious mental state— the picture of inaudible, internal deliberations, non-introspectable but retrospectable, will distract us and keep us happy.

When Freud writes '...of many of these latent states we have to assert that the only point in which they differ from states which are conscious is just in the lack of consciousness of them' (Freud, 1925 b, p. 101). 'We call a process unconscious when...it was active at a certain time, although at that time we knew nothing about it' (Freud, 1933, p. 95), he is, in these formulations, indulging in something less innocuous than a mere ontological bent and it is only by considering their relation to testability that we can understand Freud's insistence that in these occult transactions is to be found 'the true essence of what is mental'.

Consider the idioms in which Freud's interpretations are typically phrased. Symptoms, errors, etc., are not simply *caused by* but they 'announce', 'proclaim', 'express', 'realize', 'fulfil', 'gratify', 'represent', 'imitate', or 'allude to' this or that repressed impulse, thought, memory, etc. Consider the term 'allusion'. It is typically used short of its full force, in a strained sense, like that in which one might say that a hangover is an allusion to alcoholic over-indulgence or a winter sun-tan to a Mediterranean holiday. In one of his dream interpretations Freud advances the claim that the red, camellia-shaped blossoms which his patient reported carrying were 'an unmistakable allusion to menstruation' and supports this by a reference to La Dame aux Camelias who signalled the onset of her menstrual periods by replacing her usual white camellia with a red one (Freud, 1949 b, p. 319). Though we can give a sense to the statement that the dream blossoms owed their shape and colour to the dreamer's familiarity with 'La Dame aux Camelias', and that if menstrual blood were green, they too would have been green, it is not the sense which Freud requires, for it is not the kind of thing to which the dreamer could attest. She might agree but she could not corroborate. But Freud contrives by the use of such idioms as 'allusion' to get us to assimilate his explananda to a class of actions and reactions, enquiry into which naturally terminates in our receipt of the agent-subject's account of the matter, e.g. the course taken by his thoughts during a brown study. The cumulative effect of this is that, in contexts where it would otherwise be natural to demand behavioural elucidation or inductive evidence, this demand is suspended due to our conviction that it is intentional or expressive activity which is being explained; while in contexts where we normally expect an agent's candid and considered rejection sufficient to falsify or disconfirm the attribution of expressiveness or intention, this expectation is dissipated by Freud's talk of 'processes', 'mechanisms', and 'laws of the unconscious'.

Suppose we take as a paradigm of disguised expressiveness the case of the minister in *La Symphonie Pastorale*, who keeps a record of his struggle to desist from masturbating, but in order to ensure his privacy takes the precaution of writing 'smoking' for 'masturbating'. We have no difficulty in understanding this. But if we now replace this secret meaning by an *unconscious* reference to masturbation, isn't any residual feeling of understanding we may have just a shadow cast by our paradigm?

But our descent into unintelligibility is, on most occasions, concealed from us by the gentleness of the declivity, the succession of subtle dislocations of sense to which the idioms in which the interpretations have been couched were subjected. To free ourselves we have to make explicit the tacitly performed assimilations which produce the illusion of intelligibility and hide from us the extent to which the absence of the normal surroundings deprives Freud's interpretations of sense. When we do so we see these reconstructions of Freud's to be pseudo-soliloquies.

### CONCLUSION

Examination of Freud's interpretations will show that he typically proceeds by beginning with whatever content his theoretical preconceptions compel him to maintain underlies the symptoms, and then, by working back and forth between it and the explanandum, constructing persuasive but spurious links between them. It is this which enables him to find allusions to the father's coital breathing in attacks of dyspnoea, fellatio in a *tussis nervosa*, defloration in migraine, orgasm in an hysterical loss of consciousness, birth pangs in appendicitis, pregnancy wishes in hysterical vomiting, pregnancy fears in anorexia, an accouchement in a suicidal leap, castration fears in an obsessive preoccupation with hat tipping, masturbation in the practice of squeezing blackheads, the anal theory of birth in an hysterical constipation, parturition in a falling cart-horse, nocturnal emissions in bedwetting, unwed motherhood in a limp, guilt over the practice of seducing pubescent girls in the compulsion to sterilize banknotes before passing them on, etc.

It might be felt that these criticisms are biographical in character and of only limited relevance to the question of the pseudo-scientific status of psychoanalysis. This objection overlooks the fact that they do not deal with private episodes in Freud's history but with matters of public record, which we have expounded and discussed on innumerable occasions without their having aroused in us any but the most easily placated misgivings. The questions and objections which force themselves upon us when we *do* make a resolute attempt to understand Freud's claims as we would others encountered in the course of an explanatory enquiry have a blatancy which must cause us to suspect the spirit in which we originally received them. They set us problems in the phenomenology of psychoanalytic conviction.[1]

[1] Problems towards whose solution Wittgenstein has made some fruitful suggestions— see the paper, 'Wittgenstein's Freud', in *Studies in the Philosophy of Wittgenstein*, edited by Peter Winch, 1969.

If psychoanalytic claims are not hypotheses this is not because of any inspectable formal deficiencies they display but because this is not the role which they played in the lives of those who originated and transmitted them, nor of those who have since repeated, adapted or merely silently rehearsed them. We did not interpret dreams, symptoms, errors, etc., because it was discovered that they were meaningful, but we insisted that they were meaningful in order that we might interpret them. And if we reflect on the kind of thesis this is and the kind of evidence it involves, we will not find it surprising that it should prove incapable of demonstration and give rise to intractable disagreement, for it is not a question of proving of some isolated thesis of psychoanalysis that it fails to meet a particular criterion but of discerning a pattern in the total ensemble.

But if Freud's transactions with his patients' symptoms, dreams, associations, reminiscences, etc., are not explanatory what are they? To what genre do they belong? One of our difficulties is that we have no commonly accepted label for them. 'Pseudo-science', though a correct characterization, is a merely negative one which performs too general an assimilation. We can only illuminate their character by comparisons.

In this paper I have assembled reasons for concluding that whatever Dante was doing when he found a trinitarian allusion in the date of Beatrice's death; whatever the cleric encountered by Macaulay was doing when he demonstrated Bonaparte's identity with the Beast mentioned by St John; whatever Pyramidologists are doing when they discover allusions to mathematical and scientific truths in the dimensions of the Great Pyramid; whatever St Augustine was doing when he expounded the significance of St Peter's catch of 153 fish; whatever Newton was doing when he identified the subdivisions of the Western Roman Empire with the ten horns of the fourth Beast mentioned in the Book of Daniel—it is this which Freud is doing when he 'lays bare' the secret significance of his patients' dreams, symptoms, errors, memories and associations, and explains 'what the symptoms signify, what instinctual impulses lurk behind them and are satisfied by them and by what transitions the mysterious path has led from those impulses to these symptoms'.

## REFERENCES

Cioffi, F. 1969. Wittgenstein's Freud. *Studies in the Philosophy of Wittgenstein.* Ed. Peter Winch. London.

Dante. 1871. *La Vita Nuova.* Trans. Theodore Martin. Edinburgh.

Eddy, Mary Baker. 1875. *Science and Health with Key to the Scriptures.* Published in Boston by the Trustees under the Mary Baker Eddy will.

Farrell, B. A. 1961. Can psychoanalysis be refuted? *Inquiry.*

Freud, S. 1922. *Beyond the Pleasure Principle.* Trans. C. J. M. Hubback. London.

Freud, S. 1924a. *Collected Papers,* vol. 1. Trans. Joan Riviere. London.

Freud, S. 1924b. *Collected Papers,* vol. 2. Transl. Joan Riviere. London.

Freud, S. 1925a. *Collected Papers,* vol. 3. Trans. Joan Riviere. London.

Freud, S. 1925b. *Collected Papers,* vol. 4. Trans. Joan Riviere. London.

Freud, S. 1930. *Civilization and its Discontents.* Trans. Joan Riviere. London.

Freud, S. 1933. *New Introductory Lectures on Psychoanalysis.* Trans. W. J. H. Sprott. London.

Freud, S. 1938. *The Basic Writings of Sigmund Freud.* Trans. A. A. Brill. New York.

Freud, S. 1938. *Three Essays on Sexuality. Basic Writings* (see above).

Freud, S. 1949a. *An Outline of Psychoanalysis.* Trans. James Strachey. London.

Freud, S. 1949b. *The Interpretation of Dreams.* Trans. James Strachey. London.

Freud, S. 1950. *Collected Papers,* vol. 5. Trans. Joan Riviere. London.

Freud, S. 1953. *The Standard Edition of the Complete Works,* vol. 7. Trans. James Strachey. London.

Freud, S. 1956. *General Introduction to Psychoanalysis.* (American Edition of Introductory Lectures.) New York.

Freud, S. 1957. *Standard Edition of the Complete Works,* vol. 11. Trans. James Strachey. London.

Freud, S. 1961. *Inhibitions, Symptoms, and Anxiety.* Trans. Alix Strachey. London.

Freud, S. 1962. *Two Short Accounts of Psychoanalysis.* Trans. James Strachey. London.

Freud, S. 1962. Five lectures on psychoanalysis. *Two Short Accounts of Psychoanalysis* (see above).

Freud, S. 1963. *Leonardo da Vinci.* Trans. James Strachey. London.

Gardner, Martin. 1957. *Fads and Fallacies.* New York.

Hofling, Charles K. 1964. Percival Lowell and the canals of Mars. *Br. J. med. Psychol.* **37**.

Inglis, Brian. 1964. *Fringe Medicine.* London.

Marmor, Judd. 1962. Psychoanalytic therapy as an educational process. *Psychoanalytic Education,* vol. 5. New York. In the series *Science and Psychoanalysis.* (Ed. Jules H. Masserman.)

Newton, Isaac. 1841. *Observations on the Prophecies of Daniel.* London.

Pareto, V. 1963. *The Mind and Society—A Treatise on General Sociology,* vol. 1. New York.

Popper, K. 1957. Philosophy of science: a personal report. *British Philosophy in Mid-Century.* Ed. C. A. Mace. London.

Waelder, R. 1962. Psychoanalysis, scientific method and philosophy. *J. Am. psychoanal. Ass.* **10**, No. 3.

Wisdom, John. 1961. Review of *Psychoanalysis, Scientific Method and Philosophy.* Ed. Sidney Hook. *Q. Jl exp. Physiol.* **13**, No. 1.

# COMMENT

## *by* B. A. FARRELL

1.   Mr Cioffi concentrates on one 'particular historically identifiable' example of psychoanalysis, namely Freud's. He is concerned to examine both Freud's 'theses' and his 'procedures of investigation'. He argues that this example is a pseudo-science, and a successful one at that.

It is obvious that Mr Cioffi has embarked on a large and bold enterprise. It is hardly surprising that it runs into trouble at the very start.

(*a*)   It is far from clear that there is any one coherent set of theses to which Freud subscribed. For he was constantly throwing out new suggestions; and he changed his views in important ways in the course of his life. The Freud of 1896 is very different from the Freud of 1915, and this Freud is different again from the one of the thirties. It is far from clear, therefore, that there is just one historically identifiable, Freudian theory to be examined. Moreover, the fact that Freud had different views at different times *may* be of no more, or less, interest than that Darwin, say, also held different views at different times. Even if we pick on the libido theory as something Freud clung to at all times, we have to be sure that it functions in the same way throughout the changes in Freud's thought and practice. A careful look at the sources may make us doubtful whether it did do so or not. So Mr Cioffi leaves us uncertain just what particular Freudian theory, or body of theses, he is examining.

(*b*)   Much more serious—just what are the 'procedures of investigation' that Freud used? Mr Cioffi seems to suppose that there is a set of them, and that we can tell with confidence what these are by a careful study of Freud's own writings. Both parts of this supposition are open to serious doubts. It is perhaps generally agreed among students of the Freudian corpus that, even after Freud hit upon free association, he changed his procedures over the years. At the beginning he appears to have been a short-term therapist, somewhat didactic in character, and ready to produce allusive interpretations of the odd items. In later years, as the complexities of the whole business became more apparent, he seems to have grown more cautious, and to have developed procedures which are close to what may be called the classical family of analytic techniques. But many of Freud's own examples are drawn from his earlier years; and in various quarters today they would not be thought to represent good psychoanalytic procedure.

There is a further complication here, which has become clearer in recent years to students of psychotherapy. It is doubtful whether we can accept Freud's descriptions of his own activity as reliable guides to the precise and detailed nature of that activity. The important reason for this doubt is that, when Freud reports on his activity, he does so (typically) in his role of participant observer in a therapeutic situation, or dyad. It has become clear that a participant observer in such a situation, or dyad, is not in a position to give us reports that can be accepted with confidence. This is one discovery that emerges from the objective study of the therapeutic process, which psychologists have been undertaking in recent years. Hence, we cannot rely on Freud's own accounts of his interpretative activity with patients. They may be very misleading, and may do him more or less justice than he deserves. Mr Cioffi seems to suppose that Freud's own descriptions can be taken at their face value, and accepted with confidence. Here he is at one with the traditional analysts in perpetuating a widespread mistake. It is a mistake from which contemporary analysts are moving slowly to free themselves.

2. But let us ignore these difficulties. Let us suppose that Freud did offer us only one theory, and used only one set of procedures, of which he offered us adequate descriptions. There is an important and general objection which scholars of the Freudian corpus are likely to raise at once against Mr Cioffi's treatment of the whole matter.

When we set out to examine the theory and procedures advocated by a certain figure in the past, the first thing we have to do is to put on the robes of the historian, and consider his work with scholarly detachment and rigour. But Mr Cioffi does not seem to have complied with this requirement. His tone is polemical throughout, and he simply does not do justice to various important aspects of Freud's thought and work before he plunges into criticism of it.

'Consider an example', our scholar might continue, 'of such unhistorical treatment by Mr Cioffi. He emphasizes Freud's doctrine of libido, but he does so in abstraction from its context. The context is a model designed to embrace the totality of mental functioning. Part of its point is to enable us to view a person as an energy system that deals with tension by achieving equilibrium through various means and in different ways. When we put the libido doctrine back into this context, it takes on a somewhat different appearance from what it presents in Mr Cioffi's account. We are able, for example, to appreciate that it may be dispensable, and that Mr Cioffi may be flogging a dead, or dying, horse in emphasizing it. We can see that, as neurotic conflict in the system is between Ego and Id, it is just wrong to say (as Mr Cioffi does, p. 477) that "non-sexual conflict" is an expression with no use inside the Freudian story. We can appreciate, too, that the "quantitative factor" has a role to play in the equilibrating economy of the system. But, according to Mr Cioffi, the "quantitative factor" is "a device for retaining a preoccupation long after any reasonable hope of enhancing one's powers of prediction and control by means of it have been exhausted" (p. 488). To a historian, this remark is apt to seem just bizarre and comical.'

3.  But let us pass by this criticism from the historian; and let us consider one of the Freuds—say the one of the mid-twenties. Presumably, Mr Cioffi would want to argue that Freud's theory and procedures of this period constitute a pseudo-science.

What does Mr Cioffi mean by the word 'pseudo-science'? He is not, perhaps, as clear about this as he might be. He seems to apply the word both to 'theses' (or 'a body of theses'), and also to 'an activity'. As applied to an activity, part of what Mr Cioffi means seems to be this. An activity is pseudo-scientific if it satisfies three conditions conjointly.

(a) There are no states of affairs which would constitute a disconfirmation of any of the theses it purports to investigate;

(b) it would in fact prevent or obstruct the discovery of disconfirmatory states of affairs;

(c) it involves 'the habitual and wilful employment of methodologically defective procedures'.

Clearly, Mr Cioffi's notion of pseudo-science is a complex one. In addition, what is more, it seems to be connected, and in ways that are not immediately obvious, with two other claims of his.

(i) 'A host of psychoanalytic pecularities can be understood when once they are seen as manifestations of the same impulse—the need to avoid refutation'; and

(ii) there is really only one way in which 'the correctness of psycho-analytic claims may be assessed', namely, interpretation; and this 'itself ultimately proves illusory'.

Of all the things that could be said about Mr Cioffi's position here, I shall confine myself to two.

3.1.  Mr Cioffi appears to say, or imply, that the difference between a state of affairs that disconfirms and one that does not is quite clear and sharp. For he does not elucidate the notion, and he does not suggest that the notion may be a shading and not a bounding one. Mr Cioffi appears to speak in the same way about his notion of refutation. But now it seems evident that Mr Cioffi's notions of disconfirmation and refutation are a bit distant from those which have been much discussed by philosophers in recent years. We can exhibit their distance by recalling one such discussion.

In his *Conjectures and Refutations* Professor Popper (pp. 255–7) tells us about the problem he faced of '*drawing a line of demarcation* between those statements and systems of statements which could be properly described as belonging to empirical science, and others which might, perhaps, be described as "pseudo-scientific" or (in certain contexts) as "metaphysical", or which belonged, perhaps, to pure logic or to pure mathematics'. In solving this problem he was led to use the notion of degrees of testability. From here he was led, in turn, to conclude that 'the criterion of demarcation cannot be an absolutely sharp one but will itself have degrees. There will be well-testable theories, hardly testable theories, and non-testable theories. Those which are non-testable are of no interest to empirical scientists'. But, in addition, Popper goes on to say, 'we must not try to draw the line too sharply' between science and metaphysics. For 'most of our scientific theories originate in myths', which 'may develop testable components'. 'It would hardly contribute to clarity', he goes on, 'if we were to say that these theories are nonsensical gibberish in one stage of their development, and then suddenly become good sense in another.' Elsewhere Popper (1957—a full reference for the article appears on p. 499 of this volume) expressly cites Freudian theory as an example of such a myth.

It is not necessary for us here to accept Professor Popper's account of these matters. The point of recalling his discussion is simply to remind us of some of the complexities that pervade the notions of refutation, testability and confirmation, to which philosophers of science have devoted some attention in recent years. (Compare the discussions of Carnap and Hempel for similar complexities.) If, therefore, Mr Cioffi supposes that his notions of refutation and disconfirmation *are* those of a Popper or a Hempel, and hence are well known and need no further elucidation, then Mr Cioffi seems to be mistaken. His notions do not seem to be close to those under contemporary discussion, and do stand in need of more clarification than he gives them. Until he does clarify them, his concept of a pseudo-science will remain vague. He may claim that psychoanalysis, or homoeopathy, or what not, is a pseudo-science. But before any such claim will stick, he will have to make very much clearer and more precise just what he is claiming.

This is not a pedantic demand. One of the striking and puzzling features, it could be argued, of psychoanalytic history and contemporary analytic practice is the fact that psychoanalytic theory and methods *are* in some way or degree affected and controlled by empirical fact. This fact consists primarily, but not wholly, in material thrown up by analytic methods themselves. We need only think of Freud's rejection of his first instinct theory, of the large Kleinian modifications, of the development of Ego-psychology, of the effect of experience in World War II on theory and methods. It seems clear that the logical relation here between theory and fact is *not* one of 'tight' refutation or disconfirmation. But it also seems clear that there is *some* logical relation between them. The puzzle is: just what? The whole situation, both past and present, seems to be a tangled one. To disentangle it requires close historical and contemporary study of the discourse and practice of various psychoanalytic personnel, groups and bodies. When such studies have been made, and when we have before us a certain precise sense of the expression 'pseudo-science', it may be illuminating to ask whether, or how far, or in what respects, Freud's theory and procedures constituted a pseudo-science. Without these studies of psychoanalytic discourse and practice, and without any precise sense of 'pseudo-science', it is very doubtful whether it is particularly illuminating to assert or to deny, *simpliciter*, that Freudian psychoanalysis of the twenties was a pseudo-science.

3.2. For an activity to be a pseudo-science, according to Mr Cioffi, it must also involve the habitual and wilful employment of methodologically defective procedures. One of these procedures, presumably, is interpretation, which is the one way, according to Mr Cioffi, of assessing the correctness of psychoanalytic claims, and which proves illusory (p. 473).

Why does it prove to be illusory? I confess I find Mr Cioffi's argument obscure and in need of separate examination—something I cannot provide here. All I can do is to concentrate on what *seems* to be the *first* part of the argument. 'All interpretations are "allusive" statements. People such as Pyramidologists also produce such statements. These are obviously spurious. It is not possible for Freud to show that his statements differ from those by Pyramidologists in being genuine and not spurious.'

This argument is conspicuously weak.

(*a*) As we have noted, we cannot take Freud's own account of his therapeutic work at its face value; it is not easy to determine just what he did. But there is some reason to think that his mature interpretative activity resembled that of contemporary analysts in one relevant respect. His interpretations were not all of the same type, and hence were not all allusive in character. We have some reason to believe that he offered patients interpretations, for example, about patient–therapist relations and the transference, about the feelings implicit in the patient's remarks, about his unacknowledged fears and wishes, about the problem facing the patient and the *real* nature of the situation confronting him in the world, in contrast with his distorted picture of it. Consequently, the fact that Freud probably did offer interpretations of different types makes it uncertain how far Mr Cioffi's objection at this point applies to Freud's interpretative activity as a whole.

(*b*) Mr Cioffi seems to be caught here by a general inability to distinguish between the grain and the chaff in Freud. For it could be argued quite easily that Freud was mistaken, in more ways than one, about his own methods. One of the things he was mistaken about was the reliance he placed on the allusive interpretation of the slip of the tongue, the fragment from Leonardo da Vinci, and similar isolated items. To offer an interpretation of such an item is a hazardous business, and only legitimate if offered very tentatively as a possibility to be borne in mind. The general reason for this is that, though there may be a verbal or other allusion from some item of the person's conduct, etc., to the content of the interpretation, it does not follow that the item really functioned in this instance in the way suggested by the interpretation. It may have functioned in some other way, and hence have some other explanation. It is very advisable, if not necessary, to take the full context of the person's life and conduct and relations with the therapist into account before coming to a confident pronouncement about any interpretation. Freud was wrong—especially perhaps in his earlier years—in underemphasizing the therapeutic context of an interpretation.

(*c*) If, therefore, we are to take psychoanalytic interpretations seriously, we would be well advised to restrict any consideration of them to the psychoanalytic situation. Here the value of the interpretation can be assessed in the light of the subsequent conduct of the patient, and the character of the material as a whole. Even so, this conduct and material may not be anything like good enough to give psychoanalysts what they would like to have. Elsewhere I have argued that it is very doubtful whether it is good enough for their purposes. But, by putting interpretations into their therapeutic context, we give them, at least, some sort of rationale, and we provide good reasons for believing that they function very differently from the arbitrary allusions of the Pyramidologist. Mr Cioffi does not begin to do justice to the contextual defence that can be offered of psychoanalytic interpretations. Consequently, he does not succeed in showing that the use of interpretations 'involves the wilful employment of methodologically defective procedures'.

4. It seems, then, that something fundamental and pervasive has gone wrong with Mr Cioffi's criticism of Freud. What is it?

Freud may have thought that he was providing a pretty good scientific story, as we understand this expression, about human functioning. If so, he was mistaken. At best, he was making some sort of contribution towards a good scientific story. Mr Cioffi is right in trying to bring out the differences between, for example, Freud's accounts of the development of sexuality, or the aetiology of the neuroses, and what would constitute good scientific accounts of these phenomena—if we had them. Mr Cioffi is pointing to these differences when, in effect, he brings out that, for example, Freud's account of the aetiology of the neuroses does not succeed in stating the conditions in infancy necessary for a neurotic predisposition and neurotic disorder. But, then, having brought out that Freud's accounts are not good science, Mr Cioffi implies, or suggests, that they are worthless in consequence—that there is nothing to be said for them at all; that, because we cannot obtain 'determinate inferences' from them (as one can with good scientific stories), we cannot obtain any inferences from them at all; and that they merely constitute the theses of a pseudo-science.

Here Mr Cioffi reveals that he is yet another victim of the fallacy of the sharp disjunction. A set of theses and activity, which are putatively empirical in character, is either scientific or it is not. If it is the former, it is the object of approval; if it is the latter, it is intellectually disreputable. Because Mr Cioffi is the victim of this fallacy, he is led to stress at some length the differences between psychoanalysis and science, and he does not go on to deal with the complementary job that awaits him and any other enquirer. This is the job of indicating how psychoanalysis is related to a good scientific story, and in what its explanatory force resides. This job has been attempted by others (see, for example, Professor Hempel's suggestions about 'partial explanation', and my own about 'a coherent narrative'). It would have been helpful and instructive if Mr Cioffi had attempted to make a contribution to this complementary job.

Moreover, what he says about Freudian psychoanalysis seems to apply to most, or all, of contemporary psychopathology. Consequently, as he appears to argue that Freudian analysis is devoid of all explanatory power whatever, he seems forced to say the same of contemporary psychopathology. Well, Mr Cioffi *may* be right. The stories that contemporaries spin in this field may have little, or no, explanatory force. But his argument is not anywhere near good enough to establish this conclusion, and contemporary psychopathologists are unlikely to be impressed by it. On the contrary, they are more likely to regard Mr Cioffi's account as the work of a man who does not understand what they are doing, and who is obsessionally tied up with the differences between Freud's work and the activity of people in the advanced sciences.

This is all very unfortunate. It is especially unfortunate in view of the fact that it is possible for a student of psychopathology to argue that

Mr Cioffi's *own practice* is much better than his theory. 'Mr Cioffi says (as we have seen) that "there are a host of peculiarities of psychoanalytic theory and practice which are apparently gratuitous and unrelated, but which can be understood when once they are seen as manifestations of the same impulse: the need to avoid refutation" (p. 473). How does Mr Cioffi set about supporting this claim? In the same sort of way that a psycho-pathologist sets about supporting a claim that the patient, who has just been presented at the case conference in the clinic, is an obsessional with a deep and unacknowledged need to avoid mistakes and being shown to be wrong. A psychopathologist sets about this task by showing that the "host of peculiarities" in the patient's case history, which "are apparently gratuitous and unrelated", can be understood when once they are "seen as manifestations of the same impulse", namely, the need to avoid mistakes and being shown to be wrong. In other words, he sets about supporting his view of the patient by showing how this view brings order into, and makes sense out of, the material which the case conference has about the patient. Mr Cioffi seems to be arguing in the *same sort* of way about Freud. The sort of explanation and understanding he tries to give us about Freud seems to be similar to the explanation and understanding that Freud himself and contemporary psychopathologists try to give us of mental disorder and human conduct. It is a pity that Mr Cioffi's practice was only used to exhibit the limits of Freudian psychoanalysis, and was not also used to bring out its point and contribution. In particular, it is a pity that Mr Cioffi did not try to bring out how and why Freudian theory can be used to help us explain and understand—how and why it throws light on the problem of masochism, on the aetiology of the neuroses, and so on.

Of course, there is a very serious, practical difficulty in the way of Mr Cioffi trying to meet this complaint from a psychopathologist. If he were actually to try to bring out the point and contribution of Freudian analysis, he could only do so satisfactorily by going carefully into the details of particular patients. But Mr Cioffi cannot be expected to do this in the course of a single chapter, which is also concerned to uncover the limitations of Freudian psychoanalysis. Nor is it easy to do this job in a short commentary; and I have not attempted it.

A last word. I regret that the tone of this commentary as a whole has been so critical of Mr Cioffi's essay. It is possible that this tone may be accounted for, in part, by the fact that I feel I have been guilty in the past of some of the very errors of which I am now accusing Mr Cioffi.

# REPLY

## *by* FRANK CIOFFI

## I

Farrell is both unclear as to what I mean by a pseudo-science and sure that psychoanalysis is not an example of one. I shall deal with these points in that order.

In Section I of my paper I argue that it is a mistake to think that when a theory is characterized as pseudo-scientific all that is at issue is the formal character of the claims which comprise it; in particular that it is a mistake to think that what is being maintained is that these statements are 'irrefutable in principle'. I cite as illustrative of this common mistake Farrell's paper 'Can psychoanalysis be refuted?'. In this paper Farrell speaks of the 'challenge that psychoanalytic theory cannot be refuted' and 'of whether it it possible to refute it' and asks 'is it logically possible to refute this theory?' He then proceeds to show with regard to a number of psychoanalytic theses that there are states of affairs whose occurrence would be taken as disconfirming them.

A psychoanalytic generalization about infantile anal eroticism is given a clean bill of health on the grounds that 'it is open in principle to scientific investigation' and that 'there is nothing logically absurd about such a programme of investigation'. Of the claim that little girls suffer from penis envy he asks, 'Have we any grounds for asserting that it is not refutable in principle? Surely none whatever.' Of another development thesis Farrell remarks, 'And could we not *imagine* a programme of investigation, the results of which could make *us* inclined to say that the generalization had been refuted?' (both italics are Farrell's). But if the fact that a programme of investigation can be *imagined*, the results of which could make *us* inclined to say that a generalization had been refuted is to count against that generalization's being pseudo-scientific, then no generalization is pseudo-scientific. Consider the generalization that birth under Aries is more auspicious than birth under Pisces. Could we not *imagine* a programme of investigation the results of which could make *us* inclined to say that the generalization had been refuted? It is obvious that something more is wanted.

In my paper I suggest that this something more is an assurance that the theory has been subjected to as much investigation as is practicable and as much clarification as the character of the subject permits, and that its deficiencies in these respects are not due to a fear that repairing them might undermine its credibility.

Farrell finds this demand 'insufficiently precise': 'When we have before us a certain precise sense of the expression pseudo-science it may be illuminating to ask whether or how far Freud's theory and procedures constitute a pseudo-science.' Does Farrell really mean to exhort us to suspend judgment on the scientific status of astrology, Christian Science, radiesthesia, 'the Abrams Box', etc., until 'we have a precise sense of the expression pseudo-science'? If so he has provided an instructive reminder of the readiness with which logicism and obscurantism enter into alliance with one another when the need arises.

In my paper I argue that it is not mere indeterminacy, or obscurity, or ambiguity, or figurativeness, but the tendentious exploitation of these which moves us to characterize theories such as Marxism, astrology, etc., as pseudo-scientific. But Farrell, who deplores obsession with the activities of the advanced sciences, nevertheless makes it an objection to my notion of the relevant considerations that they are remote from those elaborated by Hempel and Carnap in connection with physical theory: 'His notions do not seem to be close to those under contemporary discussion and do stand in need of more clarification than he gives them. Until he does clarify them his concept of a pseudo-science will remain vague.' I am saying of the psychoanalytic investigation of the neuroses: this is not what disinterested enquiry looks like—and Farrell thinks it appropriate to demand an analysis of the notion of disinterestedness.

Let us ask: what would have to obtain for a thesis to be justly characterized as pseudo-scientific, for example the thesis that all little girls suffer from penis envy. On my view of the nature of the issue, to rebut this charge it is not sufficient to show that disconfirmatory states of affairs with respect to the thesis are conceivable—we should have further to see how disconfirmatory reports are dealt with.

In his paper on refutability Farrell notes that direct observation has failed to support the existence of penis envy. However, he does not mention that on the evidence of the psychoanalytic authorities cited in his own bibliography the reports that direct observation is even apparently inconsistent with the existence of penis envy are not taken account of (Fenichel, 1946, p. 81). Nor can this be attributed to the inevitable time lag before the results of research become disseminated, for as late as 1960 it was being asserted that '. . . direct observations in childhood have led to the conclusion, mentioned before, that the sexual development of the little girl (goes through) the stage of penis envy' and that 'there is no doubt about the ubiquity of penis envy . . . in little girls, particularly around the third and fourth year' (Waelder, 1960, p. 116). This apparent imperviousness to criticism is a phenomenon which the enquirer into the empirical status of a discipline must take into account. But because Farrell puts so narrow a construction on his task these considerations lie outside his brief. Since he is able to mention studies which he considers disconfirmatory of the existence of penis envy, its empirical status is, as far as he is concerned, settled, for he has shown that it is 'refutable in principle'. Farrell thinks he has a warrant for his procedure in Popper's remark that 'refutability is the criterion of the scientific status of a theory'. But if we consult the paper to which Farrell refers us we find that Popper's remark summarizes (too baldly, it must be said) a number of varied considerations of which Farrell's refutability in principle is only one. For example, Popper characterizes as pseudo-scientific 'some genuinely testable theories (which) when found to be false are still upheld by their admirers'. Furthermore, the examples which Popper gives of pseudo-empirical or pseudo-scientific claims are astrology and the Marxist and racial interpretations of history, and these do not fail to be empirical for the same reasons that metaphysical claims fail to be empirical and are not irrefutable in the same sense.

But in this sort of enterprise, 'irrefutability in principle' is a last desperate resort, and even then argues a want of ingenuity on the part of the advocates of an apparently falsified thesis since they have all the varieties of mystification at their disposal. Normally the qualifications and emendations to which pseudo-scientific theses are subjected when they encounter apparent refutations are only asymptotically exhaustive. They rarely, if ever, get to the point of irrefutability in principle. That they are irrefutable in principle is a (hyperbolically expressed) inference supplied by those who are attempting to characterize them.

Ernest Nagel has expressed the view that '. . . there is surely good ground for the suspicion that Freudian theory can always be manipulated so that it escapes refutation no matter what the well established facts may be'. How does Farrell think a more refined concept of testability is going to settle an issue like this? In my paper I supply the only sorts of reasons which the nature of the issue permits for believing that Nagel is correct in his suspicion and for concluding that 'Freudian theory' is a pseudo-science.

II

I will now deal with the features of psychoanalytic practice which Farrell believes absolve it from the charge of being pseudo-empirical.

'If we are to take psychoanalytic interpretations seriously we would be well advised to restrict any consideration of them to the psychoanalytic situation. Here the value of the interpretation can be assessed in the light of the subsequent conduct of the patient, and the character of the material as a whole.' But Farrell seems to have forgotten his own strictures about the heterogeneity of the notion of interpretation, and in consequence it is not clear what he means by the term 'interpretation' in this passage. If what he means is to be relevant to my criticism of Freud's reconstructions of infantile life, the expression 'interpretation' must mean 'psychoanalytic claims about the infantile mental history of patients undergoing analysis'. This gives us: 'if we are to take psychoanalytic claims about the infantile mental history of patients undergoing analysis seriously we would be well advised to restrict any consideration of them to the psychoanalytic situation (where they) can be assessed in the light of the subsequent conduct of the patient and the character of the material as a whole.'

But when we ask that the grounds for Freud's reconstructions of infantile mental life be elucidated, what we want is something which is grammatically and not just contingently related to them; i.e. we want an account of the kind of infantile behaviour which is thought to warrant the ascription to the patient of the fantasies which figure in the reconstruction. So Farrell's talk of the 'therapeutic context' of the proffered interpretation is not in order here. If I ask how sesame differs from barley you don't enlighten me by telling me that 'Open Sesame' secures access to the robbers' cave whereas 'Open Barley' does not.

But even if we restrict Farrell's proposal to interpretations which are not reconstructions, or understand him to concede that reconstructions must be investigated outside the analytic situation, it still encounters difficulties. For if it is to be relevant to the genetic claims advanced by psychoanalysts it ought to run: 'if we are to take psychoanalytic claims about the infantile vicissitudes responsible for the patient's condition seriously we would be well advised to restrict any consideration of them to the psychoanalytic situation (where they) can be assessed in the light of the subsequent conduct of the patient and the character of the material as a whole.'

It must be remembered that we are apparently being offered this rationale as an attempted elucidation of claims like the following: this patient's neurosis is due to his infantile preoccupation with the prospect of being castrated; this patient's neurosis is due to her infantile preoccupation with her lack of a penis; this patient's homosexuality is due to his infantile preoccupation with his mother's fantasied phallus.

But how do we come to learn how a patient who is conducting himself as if his neurosis were due to his infantile preoccupation with the prospect of being castrated, say, conducts himself? How can it be determined whether the infantile vicissitudes which figure in the interpretation were, in fact, the cause of the illness? Certainly not, as Farrell suggests, by restricting ourselves to the 'psychoanalytic situation'. The most natural grounds for crediting these claims would be of the kind which Freud had, or thought he had, at the time that he believed hysteria to be due to a 'passive, sexual experience before puberty'—a differential association of the attributed infantile history with the patient's particular affliction. But when Freud abandoned the seduction theory of the neuroses he did not merely abandon the veridicality of the 'reminiscences' of infantile seduction but their specificity as fantasies as well, so that not only did he no longer have a factor in the infantile history of his patients which was intervening but neither did it differentiate neurotics from non-neurotics. So that in these genetic interpretations we have theses which logically require extra-analytic validation and yet seems incapable of receiving it. Because of the oddity of designating as a cause a factor of the infantile history which is neither differentiating nor supervening I suggest in my paper that it is misleading to call the infantile sexual experiences, believed to manifest themselves in the neuroses, the historical cause of them.

We now arrive at a version of Farrell's proposal which makes it at least superficially plausible; the psychoanalytic theory of the neuroses is to be understood as advancing claims about the unconscious infantile impulses, etc., contemporary with and manifesting themselves as the neuroses. Farrell's thesis thus becomes: 'If we are to take psychoanalytic claims about the unconscious infantile impulses contemporary with, and mani-festing themselves as, the neuroses seriously, we will be well advised to restrict any consideration of them to the psychoanalytic situation (where they) can be assessed in the light of the subsequent conduct of the patient and the character of the material as a whole.' (Why is it taking in-terpretations seriously to divest them of their prophylactic and genetic implications? Isn't this just a face-saving way of abandoning the aetiological illumination and prophylactic power we were promised, and isn't Farrell here illustrating the tacit dehistoricization of psychoanalytic claims which I noted in my paper?)

But even in its present dehistoricized form Farrell's proposal repeats the error of his paper on refutability, for the issue is not whether interpretation of a patient's symptoms '*can* be assessed' in the light of the patient's subsequent conduct (since the same might with equal justification be said of the casting of horoscopes) but whether they have been so assessed. If we are to arrive at a rational assessment of the psychoanalytic theory of the neuroses we must not only have some idea of what it would be for a patient to behave as if his difficulties were *not* due to the operation of unconscious infantile impulses (or whatever other unconscious impulses the theory assigns pathogenic power to) but some assurance that he did not in fact so behave. This comes to no more than one dynamically oriented psychopathologist's requirement that if an account of the clinical data is 'to be regarded as true then there has to be some way of distinguishing genuine patterns from pseudo-patterns' (Cheshire, 1964).

In my paper I assert that the considerations which are typically advanced by psychoanalysts in support of their claims imply criteria of confirmation so unexacting as to unconscionably impede the discovery that these claims are ill-founded. In his comment Farrell denies the inadequacy of these criteria and says that to hold that they are inadequate is to display 'an obsession with the activities of the advanced sciences'. Farrell makes it difficult to deal with this claim since he neither produces specimens of what he considers good arguments in support of the psychoanalytic theory of the neuroses nor says where they are to be found. I shall therefore draw on his Leonardo essay for an example of what he there considers con-firmation—Freud's reasons for believing that Leonardo's (supposed) attitude towards his painting was determined by unconscious reminis-cences of his infantile past. Since Leonardo had:

come unconsciously to regard his father as a person who neglected him when a child, who had not cared about him, and who had left him as an infant to the sole charge of his mother . . . he came unconsciously to adopt the same attitude to his own off-spring; that is, to the products of his own work. He did not particularly care about them. He found it difficult to sustain his interest in them. He could not bother to finish them, and to see that they were properly developed and completed . . .

Hence too, Farrell adds, seeing an opportunity Freud had missed, 'the unfinished doodling character of his scientific work' (Farrell, 1963, p. 22).

Does it really argue an 'obsession with the activities of the advanced sciences' to object that if you are going to call your ability to produce this kind of thing confirming a connection between infantile experience and adult behaviour, then far from being engaged in 'a hazardous business', as Farrell maintains in his comment (p. 505), you will make it next to impossible to find instances in which there does not appear to be such a connection, and so next to impossible to discover instances in which there really is no such connection.[1]

---

[1] If Farrell wants an insight into the sources of scientism let him reflect on what it was that moved him to remind us that the relation between a therapist and his patient is one involving just two people by twice referring to it as a 'dyad'. (In any case, of the ex-amples which I employ in my paper, only that of Dora was conducted as a 'dyad'. The data relating to Hans were transmitted through his father and was thus a 'triad'. The Leonardo and Dostoevsky essays were, I suppose, 'uniads'.)

This is what Judd Marmor, a psychoanalytic practitioner and former president of the American Academy of Psychoanalysis, has to say about suggestions like Farrell's for determining the truth of interpretations 'in the light of the subsequent conduct of the patient and the character of the material as a whole':

depending upon the point of view of the analyst, the patients of each school seem to bring up precisely the kinds of phenomenological data which confirm the theories and interpretations of their analysts! Thus each theory tends to be self-validating. Freudians elicit material about the Oedipus complex and castration anxiety, Jungians about archetypes, Rankians about separation anxiety, Adlerians about masculine strivings and feelings of inferiority, Horneyites about idealized images, Sullivanians about disturbed inter-personal relationships, etc. The fact is that in so complex a transaction as the psychoanalytic therapeutic process, the impact of patient and therapist upon each other, and particularly of the latter upon the former, is an unusually profound one. What the analyst shows interest in, the kinds of questions he asks, the kind of data he chooses to react to or ignore, and the interpretations he makes, all exert a subtle but significant suggestive impact upon the patient to bring forth certain kinds of data in preference to others (Marmor, 1962).

The inference that psychoanalytic method is invalid did not have to wait on Marmor's candour but could have been drawn, on the basis of considerations analogous to those advanced in my paper, at any time during the last fifty years. Why wasn't it?

And why do those who now hold that though Freud may have been mistaken about the neuroses, his enquiry into them was nevertheless a genuine one, not ask themselves how it was possible for him and his followers to conduct investigations into the mental lives of thousands of patients without discovering that they had no adequate grounds for maintaining that 'neurotic symptoms are substitutes for instinctual satisfactions'? (Fenichel, 1946, p. 32)? My paper is an attempt to explain how.

However, in it I also stress that my conclusion, that the psychoanalyst's adherence to defective procedures was due to 'the need to avoid refutation', is not ineluctable since the judgment that a procedure or mode of argument would, with a reasonable expenditure of effort, have revealed its invalidity is not such as to be capable of the kind of rigorous demonstration which apologists think they have a right to demand.

Nevertheless, I maintain that to see in Freud's theory of the neuroses something which though it may have only produced poor explanations was still an attempt at 'a good scientific story', registers not a mere failure in logical acumen but an obtuseness to the spirit of the enterprise. It completely overlooks the role that psychoanalysis, and particularly the activity of proffering interpretations, has played in our culture.

Freud behaves neither like someone who is addressing himself to the problem of the causes and nature of the neuroses but bungles the job from incompetence or lack of methodological sophistication, nor like someone who is stymied by the intrinsic difficulties of the problem, but rather like someone who, while going through the motions of engaging in an explanatory enquiry, reveals in an enormous variety of ways that he has other ends in view. (Fine shades of misbehaviour.)

One often comes across people whose preoccupations with a putatively explanatory factor is ostensibly derived from their interest in its pathogenic potentialities, but is really intrinsic. The majority of those who speculate as to whether slums, or the decline in church-going are causes of delinquency are not really interested in delinquency, but are interested in slums and the decline in church-going. What is noteworthy in Freud is the way in which the prestige of aetiological and prophylactic enquiries is exploited in the interest of an idiosyncratic preoccupation (or, perhaps I should say, an idiosyncratically intense preoccupation).

I hope this reply has done something to clarify my reasons for maintaining 'that psychoanalysis is a pseudo-science; that the character of this claim has often been misunderstood; that when it is understood its intractability is less surprising'.

### REFERENCES

Cheshire, Neil M. 1964. On the rationale of psychodynamic argumentation. *Br. J. med. Psychol.* **37**.
Farrell, B. A. 1961. Can psychoanalysis be refuted? *Inquiry.*
Farrell, B. A. 1963. Introduction. *Leonardo da Vinci.* Ed. S. Freud. London.
Fenichel, O. 1946. *The Psychoanalytic Theory of Neurosis.* London.
Marmor, Judd. 1962. Psychoanalytic therapy as an educational process. *Psychoanalytic Education,* vol. 5. New York. In the series *Science and Psychoanalysis.* (Ed. Jules H. Nasserman.)
Waelder, R. 1960. *Basic Theory of Psychoanalysis.* New York.

# INDEX OF PERSONS

N.B. The names of contributors to this volume have not been indexed except where they occur outside their own sections

517

## DATE DUE